MW00483887

Robert Lazarsfeld

# Positivity in Algebraic Geometry II

## Positivity for Vector Bundles, and Multiplier Ideals

 Springer

Prof. Robert Lazarsfeld
Department of Mathematics
University of Michigan
Ann Arbor, MI 48109, USA
e-mail: rlaz@umich.edu

Also available as a hardcover edition as Volume 49 in our series
"Ergebnisse der Mathematik": Robert Lazarsfeld: Positivity in Algebraic
Geometry II. Positivity for Vector Bundles, and Multiplier Ideals.
Springer-Verlag 2004, ISBN 3-540-22534-X


Library of Congress Control Number: 2004109578


Mathematics Subject Classification (2000):
Primary: 14-02
Secondary: 14C20, 14F05, 14F17, 32L10, 32J99, 14J17


ISSN 0071-1136
ISBN 3-540-22531-5 Springer Berlin Heidelberg New York

This work is subject to copyright. All rights are reserved, whether the whole or part of the material
is concerned, specifically the rights of translation, reprinting, reuse of illustrations, recitation, bro-
adcasting, reproduction on microfilms or in any other ways, and storage in data banks. Duplicati-
on of this publication or parts thereof is permitted only under the provisions of the German Copyright Law of September 9, 1965, in its current version, and permission for use must always be
obtained from Springer. Violations are liable for prosecution under the German Copyright Law.

Springer is a part of Springer Science+Business Media
springeronline.com
© Springer-Verlag Berlin Heidelberg 2004
Printed in Germany

Typesetting: Computer to film by the authors' data
Printed on acid-free paper      41/3142XT - 5 4 3 2 1 0

To Lee Yen, Sarah, and John

# Preface

The object of this book is to give a contemporary account of a body of work in complex algebraic geometry loosely centered around the theme of positivity.

Our focus lies on a number of questions that grew up with the field during the period 1950–1975. The sheaf-theoretic methods that revolutionized algebraic geometry in the fifties — notably the seminal work of Kodaira, Serre, and Grothendieck — brought into relief the special importance of ample divisors. By the mid sixties a very satisfying theory of positivity for line bundles was largely complete, and first steps were taken to extend the picture to bundles of higher rank. In a related direction, work of Zariski and others led to a greatly deepened understanding of the behavior of linear series on algebraic varieties. At the border with topology the classical theorems of Lefschetz were understood from new points of view, and extended in surprising ways. Hartshorne's book [276] and the survey articles in the Arcata proceedings [281] give a good picture of the state of affairs as of the mid seventies.

The years since then have seen continued interest and activity in these matters. Work initiated during the earlier period has matured and found new applications. More importantly, the flowering of higher dimensional geometry has led to fresh perspectives and — especially in connection with vanishing theorems — vast improvements in technology. It seems fair to say that the current understanding of phenomena surrounding positivity goes fundamentally beyond what it was thirty years ago. However, many of these new ideas have remained scattered in the literature, and others up to now have not been worked out in a systematic fashion. The time seemed ripe to pull together some of these developments, and the present volumes represent an attempt to do so.

The book is divided into three parts. The first, which occupies Volume I, focuses on line bundles and linear series. In the second volume, Part Two takes up positivity for vector bundles of higher ranks. Part Three deals with ideas and methods coming from higher-dimensional geometry, in the form of multiplier ideals. A brief introduction appears at the beginning of each of the parts.

I have attempted to aim the presentation at non-specialists. Not conceiving of this work as a textbook, I haven't started from a clearly defined set of prerequisites. But the subject is relatively non-technical in nature, and familiarity with the canonical texts [280] and [248] (combined with occasional faith and effort) is more than sufficient for the bulk of the material. In places — for example, Chapter 4 on vanishing theorems — our exposition is if anything more elementary than the standard presentations.

I expect that many readers will want to access this material in short segments rather than sequentially, and I have tried to make the presentation as friendly as possible for browsing. At least a third of the book is devoted to concrete examples, applications, and pointers to further developments. The more substantial of these are often collected together into separate sections. Others appear as examples or remarks (typically distinguished by the presence and absence respectively of indications of proof). Sources and attributions are generally indicated in the body of the text: these references are supplemented by brief sections of notes at the end of each chapter.

We work throughout with algebraic varieties defined over the complex numbers. Since substantial parts of the book involve applications of vanishing theorems, hypotheses of characteristic zero are often essential. However I have attempted to flag those foundational discussions that extend with only minor changes to varieties defined over algebraically closed fields of arbitrary characteristic. By the same token we often make assumptions of projectivity when in reality properness would do. Again I try to provide hints or references for the more general facts.

Although we use the Hodge decomposition and the hard Lefschetz theorem on several occasions, we say almost nothing about the Hodge-theoretic consequences of positivity. Happily these are treated in several sources, most recently in the beautiful book [600], [599] of Voisin. Similarly, the reader will find relatively little here about the complex analytic side of the story. For this we refer to Demailly's notes [126] or his anticipated book [119].

Concerning matters of organization, each chapter is divided into several sections, many of which are further partitioned into subsections. A reference to Section 3.1 points to the first section of Chapter 3, while Section 3.1.B refers to the second subsection therein. Statements are numbered consecutively within each section: so for example Theorem 3.1.17 indicates a result from Section 3.1 (which, as it happens, appears in 3.1.B). As an aid to the reader, each of the two volumes reproduces the table of contents of the other. The index, glossary, and list of references cover both volumes. Large parts of Volume I can be read without access to Volume II, but Volume II makes frequent reference to Volume I.

**Acknowledgements.** I am grateful to the National Science Foundation, the Guggenheim Foundation and the University of Michigan for support during the preparation of these volumes.[1]

I have benefited from comments and suggestions from many students and colleagues, including: T. Bauer, G. Bini, F. Bogomolov, A. Bravo, H. Brenner, A. Chen, D. Cutkosky, M. De Cataldo, T. De Fernex, J.-P. Demailly, I. Dolgachev, L. Ein, G. Farkas, R. Friedman, T. Garrity, A. Gibney, C. Hacon, R. Hartshorne, S. Helmke, M. Hering, M. Hochster, J. Howald, C. Huneke, P. Jahnke, M. Jonsson, Y. Kawamata, D. Keeler, M. Kim, F. Knudsen, A. Küronya, J. M. Landsberg, H. Lee, M. Mustaţă, T. Nevins, M. Nori, M. Paun, T. Peternell, H. Pinkham, M. Popa, I. Radloff, M. Ramachandran, M. Reid, J. Ross, M. Roth, H. Schenck, J. Sidman, R. Smith, A. Sommese, T. Stafford, I. Swanson, T. Szemberg, B. Teissier, Z. Teitler, D. Varolin, E. Viehweg, P. Vojta, J. Winkelmann, A. Wolfe, and Q. Zhang.

This project has profited from collaborations with a number of co-authors, including Jean-Pierre Demailly, Mark Green, and Karen E. Smith. Joint work many years ago with Bill Fulton has helped to shape the content and presentation of Chapter 3 and Part Two. I likewise owe a large mathematical debt to Lawrence Ein: I either learned from him or worked out together with him a significant amount of the material in Chapter 5 and Part Three, and our collaboration has had an influence in many other places as well.

I am grateful to several individuals for making particularly valuable contributions to the preparation of these volumes. János Kollár convinced me to start the book in the first place, and Bill Fulton insisted (on several occasions) that I finish it. Besides their encouragement, they contributed detailed suggestions from careful readings of drafts of several chapters. I also received copious comments on different parts of a preliminary draft from Thomas Eckl, Jun-Muk Hwang, Steve Kleiman, and Karen Smith. Olivier Debarre and Dano Kim read through the draft in its entirety, and provided a vast number of corrections and improvements.

Like many first-time authors, I couldn't have imagined when I began writing how long and consuming this undertaking would become. I'd like to take this opportunity to express my profound appreciation to the friends and family members that offered support, encouragement, and patience along the way.

Ann Arbor                                                                 Robert Lazarsfeld
May 2004

---

[1] Specifically, this work was partially supported by NSF Grants DMS 97-13149 and DMS 01-39713.

# Contents

# Contents of Volume I

**Appendices**

# Notation and Conventions

For the most part we follow generally accepted notation, as in [280]. We do however adopt a few specific conventions:

- We work throughout over the complex numbers $\mathbf{C}$.

- A *scheme* is a separated algebraic scheme of finite type over $\mathbf{C}$. A *variety* is a reduced and irreducible scheme. We deal exclusively with closed points of schemes.

- If $X$ is a variety, a *modification* of $X$ is a projective birational mapping $\mu : X' \longrightarrow X$ from an irreducible variety $X'$ onto $X$.

- Given an irreducible variety $X$, we say that a property holds at a *general* point of $X$ if it holds for all points in the complement of a proper algebraic subset. A property holds at a *very general* point if it is satisfied off the union of countably many proper subvarieties.

- Let $f : X \longrightarrow Y$ be a morphism of varieties or schemes, and $\mathfrak{a} \subseteq \mathcal{O}_Y$ an ideal sheaf on $Y$. We sometimes denote by $f^{-1}\mathfrak{a} \subseteq \mathcal{O}_X$ the ideal sheaf $\mathfrak{a} \cdot \mathcal{O}_X$ on $X$ determined by $\mathfrak{a}$. While strictly speaking this conflicts with the notation for the sheaf-theoretic pullback of $\mathfrak{a}$, we trust that no confusion will result.

- Given a vector space or vector bundle $E$, $S^k E$ denotes the $k^{\text{th}}$ symmetric power of $E$, and $\mathrm{Sym}(E) = \oplus S^k E$ is the full symmetric algebra on $E$. The dual of $E$ is written $E^*$.

- If $E$ is a vector bundle on a scheme $X$, $\mathbf{P}(E)$ denotes the projective bundle of one-dimensional *quotients* of $E$. On a few occasions we will want to work with the bundle of one-dimensional *subspaces* of $E$: this is denoted $\mathbf{P}_{\mathrm{sub}}(E)$. Thus

$$\mathbf{P}_{\mathrm{sub}}(E) = \mathbf{P}(E^*).$$

A brief review of basic facts concerning projective bundles appears in Appendix A.

- Let $X_1$, $X_2$ be varieties or schemes. Without specific comment we write

$$\mathrm{pr}_1 : X_1 \times X_2 \longrightarrow X_1 \quad , \quad \mathrm{pr}_2 : X_1 \times X_2 \longrightarrow X_2$$

  for the two projections of $X_1 \times X_2$ onto its factors. When other notation for the projections seems preferable, we introduce it explicitly.

- Given a real-valued function $f : \mathbf{N} \longrightarrow \mathbf{R}$ defined on the natural numbers, we say that $f(m) = O(m^k)$ if

$$\limsup_{m \to \infty} \frac{|f(m)|}{m^k} < \infty.$$

Positivity for Vector Bundles

# Introduction to Part Two

The three chapters in this Part Two deal with amplitude for vector bundles of higher rank.

Starting in the early 1960s, several mathematicians — notably Grauert [232], Griffiths [245], [246], [247] and Hartshorne [274] — undertook the task of generalizing to vector bundles the theory of positivity for line bundles. One of the goals was to extend to the higher-rank setting as many as possible of the beautiful cohomological and topological properties enjoyed by ample divisors. It was not initially clear how to achieve this, and the literature of the period is marked by a certain terminological chaos as authors experimented with various definitions and approaches. With the passage of time, however, it has become apparent that Hartshorne's definition — which involves the weakest notion of positivity — does in fact lead to most of the basic results one would like. The idea is simply to pass to the associated projective bundle, where one reduces to the rank one case. This approach is by now standard, and it is the one we adopt here.

After the foundational work of the late sixties and early seventies, the geometric consequences of positivity were substantially clarified during the later seventies and the eighties. These same years brought many new examples and applications of the theory. While aspects of this story have been surveyed on several occasions — e.g. [533, Chapter V], [15, Chapter 5, §1] and [352, Chapter 3, §6] — there hasn't been a systematic exposition of the theory taking these newer developments into account. Our aim in these chapters is to help fill this gap. Although we certainly make no claims to completeness, we hope that the reader will take away the picture of a mature body of work touching on a considerable range of questions.

We start in Chapter 6 with the basic formal properties of ample and nef bundles, after which we dwell at length on examples and constructions. Chapter 7 deals with geometric properties of ample bundles: we discuss higher-rank generalizations of the Lefschetz hyperplane and Kodaira vanishing theorems, as well as the effect of positivity on degeneracy loci. Finally, we take up in Chapter 8 the numerical properties of ample bundles.

In contrast to our focus in the case of line bundles, we say very little about the various notions of "generic amplitude" for bundles that appear in the literature. Some of these — for example the weak positivity of Viehweg or Miyaoka's generic semipositivity — have led to extremely important developments. However these concepts are tied to particular applications lying beyond the scope of this volume. Our feeling was that they are best understood in vivo rather than through the sort of in vitro presentation that would have been possible here. So for the most part we (regretfully) pass over them.

# 6

# Ample and Nef Vector Bundles

This chapter is devoted to the basic theory of ample and nef vector bundles. We start in Section 6.1 with the "classical" material from [274]. In Section 6.2 we develop a formalism for twisting bundles by $\mathbf{Q}$-divisor classes, which is used to study nefness. The development parallels — and for the most part reduces to — the corresponding theory for line bundles. The next two sections constitute the heart of the chapter. In the extended Section 6.3 we present numerous examples of positive bundles arising "in nature," as well as some methods of construction. Finally, we study in Section 6.4 the situation on curves, where there is a close connection between amplitude and stability: following Gieseker [224] one obtains along the way an elementary proof of the tensorial properties of semistability for bundles on curves.

## 6.1 Classical Theory

We start by fixing notation and assumptions. Throughout this section, unless otherwise stated $X$ is a projective algebraic variety or scheme defined over $\mathbf{C}$. Given a vector bundle $E$ on $X$, $S^m E$ is the $m^{\text{th}}$ symmetric product of $E$, and

$$\pi \, : \, \mathbf{P}(E) \longrightarrow X$$

denotes the projective bundle of one-dimensional quotients of $E$. On occasion, when it seems desirable to emphasize that $\mathbf{P}(E)$ is a bundle over $X$, we will write $\mathbf{P}_X(E)$ in place of $\mathbf{P}(E)$. As usual, $\mathcal{O}_{\mathbf{P}(E)}(1)$ is the Serre line bundle on $\mathbf{P}(E)$, i.e. the tautological quotient of $\pi^* E$: thus $S^m E = \pi_* \mathcal{O}_{\mathbf{P}(E)}(m)$. We refer to Appendix A for a review of basic facts about projective bundles.

As in Chapter 1 (Definition 1.1.15), we denote by

$$N^1 \big( \mathbf{P}(E) \big) \; = \; \mathrm{Div} \big( \mathbf{P}(E) \big) / \mathrm{Num} \big( \mathbf{P}(E) \big)$$

the Néron–Severi group of numerical equivalence classes of divisors on $\mathbf{P}(E)$. Since $X$ is projective the Serre line bundle $\mathcal{O}_{\mathbf{P}(E)}(1)$ is represented by a divisor, and we write

$$\xi = \xi_E \in N^1\big(\mathbf{P}(E)\big)$$

for its numerical equivalence class: in other words, $\xi_E$ corresponds to the first Chern class of $\mathcal{O}_{\mathbf{P}(E)}(1)$.[1] Finally, given a finite-dimensional vector space $V$, $V_X = V \otimes_{\mathbf{C}} \mathcal{O}_X$ denotes the trivial vector bundle on $X$ with fibres modeled on $V$.

### 6.1.A Definition and First Properties

In the case of line bundles, essentially all notions of positivity turn out to be equivalent, but this is no longer true for vector bundles of higher rank. Consequently the early literature in the area has an experimental flavor, involving competing definitions of positivity. By the 1980s, however, the situation had stabilized, with the approach adopted by Hartshorne [274] generally accepted as the most useful.

Hartshorne's basic idea is to reduce the definition of amplitude for a bundle $E$ to the corresponding notion for divisors by passing to $\mathbf{P}(E)$:

**Definition 6.1.1. (Ample and nef vector bundles).** A vector bundle $E$ on $X$ is *ample* if the Serre line bundle $\mathcal{O}_{\mathbf{P}(E)}(1)$ is an ample line bundle on the projectivized bundle $\mathbf{P}(E)$. Similarly, $E$ is *numerically effective* (or *nef*) if $\mathcal{O}_{\mathbf{P}(E)}(1)$ is so.                                      □

Observe that if $E = L$ is a line bundle, then $\mathbf{P}(E) = X$ and $\mathcal{O}_{\mathbf{P}(E)}(1) = L$, so that at least this definition generalizes the rank one case. We will see that 6.1.1 leads to the formal properties of amplitude for which one would hope. In this subsection we focus on results that follow directly from corresponding facts for line bundles.

The first statement of the following proposition reflects the principle that "positivity increases in quotients." The second includes the fact that the restriction of an ample bundle to a closed subvariety (or subscheme) is ample.

**Proposition 6.1.2. (Quotients and finite pullbacks).** *Let $E$ be a vector bundle on the projective variety or scheme $X$.*

(i). *If $E$ is ample (or nef), then so is any quotient bundle $Q$ of $E$.*

(ii). *Let $f : X \longrightarrow Y$ be a finite mapping. If $E$ is ample (or nef), then the pullback $f^*E$ is an ample (or nef) bundle on $X$.*

*Proof.* A surjection $E \twoheadrightarrow Q$ determines an embedding

$$\mathbf{P}(Q) \subseteq \mathbf{P}(E) \qquad \text{with} \qquad \mathcal{O}_{\mathbf{P}(Q)}(1) = \mathcal{O}_{\mathbf{P}(E)}(1) \,|\, \mathbf{P}(Q),$$

---

[1] As indicated in Remark 1.1.22, one can work on arbitrary complete schemes provided that one understands $N^1\big(\mathbf{P}(E)\big)$ to be numerical equivalence classes of line bundles.

and (i) follows from the fact that the restriction of an ample (or a nef) bundle is ample (or nef). As for (ii), $f$ gives rise to a finite map $F : \mathbf{P}(f^*E) \longrightarrow \mathbf{P}(E)$ such that $\mathcal{O}_{\mathbf{P}(f^*E)}(1) = F^*\mathcal{O}_{\mathbf{P}(E)}(1)$, so again we reduce to the corresponding statement (Proposition 1.2.13) for line bundles.    □

**Example 6.1.3. (Bundles on $\mathbf{P}^1$).** Let $E$ be a vector bundle on the projective line $\mathbf{P}^1$. According to a celebrated theorem of Grothendieck [488, Chapter 1, Section 2.1], $E$ is a direct sum of line bundles, say

$$E = \mathcal{O}_{\mathbf{P}^1}(a_1) \oplus \ldots \oplus \mathcal{O}_{\mathbf{P}^1}(a_e).$$

Then $E$ is ample if and only if $a_i > 0$ for every $1 \le i \le e$. (If each $a_i$ is positive, then $\mathcal{O}_{\mathbf{P}(E)}(1)$ embeds $\mathbf{P}(E)$ as a rational normal scroll.) This is a special case of Proposition 6.1.13.    □

**Example 6.1.4. (Pullbacks of negative bundles).** Let $f : Y \longrightarrow X$ be a surjective morphism of projective varieties, with $\dim X \ge 1$, and let $E$ be an ample bundle on $X$. Then $H^0\big(Y, f^*E^*\big) = 0$. (It is enough to show that if $C \subseteq Y$ is a general curve obtained as the complete intersection of very ample divisors on $Y$ then $\mathrm{Hom}\big(f^*E \,|\, C, \mathcal{O}_C\big) = 0$. One can assume that $C$ maps finitely to $X$, and then the assertion follows from 6.1.2 upon normalizing $C$.)    □

As another illustration, we analyze the case of globally generated bundles:

**Example 6.1.5. (Globally generated bundles).** Let $V$ be a finite-dimensional vector space, and let $E$ be a quotient

$$V_X \longrightarrow E \longrightarrow 0$$

of the trivial vector bundle $V_X$ modeled on $V$. This gives rise to a morphism

$$\phi : \mathbf{P}_X(E) \longrightarrow \mathbf{P}(V) \tag{6.1}$$

from the projective bundle $\mathbf{P}_X(E) = \mathbf{P}(E)$ to the projective space of one dimensional quotients of $V$, defined as the composition

$$\mathbf{P}_X(E) \subseteq \mathbf{P}(V_X) = \mathbf{P}(V) \times X \xrightarrow{\mathrm{pr}_1} \mathbf{P}(V).$$

By construction, $\mathcal{O}_{\mathbf{P}(E)}(1) = \phi^*\mathcal{O}_{\mathbf{P}(V)}(1)$.

(i).   $E$ is ample if and only if $\phi$ is finite.

(ii).  Given $x \in X$, denote by $E(x)$ the fibre of $E$ at $x$, so that $E(x)$ is a quotient of $V$, and let

$$\mathbf{P}(E(x)) \subseteq \mathbf{P}(V)$$

be the corresponding linear subspace of $\mathbf{P}(V)$. Then $E$ is ample if and only if there are only finitely many $x \in X$ such that $\mathbf{P}(E(x))$ passes through any given point of $\mathbf{P}(V)$.    □

**Example 6.1.6. (Tautological bundle on the Grassmannian).** Let $\mathbf{G} =$ $\mathrm{Grass}(V, k)$ be the Grassmannian of $k$-dimensional quotients of a vector space $V$, and denote by $E$ the tautological rank-$k$ quotient bundle of $V_{\mathbf{G}}$. If $k \geq 2$ then $E$ is nef but *not* ample. (Use 6.1.5 (ii).) In particular, it would be incorrect to try by analogy with Definition 1.2.1 to define amplitude naively in terms of embeddings into Grassmannians.                           □

For globally generated bundles, amplitude also is equivalent to the absence of trivial quotients along curves:

**Proposition 6.1.7. (Gieseker's lemma).** *Suppose that $E$ is a globally generated bundle on an irreducible projective variety $X$. Then $E$ fails to be ample if and only if there is a curve $C \subseteq X$ such that the restriction $E \mid C$ admits a trivial quotient.*

*Proof.* If $E$ is ample, then it follows from Proposition 6.1.2 that no restriction of $E$ can have a trivial quotient. Conversely, suppose that $E$ fails to be ample. Set $V = H^0(X, E)$, and consider the morphism $\phi : \mathbf{P}(E) = \mathbf{P}_X(E) \longrightarrow \mathbf{P}(V)$ appearing in (6.1). By 6.1.5 (i), $\phi$ fails be finite, i.e. there exists a curve $C \subset \mathbf{P}(E)$ contracted by $\phi$ to a point. But $\phi$ is in any event an embedding on each fibre of $\pi : \mathbf{P}(E) \longrightarrow X$, and hence $C$ maps isomorphically to its image in $X$. Moreover
$$\mathcal{O}_{\mathbf{P}(E)}(1) \mid C \;=\; \phi^* \mathcal{O}_{\mathbf{P}(V)}(1) \mid C \,,$$
and since $\mathcal{O}_{\mathbf{P}(E)}(1)$ is a quotient of $\pi^* E$, we arrive at a trivial quotient of $E \mid C$.                           □

We refer to Section 6.3 for more substantial examples and methods of construction. Here we return to developing the general theory.

As in the case of line bundles, the first two parts of the following proposition allow one in practice to focus attention on reduced and irreducible varieties:

**Proposition 6.1.8. (Additional formal properties of amplitude).** *Let $E$ be a vector bundle on a projective variety or scheme $X$.*

(i).   *$E$ is ample (or nef) if and only if the restriction $E_{red}$ of $E$ to $X_{red}$ is ample (or nef).*

(ii).  *$E$ is ample (or nef) if and only if its restriction to each irreducible component of $X$ is ample (or nef).*

(iii). *If $f : Y \longrightarrow X$ is a surjective finite map, and if $f^* E$ is an ample vector bundle on $Y$, then $E$ is ample.*

(iv).  *If $f : Y \longrightarrow X$ is an arbitrary surjective mapping, and if $f^* E$ is a nef bundle on $Y$, then $E$ itself is nef.*

*Proof.* By passing to $\mathbf{P}(E)$, these statements again follow immediately from the corresponding facts (1.2.16, 1.2.28, 1.4.4) for line bundles.                           □

The next result expresses the strong open nature of amplitude in families:

**Proposition 6.1.9. (Amplitude in families).** *Let $f : X \longrightarrow T$ be a proper surjective mapping, and suppose that $E$ is a vector bundle on $X$. Given $t \in T$, denote by $X_t$ the fibre of $X$ over $t$, and by $E_t = E \mid X_t$ the restriction of $E$ to $X_t$. If there is a point $0 \in T$ such that $E_0$ is ample, then there is an open neighborhood $U \subset T$ of $0$ such that $E_t$ is ample for all $t \in U$.*

*Proof.* Since $\mathbf{P}(E_t)$ is the fibre of the evident map $\mathbf{P}(E) \longrightarrow T$, this yet again follows directly from the corresponding statement (Theorem 1.2.17) for line bundles. $\square$

### 6.1.B Cohomological Properties

We next analyze the asymptotic cohomological properties of ample vector bundles. The following theorem of Hartshorne [274] is the analogue for vector bundles of the theorem of Cartan–Serre–Grothendieck (Theorem 1.2.6).

**Theorem 6.1.10. (Cohomological characterization of ample vector bundles).** *Let $E$ be a vector bundle on the projective variety or scheme $X$. The following are equivalent:*

(i).  *$E$ is ample.*

(ii).  *Given any coherent sheaf $\mathcal{F}$ on $X$, there is a positive integer $m_1 = m_1(\mathcal{F})$ such that*
$$H^i\big(X, S^m E \otimes \mathcal{F}\big) = 0 \quad for \quad i > 0 , \ m \geq m_1.$$

(iii).  *Given any coherent sheaf $\mathcal{F}$ on $X$, there is a positive integer $m_2 = m_2(\mathcal{F})$ such that $S^m E \otimes \mathcal{F}$ is globally generated for all $m \geq m_2$.*

(iv).  *For any ample divisor $H$ on $X$, there is a positive integer $m_3 = m_3(H)$ such that if $m \geq m_3$ then $S^m E$ is a quotient of a direct sum of copies of $\mathcal{O}_X(H)$.*

(iv)\*.  *The statement of (iv) holds for some ample divisor $H$.*

*Proof.* (i) $\Rightarrow$ (ii). Denoting as above by $\pi : \mathbf{P}(E) \longrightarrow X$ the bundle map, assume that $E$ is ample, so that $\mathcal{O}_{\mathbf{P}(E)}(1)$ is an ample line bundle on $\mathbf{P}(E)$, let $\mathcal{F}$ be a coherent sheaf on $X$. Then by Serre's criterion (Theorem 1.2.6) there is an integer $m_1 = m_1(\mathcal{F})$ such that
$$H^i\big(\mathbf{P}(E), \mathcal{O}_{\mathbf{P}(E)}(m) \otimes \pi^*\mathcal{F}\big) = 0 \quad for \quad i > 0 , \ m \geq m_1. \qquad (*)$$

Now suppose for the moment that $\mathcal{F}$ is locally free. Then the projection formula implies that
$$\pi_*\big(\mathcal{O}_{\mathbf{P}(E)}(m) \otimes \pi^*\mathcal{F}\big) = \pi_*\big(\mathcal{O}_{\mathbf{P}(E)}(m)\big) \otimes \mathcal{F} = S^m E \otimes \mathcal{F},$$

and by the same token the higher direct images vanish provided that $m \geq 0$. Therefore

$$H^i\big(X, S^m E \otimes \mathcal{F}\big) \quad = \quad H^i\big(\mathbf{P}(E), \mathcal{O}_{\mathbf{P}(E)}(m) \otimes \pi^* \mathcal{F}\big),$$

and the required vanishings follow from (*). For an arbitrary coherent sheaf $\mathcal{F}$ we can find a (possibly non terminating) resolution

$$\cdots \longrightarrow F_2 \longrightarrow F_1 \longrightarrow F_0 \longrightarrow \mathcal{F} \longrightarrow 0$$

of $\mathcal{F}$ by locally free sheaves (Example 1.2.21). If $\dim X = n$, it is sufficient in view of Proposition B.1.2 from Appendix B to establish that

$$H^i\big(X, S^m E \otimes F_n\big) = \cdots = H^i\big(X, S^m E \otimes F_0\big) \;\; = \;\; 0 \;\; \text{for } i > 0 \text{ and } m \gg 0,$$

and this follows from the case already treated.[2]

(ii) $\Rightarrow$ (iii). Given $\mathcal{F}$, fix a point $x \in X$, with maximal ideal $\mathfrak{m}_x \subset \mathcal{O}_X$. Consider the subsheaf $\mathfrak{m}_x \mathcal{F} \subset \mathcal{F}$ of germs of sections of $\mathcal{F}$ that vanish at $x$, so that one has the exact sequence

$$0 \longrightarrow \mathfrak{m}_x \mathcal{F} \longrightarrow \mathcal{F} \longrightarrow \mathcal{F}/\mathfrak{m}_x \mathcal{F} \longrightarrow 0.$$

By (ii) there exists a positive integer $m_2(\mathcal{F}, x)$ such that $H^1(X, S^m E \otimes \mathfrak{m}_x \mathcal{F}) = 0$ for $m \geq m_2(\mathcal{F}, x)$, and so we see using the sequence above that $S^m E \otimes \mathcal{F}$ is generated by its sections at $x$. The same therefore holds in a Zariski open neighborhood of $x$, and consequently by quasi-compactness we can find a single integer $m_2 = m_2(\mathcal{F})$ such that $S^m E \otimes \mathcal{F}$ is globally generated when $m \geq m_2$.

(iii) $\Rightarrow$ (iv). Apply (iii) with $\mathcal{F} = \mathcal{O}_X(-H)$.

(iv) $\Rightarrow$ (iv)*. Tautological.

(iv)* $\Rightarrow$ (i). It follows from (iv)* that there is an ample divisor $H$ on $X$ and a positive integer $m > 0$ such that we can write $S^m E$ as a quotient of $U = \mathcal{O}_X(H)^{\oplus N+1}$ for some $N$. Note first that $U$ is ample.[3] In fact, $\mathbf{P}(U)$ is isomorphic to the product $\mathbf{P}^N \times X$ in such a way that

$$\mathcal{O}_{\mathbf{P}(U)}(1) = \mathrm{pr}_1^* \mathcal{O}_{\mathbf{P}^N}(1) \otimes \mathrm{pr}_2^* \mathcal{O}_X(H),$$

and this is an ample line bundle on $\mathbf{P}^N \times X$. Since $S^m E$ is a quotient of $U$, it follows that $S^m E$ is ample. But one has a Veronese embedding

$$\mathbf{P}(E) \hookrightarrow \mathbf{P}(S^m E) \quad \text{with} \quad \mathcal{O}_{\mathbf{P}(S^m E)}(1) \,|\, \mathbf{P}(E) \;=\; \mathcal{O}_{\mathbf{P}(E)}(m).$$

Therefore $\mathcal{O}_{\mathbf{P}(E)}(m)$ — and hence also $\mathcal{O}_{\mathbf{P}(E)}(1)$ — is ample, as required.  $\square$

---

[2] One can avoid this step by observing directly that

$$R^i \pi_* \big(\mathcal{O}_{\mathbf{P}(E)}(m) \otimes \pi^* \mathcal{F}\big) \;=\; R^i \pi_* \big(\mathcal{O}_{\mathbf{P}(E)}(m)\big) \otimes \mathcal{F}$$

for any coherent sheaf $\mathcal{F}$ on $X$: see [274, Lemma 3.1].

[3] Compare the next proposition for a more general statement.

**Remark 6.1.11.** The theorem shows in effect that amplitude of a bundle $E$ is detected by geometric properties of a high symmetric power $S^m E$. However in contrast to the situation for line bundles — where for instance one can use covering constructions (e.g. Theorem 4.1.10) to replace a divisor by a multiple — it is often not obvious how to pass from information about $S^m E$ to a statement for $E$ itself. In practice, this is the essential difficulty in working with amplitude for bundles of higher rank. □

**Remark 6.1.12.** The equivalence of statements (i), (ii), and (iii) remains valid on an arbitrary complete (but possibly non-projective) scheme over $\mathbf{C}$: see [274], (2.1), (3.2) and (3.3). On a non-complete scheme, (iii) is taken as the definition of amplitude. □

Turning to direct sums and extensions, there is the useful

**Proposition 6.1.13. (Direct sums and extensions).** *Let $E_1$ and $E_2$ be vector bundles on $X$.*

(i). *The direct sum $E_1 \oplus E_2$ is ample if and only if both $E_1$ and $E_2$ are.*

(ii). *Suppose that $F$ is an extension of $E_2$ by $E_1$:*

$$0 \longrightarrow E_1 \longrightarrow F \longrightarrow E_2 \longrightarrow 0. \qquad (6.2)$$

*If $E_1$ and $E_2$ are ample, then so is $F$.*

*Proof.* (i). If $E_1 \oplus E_2$ is ample, then so are its quotients $E_1$ and $E_2$. Conversely, assume that $E_1$ and $E_2$ are ample. The plan is to apply criterion (iii) from 6.1.10, so fix a coherent sheaf $\mathcal{F}$ on $X$. We need to show that $S^m(E_1 \oplus E_2) \otimes \mathcal{F}$ is globally generated for $m \gg 0$. But $S^m(E_1 \oplus E_2) = \sum_{p+q=m} S^p E_1 \otimes S^q E_2$, so the question is to find an integer $m_0$ such that

$$S^p E_1 \otimes S^q E_2 \otimes \mathcal{F} \quad \text{is globally generated for} \quad p + q \ge m_0. \qquad (*)$$

To this end, first use the amplitude of $E_1$ and $E_2$ to choose $t_1 > 0$ such that

$$S^t E_1 \; , \quad S^t E_2 \otimes \mathcal{F} \quad \text{are globally generated for} \quad t \ge t_1.$$

Next, for $k = 0, 1, \dots, t_1$ apply 6.1.10 (iii) to each of the coherent sheaves $S^k E_2 \otimes \mathcal{F}$ and $S^k E_1 \otimes \mathcal{F}$ to produce $t_2$ such that

$$S^t E_1 \otimes S^k E_2 \otimes \mathcal{F} \; , \quad S^k E_1 \otimes S^t E_2 \otimes \mathcal{F} \quad \text{are globally generated for} \quad t \ge t_2.$$

We claim that then (*) holds with $m_0 = t_1 + t_2$. In fact, suppose that $p + q \ge t_1 + t_2$. If $p, q \ge t_1$ then $S^p E_1$ and $S^q E_2 \otimes \mathcal{F}$ are globally generated, and hence so is their tensor product. If $p \le t_1$ then $q \ge t_2$, and so (*) holds by choice of $t_2$, and similarly if $q \le t_1$. This proves (i).

For (ii) we apply 6.1.10 (ii), so again fix a coherent sheaf $\mathcal{F}$. Now $S^m F$ has a filtration whose quotients are $S^p(E_1) \otimes S^q(E_2)$ with $p + q = m$. So (using Lemma B.1.7) it is enough to prove the vanishings

$$H^i(X, S^p E_1 \otimes S^q E_2 \otimes \mathcal{F}) = 0 \quad \text{for } i > 0 \text{ and } p + q = m \gg 0. \qquad (**)$$

But by part (i) of the proposition we already know that $E_1 \oplus E_2$ is ample, and then $(**)$ follows upon applying 6.1.10 (ii) to this bundle.      □

**Example 6.1.14.** One can also deduce statement (ii) from (i) via Proposition 6.1.9. In fact, by scaling the extension class defining (6.2), the bundle $F$ is realized as a small deformation of $E_1 \oplus E_2$, and then 6.1.9 applies.      □

At least in characteristic zero, the tensorial properties of amplitude flow from the next result.

**Theorem 6.1.15. (Symmetric products).** *A vector bundle $E$ on $X$ is ample if and only if $S^k E$ is ample for some — or equivalently, for all — $k \geq 1$.*

*Proof.* Supposing first that $S^k E$ is ample, the amplitude of $E$ follows via the Veronese embedding $\mathbf{P}(E) \subseteq \mathbf{P}(S^k E)$ as in the argument that (iv)$^*$ $\Rightarrow$ (i) in 6.1.10. Conversely, assume that $E$ is ample. We first note that $S^m E$ is ample for all $m \gg 0$. In fact, according to Theorem 6.1.10 (iv), for large $m$ the bundle in question is a quotient of a direct sum of ample line bundles. It remains to deduce that $S^k E$ is ample for any fixed $k \geq 1$. To this end, we again use a Veronese-type morphism. Specifically, we assert that (in characteristic zero!) there exists for any $\ell \geq 1$ a finite mapping

$$\nu_{\ell,k} : \mathbf{P}(S^k E) \longrightarrow \mathbf{P}(S^{k\ell} E) \qquad (6.3)$$

with $\mathcal{O}_{\mathbf{P}(S^k E)}(\ell) = \nu_{\ell,k}^* \mathcal{O}_{\mathbf{P}(S^{\ell k} E)}(1)$. Grant this for the moment. For $\ell \gg 0$ we already know that $\mathcal{O}_{\mathbf{P}(S^{k\ell} E)}(1)$ is ample, and hence so is $\mathcal{O}_{\mathbf{P}(S^k E)}(\ell)$. As before, this implies that $S^k E$ is ample, as required.

It remains to prove the existence of $\nu_{\ell,k}$. Recall to begin with that there are canonical homomorphisms

$$S^{\ell k} E \xhookrightarrow{\; i_{\ell,k} \;} S^\ell(S^k E) \xrightarrow{\; m_{\ell,k} \;} S^{\ell k} E,$$

with $m_{\ell,k}$ given by multiplication. The composition $m_{\ell,k} \circ i_{\ell,k}$ is multiplication by a fixed non-zero integer $C_{\ell,k}$, and — since we are in characteristic zero — is therefore a homothety. Now write $\pi_k : \mathbf{P}(S^k E) \longrightarrow X$ for the bundle map. Composing $\pi_k^*(i_{\ell,k})$ with the $\ell^{\text{th}}$ symmetric power of the canonical quotient $\pi_k^* S^k E \longrightarrow \mathcal{O}_{\mathbf{P}(S^k E)}(1)$ gives a vector bundle morphism

$$\pi_k^* S^{\ell k} E \longrightarrow \mathcal{O}_{\mathbf{P}(S^k E)}(\ell). \qquad (*)$$

We will show that $(*)$ is surjective: it then defines the required mapping $\nu_{\ell,k}$.

The surjectivity of $(*)$ can be checked fibre by fibre over $X$, so we assume that $E$ is a vector space. Given a non-zero functional $\phi : S^k E \longrightarrow \mathbf{C}$, and its $\ell^{\text{th}}$ symmetric power $S^\ell \phi : S^\ell(S^k E) \longrightarrow \mathbf{C}$, the issue is to show that

$$\mathrm{im}(i_{\ell,k}) \not\subseteq \ker(S^\ell \phi). \qquad (**)$$

Write $N = \ker(\phi)$, choose a one-dimensional subspace $L \subseteq S^k E$ splitting $\phi$, and denote by $L^\ell \subseteq S^{\ell k} E$ the image of $S^\ell L$ under the multiplication $m_{\ell,k}$. Then

$$\ker(S^\ell \phi) = S^\ell N + S^{\ell-1} N \cdot L + \ldots + N \cdot S^{\ell-1} L.$$

In particular, $L^\ell$ is not in the image of $\ker(S^\ell \phi)$ under $m_{\ell,k}$. By the remark at the beginning of the previous paragraph, this implies that $S^\ell \phi \circ i_{\ell,k}$ does not vanish on $L^\ell$, and $(**)$ is established, giving the surjectivity of $(*)$. A similar analysis shows that $\nu_{\ell,k}$ is one-to-one. This completes the proof.  $\square$

**Corollary 6.1.16. (Amplitude of tensor products).** *Let $E$ and $F$ be vector bundles on a projective variety or scheme $X$.*

(i). *If $E$ and $F$ are ample, then so is $E \otimes F$. Consequently any tensor power $T^q(E)$ of $E$ is ample.*

(ii). *If $E$ is ample, then so are all of its exterior powers. More generally, for any Young diagram $\lambda$ denote by $\Gamma^\lambda E$ the bundle deduced from $E$ via the representation of the general linear group corresponding to $\lambda$.[4] If $E$ is ample, then so is $\Gamma^\lambda E$.*

*Proof.* Since $E$ and $F$ are ample, so is $E \oplus F$ and consequently also $S^2(E \oplus F)$ (6.1.13, 6.1.15). But $E \otimes F$ is a summand of $S^2(E \oplus F)$, and hence is ample. The amplitude of $T^q(E)$ follows by induction. The exterior products $\Lambda^i E$ and more generally $\Gamma^\lambda E$ are quotients of $T^q(E)$ for suitable $q$, and (ii) follows.  $\square$

**Remark 6.1.17. (Positive characteristics).** All the results that have appeared so far remain valid for varieties defined over an algebraically closed field of arbitrary characteristic. The proof of Theorem 6.1.15 requires characteristic zero. However Barton [34] proved by other methods that the tensor product of ample vector bundles is ample in arbitrary characteristic, so 6.1.15 and 6.1.16 remain true. (In 6.1.16 (ii) one should define $\Gamma^\lambda E$ as in [207, Chapter 8.1], so that it is a quotient of a tensor power of $E$.)  $\square$

### 6.1.C Criteria for Amplitude

We now indicate a few criteria for amplitude and nefness parallel to some of the statements for line bundles from Sections 1.2 and 1.4.

To begin with, nefness and amplitude can be tested by pulling back to curves:

**Proposition 6.1.18. (Barton–Kleiman criterion).** *Let $E$ be a vector bundle on $X$.*

---

[4] See [210, §15.5] or Section 7.3.B.

(i). *E is nef if and only if the following condition is satisfied:*

> *Given any finite map $\nu : C \longrightarrow X$ from a smooth irreducible projective curve $C$ to $X$, and given any quotient line bundle $L$ of $\nu^*E$, one has*
>
> $$\deg L \; \geq \; 0.$$

(ii). *Fix an ample divisor class $h \in N^1(X)$ on $X$. Then $E$ is ample if and only if there exists a positive rational number $\delta = \delta_h > 0$ such that*

$$\deg L \; \geq \; \delta\left(C \cdot \nu^*h\right) \tag{*}$$

*for any $\nu : C \longrightarrow X$ and $\nu^*E \twoheadrightarrow L$ as above.*

Note that in (ii) the constant $\delta$ is required to be independent of $C$, $\nu$, and $L$.

*Proof.* Recall that giving a line bundle quotient $\nu^*E \twoheadrightarrow L$ is the same as giving a map $\mu : C \longrightarrow \mathbf{P}(E)$ commuting with the projections to $X$:

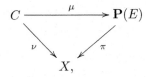

with $L = \mu^*\mathcal{O}_{\mathbf{P}(E)}(1)$. So the property stated in (i) is evidently equivalent to the nefness of $\mathcal{O}_{\mathbf{P}(E)}(1)$. For (ii), assuming that (*) holds we use Corollary 1.4.10 to establish the amplitude of $E$. In fact, as usual write

$$\xi \; = \; c_1\left(\mathcal{O}_{\mathbf{P}(E)}(1)\right) \; \in \; N^1\left(\mathbf{P}(E)\right).$$

Then (*) is equivalent to requiring that the $\mathbf{Q}$-divisor class

$$\xi - \delta \cdot \pi^*h \; \in \; N^1\left(\mathbf{P}(E)\right)_{\mathbf{Q}}$$

be nef. On the other hand, since $\mathcal{O}_{\mathbf{P}(E)}(1)$ is relatively ample with respect to $\pi$, the class $a\xi + \delta \cdot \pi^*h$ is ample for $0 < a \ll 1$ (Proposition 1.7.10). Therefore

$$\left(\xi - \delta \cdot \pi^*h\right) + \left(a\xi + \delta \cdot \pi^*h\right) \; = \; \left(1 + a\right) \cdot \xi$$

is the sum of a nef and an ample class, and hence is ample. Thus $\mathcal{O}_{\mathbf{P}(E)}(1)$ is likewise ample. We leave the converse to the reader. $\qquad\square$

The proposition shows that twisting by numerically trivial line bundles does not affect amplitude:

**Corollary 6.1.19. (Numerically trivial twists).** *Let $P$ be a numerically trivial line bundle on $X$. Then given any vector bundle $E$ on $X$, $E$ is ample or nef if and only if $E \otimes P$ is so.* $\qquad\square$

(Of course one can also argue directly. In fact, $\mathbf{P}(E \otimes P) \cong \mathbf{P}(E)$ by an isomorphism under which $\mathcal{O}_{\mathbf{P}(E \otimes P)}(1)$ corresponds to $\mathcal{O}_{\mathbf{P}(E)}(1) \otimes \pi^* P$ on $\mathbf{P}(E)$. But under the stated hypothesis on $P$, the latter bundle is numerically equivalent to $\mathcal{O}_{\mathbf{P}(E)}(1)$.)

We conclude by sketching some other criteria.

**Example 6.1.20. (Analogue of Seshadri's criterion).** Given a reduced irreducible curve $C$, write $m(C)$ for the maximum of the multiplicities of $C$ at all points $P \in C$. Now suppose that $E$ is a vector bundle on the projective variety $X$. Then $E$ is ample if and only if the following condition is satisfied:

There is a real number $\varepsilon > 0$ such that for every non-constant map

$$\nu : C \longrightarrow X$$

from a reduced and irreducible (but possibly singular) curve $C$ to $X$, and for any quotient line bundle $L$ of $\nu^* E$, one has

$$\frac{\deg_C L}{m(C)} \geq \varepsilon. \tag{*}$$

(To show that (*) implies the amplitude of $E$, fix any curve $C \subseteq \mathbf{P}(E)$ finite over $X$. As in the proof of 6.1.18, (*) implies that

$$\deg \left( \mathcal{O}_{\mathbf{P}(E)}(1) \,|\, C \right) \geq \varepsilon \cdot m(C).$$

On the other hand, if $C$ lies in a fibre of $\pi : \mathbf{P}(E) \longrightarrow X$, then

$$\deg \left( \mathcal{O}_{\mathbf{P}(E)}(1) \,|\, C \right) \geq m(C).$$

Therefore it follows from Seshadri's criterion (Theorem 1.4.13) that $\mathcal{O}_{\mathbf{P}(E)}(1)$ is ample.) $\qquad \square$

**Example 6.1.21. (Generalization of Gieseker's lemma).** Let $E$ be a vector bundle on $X$. Then $E$ is ample if and only if the following two conditions are satisfied:

(i). There is a positive integer $m_0 = m_0(E)$ such that $S^m E$ is globally generated for every $m \geq m_0$.

(ii). There is no reduced irreducible curve $C \subset X$ such that $E|C$ admits a trivial quotient.

(Assuming the conditions are satisfied, it follows from (i) that $\mathcal{O}_{\mathbf{P}(E)}(m_0)$ is globally generated and hence defines a morphism

$$\phi = \phi_{m_0} : \mathbf{P}(E) \longrightarrow \mathbf{P} = \mathbf{P} H^0(S^{m_0} E).$$

The issue is to show that $\phi$ is finite. If not there is a curve $C \subset \mathbf{P}(E)$ that $\phi$ contracts to a point. As in the proof of 6.1.7, $C$ must map isomorphically to its image in $X$, and it remains only to show that $\mathcal{O}_{\mathbf{P}(E)}(1)|C$ is trivial. But by construction $\mathcal{O}_{\mathbf{P}(E)}(m_0)|C$ is trivial, and hence

$$\mathcal{O}_{\mathbf{P}(E)}(1)|C \quad = \quad \mathcal{O}_{\mathbf{P}(E)}(m_0 + 1)|C$$

is a degree-zero line bundle on $C$ that (thanks to (i) for $m = m_0 + 1$) is generated by its global sections.)                                    □

**Example 6.1.22. (Grauert's criterion).** Let $E$ be a vector bundle on $X$, and denote by $\mathbf{F}$ the total space of the dual of $E$, so that $\mathbf{P}(E)$ is the projective bundle of one-dimensional subspaces of $\mathbf{F}$. Then $E$ is ample if and only if the zero section $\mathbf{0_F}$ can be blown down to a point. (In fact, let $\mathbf{L}$ denote the total space of the line bundle $L = \mathcal{O}_{\mathbf{P}(E)}(-1)$. One reduces to the case of line bundles by noting that the complement of the zero section in $\mathbf{F}$ is isomorphic to the complement of the zero section in $\mathbf{L}$. See [274, (3.5)] for details.)      □

**Example 6.1.23. (Big vector bundles).** A vector bundle $E$ on an irreducible projective variety $X$ is *big* if $\mathcal{O}_{\mathbf{P}(E)}(1)$ is a big line bundle on $\mathbf{P}(E)$ (Definition 2.2.1). The characterization of big divisors as sums of ample and effective classes (Corollary 2.2.7) generalizes to the bundle setting in a natural manner. Specifically, assume that $H^0(X, S^m E) \neq 0$ for some $m \geq 1$. Then for any ample line bundle $A$ on $X$, the bundle $E \otimes A$ is big. Conversely, if $E$ is big, then given any line bundle $B$ on $X$, $H^0(X, S^m E \otimes B) \neq 0$ for $m \gg 0$. (For the first statement, the hypothesis implies that $\xi = c_1\big(\mathcal{O}_{\mathbf{P}(E)}(1)\big)$ is an effective class. Then argue as in the proof of Proposition 6.1.18 or 6.2.11 that given any ample class $\alpha \in N^1(X)_{\mathbf{R}}$, $\xi + \pi^*\alpha$ is a big class on $\mathbf{P}(E)$. The second assertion follows quickly from Kodaira's Lemma 2.2.6 applied on $\mathbf{P}(E)$.) Analogous statements — which we leave to the reader — hold in the setting of $\mathbf{Q}$-twisted bundles (Section 6.2).      □

### 6.1.D Metric Approaches to Positivity of Vector Bundles

In the case of line bundles, amplitude is equivalent to the existence of a metric of positive curvature, and it is natural to attempt to generalize this approach to higher ranks. Griffiths [247] defined a differential-geometric notion of positivity that seems reasonably close to amplitude, although it is unknown whether the two concepts actually coincide. The present section is devoted to a very brief sketch of such metric definitions of positivity for bundles. We refer to [248, Chapter 0, §5], [126, §3], [381, Chapter 5], [352, Chapter III, §6] or [533, Chapter VI] for background and more details. We follow the notation and presentation of [126].[5]

Let $X$ be a complex manifold of dimension $n$, and $E$ a holomorphic vector bundle of rank $e$ on $X$ equipped with a Hermitian metric $h$. The Hermitian

---

[5] Concerning the various sign conventions in the literature, see [381, p. 136].

bundle $(E, h)$ determines a unique Hermitian connection $D_E$ compatible with the complex structures on $X$ and $E$, and $D_E$ in turn gives rise to a curvature tensor

$$\Theta(E, h) \in \mathcal{C}^\infty\big(X, \Lambda^{1,1} T_X^* \otimes \mathrm{Hom}(E, E)\big),$$

a $\mathrm{Hom}(E, E)$-valued $(1, 1)$-form on $X$. If $z_1, \ldots, z_n$ are local analytic coordinates on $X$, and if $(e_\lambda)_{(1 \le \lambda \le e)}$ is a local orthogonal frame on $E$, then one can write

$$i \cdot \Theta(E, h) = \sum_{\substack{1 \le j, k \le n \\ 1 \le \lambda, \mu \le e}} c_{jk\lambda\mu} \cdot dz_j \wedge d\bar{z}_k \otimes e_\lambda^* \otimes e_\mu,$$

where $\bar{c}_{jk\lambda\mu} = c_{kj\mu\lambda}$. This curvature tensor gives rise to a Hermitian form $\theta_E$ on the bundle $T_X \otimes E$, given locally by

$$\theta_E = \sum_{j,k,\lambda,\mu} c_{jk\lambda\mu} \cdot (dz_j \otimes e_\lambda^*) \otimes \overline{(dz_k \otimes e_\mu^*)}.$$

**Definition 6.1.24.** (i). $E$ is *Nakano-positive* if $\theta_E$ is a positive-definite Hermitian form on $TX \otimes E$.

(ii). $E$ is *Griffiths-positive* if $\theta_E$ is positive on all simple tensors in $TX \otimes E$, i.e. if at every point $x \in X$,

$$\theta_E(\xi \otimes s, \xi \otimes s) > 0 \quad \text{for all } 0 \ne \xi \in T_x X \text{ and } 0 \ne s \in E(x).$$

Nakano and Griffiths semipositivity, negativity, and seminegativity are defined analogously.    □

If we fix a holomorphic tangent vector $\xi \in T_x X$, then we can view

$$\theta_{E,\xi} = \theta_E(\xi \otimes \bullet, \xi \otimes \bullet)$$

as a Hermitian form on $E(x)$. The condition of Griffiths positivity is that $\theta_{E,\xi}$ should be positive definite for every non-zero $\xi$. Note that if $E$ has rank one, then the Nakano and Griffiths definitions both coincide with positivity in the sense of Kodaira.

The connection between these notions is given by

**Theorem 6.1.25.** *Let $(E, h)$ be a Hermitian vector bundle on a complex projective manifold $X$.*

(i). *If $(E, h)$ is Nakano-positive, then $(E, h)$ is Griffiths-positive.*

(ii). *If $(E, h)$ is Griffiths-positive, then $E$ is ample.*

Statement (i) of the theorem is clear from the definitions, and we refer to [533, Theorem 6.30] for a proof of (ii). The idea, naturally enough, is to consider the projective bundle $\pi : \mathbf{P}(E) \longrightarrow X$. The given Hermitian metric $h$ on $E$ determines one on $\pi^* E$, which in turn induces a natural Hermitian metric $k$

on its quotient $\mathcal{O}_{\mathbf{P}(E)}(1)$. Then one uses the principle that curvature increases in quotients to show that the corresponding curvature form $\Theta(\mathcal{O}_{\mathbf{P}(E)}(1), k)$ is positive, and hence that this line bundle is ample.

The Nakano condition is known to be considerably stronger than Griffiths positivity and amplitude (cf. Example 7.3.18). It is an interesting open question whether or not Griffiths-positivity is equivalent to amplitude, i.e. whether every ample bundle carries a Griffiths-positive metric. Fulton [203, p. 28] points out that a weaker condition would have many of the same consequences.

**Remark 6.1.26. (Garrity's theorem).** Garrity has given an alternate metric approach to amplitude coming from Proposition 6.1.18. A detailed account appears in [381, Chapter V, §4]. □

## 6.2 Q-Twisted and Nef Bundles

As we saw in Chapter 1, the language of **Q**-divisors greatly facilitates discussions of positivity for line bundles. In the present section, we start by developing an analogous formalism involving twists of vector bundles by **Q**-divisor classes. This device was initiated by Miyaoka [429], and its utility was re-emphasized by some related constructions in [133]. As in the rank-one case, the formalism will allow us first of all to treat nef bundles as limits of ample ones. Beyond that, it absorbs many of the branched covering arguments that arise in previous developments of the theory. In the second subsection, we apply the formalism to establish the basic properties of nef bundles.

Throughout this section, $X$ denotes a complex projective variety or scheme. Concerning divisors, we adhere to the notation and conventions introduced in Sections 1.1 and 1.3. In particular, $\mathrm{Div}(X)$ denotes the group of Cartier divisors on $X$, and by a **Q**-divisor we understand an element of $\mathrm{Div}_{\mathbf{Q}}(X) = \mathrm{Div}(X) \otimes_{\mathbf{Z}} \mathbf{Q}$.

### 6.2.A  Twists by Q-Divisors

Whereas the definition of a **Q**-divisor did not present any difficulties, it is not really clear what one might mean by a **Q**-vector bundle. However, for many purposes — notably for questions of positivity and numerical properties — there is a natural way to make sense of twisting a bundle by a **Q**-divisor (class).

We start by defining formally the objects with which we shall deal:

**Definition 6.2.1. (Q-twisted bundles).** A **Q**-*twisted vector bundle*

$$E\langle\delta\rangle$$

on $X$ is an ordered pair consisting of a vector bundle $E$ on $X$, defined up to isomorphism, and a numerical equivalence class $\delta \in N^1(X)_{\mathbf{Q}}$. If $D \in \mathrm{Div}_{\mathbf{Q}}(X)$ is a $\mathbf{Q}$-divisor, we write $E{<}D{>}$ for the twist of $E$ by the numerical equivalence class of $D$.                                                                                               □

So in other words, $E{<}D{>}$ and $E{<}\delta{>}$ are just formal symbols, but the notation is intended to suggest that we are twisting $E$ by the $\mathbf{Q}$-divisor $D$ or class $\delta$. This leads to a natural notion of equivalence among such pairs:

**Definition 6.2.2. (Q-isomorphism).** We define **Q-isomorphism** of **Q**-twisted bundles to be the equivalence relation generated by declaring that $E{<}A + D{>}$ be equivalent to

$$\big(E \otimes \mathcal{O}_X(A)\big){<}D{>}$$

whenever $A$ is an integral Cartier divisor on $X$ and $D \in \mathrm{Div}_{\mathbf{Q}}(X)$.      □

We generally identify **Q**-isomorphic **Q**-twisted bundles without explicit mention. Therefore we will take care that further definitions respect this equivalence relation.

A "classical" (untwisted) bundle $E$ may be considered in the natural way as a $\mathbf{Q}$-twisted bundle, viz. as $E{<}0{>}$. Note that then $E_1$ and $E_2$ determine the same $\mathbf{Q}$-twisted bundle — i.e. $E_1{<}0{>}$ and $E_2{<}0{>}$ are $\mathbf{Q}$-isomorphic — if $E_1 = E_2 \otimes P$ for a numerically trivial line bundle $P$. But as long as we are dealing with positivity or numerical properties, this shouldn't cause undue confusion.

We have already had occasion to note that if $E$ is a vector bundle and $A$ is a line bundle on $X$, then $\mathbf{P}(E \otimes A) \cong \mathbf{P}(E)$ by an isomorphism under which $\mathcal{O}_{\mathbf{P}(E \otimes A)}(1)$ corresponds to the bundle

$$\mathcal{O}_{\mathbf{P}(E)}(1) \otimes \pi^* A$$

on $\mathbf{P}(E)$, where $\pi : \mathbf{P}(E) \longrightarrow X$ denotes as usual the bundle map. This motivates

**Definition 6.2.3. (Ample and nef Q-twisted bundles).** A $\mathbf{Q}$-twisted vector bundle $E{<}\delta{>}$ is *ample* (or *nef*) if

$$\xi_E + \pi^* \delta \; \in \; N^1\big(\mathbf{P}(E)\big)_{\mathbf{Q}}$$

is an ample (or nef) $\mathbf{Q}$-divisor class on $\mathbf{P}(E)$. Here $\xi_E = c_1\big(\mathcal{O}_{\mathbf{P}(E)}(1)\big)$ is the class of a divisor representing the Serre line bundle $\mathcal{O}_{\mathbf{P}(E)}(1)$.      □

Observe that $E$ is ample or nef in the usual sense if and only if it so considered as the $\mathbf{Q}$-twisted bundle $E{<}0{>}$. The remarks preceding the definition show that the notions of amplitude and nefness just introduced respect $\mathbf{Q}$-isomorphism.

The same is true for the natural definitions of tensor products and pullbacks:

**Definition 6.2.4. (Tensorial operations and pullbacks).** Let $X$ be a projective variety or scheme, and let $E$, $E_1$, $E_2$ be vector bundles on $X$.

(i).   Tensor, symmetric, and exterior powers of $\mathbf{Q}$-twisted bundles are defined via the rules

$$E_1 <\delta_1> \ \otimes \ E_2 <\delta_2> \ = \ (E_1 \otimes E_2) <\delta_1 + \delta_2>, \qquad (6.4a)$$
$$S^m ( E<\delta> ) \ = \ (S^m E) <m\delta>, \qquad (6.4b)$$
$$\Lambda^m ( E<\delta> ) \ = \ (\Lambda^m E) <m\delta> . \qquad (6.4c)$$

(ii).  If $f : Y \longrightarrow X$ is a morphism, then we define the pullback of a $\mathbf{Q}$-twisted bundle $E<\delta>$ on $X$ to be the $\mathbf{Q}$-twisted bundle

$$f^* (E<\delta>) \ = \ ( f^* E ) <f^* \delta> \qquad (6.5)$$

on $Y$.    □

Note that it follows from the definition that pullback commutes with each of the products in (i).

There is also a natural notion of quotients, sums, and ranks:

**Definition 6.2.5. (Quotients, sums, and ranks).** Keep notation as above.

(i).   A *quotient* of a $\mathbf{Q}$-twisted bundle $E<\delta>$ is a $\mathbf{Q}$-twisted bundle $Q<\delta>$, where $Q$ is a quotient of $E$. Sub-bundles and locally free subsheaves of $E<\delta>$ are defined similarly.

(ii).  The *direct sum* of two $\mathbf{Q}$-twisted bundles $E_1<\delta>$ and $E_2<\delta>$ with the same twisting class is the $\mathbf{Q}$-twisted bundle $(E_1 \oplus E_2)<\delta>$. Extensions are defined similarly.

(iii). The *rank* of $E<\delta>$ is simply $\mathrm{rank}(E)$.    □

**Remark 6.2.6.** Note that we do not attempt to define direct sums or extensions of two vector bundles twisted by different numerical equivalence classes.    □

We will discuss Chern classes of $\mathbf{Q}$-twisted bundles later in Section 8.1.A. However a special case will be useful before then:

**Definition 6.2.7. (Degrees of Q-twists on curves).** If $E<\delta>$ is a $\mathbf{Q}$-twisted bundle of rank $e$ on a curve $X$, then its *degree* is the rational number

$$\deg E + e \cdot \deg \delta. \quad □$$

The basic facts about $\mathbf{Q}$-twisted bundles are summarized in the next lemma. The first two statements allow one to pass to covers to reduce to the case of integral divisors.

**Lemma 6.2.8. (Formal properties of Q-twists).** *Let $E$ be a vector bundle on $X$ and $\delta \in N^1(X)_{\mathbf{Q}}$ a numerical equivalence class.*

(i). *The definitions above respect the relation of **Q**-isomorphism. In particular, if $A$ is an integral Cartier divisor on $X$, then $E{<}A{>}$ is ample (or nef) if and only if $E \otimes \mathcal{O}_X(A)$ is.*

(ii). *If $f : Y \longrightarrow X$ is a finite surjective map, then $f^*(E{<}\delta{>})$ is ample (or nef) on $Y$ if and only if $E{<}\delta{>}$ is ample (or nef) on $X$.*

(iii). *$E{<}\delta{>}$ is ample if and only if $S^k(E{<}\delta{>})$ is ample for some — or equivalently, for all — integers $k \geq 1$.*

(iv). *If $E_1{<}\delta_1{>}$ and $E_2{<}\delta_2{>}$ are ample **Q**-twisted bundles on $X$, then*

$$E_1{<}\delta_1{>} \otimes E_2{<}\delta_2{>}$$

*is also ample.*

(v). *If $E_1{<}\delta{>}$ and $E_2{<}\delta{>}$ are ample, then so too is their direct sum (or any extension of one by the other).*

(vi). *Suppose that $E{<}\delta{>}$ is ample, and let $\delta'$ be any **Q**-divisor class on $X$. Then $E{<}\delta + \varepsilon \cdot \delta'{>}$ is ample for all sufficiently small rational numbers $0 < \varepsilon \ll 1$.*

*Proof.* The first assertion has already been observed. The second statement follows from the fact that amplitude or nefness for **Q**-divisor classes can be tested after pulling back by a finite surjective map (1.2.28, 1.4.4 (ii)). For (iii), use the covering constructions from Section 4.1 (e.g. Theorem 4.1.10) to form a branched covering $f : Y \longrightarrow X$ such that $f^*\delta$ is an integral class on $Y$. By (ii) it is equivalent to prove the statement after pulling back to $Y$. But thanks to (i), here we are reduced to Theorem 6.1.15. For (iv) and (v) one reduces in a similar manner to the case in which $\delta_1$ and $\delta_2$ are integral, and then 6.1.16 and 6.1.13 apply. Finally, (vi) is a consequence of the fact (1.3.7) that the ample cone in $N^1(\mathbf{P}(E))_{\mathbf{Q}}$ is open. $\qquad\square$

**Remark 6.2.9.** After the evident modifications, Proposition 6.1.18 and Example 6.1.20 extend to **Q**-twisted bundles. We leave the statements to the reader. $\qquad\square$

**Remark 6.2.10. (R-twists).** With some extra care, it would be possible to define and work out the elementary properties of **R**-twisted bundles. However, except at one point in Chapter 8 — viz. Theorem 8.2.1, where we will proceed ad hoc — we do not require this formalism, and it seemed simplest to bypass it. $\qquad\square$

## 6.2.B Nef Bundles

With the formalism of **Q**-twists in hand, the basic properties of nef bundles follow easily from the corresponding facts for amplitude.

We start with the analogue of Corollary 1.4.10, showing that a bundle is nef if and only if it is a limit of ample **Q**-twisted bundles:

**Proposition 6.2.11. (Ample twists of nef bundles).** *A vector bundle $E$ on $X$ is nef if and only if the **Q**-twisted bundle $E{<}h{>}$ is ample for every ample class $h \in N^1(X)_{\mathbf{Q}}$. Similarly, if $E{<}\delta{>}$ is a **Q**-twisted bundle, then $E{<}\delta{>}$ is nef if and only if $E{<}\delta + h{>}$ is ample for any such $h$.*

*Proof.* If $E{<}h{>}$ is ample for every ample class $h$, then $\xi = \xi_E$ is the limit as $h \to 0$ of the ample classes $\xi + \pi^*h$, and hence is nef. Assuming conversely that $E$ is nef, we argue as in the proof of 6.1.18. Specifically, since $\xi$ is ample for $\pi$, $a\xi + \pi^*h$ is an ample class on $\mathbf{P}(E)$ for $0 < a \ll 1$ (Proposition 1.7.10). Therefore

$$\xi + \pi^*h \; = \; \big((1-a)\xi\big) + \big(a\xi + \pi^*h\big),$$

being the sum of a nef and an ample class, is ample (Corollary 1.4.10). Hence $E{<}h{>}$ is ample. The statement for **Q**-twists is similar.     □

It is now immediate to establish the analogues for nef bundles of the formal properties of ample bundles given earlier. We state them all explicitly for ease of reference.

**Theorem 6.2.12. (Formal properties of nef vector bundles).** *Let $X$ be a projective variety or scheme.*

(i).   *Quotients and arbitrary pullbacks of nef bundles on $X$ are nef. Given a vector bundle $E$ on $X$, and a surjective morphism $f : Y \longrightarrow X$ from a projective variety (or scheme)) to $X$, if $f^*E$ is a nef bundle on $Y$, then $E$ is nef.*

(ii).   *Direct sums and extensions of nef bundles are nef.*

(iii).   *A vector bundle $E$ on $X$ is nef if and only if $S^k E$ is nef for any — or equivalently for all — $k \geq 1$.*

(iv).   *Any tensor or exterior product of nef bundles is nef. If $E$ is nef and $F$ is ample, then $E \otimes F$ is ample.*

(v).   *All of these statements hold if the "classical" bundles involved are replaced by **Q**-twists (provided in (ii) that the sum or extension is defined).*

*Proof.* Statement (i) follows immediately from the corresponding facts for nef line bundles. For the first assertion in (ii), suppose that $E_1$ and $E_2$ are nef, and fix any ample **Q**-divisor class $h \in N^1(X)_{\mathbf{Q}}$. By the previous proposition, it suffices to prove that $(E_1 \oplus E_2){<}h{>}$ is an ample **Q**-twisted bundle, and this follows from Lemma 6.2.8 (v). Extensions are treated in the same manner. Turning to (iii), note that

$$\left(S^k E\right)<\delta> = \ S^k\big(E<\tfrac{1}{k}\delta>\big),$$

and so the amplitude of $\left(S^k E\right)<h>$ for all ample **Q**-divisor classes $h$ is equivalent to the amplitude of $S^k\big(E<h'>\big)$ for all ample classes $h'$. Therefore we are reduced by the previous proposition to Lemma 6.2.8 (iii). The first assertion of (iv) is similar, and in characteristic zero the nefness of tensor powers implies in the usual way the nefness of exterior products. For the last statement in (iv), suppose that $E$ is nef and $F$ is ample, and let $h$ be any ample class on $X$. Then $E<\varepsilon h>$ is ample for all $\varepsilon > 0$ by 6.2.11, and $F<-\varepsilon h>$ is ample for $0 < \varepsilon \ll 1$ thanks to 6.2.8 (vi). Therefore 6.2.8 (iv) applies to show that

$$E \otimes F \ = \ E<\varepsilon h> \, \otimes \, F<-\varepsilon h>$$

is ample. We leave the extensions to **Q**-twisted bundles to the reader.    □

**Example 6.2.13. (Fixed twists of large symmetric powers).** Let $E$ be a vector bundle and $B$ an ample divisor on $X$. If $S^m E \otimes \mathcal{O}_X(B)$ is nef for all $m \gg 0$, then $E$ itself is nef. Similarly, the nefness of the $m$-fold tensor power $T^m E \otimes \mathcal{O}_X(B)$ for every $m \gg 0$ implies the nefness of $E$. (Replacing $B$ by $2B$, there is no loss in generality in supposing that $S^m E \otimes \mathcal{O}_X(B)$ is ample for all $m \gg 0$. But considered as a **Q**-twisted bundle, $S^m E \otimes \mathcal{O}_X(B)$ is **Q**-isomorphic to

$$S^m\big(E<\tfrac{1}{m}B>\big).$$

Therefore $E<\tfrac{1}{m}B>$ is ample for all $m \gg 0$, and hence $E$ is nef. The analogous assertion for tensor powers follows from this since in characteristic zero $S^m E$ is a summand of $T^m E$.)    □

**Example 6.2.14. (The Barton invariant of a bundle).** Let $E$ be a vector bundle and $h$ an ample divisor class on the projective variety $X$. Define the *Barton invariant* of $E$ with respect to $h$ to be the real number

$$\delta(X, E, h) \ = \ \sup\big\{t \in \mathbf{Q} \mid E<-t \cdot h> \ \text{is nef}\,\big\}.$$

(i).  $E$ is ample if and only if $\delta(X, E, h) > 0$, and $E$ is nef if and only if $\delta(X, E, h) \geq 0$. (Compare Proposition 6.1.18.)

(ii).  For every integer $m > 0$, $\delta(X, E, m \cdot h) = \tfrac{1}{m}\delta(X, E, h)$.

(iii).  If $f : Y \longrightarrow X$ is a finite surjective mapping, then

$$\delta(Y, f^* E, f^* h) = \delta(X, E, h).$$

Given an ample divisor $D$ or line bundle bundle $B$, we write $\delta(X, E, D)$ or $\delta(X, E, B)$ for the quantity defined by using $D$ or $B$ in place of $h$.    □

**Example 6.2.15. (Irrational Barton invariants).** It is possible for the Barton invariant to be an irrational number. For a simple example, take $X = C \times C$ to be the product of an elliptic curve with itself, so that $\mathrm{Nef}(X)$ is a circular cone (Example 1.5.4). Given ample divisors $A$ and $H$ on $X$, put $E = \mathcal{O}_X(A)$. Then the Barton invariant $\delta(X, E, H)$ of $E$ with respect to $H$ is the smallest root of the quadratic polynomial $s(t) = ((A - tH)^2)$, and for general choices of $A$ and $H$ this will be irrational (compare Section 2.3.B). Peternell (private communication) has constructed more interesting examples involving bundles of higher rank.    □

**Remark 6.2.16. (Positive characteristics).** Except for the second assertion of Example 6.2.13, all of the material appearing so far in this section remains valid for varieties defined over an algebraically closed field of arbitrary characteristic.    □

By analogy with the corresponding notion for line bundles (Section 2.1.B), it is natural to define semiamplitude for vector bundles.

**Definition 6.2.17. (Semiample vector bundles).** A vector bundle $E$ on a complete variety or scheme is *semiample* if $\mathcal{O}_{\mathbf{P}(E)}(m)$ is globally generated for some $m > 0$.    □

The condition is satisfied for example if $S^m(E)$ is globally generated for some $m > 0$. Evidently a semiample bundle is nef.

**Remark 6.2.18. (k-amplitude).** Sommese [551] has introduced a quantitative measure of how close a semiample bundle comes to being ample. Specifically, let $E$ be a semiample bundle – so that $\mathcal{O}_{\mathbf{P}(E)}(m)$ is free for some $m > 0$ – and let

$$\phi = \phi_m : \mathbf{P}(E) \longrightarrow \mathbf{P}\big(H^0\big(\mathbf{P}(E), \mathcal{O}_{\mathbf{P}(E)}(m)\big)\big)$$

be the corresponding map. Sommese defines $E$ to be *k-ample* if every fibre of $\phi$ has dimension $\leq k$. Thus $E$ is ample if and only if it is 0-ample. Many of the basic facts about amplitude extend with natural modifications to this more general setting. We shall point out some of these as we go along, and the reader can consult [551] for a more complete survey.    □

**Example 6.2.19. (Some formal properties of k-amplitude).** We indicate how some of the results of this and the previous section extend to $k$-ample bundles on a projective scheme $X$ (Remark 6.2.18).

(i).    Any quotient of a $k$-ample bundle is $k$-ample.

(ii).   If $E$ is a $k$-ample bundle on $X$ and $f : Y \longrightarrow X$ is a projective morphism all of whose fibres have dimension $\leq m$, then $f^*E$ is $(k + m)$-ample.

(iii). If $E$ is $k$-ample then for any coherent sheaf $\mathcal{F}$ on $X$ there is an integer $m(\mathcal{F})$ such that

$$H^i\big(X, S^m E \otimes \mathcal{F}\big) \;=\; 0 \quad \text{for} \;\; m \geq m(\mathcal{F}) \;\; \text{and} \;\; i > k;$$

the converse holds provided that $E$ is semiample.

(iv). If $E$ and $F$ are globally generated and $k$-ample, then $E \otimes F$ and $E \oplus F$ and any extension of $E$ by $F$ are $k$-ample.

(See [551, §1, (1.7), (1.9), and (1.10)].) $\hfill\square$

## 6.3 Examples and Constructions

In order to add substance to the general theory, we present in this section several examples and constructions of ample and nef vector bundles. Our hope is to convey a sense of some of the many settings in which positivity of bundles arises "in nature," and to illustrate a few of the methods that have been used to detect and exploit it.

In the first two subsections, we discuss the geometric consequences of positivity conditions on tangent, cotangent, and normal bundles. We then consider the Picard bundles on the Jacobian of a curve (and their analogues for irregular varieties of arbitrary dimension), and prove the positivity of a vector bundle associated to a branched covering of projective space. Direct images of canonical bundles are discussed briefly in 6.3.E, while in 6.3.F we indicate some methods of construction. On several occasions we call on results to be established later, but we felt that the value of presenting early on some non-trivial illustrations of the theory outweighs any lapses in strict logical development.

### 6.3.A Normal and Tangent Bundles

Historically, an important motivation for developing the theory of ample vector bundles was the desire to generalize to higher codimensions some of the positivity properties enjoyed by ample divisors. If $M$ is a smooth projective variety, and if $X \subset M$ is a non-singular ample divisor, then of course its normal bundle $N_{X/M} = \mathcal{O}_M(X)|X$ is an ample line bundle on $X$. Given $X \subset M$ of arbitrary codimension, one would like to think of the amplitude of $N_{X/M}$ as reflecting the intuition that $X$ is "positively embedded" in $M$. While this has never been made precise, we present here some of the basic examples and results supporting this heuristic. We start by considering subvarieties of projective space and abelian varieties, and then briefly discuss what is known on arbitrary smooth ambient varieties. Hartshorne's book [276] remains a valuable source of information on these matters.

**Projective space and its subvarieties.** The positivity of projective space is manifested in the amplitude of its tangent bundle and the normal bundles of all smooth subvarieties.

**Proposition 6.3.1.** (i). *The tangent bundle $T\mathbf{P}^n$ of $n$-dimensional projective space is ample.*

(ii). *If $X \subseteq \mathbf{P}^n$ is any smooth subvariety, then the normal bundle $N_{X/\mathbf{P}}$ to $X$ in $\mathbf{P} = \mathbf{P}^n$ is ample.*

*Proof.* Let $V$ be a vector space of dimension $n+1$, so that $\mathbf{P}(V) = \mathbf{P}^n$. Then one has the Euler sequence [280, II.8.13]

$$0 \longrightarrow \mathcal{O}_{\mathbf{P}^n} \longrightarrow V^* \otimes_k \mathcal{O}_{\mathbf{P}^n}(1) \longrightarrow T\mathbf{P}^n \longrightarrow 0. \tag{6.6}$$

So $T\mathbf{P}^n$ is a quotient of the $(n+1)$-fold direct sum of copies of $\mathcal{O}_{\mathbf{P}}(1)$, and hence is ample. The second statement then follows from the normal bundle sequence

$$0 \longrightarrow TX \longrightarrow T\mathbf{P}^n|X \longrightarrow N_{X/\mathbf{P}} \longrightarrow 0. \tag{6.7}$$

In fact, the restriction $T\mathbf{P}^n|X$ is ample, and hence so it its quotient $N_{X/\mathbf{P}}$. □

**Remark 6.3.2. (Mori's theorem).** A very fundamental theorem of Mori [437] states that projective space is the only smooth projective variety with ample tangent bundle. This had been conjectured by Hartshorne [276], and Siu and Yau [542] proved a complex geometric analogue at essentially the same time as Mori's theorem. In the course of his argument, Mori introduced several spectacular ideas that soon led to a flowering of higher-dimensional geometry. We refer to [135] for an excellent overview of Mori's proof, and to [363] for an account of the developments growing in part out of his techniques. The papers [80], [81], [6], [91], and [603] contain some results and conjectures concerning other characterizations of projective space involving positive vector bundles. □

**Remark 6.3.3. (Nef tangent bundles).** Continuing the train of thought of the previous remark, it is natural to ask whether one can characterize complex varieties whose tangent bundles satisfy weaker positivity properties. The first results in this direction were obtained by Mok [435], who classified all compact Kähler manifolds with semi-positive bisectional curvature. Following work of Campana and Peternell, [79], Demailly, Peternell and Schneider [133] studied compact Kähler manifolds $X$ with nef tangent bundles.[6] They showed that $X$ admits a finite étale cover $\tilde{X} \longrightarrow X$ having the property that the Albanese mapping $\tilde{X} \longrightarrow \mathrm{Alb}(\tilde{X})$ is a smooth fibration whose fibres are Fano manifolds with nef tangent bundles. It is conjectured in [79] that a complex Fano variety

---

[6] On possibly non-algebraic Kähler manifolds, nefness of a bundle $E$ is defined by asking that $\mathcal{O}_{\mathbf{P}(E)}(1)$ carry a metric satisfying the condition of Remark 1.4.7. See [133, §1B].

with nef tangent bundle must be a rational homogeneous space $G/P$. If true, this would give a complete picture, up to étale covers, of Kähler manifolds whose tangent bundles are nef. An amusing numerical property of varieties with nef tangent bundles appears in Corollary 8.4.4.    □

**Remark 6.3.4. (Bundles of differential operators).** In a somewhat related direction, Ran and Clemens [512] use very interesting considerations of positivity and stability for sheaves of differential operators to study the geometry of Fano manifolds of Picard number one.    □

While every smooth subvariety of projective space has ample normal bundle, it was observed in [211, Remark 7.5] that a twist of that bundle carries additional geometric information:

**Proposition 6.3.5. (Amplitude of N(-1)).** *Given a smooth subvariety* $X \subset \mathbf{P} = \mathbf{P}^n$ *not contained in any hyperplane, the twisted normal bundle* $N_{X/\mathbf{P}}(-1)$ *is ample if and only if every hyperplane* $H \subset \mathbf{P}$ *that is tangent to* $X$ *is tangent at only finitely many points.*

The condition is equivalent to asking that for every hyperplane $H$, $X \cap H$ have at most finitely many singular points.

*Proof of 6.3.5.* Restricting the Euler sequence to $X$, and combining it with the normal bundle sequence, one arrives at a bundle surjection

$$V_X^* = V^* \otimes \mathcal{O}_X \longrightarrow N_{X/\mathbf{P}}(-1) \longrightarrow 0.$$

This gives rise to an embedding

$$\mathbf{P}\big(N_{X/\mathbf{P}}(-1)\big) \subseteq \mathbf{P}\big(V_X^*\big) = X \times \mathbf{P}(V)^*,$$

and in $X \times \mathbf{P}(V)^*$, $\mathbf{P}\big(N_{X/\mathbf{P}}(-1)\big)$ is identified with the locus

$$\big\{ (x, H) \mid x \in X, \ H \subset \mathbf{P}^n \text{ a hyperplane tangent to } X \text{ at } x \ \big\}.$$

So the condition in the proposition is equivalent to the finiteness of the projection $\mathbf{P}(N_{X/\mathbf{P}}(-1)) \longrightarrow \mathbf{P}(V)^*$, which in turn is equivalent to the amplitude of $N_{X/\mathbf{P}}(-1)$.    □

**Example 6.3.6. (Tangencies to complete intersections).** Let $X \subseteq \mathbf{P}^n$ be a smooth complete intersection of hypersurfaces of degrees $\geq 2$. Then a hyperplane can be tangent to $X$ at only a finite number of points.    □

**Example 6.3.7.** Given a smooth non-degenerate subvariety $X \subseteq \mathbf{P} = \mathbf{P}^n$, the twisted normal bundle $N_{X/\mathbf{P}}(-1)$ is $k$-ample in the sense of Sommese (Remark 6.2.18) if and only if no hyperplane is tangent to $X$ along a subset of dimension $\geq k + 1$. It follows from Zak's theorem on tangencies (Theorem 3.4.17) that in fact this always holds with $k = \mathrm{codim}(X, \mathbf{P}) - 1$.    □

**Example 6.3.8. (Tangencies along hypersurfaces).** Proposition 6.3.5 admits a partial generalization to other twists. Specifically let $X \subseteq \mathbf{P} = \mathbf{P}^n$ be a smooth subvariety, and let $S = S_d \subseteq \mathbf{P}^n$ be a non-singular hypersurface of degree $d$, not containing $X$. If the twisted normal bundle $N_{X/\mathbf{P}}(-d)$ is ample, then $S$ cannot be tangent to $X$ along a curve.[7] This applies for example when $X$ is the complete intersection of hypersurfaces of degrees $> d$. (In fact, suppose to the contrary that there exists a reduced and irreducible curve $C \subset X \cap S$ such that $TX|C \subseteq TS|C$. This gives rise to a surjective homomorphism

$$N_{X/\mathbf{P}}|C \longrightarrow N_{S/\mathbf{P}}|C$$

and hence a surjection $N_{X/\mathbf{P}}(-d)|C \twoheadrightarrow \mathcal{O}_C$. But an ample bundle on a curve does not admit a trivial quotient.) This result appears in [186, Proposition 4.3.6].    □

**Remark 6.3.9. (Normal bundles to local complete intersection subvarieties).** If $X \subset \mathbf{P}^n$ is a singular local complete intersection subvariety, then the normal bundle $N_{X/\mathbf{P}}$ is still defined, but one no longer has the normal bundle sequence (6.7). Therefore one can no longer conclude the amplitude of $N_{X/\mathbf{P}}$. This is discussed in Fritzsche's paper [189, §3]    □

**Subvarieties of abelian varieties.** Let $A$ be an abelian variety of dimension $n$. The tangent bundle $TA$ of $A$ is trivial, so the first interesting question is the amplitude of normal bundles. This was analyzed by Hartshorne [277]:

**Proposition 6.3.10.** *Let $X \subseteq A$ be a smooth subvariety, and denote by $N = N_{X/A}$ the normal bundle to $X$ in $A$. Then:*

(i). *$N$ is ample if and only if for every regular one-form $\omega \in \Gamma(A, \Omega_A^1)$, the restriction*

$$\omega \mid X \;\in\; \Gamma(X, \Omega_X^1)$$

*of $\omega$ to $X$ vanishes on at most a finite set.*

(ii). *If $N$ fails to be ample, then there is a reduced and irreducible curve $C \subseteq X$ that lies in a proper abelian subvariety of $A$.*

**Corollary 6.3.11. (Subvarieties of simple abelian varieties).** *If $A$ is simple, i.e. if it contains no proper abelian subvarieties, then every smooth subvariety of $A$ has ample normal bundle.*    □

*Proof of Proposition 6.3.10.* Write $TA = V \otimes \mathcal{O}_A$ where $V = T_0 A$ is a vector space of dimension $n$, which is canonically identified with the tangent space to $A$ at the origin $0 \in A$. By Gieseker's Lemma 6.1.7, $N$ fails to be ample if and only if there is a curve $C \subseteq X$ such that $N \mid C$ admits a trivial quotient. The normal bundle sequence

---

[7] By definition, $S$ is tangent to $X$ at a point $x \in S \cap X$ if $T_x X \subset T_x S$.

$$0 \longrightarrow TX \longrightarrow V_X = TA|X \longrightarrow N \longrightarrow 0$$

shows that this is equivalent to the existence of a one-form $\omega \in V^* = \Gamma\left(A, \Omega^1_A\right)$ whose restriction $\omega \,|\, X \in \Gamma\left(X, \Omega^1_X\right)$ vanishes along $C$. This proves the first assertion. Assuming such a curve exists, let $C'$ be its normalization, and denote by $\nu : C' \longrightarrow X$ the natural (finite) mapping. Then $\nu^*(\omega) = 0 \in \Gamma\left(C', \nu^* \Omega^1_X\right)$ and consequently

$$(d\nu)^* \omega \;=\; 0 \;\in\; \Gamma\left(C', \Omega^1_{C'}\right).$$

But this implies that the Jacobian $\mathrm{Jac}(C')$ — and hence also $C'$ itself — maps to a proper abelian subvariety of $A$. □

**Example 6.3.12.** The converse of the second statement of Proposition 6.3.10 is not true in general. (For instance, let $A$ be an abelian variety of dimension $n \ge 3$ that contains an elliptic curve $C \subseteq A$. Then there exist smooth ample divisors $X \subset A$ with $C \subset X$, but of course $N_{X/A} = \mathcal{O}_X(X)$ is ample.) □

**General ambient manifolds.** We now survey some of the geometric consequences of the amplitude of normal bundles in a general ambient manifold, referring to [276] for more information. Throughout this discussion, $M$ denotes a non-singular complex quasi-projective complex variety of dimension $n$, and $X \subseteq M$ is a smooth irreducible subvariety of dimension $d$. Unless otherwise stated we do not assume that $M$ is complete, but we always suppose that $X$ is projective. We denote by $N = N_{X/M}$ the normal bundle of $X$ in $M$.

As suggested above, the basic goal is to give some substance to the intuition that the amplitude of $N$ reflects the fact that $X$ is "positively embedded" in $M$. A first idea in this direction is to study the ring of formal functions along $X$, or more generally the sections of the formal completion along $X$ of a locally free sheaf on $M$. Along these lines one has:

**Proposition 6.3.13. (Formal functions along ample subvarieties).** *Assume that $d = \dim X \ge 1$ and that $N = N_{X/M}$ is ample, and consider the formal completion $\widehat{M} = \widehat{M}_{/X}$ of $M$ along $X$.*

(i). *The only formal holomorphic functions along $X$ are constants, i.e.*

$$H^0\left(\widehat{M}, \mathcal{O}_{\widehat{M}_{/X}}\right) \;=\; \mathbf{C}.$$

(ii). *Given a locally free sheaf $E$ on $M$, denote by $\widehat{E}$ its completion along $X$. Then $H^0\left(\widehat{M}, \widehat{E}\right)$ is finite dimensional.*

Recall that if $U \supseteq X$ is any connected neighborhood of $X$ in $M$, then the natural map

$$H^0\left(U, E|U\right) \longrightarrow H^0\left(\widehat{M}, \widehat{E}\right)$$

is injective. The same is true if $U$ is a connected neighborhood in the classical topology and the group on the left is replaced by the space $H^0(U_{an}, E_{an})$ of holomorphic sections of $E$ on $U$. So it follows that both these spaces of sections are finite dimensional. We refer to [276, Chapter 5] for a discussion of some related results of Hartshorne, Hironaka, Matsumura, and others concerning formal rational functions.

**Remark 6.3.14. (Contractible subvarieties).** To appreciate why this sort of statement is suggestive of the positivity of $X$ in $M$, consider by contrast the "opposite" case, when $X$ contracts. Specifically, suppose that $X$ is the fibre over a point $p \in N$ of a surjective mapping $f : M \longrightarrow N$ of $M$ onto a variety $N$ of dimension $\geq 1$. Then by the theorem on formal functions [280, III.11] the group appearing in 6.3.13 (i) is isomorphic to the completion of the stalk at $p$ of the sheaf $f_* \mathcal{O}_M$:

$$H^0(\widehat{M}, \mathcal{O}_{\widehat{M}/X}) = (\widehat{f_* \mathcal{O}_X})_p.$$

In particular, the group in question is infinite dimensional. Similarly, if $E = f^* F$ for some bundle $F$ on $N$, the finiteness in (ii) will also fail.    □

*Proof of Proposition 6.3.13.* We focus on statement (ii). Let $\mathcal{I}$ denote the ideal sheaf of $X$ in $M$, and let $X_k$ be the $k^{\text{th}}$ infinitesimal neighborhood of $X$ in $M$, i.e. the subscheme of $M$ defined by $\mathcal{I}^k$. Then

$$H^0(\widehat{M}, \widehat{E}) = \varprojlim H^0(M, E \otimes \mathcal{O}_{X_k}).$$

For every $k \geq 1$ there is an exact sequence

$$0 \longrightarrow E \otimes \mathcal{I}^k / \mathcal{I}^{k+1} \longrightarrow E \otimes \mathcal{O}_{X_{k+1}} \longrightarrow E \otimes \mathcal{O}_{X_k} \longrightarrow 0,$$

and one has the isomorphism $\mathcal{I}^k / \mathcal{I}^{k+1} = S^k N^*$ of $\mathcal{O}_X$-modules. It is enough for (ii) to show that $H^0(X, S^k N^* \otimes E) = 0$ if $k \gg 0$: for then $H^0(\widehat{M}, \widehat{E})$ injects into $H^0(X_k, E \otimes \mathcal{O}_{X_k})$ for some fixed $k \gg 0$, and hence is finite-dimensional. But since we are in characteristic zero, $S^k N^* = (S^k N)^*$. So for $k \gg 0$, $S^k N^* \otimes E$ is the dual of an ample bundle and hence has no non-vanishing sections, as required. By the same token, $H^0(X, S^k N^*) = 0$ for all $k \geq 1$, which in a similar fashion yields (i).    □

**Remark 6.3.15.** Since we only need to control $H^0$ in the argument just completed, Proposition 6.3.13 holds under considerably weaker conditions than the amplitude of $N = N_{X/M}$. For example, it suffices to assume the following:

> If $C \subseteq X$ is a curve arising as a general complete intersection of very ample divisors on $X$, then the restriction $N \mid C$ of $N$ to $C$ is ample.[8]

---

[8] This is analogous to the condition of generic semipositivity figuring in the theorem of Miyaoka described in Remark 6.3.34.

For then, keeping notation as in the previous proof, one has

$$H^0\big(C, (S^k N^* \otimes E) \mid C\big) \; = \; 0$$

for $k \gg 0$. Since this holds for the general member of a family of curves that covers $X$, it follows that $H^0\big(X, S^k N^* \otimes E\big) = 0$.    □

**Remark 6.3.16. (Finiteness of formal cohomology).** By a similar argument, Hartshorne [275, Theorem 5.1] proves in the situation of the Proposition that the formal cohomology groups $H^i\big(\widehat{M}, \widehat{E}\big)$ are finite dimensional for all $i < d$. Combining this with formal duality, he deduces [275, Corollary 5.5] that if in addition $M$ is projective, then for every coherent sheaf $\mathcal{F}$ on $M$, the cohomology group $H^i\big(M - X, \mathcal{F}\big)$ is finite-dimensional whenever $i \geq n - d$.    □

There have been a number of attempts to find global geometric consequences of the amplitude of the normal bundle $N = N_{X/M}$. Hartshorne [276, III.4.2] showed that if $X$ is a smooth divisor in $M$ whose normal bundle $\mathcal{O}_X(X)$ is ample, then some large multiple of $X$ moves in a free linear series, and hence meets any curve with ample normal bundle (Example 1.2.30). This led him to make two conjectures [276, Conjectures 4.4 and 4.5] concerning what one might expect in higher codimension:

**Hartshorne's Conjecture A.** If $X \subseteq M$ is a smooth subvariety with ample normal bundle, then a sufficiently high multiple of $[X]$ should move (as a cycle) in a large algebraic family.

**Hartshorne's Conjecture B.** Let $X$ , $Y \subseteq M$ be smooth complete subvarieties having ample normal bundles. If $\dim X + \dim Y \geq \dim M$, then $X$ and $Y$ must meet in $M$.

It was observed by Fulton and the author in [213] that Conjecture A would imply B: see Remark 6.3.18 or Corollary 8.4.3. Conjecture B remains open, although it has been verified in some special cases (cf. [213], [25], [27], [26]). However, it was also shown in [213] that Conjecture A can fail:

**Example 6.3.17. (Counterexample to Conjecture A).** There exists for $d \gg 0$ a rank-2 ample vector bundle $E$ on $\mathbf{P}^2$ sitting in an exact sequence

$$0 \longrightarrow \mathcal{O}_{\mathbf{P}^2}(-d)^2 \longrightarrow \mathcal{O}_{\mathbf{P}^2}(-1)^4 \longrightarrow E \longrightarrow 0. \tag{*}$$

Such bundles were originally constructed by Gieseker [223] via a reduction to characteristic $p$; we exhibit them as a special case of a general construction in Example 6.3.67 below. For our example we take $M = \mathbf{E}$ to be the total space of $E$, and $X \subseteq M$ to be the zero-section. Thus $X = \mathbf{P}^2$ and $N_{X/M} = E$. We will show that the only projective surface $Y \subseteq M$ is the zero-section $X$ itself, and so in particular no multiple of $X$ can move. In fact, let $Y' \longrightarrow Y$ be a resolution of singularities, and denote by $f : Y' \longrightarrow \mathbf{P}^2$ the natural map. The inclusion $Y \hookrightarrow M$ gives rise to a mapping $Y' \longrightarrow M$, which in turn determines a "tautological" section $s \in H^0\big(Y', f^*E\big)$: for $y \in Y'$, $s(y) \in E\big(f(y)\big)$ is the point of the fibre of $\mathbf{E} \longrightarrow \mathbf{P}^2$ over $f(y)$ to which $y$ maps. But it follows from vanishing for the big and nef bundle $f^*\mathcal{O}_\mathbf{P}(d)$ (Theorem 4.3.1) that

$$H^1\big(Y', f^*\mathcal{O}_\mathbf{P}(-d)\big) = 0,$$

and by pulling back (*) we deduce that $H^0\big(Y', f^*E\big) = 0$. Therefore $Y'$ must map to the zero-section, as claimed.    $\square$

**Remark 6.3.18. (Numerical consequences of positive normal bundles).** With $X \subseteq M$ as above, the positivity of the normal bundle $N = N_{X/M}$ has numerical consequences. Specifically, if $N$ is nef then for every subvariety $Y$ of dimension complementary to $X$, the intersection number of $X$ and $Y$ satisfies $(X \cdot Y) \geq 0$. If $N$ is ample then strict inequality holds provided that $Y$ is homologous to an effective algebraic cycle that meets $X$. This appears as Corollary 8.4.3.    $\square$

Finally, in the spirit of the Lefschetz hyperplane theorem, one can attempt to compare the topology of $X$ and $M$. Assume now that $M$ is projective. Napier and Ramachandran [473] used $L^2$ methods to prove that if $N = N_{X/M}$ is ample, then the image of the map

$$\pi_1(X) \longrightarrow \pi_1(M)$$

on fundamental groups has finite index in $\pi_1(M)$. To give a taste of the argument, we will establish a somewhat weaker algebro-geometric assertion:

**Theorem 6.3.19. (Analogue of theorem of Napier–Ramachandran).** *Let $M$ be a connected complex projective manifold, and let*

$$X \subseteq M$$

*be a smooth irreducible subvariety with $N = N_{X/M}$ ample. Then there is a positive constant $\ell = \ell(X, M)$ depending only on $X$ and $M$ with the following property:*

> *If $f : M' \longrightarrow M$ is any finite connected étale covering that admits a section over $X$, then $\deg f \leq \ell$.*

*Equivalently, there do not exist subgroups of arbitrarily large finite index in $\pi_1(M)$ which contain the image of $\pi_1(X)$.*

**Remark 6.3.20.** One can deduce this from the results of Hironaka, Matsumura, et al. on formal rational functions (see [276, Chapter V]). However, we prefer an argument based on the very nice approach of Napier and Ramachandran. In fact, the proof will show — thanks to Remark 6.3.15 — that it is sufficient to assume that the restriction of $N$ to a general complete intersection curve $C \subset M$ is ample. A related cohomological result due to Sommese appears later as Proposition 7.1.12.    $\square$

*Proof of Theorem 6.3.19.* Suppose to the contrary that one can find an infinite sequence $f_k : M_k \longrightarrow M$ of connected (and hence irreducible) étale coverings, with $\deg f_k \to \infty$, each of which admits a section $\mu_k : X \longrightarrow M_k$ over $X$:

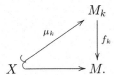

We view $X$ as a subvariety both of $M$ and of $M_k$. Observe that $\mu_k$ extends to an isomorphism $U \cong U_k$ between small (classical) neighborhoods of $X$ in $M$ and $M_k$ respectively. It follows in particular that the formal completions $\widehat{M} = \widehat{M}_{/X}$ and $\widehat{M}_k = \widehat{M}_{k/X}$ of $M$ and $M_k$ respectively along $X$ are isomorphic.

Now fix any line bundle $L$ on $M$, and set $L_k = f_k^* L$. Then $L \mid U \cong L_k \mid U_k$ (as holomorphic line bundles) and therefore

$$H^0\big(\widehat{M}_k, \widehat{L}_k\big) \;\cong\; H^0\big(\widehat{M}, \widehat{L}\big)$$

(since by GAGA we can compute these completions analytically). But quite generally $H^0(M_k, L_k)$ injects into $H^0\big(\widehat{M}_k, \widehat{L}_k\big)$, and so we conclude that

$$h^0\big(M_k, L_k\big) \;\leq\; h^0\big(\widehat{M}, \widehat{L}\big),$$

the right hand side being finite thanks to Proposition 6.3.13 and the hypothesis on $N_{X/M}$. So to get the required contradiction, it is sufficient to exhibit any line bundle $L$ on $M$ such that the dimension $h^0\big(M_k, f_k^* L\big)$ goes to infinity with $k$.

But this is easily achieved. In fact, fix an ample divisor $H$ on $M$ and choose any positive integer $b \gg 0$ large enough so that

$$\chi\big(M, \mathcal{O}_M(K_M + bH)\big) \;=\; h^0\big(M, \mathcal{O}_M(K_M + bH)\big) \;\neq\; 0$$

(the equality coming from the Kodaira vanishing theorem). Then

$$f_k^* \mathcal{O}_M\big(K_M + bH\big) \;=\; \mathcal{O}_{M_k}\big(K_{M_k} + f_k^*(bH)\big)$$

since $f$ is étale, and

$$h^0\big(M_k \,,\, f_k^* \mathcal{O}_M\big(K_M + bH\big)\big) \;=\; \chi\big(M_k \,,\, f_k^* \mathcal{O}_M\big(K_M + bH\big)\big)$$

thanks again to Kodaira vanishing. But since Euler characteristics are multiplicative in étale covers (Proposition 1.1.28) we conclude that

$$h^0\big(M_k, f_k^* \mathcal{O}_M(K_M + bH)\big) \;=\; \deg f_k \cdot h^0\big(M, \mathcal{O}_M(K_M + bH)\big),$$

so setting $L = \mathcal{O}_M(K_M + bH)$ we are done.    $\square$

**Example 6.3.21. (Hironaka's example).** A construction of Hironaka shows that one cannot expect a surjection $\pi_1(X) \twoheadrightarrow \pi_1(M)$ on fundamental groups in the setting of Theorem 6.3.19. Let $f : M' \longrightarrow M$ be a non-trivial connected étale covering between smooth projective varieties of dimension $\geq 3$. If $Y \subset M'$ is a sufficiently general complete intersection curve, then the restriction of $f$ will determine an embedding of $Y$ in $M$: let $X = f(Y) \subset M$ denote its image. But then $f$ splits over $X$, since by construction $f^{-1}(X)$ contains $Y$ as a connected component.                                        $\square$

**Remark 6.3.22. (Concavity and convexity of complements).** From an analytic viewpoint, a natural way to measure the positivity properties of an embedding $X \subseteq M$ is to study the (pseudo)-concavity or convexity of the complement $M - X$ in the sense of Andreotti and Grauert [8]. Precise definitions and statements would take us too far afield here: see e.g. [276, Chapter 6, §1 (iv)] for a quick overview. Suffice it to say that Barth [29] obtained some estimates on the concavity and convexity of $\mathbf{P}^r - X$ when $X \subseteq \mathbf{P}^r$ is a closed submanifold: as explained in [276], these are related to Barth's Theorem 3.2.1 on the cohomology of low-codimensional subvarieties of projective space. Sommese [550], [554] extended some of these results to subvarieties of other homogeneous varieties. Assuming that $M$ is projective, Sommese [554, Corollary 1.4] also proves a convexity estimate for $M - X$ when $X \subseteq M$ is any smooth subvariety whose normal bundle is ample and globally generated.                                        $\square$

## 6.3.B Ample Cotangent Bundles and Hyperbolicity

We now consider smooth projective varieties with ample cotangent bundles. Such varieties are hyperbolic, and the theme is that they exhibit strong forms of properties known or expected for hyperbolic varieties. In the first part of this subsection we summarize some of the basic geometric facts. In the second, we discuss methods of construction.

**Geometric properties.** We begin by recalling two notions of hyperbolicity:

**Definition 6.3.23. (Hyperbolicity).** Let $X$ be a smooth complex projective variety, and let $h \in N^1(X)$ be an ample divisor class on $X$.

(i). $X$ is *algebraically hyperbolic* if there is a positive real number $\varepsilon > 0$ with the following property:

> For every finite map $\nu : C \longrightarrow X$ from a smooth curve $C$ to $X$ one has the inequality

$$\bigl(2g(C) - 2\bigr) \geq \varepsilon \cdot \bigl(C \cdot \nu^* h\bigr), \tag{6.8}$$

where as usual $g(C)$ denotes the genus of $C$.

(ii). Viewed as a complex manifold, $X$ is *Kobayashi hyperbolic* if there are no non-constant entire holomorphic mappings $g : \mathbf{C} \longrightarrow X$.    □

Algebraic hyperbolicity was introduced and studied by Demailly in [127]. He actually requires (6.8) to hold only for the normalizations of embedded curves, but it is easily seen using Riemann–Hurwitz that this is equivalent to the condition stated above. Similarly, the absence of entire holomorphic mappings is usually not taken as the definition of hyperbolicity, but for compact targets it is equivalent to the standard definition thanks to a theorem of Brody. (See [127, Corollary 1.2] or [381, Chapter III].)

**Example 6.3.24. (Properties of hyperbolic varieties).** Keep assumptions as above.

(i).    The definition of algebraic hyperbolicity is independent of the ample class $h$.

(ii).    If $X$ is algebraically hyperbolic, then $X$ does not contain any rational or elliptic curves.

(iii).    If $X$ is algebraically hyperbolic, then there are no non-constant maps $f : A \longrightarrow X$ from an abelian variety $A$ to $X$.

(iv).    If $X$ is Kobayashi hyperbolic, then $X$ is algebraically hyperbolic.

(The first two statements are clear. For (iii), suppose that $f : A \longrightarrow X$ is non-constant, and consider $\nu_k = f \circ m_k$, where $m_k : A \longrightarrow A$ is multiplication by $k$. If $C \subset A$ is (say) a general complete intersection curve, then as $k \to \infty$, $\nu_k$ will eventually violate (6.8). For (iv), see [127, Theorem 2.1].)    □

**Remark 6.3.25. (Hypersurfaces of large degree).** Let $X \subseteq \mathbf{P}^{n+1}$ be a very general hypersurface of degree $d \geq 2(n+1)$. Then $X$ is algebraically hyperbolic. This is proved (but not explicitly stated) by Voisin [597, §1], building on earlier work of Clemens [93] and Ein [143].    □

It was established by Kobayashi [351] that compact manifolds with negative tangent bundles are hyperbolic:

**Theorem 6.3.26. (Kobayashi's theorem).** *Let $X$ be a smooth projective variety whose cotangent bundle $\Omega_X^1$ is ample. Then $X$ is algebraically hyperbolic. In fact, $X$ is hyperbolic in the sense of Kobayashi.*

*Partial Proof.* For the Kobayashi hyperbolicity we refer to [127, (3.1)], or [381, III.3]. We prove the first statement using results from Section 6.4 concerning amplitude of bundles on curves. Assuming then that $\Omega_X^1$ is ample, fix an ample class $h$ on $X$ and a positive number $\varepsilon > 0$ sufficiently small so that the $\mathbf{Q}$-twisted bundle $\Omega_X^1 <\!-\varepsilon h\!>$ remains ample (Lemma 6.2.8.vi). Given a finite mapping $\nu : C \longrightarrow X$ from a smooth curve to $X$, the pullback $\nu^* (\Omega_X^1 <\!-\varepsilon h\!>)$ is then an ample $\mathbf{Q}$-twisted bundle on $C$. On the other hand, the derivative of $\nu$ determines a generically surjective homomorphism $\nu^* \Omega_X^1 \longrightarrow \Omega_C^1$, and it then follows from Example 6.4.17 below that $\Omega_C^1 <-\varepsilon h>$ is also ample. Therefore

$$\left(2g(C) - 2\right) - \varepsilon \left(C \cdot \nu^* h\right) \ = \ \deg\left(\Omega_C^1 <-\varepsilon h>\right) \ > \ 0$$

thanks to Lemma 6.4.10.                                                                                       □

**Remark 6.3.27.** The converse of Theorem 6.3.26 can easily fail. For example, if $B$ is a curve of genus $\geq 2$ then $X = B \times B$ is Kobayashi hyperbolic since it is uniformized by the product of two discs. But $\Omega_X^1$ is evidently not ample, since its restriction to $B \times \{pt\}$ admits a trivial quotient.

**Example 6.3.28. (Subvarieties).** Let $X$ be a smooth complex projective variety with ample cotangent bundle. Then every irreducible subvariety of $X$ is of general type. (In fact, let $Y_0 \subset X$ be an irreducible subvariety of dimension $d$, and let $\mu : Y \longrightarrow Y_0$ be a resolution of singularities. Then there is a generically surjective homomorphism $\mu^* \Omega_X^d \longrightarrow \Omega_Y^d = \mathcal{O}_Y(K_Y)$. Since $\Omega_X^d$ is ample, this implies upon taking symmetric powers after twisting by a small negative multiple of an ample class that $\mathcal{O}_Y(K_Y)$ is big.) It is conjectured by Lang (cf. [127, (3.8)]) that $X$ is hyperbolic if and only if every subvariety of $X$ (including of course $X$ itself) is of general type.                                    □

There are a number of interesting finiteness theorems in the literature for mappings to varieties with ample cotangent bundles: a nice survey appears in [618]. Here we use some ideas from Part One to prove a general boundedness statement for mappings into algebraically hyperbolic varieties.

**Theorem 6.3.29. (Boundedness of regular mappings).** *Let $X$ be an algebraically hyperbolic variety, and let $Y$ be any irreducible projective variety of dimension $m > 0$. Then the set $\mathrm{Hom}(Y, X)$ of morphisms from $Y$ to $X$ forms a bounded family, i.e. all such morphisms are parameterized by finitely many irreducible varieties.*

*Sketch of Proof.* In fact, fix very ample divisors $H$ and $D$ on $Y$ and $X$ respectively, and let $f : Y \longrightarrow X$ be any morphism. By standard finiteness results, it is sufficient to show that there is a positive integer $a > 0$ such that the degree of the graph $\Gamma_f \subset Y \times X$ with respect to the ample divisor $\mathrm{pr}_1^* aH + \mathrm{pr}_2^* D$ is bounded independent of $f$, i.e. we need to bound from above the intersection number

$$\int_Y c_1\left(\mathcal{O}_Y(aH + f^* D)\right)^m \tag{$*$}$$

independently of $f$. To this end, we may assume that $f$ is non-constant. In this case the intersection number $\left(H^{m-1} \cdot f^* D\right)$ computes the degree (with respect to $D$) of the image in $X$ of a curve obtained as the complete intersection of $(m-1)$ divisors in the linear series $|H|$. Therefore the algebraic hyperbolicity of $X$ implies that there is a uniform upper bound on $\left(H^{m-1} \cdot f^* D\right)$. Then we can fix a positive integer $a \gg 0$, independent of $f$, such that

$$((aH)^m) \; > \; m((aH)^{(m-1)} \cdot f^*D).$$

As $H$ and $f^*D$ are nef, it follows from Theorem 2.2.15 that $(aH - f^*D)$ is big. Therefore some large multiple (possibly depending on $f$) of $(aH - f^*D)$ is effective. Thanks again to the nefness of $H$ and $f^*D$, one concludes that

$$\left((aH - f^*D) \cdot (aH)^{(m-1-i)} \cdot (f^*D)^i\right) \; \geq \; 0$$

for all $0 \leq i \leq n - 1$. This in turn leads to the inequalities

$$a^m(H^m) \; \geq \; a^{m-1}(H^{m-1} \cdot f^*D) \; \geq \; a^{m-2}(H^{m-2} \cdot f^*D^2) \geq \cdots$$
$$\geq (f^*D^m) \; \geq 0, \quad (6.9)$$

from which it follows that the quantity in $(*)$ is bounded above by $2^m a^m (H^m)$.
□

As a consequence, we get a quick proof of a result of Kalka, Shiffman, and Wong from [309, Corollary 4]:

**Corollary 6.3.30. (Finiteness of regular mappings).** *Let $X$ be a smooth projective variety whose cotangent bundle $\Omega^1_X$ is ample, and let $Y$ be any irreducible projective variety. Then the set $\mathrm{Hom}_*(Y, X)$ of non-constant morphisms from $Y$ to $X$ is finite.*

*Proof.* The amplitude of $\Omega^1_X$ implies by 6.1.4 that $H^0(Y, f^*TX) = 0$, so in any event the Hom scheme in question is discrete (cf. [114, Proposition 2.4]). On the other hand, it follows from the previous result that the set of all maps $f : Y \longrightarrow X$ is parametrized by finitely many irreducible varieties. Putting these facts together, it follows that there are only finitely many such maps.   □

**Remark 6.3.31. (Finiteness of rational mappings).** A related result of Noguchi and Sunada [480] states that with $X$ and $Y$ as in Corollary 6.3.30, the set $\mathrm{Rat}_*(Y, X)$ of non-constant rational maps from $Y$ to $X$ is also finite.   □

**Remark 6.3.32. (Rational points over function fields).** Another interesting avenue of investigation concerns the diophantine properties of varieties with ample cotangent bundles defined over function fields. In this setting, a number of authors have obtained Mordell-type statements. For example, suppose that $L$ is an algebraic function field over an algebraically closed groundfield $K$ of characteristic zero, and suppose that $X$ is a smooth, projective, and geometrically integral variety over $L$ with ample cotangent bundle $\Omega^1_{X/L}$. Inspired by theorems of Grauert [233] and Manin [415] in the one-dimensional case, Martin-Deschamps [418] proves that if the set of $L$-rational points of $X$ is Zariski dense, then $X$ is isotrivial over $L$. There are related results due to Noguchi [479] and Moriwaki [441].   □

**Remark 6.3.33. (Rational points over number fields).** If $X$ is a smooth projective variety defined over a number field $L$ that has ample cotangent bundle, then it is a conjecture of Lang [380] that the set of $L$-rational points of $X$ is finite. Moriwaki [442] remarks that this follows from work of Faltings [178], [179] if one assumes in addition that the cotangent bundle of $X$ is globally generated.                                                                      □

**Remark 6.3.34. (Miyaoka's theorem on generic semipositivity).** A basic theorem of Miyaoka [430] shows that the cotangent bundle of a projective variety satisfies a weak positivity property in very general circumstances. Specifically, let $X$ be a smooth complex projective variety of dimension $n$, and let $H$ be an ample divisor on $X$. Suppose that $X$ is not uniruled, i.e. assume that $X$ is not covered by rational curves. Miyaoka's theorem states that if $C \subset X$ is a sufficiently general curve arising as the complete intersection of $n-1$ divisors in the linear series $|mH|$ for $m \gg 0$, then the restriction of $\Omega_X^1$ to $C$ is nef. We refer to [432] and Shepherd-Barron's exposition in [360, Chapter 9] for proofs and a discussion of some of the applications.        □

**Remark 6.3.35. (Varieties of general type).** There are a number of very interesting results and conjectures concerning finiteness properties for varieties of general type. Bogomolov [61] proved that if $X$ is a surface of general type satisfying the inequality $c_1(X)^2 > c_2(X)$, then the family of curves on $X$ of fixed geometric genus is bounded. Martin-Deschamps gives a nice account of this work in [138]. One can view Bogomolov's theorem as going in the direction of conjectures of Lang concerning the diophantine and geometric properties of varieties of general type. These conjectures predict, for example, that if $X$ is a projective variety of general type, then there exists a proper Zariski-closed subset $Z \subsetneq X$ having the property that the image of any non-constant morphism $f : G \longrightarrow X$ from an algebraic group to $X$ must lie in $Z$: in particular, $Z$ must contain all rational curves on $X$. We refer to [382, Chapter 1] for a pleasant discussion of this circle of ideas.        □

**Constructions.** Although one expects that varieties with ample cotangent bundle should be reasonably plentiful, until recently relatively few explicit constructions appeared in the literature except in the case of surfaces.

**Construction 6.3.36. (Ball quotients).** If $X$ is a smooth complex projective variety that is uniformized by the ball $\mathbf{B}^n \subset \mathbf{C}^n$, then the Bergman metric on $\mathbf{B}^n$ descends to a metric on $X$ with negative holomorphic sectional curvature, and hence $\Omega_X^1$ is ample (cf. [618, Example 2, p. 147]).        □

**Construction 6.3.37. (Surfaces).** Yau raised the question of classifying all surfaces with positive cotangent bundles, and motivated in part by this several authors have given constructions of such surfaces.

- **Miyaoka's examples.** Building on ideas of Bogomolov, Miyaoka [428] showed that if $X$ is a smooth complex projective surface of general type with $c_1(X)^2 > 2c_2(X)$, then the cotangent bundle $\Omega_X^1$ is "almost everywhere ample," which very roughly means that it fails to be ample only along finitely many curves. Using this, he deduces that if $X_1$ and $X_2$ are two such surfaces, then a complete intersection of two general sufficiently positive divisors in $X_1 \times X_2$ is a surface $X$ with $\Omega_X^1$ ample.

- **Kodaira surfaces.** Martin-Deschamps [418] established the amplitude of the cotangent bundles of certain Kodaira surfaces, i.e. surfaces that admit a smooth map to a non-singular curve. A similar result was proved independently by Schneider and Tancredi [524], who also generalize Miyaoka's construction.

- **Hirzebruch–Sommese examples.** Hirzebruch found some interesting surfaces by desingularizing Kummer coverings of $\mathbf{P}^2$ branched over line arrangements, and Sommese [555] classified which of these have ample cotangent bundles.                                          □

In higher dimensions, the most general constructions are due to Bogomolov and Debarre. We recommend Debarre's nice paper [115] for more information.

**Construction 6.3.38. (Bogomolov's construction).** Let $Y_1, \ldots, Y_m$ be smooth projective varieties of dimension $d \geq 1$, each having big cotangent bundle,[9] and let

$$ X \subseteq Y_1 \times \ldots \times Y_m $$

be a general complete intersection of sufficiently high multiples of an ample divisor. Bogomolov proves that if

$$ \dim X \ \leq \ \frac{d(m+1)+1}{2(d+1)}, $$

then $X$ has ample cotangent bundle. Bogomolov and Debarre deduce from this that there exists a projective variety $X$ having ample cotangent bundle with the additional property that $\pi_1(X)$ can be any fixed group that arises as the fundamental group of a smooth projective variety: in particular, $X$ can be simply connected. Wong [612] employed a similar construction in a differential-geometric context. A detailed description and verification of Bogomolov's construction appears in [115, §3]: we will work through an elementary special case in Construction 6.3.42.                                          □

**Construction 6.3.39. (Complete intersections in abelian varieties).** Debarre [115] recently proved that if $X$ is the complete intersection of $e \geq n$ sufficiently ample general divisors in a simple abelian variety of dimension $n + e$, then the cotangent bundle $\Omega_X^1$ is ample.                                          □

---

[9] Recall from Example 6.1.23 that a vector bundle $E$ on a projective variety $V$ is big if $\mathcal{O}_{\mathbf{P}(E)}(1)$ is a big line bundle on $\mathbf{P}(E)$. Bogomolov shows that if $Y$ is a surface of general type satisfying $c_1(Y)^2 > c_2(Y)$, then the cotangent bundle of $Y$ is big.

**Remark 6.3.40. (Complete intersections in projective space).** De-barre [115, §2.2] conjectures that if $X \subseteq \mathbf{P}^r$ is the complete intersection of $e \geq r/2$ hypersurfaces of sufficiently high degree, then the cotangent bundle of $X$ is ample. $\qquad\qquad\square$

**Remark 6.3.41. (Nef cotangent bundles).** It is also very interesting to ask for examples of projective manifolds whose cotangent bundles are numerically effective. The class of all such is evidently closed under taking products, subvarieties, and finite unramified covers, and it includes smooth subvarieties of abelian varieties. A nice theorem of Kratz [372, Theorem 2] states that if $X$ is a complex projective variety whose universal covering space is a bounded domain in $\mathbf{C}^n$ or in a Stein manifold, then $\Omega^1_X$ is nef. We refer to Theorem 7.2.19 for a result about projective embeddings of such varieties. $\qquad\square$

As Miyaoka suggested, a special case of Bogomolov's construction is particularly elementary. We devote the rest of this subsection to working this case out explicitly.

**Construction 6.3.42. (Complete intersections in curve products).** Start with smooth projective curves $T_1, \ldots, T_{n+e}$ of genus $\geq 2$, and set $T = T_1 \times \cdots \times T_{n+e}$, with projections $p_i : T \longrightarrow T_i$. Fix next very ample line bundles $A_i$ on $T_i$ and for each $d > 0$ put

$$A = p_1^* A_1 \otimes \ldots \otimes p_{n+e}^* A_{n+e} \quad , \quad L = L_d = A^{\otimes d}.$$

Choose finally $e$ general divisors $D_1, \ldots, D_e \in |L_d|$, and set

$$X = D_1 \cap \cdots \cap D_e \subseteq T.$$

Assuming that $e \geq 2n - 1$ and $d \geq n$, we will now verify that $X$ is a smooth projective $n$-fold whose cotangent bundle $\Omega^1_X$ is ample.

*Sketch of Verification of Construction 6.3.42.* The fact that $X$ is a smooth $n$-fold is clear, and the issue is to establish the amplitude of its cotangent bundle. To this end, we will consider projections of $X$ onto various products of the $T_i$. As a matter of notation, for any multi-index $I = \{i_1, \ldots, i_k\}$ ($1 \leq i_1 < \cdots < i_k \leq n + e$), write

$$T_I = T_{i_1} \times \cdots \times T_{i_k},$$

and denote by $p_I : T \longrightarrow T_I$ the corresponding projection. Somewhat abusively, we will also write $p_I$ for the restriction of this projection to subvarieties of $T$. The first point is to check that one can arrange by choosing the $D_i$ generally enough that $X$ satisfies two genericity conditions:

(i). For every $I$ of length $2n$, the projection $p_I : X \longrightarrow T_I$ is unramified.

(ii). For every $J$ of length $n$, the projection $p_J : X \longrightarrow T_J$ is finite.

Property (i) is verified by a standard dimension count as in [280, II.8.18]. The second follows from a general finiteness statement (Lemma 6.3.43) formulated and proved at the end of this subsection: this is where the hypothesis $d \geq n$ is used.

Assuming that we have arranged for $X$ to satisfy the two properties just discussed, we verify that $\Omega_X^1$ is ample. Since $\Omega_X^1$ — being a quotient of $\Omega_T^1 \,|\, X$ — is globally generated, it is enough by Gieseker's lemma (Proposition 6.1.7) to show that it does not admit a trivial quotient along any curve. Suppose then that $C \subseteq X$ is a reduced and irreducible curve. It follows from property (ii) that there can be at most $n-1$ indices $i \in [1, n+e]$ such that the projection $p_i : X \longrightarrow T_i$ maps $C$ to a point. Since $n + e \geq 3n - 1$ we may assume after reindexing that $p_i | C$ is finite for $1 \leq i \leq 2n$. Set $I_0 = \{1, 2, \ldots, 2n\}$, and for any $J \subset I_0$ of length $n$ denote by $R_J \subset X$ the ramification divisor of the branched covering $p_J : X \longrightarrow T_J$. It follows by a simple argument from property (i) that

$$\bigcap_{\substack{J \subset I_0 \\ |J| = n}} \mathrm{supp}(R_J) \; = \; \varnothing. \tag{*}$$

Therefore we can choose $J \subset I_0$ so that $C \not\subset \mathrm{supp}(R_J)$. Then the derivative $dp_J$ gives rise to an exact sequence

$$0 \longrightarrow p_J^* \left( \Omega_{T_J}^1 \right) | C \longrightarrow \Omega_X^1 | C \longrightarrow \tau \longrightarrow 0,$$

where $\tau$ is a torsion sheaf supported on $C \cap \mathrm{supp}(R_J)$. But the bundle on the left is ample, and it follows right away that $\Omega_X^1 | C$ has no trivial quotients, as required.    □

Finally, we state and prove the finiteness lemma that was used in the course of the argument just completed. We will have occasion to refer to it also in Section 6.3.F.

**Lemma 6.3.43. (Finiteness lemma).** *Let $Y$ and $T$ be irreducible projective varieties of dimensions $e$ and $n$ respectively. Let $A$ and $B$ be very ample line bundles on $Y$ and $T$, and for $d \geq 1$ set*

$$L \; = \; L_d \; = \; \mathrm{pr}_1^* A^{\otimes d} \otimes \mathrm{pr}_2^* B.$$

*Consider $e$ general divisors $D_1, \ldots, D_e \in |L_d|$ in the indicated linear series on $Y \times T$. If $d \geq n$, then the intersection $D_1 \cap \cdots \cap D_e$ is finite over $T$.*

*Proof.* Fix some $0 \leq k < e$ and consider

$$X \; = \; D_1 \cap \cdots \cap D_k \; \subseteq \; Y \times T.$$

(If $k = 0$ take $X = Y \times T$.) We assume inductively that every fibre of the projection $X \longrightarrow T$ has pure dimension $e - k$, and we will show that one can arrange for every fibre of $X \cap D_{k+1} \longrightarrow T$ to have pure dimension $e - (k+1)$.

To this end, fix $t \in T$, and denote by $X_t$ the fibre of $X$ over $t$. Consider the set of "bad" divisors at $t$:

$$Z_t = \{ D \in |L_d| \mid D \text{ contains one or more components of } X_t \}.$$

We claim:

$$Z_t \text{ has codimension} > d \text{ in the linear series } |L_d|. \tag{*}$$

Granting this, it follows that the set

$$Z = \{ D \in |L_d| \mid \text{ some fibre of } X \cap D \longrightarrow T \text{ has dim} \ge e - k \}$$

has codimension $> d - n$ in $|L_d|$. Hence if $d \ge n$ we can find $D_{k+1} \notin Z$.

Turning to (*), since $H^0(Y \times T, L_d)$ maps surjectively onto the fibre-wise space of sections $H^0(Y, A^{\otimes d})$ over $t$, and since $A$ is very ample, it is enough to verify the following assertion:

Let $V \subseteq \mathbf{P}$ be any algebraic subset of positive dimension in some projective space $\mathbf{P}$, and let

$$Z_V = \{ E \in |\mathcal{O}_{\mathbf{P}}(d)| \mid E \text{ contains } V \}.$$

Then $Z_V$ has codimension $> d$ in $|\mathcal{O}_{\mathbf{P}}(d)|$.

But this follows from the elementary and well-known fact that any $d+1$ points on $V$ impose independent conditions on hypersurfaces of degree $d$ in $\mathbf{P}$. □

### 6.3.C Picard Bundles

When $C$ is a smooth projective curve of genus $g \ge 1$, the Jacobian of $C$ carries some interesting bundles, whose projectivizations are the symmetric products of $C$. It was established by Fulton and the author in [212] that these so-called Picard bundles are negative, a fact that was used there to study the varieties of special divisors on $C$. Here we follow the same arguments to prove the negativity of the analogous bundles on the Picard variety of any irregular smooth projective variety. The application to special divisors appears as Theorem 7.2.12.

**Convention 6.3.44.** In the present subsection, it is most natural to deal with the projective bundle of one-dimensional **sub**-bundles of a given bundle $F$ on a variety $Y$. We denote this projectivization by $\mathbf{P}_{\text{sub}}(F)$, with $\pi : \mathbf{P}_{\text{sub}}(F) \longrightarrow Y$ the projection. So $\mathbf{P}_{\text{sub}}(F) = \mathbf{P}(F^*)$. On $\mathbf{P}_{\text{sub}}(F)$ one has the tautological line sub-bundle $\mathcal{O}_{\mathbf{P}_{\text{sub}}}(-1) \subseteq \pi^* F$. We say that $F$ is *negative* if $F^*$ is ample. Thus $F$ is negative if and only if the tautological line bundle $\mathcal{O}_{\mathbf{P}_{\text{sub}}}(1)$ on $\mathbf{P}_{\text{sub}}(F)$ is ample. □

We start by constructing the Picard bundles. Throughout this subsection, $X$ is a smooth projective variety of dimension $n$. Fix an algebraic equivalence class $\lambda$ on $X$ and denote by $\mathrm{Pic}^\lambda(X)$ the component of the Picard variety parameterizing bundles in the chosen class. Thus $\mathrm{Pic}^\lambda(X)$ is a torus of dimension $q(X) = \dim H^1(X, \mathcal{O}_X)$. Given a point $t \in \mathrm{Pic}^\lambda(X)$ we denote by $L_t$ the corresponding line bundle on $X$. Choosing a base point $0 \in X$, there is a Poincaré line bundle $\mathcal{L}$ on $X \times \mathrm{Pic}^\lambda(X)$, characterized by the properties

$$\mathcal{L} \mid \big(X \times \{t\}\big) \;=\; L_t \quad \forall\, t \in \mathrm{Pic}^\lambda(X);$$
$$\mathcal{L} \mid \big(\{0\} \times \mathrm{Pic}^\lambda(X)\big) \;=\; \mathcal{O}_{\mathrm{Pic}^\lambda(X)}.$$

Our object is to realize the groups $H^0(X, L_t)$ for $t \in \mathrm{Pic}^\lambda(X)$ as the fibres of a vector bundle $E_\lambda$ on $\mathrm{Pic}^\lambda(X)$. In order for this to work smoothly we will suppose that the class $\lambda$ is sufficiently positive so that

$$H^i(X, L_t) \;=\; 0 \quad \text{for all } i > 0 \text{ and all } t \in \mathrm{Pic}^\lambda(X), \tag{6.10}$$
$$H^0(X, L_t) \;\neq\; 0 \quad \text{for all } t \in \mathrm{Pic}^\lambda(X). \tag{6.11}$$

It follows from (6.10) by the theorems on cohomology and base-change that the direct image

$$E_\lambda \;=_{\mathrm{def}}\; \mathrm{pr}_{2,*}\big(\mathcal{L}\big)$$

under the second projection $\mathrm{pr}_2 : X \times \mathrm{Pic}^\lambda(X) \longrightarrow \mathrm{Pic}^\lambda(X)$ is a vector bundle on $\mathrm{Pic}^\lambda(X)$, which we call the *Picard bundle* corresponding to the class $\lambda$. It is non-zero by (6.11). Furthermore, push-forwards of $\mathcal{L}$ commute with base-change, so in particular one has a canonical isomorphism

$$E_\lambda(t) \;=\; H^0\big(X, L_t\big) \tag{6.12}$$

of the fibres of $E_\lambda$ with the corresponding cohomology groups on $X$.

As in the case of curves the key to analyzing the properties of these Picard bundles is to interpret their projectivizations as spaces of divisors. Specifically, let $\mathrm{Div}^\lambda(X)$ be the Hilbert scheme parameterizing all effective divisors in the algebraic equivalence class $\lambda$, and denote by

$$u : \mathrm{Div}^\lambda(X) \longrightarrow \mathrm{Pic}^\lambda(X)$$

the Abel–Jacobi mapping that sends a divisor to its linear equivalence class. Given a point $s \in \mathrm{Div}^\lambda(X)$ we denote by $D_s$ the corresponding divisor on $X$.

**Lemma 6.3.45.** *Still assuming that $\lambda$ satisfies (6.10) and (6.11), one has a canonical isomorphism*

$$\mathbf{P}_{\mathrm{sub}}(E_\lambda) \;=\; \mathrm{Div}^\lambda(X)$$

*under which the bundle projection $\pi : \mathbf{P}_{\mathrm{sub}}(E_\lambda) \longrightarrow \mathrm{Pic}^\lambda(X)$ corresponds to the Abel–Jacobi mapping $u$.*

*Idea of Proof.* The essential point is simply that one has natural identifica-
tions

$$\pi^{-1}(t) \ = \ \mathbf{P}_{\mathrm{sub}}\big(E_\lambda(t)\big) \ = \ \mathbf{P}_{\mathrm{sub}}\big(H^0(X, L_t)\big) \ = \ u^{-1}(t)$$

coming from (6.12). We leave it to the reader to use the universal property of
$\mathrm{Div}^\lambda$ to construct the stated isomorphism globally.     □

**Remark 6.3.46.** When $X$ is a smooth curve, the Picard bundles have been
intensively studied. For instance, their Chern classes are given by a formula
of Poincaré ([15, I.5] and [208, 4.3.3]), and they are stable with respect to the
canonical polarization on $\mathrm{Jac}(X)$ ([150]).     □

Homomorphisms defined by restriction to subsets also play an important
role. Let $Z$ be a fixed finite subscheme of $X$. Then the evaluation maps

$$\sigma_t : H^0(X, L_t) \longrightarrow H^0(X, L_t \otimes \mathcal{O}_Z)$$

globalize to a morphism $\sigma = \sigma_Z : E_\lambda \longrightarrow \Sigma_Z$ of vector bundles on $\mathrm{Pic}^\lambda(X)$.
In fact, consider the subscheme $Z \times \mathrm{Pic}^\lambda(X) \subseteq X \times \mathrm{Pic}^\lambda(X)$, and define

$$\Sigma_Z \ = \ \mathrm{pr}_{2,*}\big(\mathcal{L} \mid Z \times \mathrm{Pic}^\lambda(X)\,\big).$$

Taking direct images of the restriction $\mathcal{L} \longrightarrow \mathcal{L} \otimes \mathcal{O}_{Z \times \mathrm{Pic}^\lambda(X)}$ gives rise to $\sigma$.
Observe that if $Z = \{x_1, \ldots, x_w\}$ is a reduced scheme consisting of $w$ distinct
points of $X$, then

$$\Sigma_Z = P_{x_1} \oplus \cdots \oplus P_{x_w},$$

where for $x \in X$, $P_x = \mathcal{L} \mid \big(\{x\} \times \mathrm{Pic}^\lambda(X)\big)$ is a line bundle on $\mathrm{Pic}^\lambda(X)$ that
is a deformation of the trivial bundle $P_0 = \mathcal{O}_{\mathrm{Pic}^\lambda(X)}$. For an arbitrary finite
subscheme $Z \subset X$, $\Sigma_Z$ has a filtration whose quotients are line bundles of
this type.

The maps $\sigma_Z$ have a simple meaning in terms of the isomorphism in 6.3.45.
In fact, consider on $\mathbf{P}_{\mathrm{sub}}(E_\lambda)$ the composition

defining $s_Z$. Viewing $s_Z$ as a section

$$s_Z \ \in \ \Gamma\Big(\mathbf{P}_{\mathrm{sub}}(E_\lambda), \, \pi^*\Sigma_Z \otimes \mathcal{O}_{\mathbf{P}_{\mathrm{sub}}(E_\lambda)}(1)\Big),$$

it is immediate to verify

**Lemma 6.3.47.** *Under the identification* $\mathbf{P}_{\mathrm{sub}}(E_\lambda) = \mathrm{Div}^\lambda(X)$, *the zero locus
of* $s_Z$ *consists of all* $s \in \mathrm{Div}^\lambda(X)$ *such that the corresponding divisor* $D_s$
*contains* $Z$.     □

A basic fact is that Picard bundles are negative:[10]

**Theorem 6.3.48. (Negativity of the Picard bundle).** *Let $\lambda$ be any class satisfying (6.10) and (6.11). Then the Picard bundle $E_\lambda$ on $\mathrm{Pic}^\lambda(X)$ is negative, i.e. its dual $E_\lambda^*$ is ample. Moreover, if $Z \subseteq X$ is any finite subscheme, then the bundle*

$$\mathrm{Hom}(E_\lambda, \Sigma_Z) = E_\lambda^* \otimes \Sigma_Z$$

*is ample.*

*Proof.* For the first statement, we apply Nakai's criterion to establish the amplitude of the tautological bundle $\mathcal{O}_{\mathbf{P}_{\mathrm{sub}}(E_\lambda)}(1)$ on $\mathbf{P}_{\mathrm{sub}}(E_\lambda)$. Suppose then that $V \subseteq \mathbf{P}_{\mathrm{sub}}(E_\lambda)$ is any irreducible subvariety of dimension $k \geq 1$, and let $\xi$ denote the numerical equivalence class of $c_1\big(\mathcal{O}_{\mathbf{P}_{\mathrm{sub}}(E_\lambda)}(1)\big)$. The positivity of $\int_V \xi^k$ will follow by induction if we show that $\xi \cap [V]$ is represented (in numerical equivalence) by a non-zero effective $(k-1)$-cycle. To this end, fix a general point $x \in X$, and consider in $\mathrm{Div}^\lambda(X)$ the divisor

$$I_x = \{s \in \mathrm{Div}^\lambda(X) \mid D_s \ni x\}.$$

It follows from Lemma 6.3.47 that under the identification $\mathrm{Div}^\lambda(X) = \mathbf{P}_{\mathrm{sub}}(E_\lambda)$, $I_x$ arises as the zeroes of a section of $\mathcal{O}_{\mathbf{P}_{\mathrm{sub}}(E_\lambda)}(1) \otimes P_x$, where $P_x$ is the deformation of $\mathcal{O}_{\mathrm{Pic}^\lambda(X)}$ introduced above. Hence $I_x \equiv_{\mathrm{num}} \xi$. So it suffices to show that for general $x$, $I_x$ meets $V$ in a non-empty proper subset of $V$. But this is clear: given any positive-dimensional family of effective divisors, those passing through a given general point form a non-empty proper subfamily. Turning to the amplitude of $E_\lambda^* \otimes \Sigma_Z$ we focus on the case in which $Z$ is the reduced subscheme consisting of distinct points $\{x_1, \ldots, x_w\}$, leaving the general assertion to the interested reader. Then the Hom bundle in question is a direct sum of the bundles $E_\lambda^* \otimes P_{x_i}$, $P_{x_i}$ being a deformation of $\mathcal{O}_{\mathrm{Pic}^\lambda(X)}$. But by Corollary 6.1.19, the amplitude of $E_\lambda^* \otimes P_{x_i}$ is equivalent to that of $E_\lambda^*$, which we have just treated. $\square$

### 6.3.D The Bundle Associated to a Branched Covering

Following the author's paper [387] we discuss a vector bundle that is associated to a branched covering of smooth varieties, and establish in particular that it is ample for coverings of projective space. This will be used in Section 7.1.C to prove a Barth-type theorem for such coverings.

Let $X$ and $Y$ be smooth varieties of dimension $n$, with $Y$ irreducible, and let

$$f : X \longrightarrow Y$$

---

[10] The second statement of the theorem will be needed later, in Section 7.2.C.

be a branched covering (i.e. a finite surjective mapping) of degree $d$. Then $f$ is flat, and consequently the direct image sheaf $f_*\mathcal{O}_X$ is locally free of rank $d$ on $Y$. Moreover, as we are in characteristic zero the natural inclusion $\mathcal{O}_Y \longrightarrow f_*\mathcal{O}_X$ splits via the trace

$$\mathrm{Tr}_{X/Y} : f_*\mathcal{O}_X \longrightarrow \mathcal{O}_Y.$$

Let $F = \ker \mathrm{Tr}_{X/Y}$, so that $F$ is a bundle of rank $d-1$ on $Y$ that appears in a canonical decomposition $f_*\mathcal{O}_X = \mathcal{O}_Y \oplus F$. We consider (for reasons that will become apparent) the dual bundle

$$E = E_f = F^*,$$

which we call the *bundle associated to the covering* $f$.

**Proposition 6.3.49.** *Denote by* $\mathbf{E}$ *the total space of* $E$, *and by* $p : \mathbf{E} \longrightarrow Y$ *the bundle projection. Then the covering* $f$ *canonically factors through an embedding of* $X$ *into* $\mathbf{E}$:

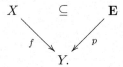

*Proof.* Recall that

$$\mathbf{E} = \mathrm{Spec}_{\mathcal{O}_Y} \mathrm{Sym}(E^*) = \mathrm{Spec}_{\mathcal{O}_Y} \mathrm{Sym}(F).$$

The natural inclusion $F \subseteq f_*\mathcal{O}_X$ gives rise to a surjection $\mathrm{Sym}(F) \longrightarrow f_*\mathcal{O}_X$ of $\mathcal{O}_Y$-algebras, which in turn defines the required embedding of $X$ into $\mathbf{E}$ over $Y$.    □

**Example 6.3.50. (Double covers).** Let $L$ be a line bundle on $Y$ with the property that there exists a smooth divisor $D \in |2L|$. In Proposition 4.1.6 we constructed a degree-two cyclic covering $f : X \longrightarrow Y$ branched over $D$, realizing $X$ as a subvariety of the total space of $L$. Then $E_f = L$, and the embedding of $X$ into $L$ is a special case of Proposition 6.3.49.    □

**Remark 6.3.51. (Triple covers).** Miranda [425] has given a quite complete description of the data involved in specifying a triple cover with given bundle.    □

**Example 6.3.52.** Let $f : X \longrightarrow \mathbf{P}^n$ be the degree $(n+1)$ covering constructed in Example 3.4.13. Then $E_f$ is isomorphic to the tangent bundle $T\mathbf{P}^n$.    □

**Example 6.3.53.** There is a canonical isomorphism

$$(f_*\mathcal{O}_X)^* = f_*\omega_{X/Y},$$

where $\omega_{X/Y} = \mathcal{O}_X(K_X - f^*K_Y)$ is the relative canonical bundle of $X$ over $Y$. (This is a special case of duality for a finite flat morphism: cf. [280, Exercises III.6.10, III.7.2].)                                                    □

**Example 6.3.54. (Branch divisor).** The algebra structure on $f_*\mathcal{O}_X$ defines a mapping $f_*\mathcal{O}_X \otimes f_*\mathcal{O}_X \longrightarrow f_*\mathcal{O}_X$ of bundles on $Y$, which composed with the trace determines a homomorphism $f_*\mathcal{O}_X \otimes f_*\mathcal{O}_X \longrightarrow \mathcal{O}_Y$ and hence

$$\delta : (f_*\mathcal{O}_X) \longrightarrow (f_*\mathcal{O}_X)^*.$$

Then $\delta$ drops rank precisely on the branch divisor $B \subset Y$ of $f$, and in particular

$$-2 \cdot c_1(f_*\mathcal{O}_X) \equiv_{\mathrm{lin}} [B].$$

(Locally $\det \delta$ is the classical discriminant of $f : Y \longrightarrow X$, cf. [3, Chapt. 6, §3]. Note that the divisor structure on $B$ is determined by taking it to be the divisor-theoretic push-forward $f_*[R]$ of the ramification divisor $R \subset X$.)    □

We now turn to branched coverings of projective space. Here the associated bundles $E_f$ satisfy a very strong positivity property:

**Theorem 6.3.55. (Coverings of projective space).** *Let $f : X \longrightarrow \mathbf{P}^n$ be a branched covering of projective space by a smooth irreducible complex projective variety $X$, and let $E = E_f$ be the corresponding bundle on $\mathbf{P}^n$. Then $E(-1)$ is globally generated. In particular, $E$ is ample.*

*Proof.* We use the basic criterion of Castelnuovo–Mumford regularity that if $\mathcal{F}$ is a coherent sheaf on $\mathbf{P}^n$ such that $H^i(\mathbf{P}^n, \mathcal{F}(-i)) = 0$ for all $i > 0$, then $\mathcal{F}$ is globally generated (Theorem 1.8.3). The plan is to apply this to the bundle $E(-1)$.

Example 6.3.53 implies that

$$E \oplus \mathcal{O}_{\mathbf{P}^n} = (f_*\mathcal{O}_X)^* = (f_*\omega_X)(n+1). \qquad (*)$$

Since $f$ is finite, one has isomorphisms

$$H^i(\mathbf{P}^n, f_*\omega_X(k)) = H^i(X, \omega_X \otimes f^*\mathcal{O}_{\mathbf{P}^n}(k)) \quad \text{for all } i, k.$$

It then follows from (*) and Kodaira vanishing (Theorem 4.2.1) that

$$H^i(\mathbf{P}^n, E(-1-i)) = 0 \quad \text{for } 1 \le i \le n-1.$$

On the other hand, $H^n(\mathbf{P}^n, \mathcal{O}_{\mathbf{P}^n}(-n-1)) = H^n(X, \omega_X) = \mathbf{C}$, so it also follows from (*) that $H^n(\mathbf{P}^n, E(-n-1)) = 0$. Thus $E(-1)$ satisfies the required vanishings, and hence is globally generated.    □

**Example 6.3.56. (Consequences of amplitude).** Let $f : X \longrightarrow Y$ be a branched covering of smooth projective varieties of dimension $n$, and $E = E_f$ the corresponding bundle on $Y$. The amplitude of $E$ has some interesting geometric consequences analogous to those deduced when $Y = \mathbf{P}^n$ from the Fulton–Hansen connectedness theorem (Section 3.4).

(i). Let $S$ be a (possibly singular) reduced and irreducible projective variety of dimension $\geq 1$, and let $g : S \longrightarrow Y$ be a finite morphism. If $E_f$ is ample, then the fibre product $Z = X \times_Y S$ is connected.

(ii). Denote by $e_f(x)$ the local degree of $f$ at a point $x \in X$ (Definition 3.4.7). Then there exists at least one point $x \in X$ at which $e_f(x) \geq \min\{\deg f , \ n + 1\}$.

(For (i), the amplitude of $E_f$ implies that $H^0(S, g^*E^*) = 0$. Writing $f' : Z \longrightarrow S$ for the induced map, it follows from this that $H^0(S, f'_*\mathcal{O}_Z) = \mathbf{C}$. Statement (ii) is then deduced as in the proof of Theorem 3.4.8. See [387, Proposition 1.3] for details.) $\hfill\square$

**Example 6.3.57.** Given a branched covering $f : X \longrightarrow Y$ of smooth projective varieties, it is not true in general that the associated bundle $E_f$ is ample or even nef. Examples may be constructed for instance by observing that if $f$ is the double covering associated to a smooth divisor $D \in |2L|$ as in 6.3.50, then $E_f = L$ is ample or nef if and only if $L$ is. On the other hand, if $C \subset Y$ is any curve not contained in the branch locus of $f$, then $E \mid C$ is nef. (It is enough to test this after pulling back by any cover $C' \longrightarrow C$, and then one reduces to the case of a branched covering $D' \longrightarrow C'$ of curves with the property that every irreducible component of $D'$ maps isomorphically to $C'$. See [504, Appendix] for details.)

**Remark 6.3.58. (Rational homogeneous spaces and other varieties).** Kim and Manivel [336], [335], [337], [417] have studied the bundle associated to a branched covering $f : X \longrightarrow Y$ for certain rational homogeneous spaces $Y$. In the cases they treat, they prove that for any $X$ and $f$, the bundle $E_f$ is always spanned, and even ample when $b_2(Y) = 1$. They conjecture that this is true for any rational homogeneous space $Y$ with $b_2(Y) = 1$. Some other results concerning these bundles — notably when $Y$ is a curve or a Del Pezzo (or more general Fano) manifold — appear in the papers [504], [503] of Peternell and Sommese. $\hfill\square$

**Example 6.3.59. (Coverings of abelian varieties).** Let

$$f : X \longrightarrow A$$

be a branched covering of an abelian variety $A$ by a smooth (but possibly disconnected) projective variety $X$ of dimension $n$. Then $E_f$ is nef. This is a result of Peternell–Sommese [504], extending earlier work of Debarre [109]. (Choose a very ample divisor $B$ on $A$, and argue first as in the proof of Theorem 6.3.55 that $(f_* \omega_X) \otimes \mathcal{O}_A((n + 1)B)$ is globally generated. This gives a lower bound on the Barton invariant (Example 6.2.14) of $f_* \omega_X$:

$$\delta(A, f_* \omega_X, B) \ \geq \ -(n + 1). \tag{*}$$

Now fix $k > 0$ and consider the map $\nu = \nu_k : A \longrightarrow A$ given by multiplication by $k$. Applying (*) to the pulled-back covering $f' : X' = X \times_A A \longrightarrow A$ one finds that

$$\delta\big(A\;,\;f_*\,\omega_X\;,\;B\big)\;=\;\delta\big(A\;,\;\nu^*f_*\,\omega_X\;,\;\nu^*B\big)$$
$$=\;\delta\big(A\;,\;\nu^*f_*\,\omega_X\;,\;k^2\cdot B\big)$$
$$=\;\tfrac{1}{k^2}\cdot\delta\big(A\;,\;\nu^*f_*\,\omega_X\;,\;B\big)$$
$$=\;\tfrac{1}{k^2}\cdot\delta\big(A\;,\;f'_*\,\omega_{X'}\;,\;B\big)$$
$$\geq\;\frac{-(n+1)}{k^2}.$$

Letting $k\to\infty$ it follows that $\delta(A,f_*\,\omega_X,B)\geq 0$, as required.) Debarre conjectures that if $f:X\longrightarrow A$ is a non-trivial branched covering of a *simple* abelian variety $A$ by a smooth irreducible variety, then $E_f$ is ample provided that $f$ does not factor through an étale covering of $A$.                                   □

### 6.3.E  Direct Images of Canonical Bundles

Here we discuss very briefly one more instance where positivity properties of vector bundles have proven to be of great importance. Our modest intention is to convey something of the flavor of a large and imposing body of work through a couple of highly oversimplified statements.

In 1978, Fujita [192] proved an important and suggestive result about the direct images of the relative canonical bundles of fibre spaces over curves:[11]

**Theorem. (Fujita's theorem).** *Let $X$ be a smooth projective variety of dimension $n$, and suppose given a surjective mapping $f:X\longrightarrow C$ with connected fibres from $X$ to a smooth projective curve. Denote by $\omega_{X/C}$ the relative canonical bundle of $X$ over $C$. Then $f_*\omega_{X/C}$ is a nef vector bundle on $C$.*                                   □

While we do not attempt to reproduce the calculations here, the rough strategy of Fujita's proof is easily described. Specifically, at least away from the finitely many points $t\in C$ over which $F_t=f^{-1}(t)$ is singular, there is a natural identification of the fibres of the bundle in question:

$$\big(f_*\omega_{X/C}\big)(t)\;=\;H^0\big(F_t,\omega_{F_t}\big)\;=\;H^{n-1,0}(F_t).$$

The space on the right carries a natural Hermitian metric defined by integration, which in fact extends over the singular fibres to define a Hermitian metric on the bundle $f_*\omega_{X/C}$. Fujita then deduces the statement by an explicit curvature calculation.

The interest in such a result is that it can be used to study the geometry of $f$, which is the simplest example of the sort of fibre space that arises frequently in the approach to birational geometry pioneered by Iitaka and his school. For example, Fujita ([192], Corollary 4.2) uses it to re-prove a statement of Ueno concerning additivity of Kodaira dimension in the case at hand:

---

[11] Recall (Definition 2.1.11) that for a surjective mapping between smooth projective varieties to be a fibre space means simply that it has connected fibres.

**Example 6.3.60. (Fibre spaces over curves).** In the setting of Fujita's theorem, assume that $C$ has genus $g \geq 2$ and that a general fibre $F$ of $f$ has positive geometric genus $p_g(F) =_{\text{def}} h^{n-1,0}(F) > 0$. Then

$$\kappa(X) = \kappa(F) + 1.$$

(The inequality $\kappa(X) \leq \kappa(F) + 1$ holds quite generally, so the issue is to show that $\kappa(X) \geq \kappa(F) + 1$. To this end, observe first that $f_*\omega_{X/C} \neq 0$ thanks to the hypothesis on $p_g(F)$. Moreover, since $g(C) \geq 2$ it follows from Fujita's theorem that

$$f_*\omega_X = f_*\omega_{X/C} \otimes \omega_C$$

is ample. Fixing a very ample divisor $H$ on $C$, this implies that $S^m(f_*\omega_X) \otimes \mathcal{O}_C(-H)$ is globally generated for all $m \gg 0$. Using the natural map $S^m(f_*\omega_X) \longrightarrow f_*(\omega_X^{\otimes m})$, one then deduces that

$$H^0(X, \omega_X^{\otimes m} \otimes f^*\mathcal{O}_C(-H)) = H^0(C, f_*(\omega_X^{\otimes m})(-H)) \neq 0,$$

and hence $H^0(X, \mathcal{O}_X(f^*H))$ is realized as a subspace of $H^0(X, \mathcal{O}_X(mK_X))$ for all $m \gg 0$. This implies that $f$ factors as a composition of rational maps:

$$X \overset{\rho}{\dashrightarrow} V \dashrightarrow C,$$

where $\rho$ is the Iitaka fibration of $X$ associated to the canonical bundle $\mathcal{O}_X(K_X)$ (Section 2.1.C). In particular, if $G \subset X$ is a general fibre of $\rho$, then $\kappa(G, K_X|G) = \kappa(G, K_F|G) = 0$, from which it follows that $\kappa(F) \leq \dim F - \dim G = \kappa(X) - 1$. See [192, Propositions 1 and 2] for details.) □

In the years since [192], these ideas have been greatly developed by a number of authors, notably Viehweg [588], [590], [591], [592], [593], Kawamata [315], and Kollár [358], [356], [357], to study the positivity properties of direct images of dualizing sheaves for fibre spaces $f : X \longrightarrow Y$ of projective varieties under various smoothness hypotheses. We refer to [315], [358], and [594] for further references and precise statements of the results that have been obtained, which are necessarily somewhat involved and technical. Besides consequences for the geometry of fibre spaces (e.g. questions involving additivity of Kodaira dimension), this machine has found important applications to proving projectivity or quasi-projectivity of moduli spaces. Kawamata [321] [322] has recently applied positivity theorems for direct images of canonical bundles to study linear series on higher-dimensional varieties: see Remark 10.4.9.

As in the paper of Fujita, the work of Kawamata [315] and Kollár [358] analyzed the direct image bundles via metrics arising from Hodge theory. However, Kollár showed in [356, 357] that one could replace some of these arguments with vanishing theorems. To give the flavor, we conclude this subsection with a "toy" special case of [356, Corollary 3.7]:

**Proposition 6.3.61.** *Let $f : X \longrightarrow Y$ be a morphism between smooth projective varieties, and assume that $f$ is smooth, i.e. that the derivative of $f$ is everywhere surjective. Then $f_* \omega_{X/Y}$ is nef.*

It goes without saying that the hypothesis that $f$ is smooth is unrealistic in practice, and the assumptions in [356] are much weaker.

*Proof of Proposition 6.3.61.* We will use the theorem of Kollár stated in 4.3.8 that if $\pi : V \longrightarrow W$ is any surjective projective mapping with $V$ smooth and projective, and if $L$ is any ample line bundle on $W$, then

$$H^j(W, L \otimes R^i \pi_* \omega_V) = 0 \text{ for any } i \geq 0 \text{ and } j > 0. \tag{*}$$

Fix $s > 0$, and consider the $s$-fold fibre product $f^{(s)} : X^{(s)} = X \times_Y \cdots \times_Y X \longrightarrow Y$ of $X$ over $Y$. The smoothness hypothesis on $f$ guarantees that $X^{(s)}$ is still smooth, and one has

$$f_*^{(s)} \omega_{X^{(s)}/Y} = \left( f_* \omega_{X/Y} \right)^{\otimes s}.$$

Now suppose that $\dim Y = d$, and let $B$ be a very ample line bundle on $Y$ that is sufficiently positive so that $B \otimes \omega_Y^*$ is ample. Applying Kollár's vanishing theorem (*) to $f^{(s)}$, we deduce that

$$H^j\left(Y, (f_* \omega_{X/Y})^{\otimes s} \otimes B^{\otimes(d+1-j)}\right) = H^j\left(Y, f_*^{(s)}(\omega_{X^{(s)}}) \otimes \omega_Y^* \otimes B^{\otimes(d+1-j)}\right)$$
$$= 0$$

for $j > 0$. By Castelnuovo–Mumford regularity (Theorem 1.8.5), this implies that the vector bundle

$$\left(f_* \omega_{X/Y}\right)^{\otimes s} \otimes B^{\otimes(d+1)}$$

is globally generated. But since this holds for all $s > 0$, the bundle in question must be nef thanks to 6.2.13. □

### 6.3.F Some Constructions of Positive Vector Bundles

We conclude by presenting a couple of methods of construction of ample bundles.

**Pulling back bundles on $\mathbf{P}^n$.** We discuss an "amplification" process for bundles on $\mathbf{P}^n$ suggested by Barton's use of the Frobenius in [34, Proposition 3.1]. Fix for each $k \geq 1$ a branched covering

$$\nu_k : \mathbf{P}^n \longrightarrow \mathbf{P}^n \quad \text{with} \quad \nu_k^* \mathcal{O}_{\mathbf{P}^n}(1) = \mathcal{O}_{\mathbf{P}^n}(k).$$

For example, one might take a Fermat-type covering $\nu_k([a_0, \ldots, a_n]) = [a_0^k, \ldots, a_n^k]$, but the next proposition holds for any choice of $\nu_k$.

**Proposition 6.3.62.** *Let $E$ be an ample vector bundle on $\mathbf{P}^n$, and let $F$ be an arbitrary bundle on $\mathbf{P}^n$. Then there is a positive integer $k_0 = k_0(E, F)$ such that*

$$\nu_k^* E \otimes F \quad \text{is ample for all} \quad k \geq k_0.$$

*Proof.* For suitable $a \in \mathbf{Z}$ we can realize $F$ as a quotient of a direct sum of copies of $\mathcal{O}_{\mathbf{P}^n}(a)$. So it suffices to treat the case $F = \mathcal{O}_{\mathbf{P}^n}(a)$. Let $\delta = \delta(\mathbf{P}^n, E, \mathcal{O}_{\mathbf{P}^n}(1))$ be the Barton invariant of $E$ with respect to $\mathcal{O}_{\mathbf{P}^n}(1)$ (Example 6.2.14). In other words,

$$\delta = \sup \left\{ t \in \mathbf{Q} \mid E{<}{-}tH{>} \quad \text{is nef} \right\},$$

where $H$ denotes the hyperplane divisor on $\mathbf{P}^n$. Then by part (iii) of the example just cited,

$$
\begin{aligned}
\delta(\mathbf{P}^n, \nu_k^* E, \mathcal{O}_{\mathbf{P}^n}(1)) &= k \cdot \delta(\mathbf{P}^n, \nu_k^* E, \mathcal{O}_{\mathbf{P}^n}(k)) \\
&= k \cdot \delta(\mathbf{P}^n, \nu_k^* E, \nu_k^* \mathcal{O}_{\mathbf{P}^n}(1)) \\
&= k \cdot \delta(\mathbf{P}^n, E, \mathcal{O}_{\mathbf{P}^n}(1)) \\
&= k \cdot \delta.
\end{aligned}
$$

Therefore

$$\delta(\mathbf{P}^n, \nu_k^* E \otimes \mathcal{O}_{\mathbf{P}^n}(a), \mathcal{O}_{\mathbf{P}^n}(1)) = k\delta + a. \qquad (*)$$

But $\delta > 0$ since $E$ is ample, so if $k \gg 0$ the right-hand side of $(*)$ is positive, and $\nu_k^* E \otimes \mathcal{O}_{\mathbf{P}^n}(a)$ is ample. $\qquad\square$

**Example 6.3.63.** An analogous statement holds if $A$ is an abelian variety, and $\nu_k : A \longrightarrow A$ is the isogeny determined by multiplication by $k$. (Compare Example 6.3.59.) $\qquad\square$

**Generic cokernels.** In [223] Gieseker used a reduction to characteristic $p > 0$ to produce some interesting ample bundles on $\mathbf{P}^2$. In this paragraph we construct analogous bundles on an arbitrary projective variety.

**Proposition 6.3.64. (Generic cokernels, I).** *Let $X$ be an irreducible projective variety of dimension $n$, let $H$ be a very ample divisor on $X$, and let $V$ be a vector space of dimension $n + e$ with $e \geq n$. Then for $d \geq n + e$ the cokernel of a general vector bundle map*

$$\mathcal{O}_X(-dH)^{\oplus n} \xrightarrow{\ u\ } V_X$$

*is an ample vector bundle of rank $e$ on $X$.*

*Proof.* The condition $e \geq n$ guarantees that a general map $u$ has constant rank $n$ on $X$, and hence that $E_u = \operatorname{coker}(u)$ is indeed a bundle of rank $e$. Since ampleness is an open condition in a family of bundles (Proposition 6.1.9), it is enough to show that $E_u$ is ample for some $u$. Now $u$ is defined by an $n \times (n+e)$

matrix of sections in $\Gamma(X, \mathcal{O}_X(dH))$, and we are then free to assume that these are pulled back from $\Gamma(\mathbf{P}^n, \mathcal{O}_{\mathbf{P}^n}(d))$ under a branched covering $f : X \longrightarrow \mathbf{P}^n$ with $f^*\mathcal{O}_{\mathbf{P}^n}(1) = \mathcal{O}_X(H)$. Then $u$ and $E_u$ are themselves pulled back from $\mathbf{P}^n$, and since amplitude is preserved under finite coverings (Proposition 6.1.8), we are reduced to the case $X = \mathbf{P}^n$.

Set $r = n + e - 1$. Fixing $u$, we have an exact sequence

$$0 \longrightarrow \mathcal{O}_{\mathbf{P}^n}(-d)^{\oplus n} \xrightarrow{u} V_{\mathbf{P}^n} \longrightarrow E_u \longrightarrow 0,$$

and the amplitude of $E_u$ is equivalent to the assertion that the natural map

$$\mathbf{P}(E_u) \longrightarrow \mathbf{P}(V) = \mathbf{P}^r$$

is finite. But $\mathbf{P}(E_u) \subset \mathbf{P}(V_{\mathbf{P}^n}) = \mathbf{P}^n \times \mathbf{P}^r$ is the complete intersection of $n$ divisors in the linear series $|\mathrm{pr}_1^*\mathcal{O}_{\mathbf{P}^n}(d) \otimes \mathrm{pr}_2^*\mathcal{O}_{\mathbf{P}^r}(1)|$ determined by the vanishing of the composition

$$\mathrm{pr}_1^*\mathcal{O}_{\mathbf{P}^n}(-d)^{\oplus n} \xrightarrow{\mathrm{pr}_1^* u} V_{\mathbf{P}^n \times \mathbf{P}^r} \longrightarrow \mathrm{pr}_2^*\mathcal{O}_{\mathbf{P}^r}(1),$$

the second map being the pullback under the projection $\mathrm{pr}_2$ of the evaluation $V_{\mathbf{P}^r} \longrightarrow \mathcal{O}_{\mathbf{P}^r}(1)$. The required finiteness is then a consequence of the finiteness lemma (Lemma 6.3.43) established above.                                        $\square$

The previous proposition generalizes to quotients of an arbitrary bundle:

**Theorem 6.3.65. (Generic cokernels, II).** *Let $X$ be an irreducible projective variety of dimension $n$, let $H$ be a very ample divisor on $X$, and let $F_0$ be any vector bundle on $X$ of rank $n + f$ with $f \geq n$. Then for any $d \gg 0$ the cokernel of a sufficiently general vector bundle map*

$$u : \mathcal{O}_X(-dH)^{\oplus n} \longrightarrow F_0$$

*is an ample vector bundle of rank $f$.*

**Remark 6.3.66.** Demailly informs us that he has proven a similar result via a metric argument.                                        $\square$

**Example 6.3.67.** ([223]). Taking $X = \mathbf{P}^2$, for $d \gg 0$ there is an ample vector bundle $E$ having a presentation of the form

$$0 \longrightarrow \mathcal{O}_{\mathbf{P}^2}(-d)^2 \longrightarrow \mathcal{O}_{\mathbf{P}^2}(-1)^4 \longrightarrow E \longrightarrow 0. \quad \square$$

*Proof of Theorem 6.3.65.* We start with some reductions analogous to those in the proof of Proposition 6.3.64. Specifically, in the first place we will require that $d$ be large enough so that the bundle $\mathrm{Hom}(\mathcal{O}_X(-dH)^n, F_0)$ is globally generated. This guarantees that the general cokernel is at least a vector bundle of the stated rank. Note also that as before, it is enough to establish the amplitude in question for some $u$ having a locally free cokernel.

We next argue that we can reduce to the case in which $F_0$ is of the form $V_X \otimes O_X(-mH)$ for some vector space $V$ of rank $\geq 2n + f$. In fact, choose $m_0 \gg 0$ so that $F_0(mH)$ is globally generated for every $m \geq m_0$. Then we can fix for each $m \geq m_0$ a surjective map $v_m : V_X \otimes O_X(-mH) \longrightarrow F_0$. Now for $d \geq m$ consider a general homomorphism

$$\tilde{u}_{d,m} : O_X(-dH)^n \longrightarrow V_X \otimes O_X(-mH),$$

and set $u_{d,m} = v_m \circ \tilde{u}_{d,m}$:

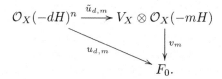

Then $\mathrm{coker}(u_{d,m})$ is a quotient of $\mathrm{coker}(\tilde{u}_{d,m})$, so it suffices to establish the amplitude of the latter. Moreover, as in the proof of 6.3.64 we may further restrict attention to the case when $\tilde{u}_{d,m}$ is pulled back from $\mathbf{P}^n$ under a branched covering $f : X \longrightarrow \mathbf{P}^n$. So finally we are reduced to considering $X = \mathbf{P}^n$, and a trivial vector bundle $V_{\mathbf{P}^n}$ of fixed large rank $\geq 2n$. Given an integer $m_0$, we need to show that for every sufficiently large $d \gg 0$ there is some $m \geq m_0$ such that the cokernel of a general map $O_{\mathbf{P}^n}(-d)^n \longrightarrow V_{\mathbf{P}^n}(-m)$ is ample.

To this end, first apply Proposition 6.3.64 to choose a natural number $d_0$, depending only on $n$ and $\dim V$, plus a map $u_0 : O_{\mathbf{P}^n}(-d_0)^n \longrightarrow V_{\mathbf{P}^n}$ such that $E_0 = \mathrm{coker}\, u_0$ is ample. By Proposition 6.3.62, there is a large integer $k_0$ such that whenever we pull back by a covering $\nu_k : \mathbf{P}^n \longrightarrow \mathbf{P}^n$ defined by $O_{\mathbf{P}^n}(k)$ with $k \geq k_0$, then for $0 \leq i \leq d_0 - 1$ each of the bundles

$$E_{k,i} = \nu_k^* E_0 \otimes O_{\mathbf{P}^n}(-m_0 - i)$$

is ample. Note that $E_{k,i}$ sits in an exact sequence

$$0 \longrightarrow O_{\mathbf{P}^n}(-kd_0 - m_0 - i)^n \longrightarrow V \otimes O_{\mathbf{P}^n}(-m_0 - i) \longrightarrow E_{k,i} \longrightarrow 0.$$

But every $d \gg 0$ is of the form $kd_0 + m_0 + i$ for some $k \geq k_0$ and $0 \leq i \leq d_0 - 1$, so we are done.    $\square$

## 6.4 Ample Vector Bundles on Curves

The object of this section is to study ample vector bundles on smooth curves, and in particular to give Hartshorne's characterization (Theorem 6.4.15) of such bundles. It was observed already in [277] that amplitude on curves is closely related to the concept of stability, and we emphasize that connection here. In order to control the stability of tensor products, Hartshorne originally drew on results of Narasimhan and Seshadri giving an essentially analytic characterization of stability. Gieseker [224] and Miyaoka [429] later realized that one can in effect reverse the process, and we follow their approach. In particular, along the way to Hartshorne's theorem we will use the theory of ample bundles to recover in an elementary fashion the tensorial properties of stability (Corollary 6.4.14).

Throughout this section, $C$ denotes a smooth irreducible complex projective curve.

## 6.4.A Review of Semistability

For the convenience of the reader we recall in this subsection the basic facts and definitions surrounding semistability of bundles on curves.

It is classical that much of the geometry associated to a line bundle on a curve is governed by the degree of the bundle. However, it was recognized early on that the degree is a less satisfactory invariant for vector bundles of higher rank. The pathologies stem from the fact that bundles of a given degree can become arbitrarily "unbalanced": for instance, if $L_n$ is a line bundle of degree $n$ on $C$, then $E_n = L_n \oplus L_n^*$ has degree zero, but most of the properties of $E_n$ depend on $n$. The condition of semistability in effect rules out this sort of problem, and the Harder–Narasimhan filtration expresses an arbitrary bundle as a successive extension with semistable quotients.

**Definition 6.4.1. (Slope and semistability).** Let $E$ be a vector bundle on the smooth projective curve $C$. The *slope* of $E$ is the rational number

$$\mu(E) \;=\; \frac{\deg(E)}{\operatorname{rank}(E)},$$

where as usual the degree of $E$ is the integer $\deg(E) = \int c_1(E)$. One says that $E$ is *semistable* if

$$\mu(F) \;\leq\; \mu(E) \text{ for every sub-bundle } F \subseteq E. \tag{*}$$

$E$ is *unstable* if it is not semistable.    □

Thus the slope of $E$ measures "degree per unit rank," and the condition of semistability means that $E$ cannot have any inordinately positive sub-bundles. There is a related notion of *stability*, for which one requires strict inequality in (*), but we will not require this.

We collect some elementary but useful observations that the reader may check:

**Example 6.4.2.** (i).   If $E = E_1 \oplus E_2$ is a direct sum of two bundles, then $E$ is semistable if and only if $E_1$ and $E_2$ are semistable with $\mu(E_1) = \mu(E_2)$.

(ii).  $E$ is semistable if and only if $\mu(Q) \geq \mu(E)$ for all quotients $Q$ of $E$.

(iii). In the definition of semistability, it is equivalent to work with sub-*sheaves* $F \subseteq E$ (so $F$ is still locally free, but one can allow $E/F$ to have torsion).

(iv). If $E$ and $F$ are any two vector bundles on $C$, then $\mu(E \otimes F) = \mu(E) + \mu(F)$.

(v).  Given any divisor $D$ on $C$, $E$ is semistable iff $E \otimes \mathcal{O}_C(D)$ is.        $\square$

Facts (iv) and (v) of the previous example give two equivalent ways to extend these concepts to **Q**-twisted bundles. Suppose that $E$ is a vector bundle on $C$, and that $\delta$ is a **Q**-divisor class on $C$. Recall that in Definition 6.2.7 we defined the degree of the **Q**-twisted bundle $E{<}\delta{>}$ on $C$ to be the rational number

$$\deg\left(E{<}\delta{>}\right) \;=\; \deg E \;+\; \operatorname{rank}(E) \cdot \deg \delta.$$

Recalling also that a sub-bundle of $E{<}\delta{>}$ is the **Q**-twisted bundle $F{<}\delta{>}$ where $F \subseteq E$ is a sub-bundle of $E$, we are led to make

**Definition 6.4.3.** (i).  The *slope* of $E{<}\delta{>}$ is defined to be

$$\mu\left(E{<}\delta{>}\right) \;=\; \frac{\deg\left(E{<}\delta{>}\right)}{\operatorname{rank}\left(E{<}\delta{>}\right)} \;=\; \mu(E) + \deg \delta.$$

(ii).  $E{<}\delta{>}$ is *semistable* if $\mu\left(F{<}\delta{>}\right) \leq \mu\left(E{<}\delta{>}\right)$ for every **Q**-twisted sub-bundle $F{<}\delta{>}$ of $E{<}\delta{>}$. In the contrary case, $E{<}\delta{>}$ is *unstable*.        $\square$

This definition evidently agrees with 6.4.1 in case $\delta$ is the class of an integral divisor, i.e. it respects **Q**-isomorphism. As one expects in light of 6.4.2 (v):

**Lemma 6.4.4.** *The* **Q**-*twisted bundle* $E{<}\delta{>}$ *is semistable if and only if* $E$ *itself is.*        $\square$

**Remark 6.4.5.** All the statements of Example 6.4.2 remain valid for **Q**-twisted bundles, with the proviso that in (i) one deals with two bundles twisted by the same class, so that their sum is defined.        $\square$

A basic fact for our purposes is that an unstable bundle has a canonical filtration with semistable graded pieces (Proposition 6.4.7). We start by establishing the following

**Lemma 6.4.6.** *Let* $E$ *be a vector bundle on* $C$. *Then the set of slopes*

$$\left\{\mu(F) \mid F \subseteq E\right\}$$

*of sub-bundles of* $E$ *is bounded from above. Moreover, if* $E$ *is unstable, there is a unique maximal sub-bundle* $U \subseteq E$ *of largest slope.*

The sub-bundle $U \subseteq E$ is called the *maximal destabilizing sub-bundle* of $E$.

*Proof of Lemma 6.4.6.* The first statement is clear for sub-bundles of a trivial bundle $\mathcal{O}_C^{\oplus N}$, and hence also for sub-bundles of $\mathcal{O}_C(H)^{\oplus N}$ for any divisor $H$ thanks to Example 6.4.2 (iv). But we can realize any bundle $E$ as a sub-bundle of $\mathcal{O}_C(H)^{\oplus N}$ for some sufficiently positive $H$, and the first statement follows. For the second, let $E_1, E_2 \subset E$ be sub-bundles having maximal slope $\mu$, and let $E_1 + E_2 \subseteq E$ be the sub-sheaf they generate. It is sufficient to prove that $\mu(E_1 + E_2) = \mu$. But this follows from the exact sequence

$$0 \longrightarrow E_1 \cap E_2 \longrightarrow E_1 \oplus E_2 \longrightarrow E_1 + E_2 \longrightarrow 0.$$

Indeed, by maximality $\mu(E_1 \cap E_2) \leq \mu$, and then it follows with a computation that $\mu(E_1 + E_2) \geq \mu(E_1 \oplus E_2) = \mu$. $\qquad\square$

One then obtains:

**Proposition 6.4.7. (Harder–Narasimhan filtration).** *Any vector bundle $E$ on $C$ has a canonically defined filtration*

$$\mathrm{HN}_\bullet(E) \;:\; 0 = \mathrm{HN}_\ell(E) \subset \mathrm{HN}_{\ell-1}(E) \subset \ldots \subset \mathrm{HN}_1(E) \subset \mathrm{HN}_0(E) = E$$

*by sub-bundles, characterized by the properties that if*

$$\mathrm{Gr}_i \;=\; \mathrm{HN}_i(E) \,/\, \mathrm{HN}_{i+1}(E)$$

*is the $i^{th}$ associated graded bundle, then each of the bundles $\mathrm{Gr}_i$ is semistable, and*

$$\mu(\mathrm{Gr}_{\ell-1}) \;>\; \ldots \;>\; \mu(\mathrm{Gr}_1) \;>\; \mu(\mathrm{Gr}_0).$$

*Proof.* In fact, if $E$ is semistable, take $\ell = 1$. Otherwise, let $\mathrm{HN}_1(E) \subset E$ be the maximal destabilizing sub-bundle of $E$, and continue inductively. $\qquad\square$

**Remark 6.4.8. (Extension to Q-twists).** Again the previous results extend in a natural way to **Q**-twisted bundles. In the situation of Definition 6.4.3, we define the maximal destabilizing sub-bundle of a **Q**-twisted bundle $E{<}\delta{>}$ to be the **Q**-twisted sub-bundle $U{<}\delta{>}$ of $E{<}\delta{>}$, where $U \subset E$ is the maximal destabilizing subsheaf of $E$. In a similar fashion, $E{<}\delta{>}$ has the Harder–Narasimhan filtration $\mathrm{HN}_\bullet(E{<}\delta{>}) = \mathrm{HN}_\bullet(E){<}\delta{>}$. $\qquad\square$

**Remark 6.4.9.** In the present discussion, stability appears as a technical tool for studying positivity. However, it originally arose in connection with the construction of moduli spaces of bundles (see [477] for a very readable account and [47] for a survey of later developments). More recently, the concept of stability has proven to be very fundamental from many points of view; see for example [474], [139], [579], [180], [181]. $\qquad\square$

## 6.4.B Semistability and Amplitude

The result for which we are aiming states that the positivity of a bundle $E$ on a curve is characterized in terms of the degrees of $E$ and its quotients. One direction is elementary:

**Lemma 6.4.10.** (i). *Let $E$ be a vector bundle of rank $e$ on $C$. If $E$ is nef then $\deg E \geq 0$ and if $E$ is ample then $\deg E > 0$.*

(ii). *The same statement holds more generally if $E$ is replaced by a $\mathbf{Q}$-twisted bundle $E{<}\delta{>}$.*

*Proof.* We treat (ii). Consider the projective bundle $\pi : \mathbf{P}(E) \longrightarrow C$, and let

$$\xi \;=\; \xi_E + \pi^*\delta \;\in\; N^1\big(\mathbf{P}(E)\big)_{\mathbf{Q}},$$

where as usual $\xi_E$ represents (the first Chern class of) $\mathcal{O}_{\mathbf{P}(E)}(1)$. Then

$$\deg\, E{<}\delta{>} \;=\; \int_{\mathbf{P}(E)} \xi^e.$$

But if $E{<}\delta{>}$ is ample then by definition $\xi$ is an ample class on $\mathbf{P}(E)$, and consequently the degree in question is strictly positive. Similarly, Kleiman's theorem (Theorem 1.4.9) shows that it is non-negative if $E{<}\delta{>}$ is nef.      □

We now wish to consider how to pass from numerical properties to statements about amplitude and stability. Let $E$ be a vector bundle of rank $e$ on the smooth curve $C$, and let $\Delta$ be a divisor representing $\det(E)$. It will be convenient to work with the $\mathbf{Q}$-twisted bundle

$$E_{\mathrm{norm}} \;=\; E{<}-\tfrac{1}{e}\Delta{>}\,.$$

The point of this normalization is that it reduces one to the case of bundles of degree zero: it follows from Definition 6.2.7 that $\deg E_{\mathrm{norm}} = 0$. Since by 6.4.4 $E_{\mathrm{norm}}$ is semistable if and only if $E$ is, we see using Example 6.4.2 (ii) that $E$ is semistable if and only if every quotient of $E_{\mathrm{norm}}$ has degree $\geq 0$ in the sense that

$$\deg\big(Q{<}-\tfrac{1}{e}\Delta{>}\big) \;\geq\; 0$$

for any quotient bundle $Q$ of $E$. Analogous remarks hold starting from a $\mathbf{Q}$-twisted bundle $E{<}\delta{>}$: indeed, if one mirrors the definition above one finds that in fact

$$\big(E{<}\delta{>}\big)_{\mathrm{norm}} = E_{\mathrm{norm}}.$$

The basic link between stability and positivity is then given by

**Proposition 6.4.11. (Semistability and nefness).** *$E$ is semistable if and only if $E_{\mathrm{norm}}$ is nef.*

We start with a useful lemma:

**Lemma 6.4.12.** *Let $E$ be a vector bundle on $C$, and let $f : C' \longrightarrow C$ be any branched covering of $C$ by a smooth irreducible projective curve $C'$. Then $E$ is semistable if and only if $f^*E$ is semistable.*

*Proof.* If $E$ is unstable, then certainly $f^*E$ is as well since the pullback of a destabilizing sub-bundle $U \subseteq E$ will destabilize $f^*E$. Conversely, suppose for a contradiction that $E$ is semistable but that $f^*E$ is unstable. By what we have just observed the pullback of $f^*E$ under a covering $C'' \longrightarrow C'$ will remain unstable, so we may assume without loss of generality that $f$ is Galois, with group $G = \mathrm{Gal}(C'/C)$. Let $V \subseteq f^*E$ be the maximal destabilizing sub-bundle of $f^*E$. Now $G$ acts in the natural way on $f^*E$ and hence also on the collection of sub-bundles of $f^*E$. It follows from the uniqueness of the maximal destabilizing sub-bundle that $V$ is $G$-stable. Hence $V = f^*U$ for a sub-bundle $U \subset E$, and one checks right away that $U$ must destabilize $E$. $\square$

*Proof of Proposition 6.4.11.* Suppose first that $E_{\mathrm{norm}}$ is nef. Then thanks to Theorem 6.2.12 (i), all of its quotients are nef, and hence have non-negative degree. Therefore by the remarks preceding the statement of Proposition 6.4.11, $E$ is semistable. Conversely, suppose that $E$ is semistable, but that $E_{\mathrm{norm}}$ is not nef. It follows from the Barton–Kleiman criterion (Proposition 6.1.18 and Remark 6.2.9) that there is a finite map $f : C' \longrightarrow C$ from a smooth irreducible curve $C'$ to $C$ such that $f^*E_{\mathrm{norm}}$ has a rank one quotient of negative degree. Then $f^*E_{\mathrm{norm}}$ is unstable. On the other hand, since $C$ is smooth $f$ must be a branched covering. Thus we arrive at a contradiction to 6.4.12, which completes the proof. $\square$

**Remark 6.4.13.** It follows from 6.4.4 that Proposition 6.4.11 and Lemma 6.4.12 both remain valid if the "classical" bundle $E$ is replaced by a **Q**-twist $E{<}\delta{>}$. $\square$

We pause to note that as an application one obtains the following fundamental result, which is traditionally established via the theorem of Narasimhan and Seshadri [474] characterizing stable bundles in terms of representations of the fundamental group of $C$. The present much more elementary approach is due to Gieseker [224], rendered particularly transparent via **Q**-twists in [429].

**Corollary 6.4.14. (Semistability of tensor products).** *The tensor product of two semistable bundles on a smooth curve is semistable. Consequently, if $E$ is semistable, then so is $S^m E$ for every $m \geq 0$. The same statements hold for **Q**-twisted bundles.*

*Proof.* Suppose that $E$ and $F$ are semistable. Then $E_{\mathrm{norm}}$ and $F_{\mathrm{norm}}$ are nef, and hence

$$\left( E \otimes F \right)_{\mathrm{norm}} = E_{\mathrm{norm}} \otimes F_{\mathrm{norm}}$$

is also nef thanks to Theorem 6.2.12 (iv) and (v). The semistability of $E \otimes F$ then follows from the previous proposition. By induction, if $E$ is semistable then so is any tensor power $T^q(E)$. In characteristic zero, $S^m E$ is a summand of $T^m E$, so its semistability follows from 6.4.2 (i). The extension to **Q**-twists is evident. □

Finally, we turn to a theorem of Hartshorne [277] giving a very pleasant characterization of nef and ample vector bundles on a curve.

**Theorem 6.4.15. (Hartshorne's theorem).** *A vector bundle $E$ on $C$ is nef if and only if $E$ and every quotient bundle of $E$ has non-negative degree, and $E$ is ample if and only if $E$ and every quotient has strictly positive degree. The same statements hold if $E$ is replaced by a **Q**-twisted bundle $E{<}\delta{>}$.*

*Proof.* One direction is immediate: if $E$ is ample (or nef), then every quotient is ample (or nef) and hence has positive (or non-negative) degree thanks to Lemma 6.4.10. Conversely, the essential point will be to treat the case in which $E$ is semistable:

> *Main Claim:* Let $E$ be a semistable bundle on $C$. If $E$ has non-negative degree then $E$ is nef, and if $E$ has positive degree then $E$ is ample.

Granting this for the moment, we complete the proof.

Suppose then that $E$ is not nef. We need to show that $E$ itself or a quotient has negative degree. If $\deg E < 0$ there is nothing further to prove, so we may suppose that $\deg E \geq 0$. It then follows from the Claim that $E$ must be unstable. Consider its Harder–Narasimhan filtration

$$\mathrm{HN}_\bullet(E) : \quad 0 = \mathrm{HN}_\ell(E) \subset \mathrm{HN}_{\ell-1}(E) \subset \ldots \subset \mathrm{HN}_1(E) \subset \mathrm{HN}_0(E) = E$$

and as before set $\mathrm{Gr}_i = \mathrm{HN}_i(E)/\mathrm{HN}_{i+1}(E)$. Since an extension of nef bundles is nef, it follows again from the Claim that at least one of these graded pieces — say $\mathrm{Gr}_k$ — must have negative degree. Therefore

$$\mu(\mathrm{Gr}_0) < \mu(\mathrm{Gr}_1) < \ldots < \mu(\mathrm{Gr}_k) < 0.$$

But then $\deg\big(E/\mathrm{HN}_{k+1}(E)\big) < 0$, and we have produced the desired quotient. A similar argument shows that if $E$ is not ample, then $E$ or some quotient has degree $\leq 0$.

Turning to the claim, the point is to apply Proposition 6.4.11. In fact, assume that $E$ is semistable. Then $E_{\mathrm{norm}}$ is nef. But

$$E = E_{\mathrm{norm}}{<}\tfrac{1}{e}\Delta{>} = E_{\mathrm{norm}} \otimes \mathcal{O}_C{<}\tfrac{1}{e}\Delta{>},$$

where as above $\Delta$ is a divisor class representing $\det(E)$. If $\deg(E) \geq 0$ then $\deg(\Delta) \geq 0$, whence $\mathcal{O}_C{<}\tfrac{1}{e}\Delta{>}$ is nef, and if $\deg(E) > 0$ then $\mathcal{O}_C{<}\tfrac{1}{e}\Delta{>}$ is ample. Therefore thanks to Theorem 6.2.12, $E$ itself is nef in the first case and ample in the second, as asserted.

Finally, the extension to **Q**-twists presents no difficulties, and is left to the reader. □

**Example 6.4.16. (Higher cohomology and amplitude).** Suppose that $C$ has genus $g \geq 2$. If $E$ is a bundle on $C$ such that $H^1(C, E) = 0$, then $E$ is ample. (If $Q$ is a quotient of $E$ having degree $\leq 0$, then it follows from Riemann–Roch that $H^1(C, Q) \neq 0$, and hence that $H^1(C, E) \neq 0$.) This result is due to Fujita [192, Lemma 3].    □

**Example 6.4.17. (Generically surjective morphisms).** Let $E$ and $F$ be vector bundles on $C$, and suppose that $u : E \longrightarrow F$ is a homomorphism that is surjective away from finitely many points of $C$. If $E$ is ample then so too is $F$, and the analogous statements hold also for $\mathbf{Q}$-twists. (Any quotient $Q$ of $F$ gives rise to a quotient $Q'$ of $E$ with $\deg Q' \leq \deg Q$.)    □

**Example 6.4.18. (Singular curves).** Fulton [201, Proposition 4] gave an example to show that the conclusion of Theorem 6.4.15 can fail on a singular curve $C$. In fact, fixing a non-singular point $P \in C$, one can construct a non-split extension

$$0 \longrightarrow \mathcal{O}_C \longrightarrow E \longrightarrow \mathcal{O}_C(P) \longrightarrow 0$$

which splits when pulled back to the normalization $\nu : C' \longrightarrow C$ of $C$. Then every quotient of $E$ has positive degree, but $\nu^* E$ is not ample. Serre had used a similar construction to show that Hartshorne's theorem fails in positive characteristics.    □

**Remark 6.4.19. (Higher-dimensional varieties).** Examples suggest that there cannot be a clean numerical criterion for the amplitude of bundles on higher-dimensional varieties analogous to Hartshorne's theorem: see Remark 8.3.14.    □

**Example 6.4.20. (Hartshorne's proof of Theorem 6.4.15).** The original proof of Theorem 6.4.15 in [277] is quite interesting, and we indicate the idea. As in the argument above, the essential point is to show that if $E$ is a semistable bundle of non-negative degree on $C$, then $E$ is nef. Supposing this is false, there is a reduced irreducible curve $\Gamma \subset \mathbf{P}(E)$ such that

$$\int_\Gamma c_1(\mathcal{O}_{\mathbf{P}(E)}(1)) \; < \; 0. \tag{*}$$

Evidently $\Gamma$ cannot lie in a fibre of the projection $\pi : \mathbf{P}(E) \longrightarrow C$, and so $\Gamma$ is flat over $C$, say of degree $d$. Consider the exact sequence

$$0 \longrightarrow \mathcal{I}_{\Gamma/\mathbf{P}(E)}(m) \longrightarrow \mathcal{O}_{\mathbf{P}(E)}(m) \longrightarrow \mathcal{O}_\Gamma(m) \longrightarrow 0.$$

For $m \gg 0$ — so that $R^1 \pi_* \mathcal{I}_{\Gamma/\mathbf{P}(E)}(m) = 0$ — this gives rise to a surjection

$$S^m E \longrightarrow \pi_* \mathcal{O}_\Gamma(m) \longrightarrow 0.$$

But $\deg \mathcal{O}_\Gamma(m) \leq -m$ by virtue of (*), and it follows from Riemann–Roch on the normalization $\Gamma'$ of $\Gamma$ that $\chi(\Gamma, \mathcal{O}_\Gamma(m))$ becomes increasingly negative as

$m$ grows. On the other hand, the bundle $G_m = \pi_* \mathcal{O}_\Gamma(m)$ has fixed rank $d$, and satisfies $h^i(C, G_m) = h^i(\Gamma, \mathcal{O}_\Gamma(m))$. Applying the Riemann–Roch formula

$$\chi(C, G_m) \;=\; \deg G_m + d \cdot (1 - \operatorname{genus}(C))$$

on $C$, we see that $\deg G_m \ll 0$ for $m \gg 0$. In other words, we have established that if $m \gg 0$, then $S^m E$ has a quotient of negative degree. But this is impossible: for $E$ is semistable of non-negative slope, and hence so is $S^m E$.  $\square$

# Notes

Section 6.1 largely follows [274], although the use of a Veronese mapping to establish the tensorial properties of ample bundles in characteristic zero (Theorem 6.1.15) — which simplifies earlier approaches — is new. The presentation of this argument follows some suggestions of Fulton. Gieseker's lemma (Proposition 6.1.7) appears in [223].

A formalism of twisting bundles by **Q**-divisors was initiated by Miyaoka in [429], and its utility was re-emphasized by an analogous construction in [133]. Nef vector bundles have come into focus in recent years with the flowering of higher-dimensional geometry, where nef line bundles play a critical role. The Barton invariant (Example 6.2.14) appears implicitly in [34], where Barton uses the Frobenius to establish the amplitude of tensor products in positive characteristics.

Section 6.3 draws on many sources, most of which are cited in the text and won't be repeated here. The proof of Theorem 6.3.19 is adapted from [473], while Theorem 6.3.29 is new. The discussion of Picard bundles closely follows [212], although it seems to have been overlooked that the arguments for curves work without change on irregular varieties of all dimensions. Section 6.3.D follows [387]. The proof of Proposition 6.3.61 was explained to me by Ein. The material in Section 6.3.F is new, although as noted Demailly has obtained some similar results by different methods.

As we have indicated, the approach of Section 6.4 originates with Gieseker [224] and Miyaoka [429]. We have drawn on the very nice exposition in [432, Lecture III, §2].

# Geometric Properties of Ample Vector Bundles

This chapter is devoted to some geometric properties of positive vector bundles. We start in Section 7.1 with analogues and extensions of the theorems of Lefschetz and Barth. The next section concerns degeneracy loci: we prove that under suitable positivity hypotheses a map of bundles must drop rank (on a connected locus) whenever it is dimensionally predicted to do so, and give some applications. Finally, in the last section we briefly take up vanishing theorems for ample bundles.

## 7.1 Topology

This section focuses on the topological properties of ample bundles. We begin with Sommese's analogue of the Lefschetz hyperplane theorem. Then we present a generalization, due to Bloch and Gieseker, of the hard Lefschetz theorem. The section concludes with some Barth-type results of the author for branched coverings.

### 7.1.A Sommese's Theorem

Let $X$ be an irreducible complex projective variety of dimension $n$, and $E$ a vector bundle of rank $e$ on $X$. Given a section $s \in \Gamma(X, E)$ of $E$, denote by

$$Z = \text{Zeroes}(s) = \{ x \in X \mid s(x) = 0 \}$$

the zero-locus of $s$. Thus $Z \subseteq X$ is a closed algebraic subset of $X$, of expected codimension $e$.

In the spirit of the Lefschetz hyperplane theorem (Theorem 3.1.17) one wishes to compare the topology of $Z$ with that of $X$ under suitable positivity hypotheses on $E$. The cleanest result in this direction is due to Sommese:

**Theorem 7.1.1. (Sommese's theorem).** *Assume that $X$ is non-singular and that $E$ is ample. Then*

$$H^i(X, Z; \mathbf{Z}) \;=\; 0 \quad \text{for } i \le n - e.$$

*In particular, the restriction map $H^i(X; \mathbf{Z}) \longrightarrow H^i(Z; \mathbf{Z})$ is an isomorphism for $i < n - e$, and injective when $i = n - e$.*

Note that the theorem does not require that $s$ vanish transversely, or even that $Z$ have the expected dimension $n - e$.

*Proof of Theorem 7.1.1.* We will show that

$$H_i(X - Z; \mathbf{Z}) \;=\; 0 \quad \text{for } i \ge n + e. \tag{7.1}$$

Since $X$ is non-singular, the theorem as stated follows by Lefschetz duality $H^j(X, Z; \mathbf{Z}) = H_{2n-j}(X - Z; \mathbf{Z})$ (cf. [71, Corollary 8.4]).

For (7.1), pass to the projective bundle $\pi : \mathbf{P}(E) \longrightarrow X$ and consider the pullback of $s$ and the tautological quotient of $\pi^* E$:

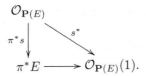

Denote by

$$s^* \;\in\; \Gamma\big(\mathbf{P}(E), \mathcal{O}_{\mathbf{P}(E)}(1)\big)$$

the indicated composition, and by $Z^* \subseteq \mathbf{P}(E)$ the zero-locus of $s^*$. Thinking of $\mathbf{P}(E)$ as the bundle of one-dimensional quotients of $E$:

$$\mathbf{P}(E) \;=\; \big\{(x, \lambda) \mid \lambda : E(x) \twoheadrightarrow \mathbf{C}\big\},$$

one sees that

$$Z^* \;=\; \big\{(x, \lambda) \mid (\lambda \circ s)(x) = 0\big\}. \tag{7.2}$$

Given $x \in X$, the condition on the right in (7.2) defines a hyperplane in $\mathbf{P}\big(E(x)\big) = \pi^{-1}(x)$ if $x \in X - Z$, while if $x \in Z$ there is no condition on $\lambda$. In other words, $\pi$ restricts to a mapping

$$p : \big(\mathbf{P}(E) - Z^*\big) \longrightarrow \big(X - Z\big),$$

and $p$ is in fact a $\mathbf{C}^{e-1}$-bundle (but not in general a vector bundle, i.e. $p$ may not section). In particular

$$H_i\big(\mathbf{P}(E) - Z^*; \mathbf{Z}\big) \;=\; H_i\big(X - Z; \mathbf{Z}\big) \quad \text{for all } i.$$

On the other hand, $\mathcal{O}_{\mathbf{P}(E)}(1)$ is an ample line bundle on $\mathbf{P}(E)$ thanks to the amplitude of $E$, and hence $\mathbf{P}(E) - Z^*$ is an affine variety, of dimension

$n + e - 1$. Therefore the theorem of Andreotti and Frankel (Theorem 3.1.1) implies that

$$H_i\big(\mathbf{P}(E) - Z^*; \mathbf{Z}\big) \;=\; 0 \quad \text{for} \quad i \;\geq\; \dim_{\mathbf{C}} \mathbf{P}(E) + 1 = n + e,$$

and (7.1) follows.                                                                □

**Example 7.1.2.** In the situation of Sommese's theorem, the cokernel of the restriction map $H^{n-e}(X; \mathbf{Z}) \longrightarrow H^{n-e}(Z; \mathbf{Z})$ is torsion-free. (In fact,

$$H^{n-(e-1)}\big(X, Z; \mathbf{Z}\big) \;=\; H_{n+(e-1)}\big(X - Z; \mathbf{Z}\big) \;=\; H_{n+e-1}\big(\mathbf{P}(E) - Z^*; \mathbf{Z}\big)$$

is torsion-free thanks to 3.1.3.)                                                □

**Remark 7.1.3. (Singular varieties).** Recalling (3.1.4, 3.1.14) that the cohomology of an arbitrarily singular affine variety of (complex) dimension $d$ still vanishes in degrees $> d$, this proof shows that the vanishing

$$H_i\big(X - Z; \mathbf{Z}\big) \;=\; 0 \quad \text{for} \quad i \;\geq\; n + e \tag{7.3}$$

from (7.1) holds without any non-singularity assumptions on $X$.            □

**Example 7.1.4. (Non-emptiness and connectedness of zero loci).** Suppose that $X$ is a smooth irreducible projective variety of dimension $n$, and that $E$ is an ample vector bundle of rank $e$ on $X$. Let $s \in \Gamma(X, E)$ be a section of $E$, with zero-locus $Z = \mathrm{Zeroes}(s) \subseteq X$.

(i).  If $e \leq n$ then $Z$ is non-empty, and if $e \leq n - 1$ then $Z$ is connected.

(ii).  The statements in (i) continue to hold even if $X$ is singular.

(In fact, (i) follows immediately from Sommese's theorem. If $X$ is singular then the non-emptiness of $Z$ when $e \leq n$ is a consequence of the previous remark. For the connectedness one can assume that $X$ is normal, and in this case it follows from the Zeeman spectral sequence that there is an injection $H^1(X, Z; \mathbf{Q}) \hookrightarrow H_{2n-1}(X - Z; \mathbf{Q})$. See [212, Lemma 1.3] for details.)    □

**Example 7.1.5. (Nonsingular zero loci).** Let $X$ be a smooth projective variety of dimension $n$, let $E$ be an ample bundle of rank $e$ on $X$, and let $s \in \Gamma(X, E)$ be a section whose zero locus $Y = \mathrm{Zeroes}(s) \subseteq X$ is non-singular.

(i).  The restriction maps $H^q(X, \Omega_X^p) \longrightarrow H^q(Y, \Omega_Y^p)$ are isomorphisms for $p + q < n - e$. In particular, one has an equality of Hodge numbers $h^{p,q}(Y) = h^{p,q}(X)$ in the indicated range.

(ii).  If $e < n - 2$, then the canonical map $\mathrm{Pic}(X) \longrightarrow \mathrm{Pic}(Y)$ on Picard groups is an isomorphism.

Note that we do not assume here that $Y$ actually has the expected codimension $n - e$. (The isomorphisms in (i) follow from the theorem via the Hodge decomposition. For (ii), use the exponential sequence.) $\qquad\Box$

**Example 7.1.6. (Set-theoretic complete intersections).** Since it does not involve any transversality hypotheses, Sommese's theorem applies to a direct sum of line bundles to give obstructions to realizing a projective variety as a set-theoretic complete intersection.

(i). Let $X \subset \mathbf{P}^n$ be an algebraic subset defined set-theoretically as the zero-locus of $e$ homogeneous polynomials. Then $b_i(X) = b_i(\mathbf{P}^n)$ for $i < n - e$.

(ii). The Segre variety $\mathbf{P}^1 \times \mathbf{P}^2 \subset \mathbf{P}^5$ is not a set-theoretic complete intersection. $\qquad\Box$

**Example 7.1.7. (Realizing cohomology by affine varieties).** Given any projective variety $X$, there is an affine variety $Y$, together with a mapping $p : Y \longrightarrow X$ inducing an isomorphism on integral homology and cohomology. (Apply the argument in the proof of 7.1.1 to a suitable bundle and section.) $\quad\Box$

**Remark 7.1.8. (Higher homotopy).** It is natural to ask whether in the situation of the theorem one has the vanishing of the relative homotopy groups

$$\pi_i(X, Z) \;=\; 0 \quad \text{for } i \le n - e. \tag{7.4}$$

This seems to be unknown in general. However Okonek [487, Corollary 2.2] has established that (7.4) holds if $E$ is ample and globally generated, and if in addition $Z$ has the expected codimension $e$: in fact, Okonek allows $X$ to have local complete intersection singularities. The vanishing (7.4) is proved in [389, Theorem 3.5] without assuming that $\dim Z = n - e$ under the stronger positivity hypothesis that $E \otimes B^*$ is globally generated for some ample line bundle $B$ that is itself globally generated. $\qquad\Box$

**Remark 7.1.9. (k-ample bundles).** Sommese's theorem extends to $k$-ample bundles (Remark 6.2.18). In fact, if $E$ is a $k$-ample bundle of rank $e$ on a smooth projective variety of dimension $n$, and $Z = \text{Zeroes}(s)$ is the zero-locus of a section of $E$, then $H^i(X, Z; \mathbf{Z}) = 0$ for $i \le n - e - k$. (The proof of 7.1.1 reduces the question to the case in which $E$ is a line bundle, and there one cuts down by general hyperplane sections to arrive at the case of ample bundles. See [551, Proposition 1.16].) $\qquad\Box$

### 7.1.B  Theorem of Bloch and Gieseker

We turn next to a result of Bloch and Gieseker [60] that gives an extension of the hard Lefschetz theorem (Theorem 3.1.39 and Corollary 3.1.40) to bundles of higher rank. We will use this statement on several occasions.

**Theorem 7.1.10. (Hard Lefschetz for ample bundles).** *Let $X$ be a smooth projective variety of dimension $n$, and let $E$ be an ample vector bundle on $X$ of rank $e \le n$. Consider for each $i$ the map*

$$L : H^i(X; \mathbf{C}) \longrightarrow H^{i+2e}(X; \mathbf{C})$$

*determined by cup-product with the top Chern class $c_e(E) \in H^{2e}(X; \mathbf{C})$. Then $L$ is injective if $i \le n - e$ and $L$ is surjective when $i \ge n - e$. In particular, the mapping*

$$H^{n-e}(X; \mathbf{C}) \xrightarrow{\;\;\cdot c_e(E)\;\;} H^{n+e}(X; \mathbf{C})$$

*is an isomorphism.*

*Proof.* The idea, naturally enough, is to pass to the projective bundle

$$\pi : \mathbf{P}(E) \longrightarrow X,$$

and apply the classical theorem to the first Chern class of the line bundle $\mathcal{O}_{\mathbf{P}(E)}(1)$. Specifically, set $\xi = c_1(\mathcal{O}_{\mathbf{P}(E)}(1))$, and write $H^*(\ )$ for cohomology with complex coefficients. Viewing $H^*(\mathbf{P}(E))$ as a module over $H^*(X)$ via $\pi^*$, recall that one has the presentation

$$H^*(\mathbf{P}(E), \mathbf{C}) \;=\; \frac{H^*(X, \mathbf{C})[\xi]}{\left(\xi^e - c_1(E) \cdot \xi^{e-1} + \cdots + (-1)^e \cdot c_e(E)\right)}. \qquad (7.5)$$

Fix a class $a \in H^i(X)$ such that $c_e(E) \cdot a = 0 \in H^{i+2e}(X)$, and set

$$b = a \cdot \left(\xi^{e-1} - c_1(E) \cdot \xi^{e-2} + \cdots + (-1)^{e-1} c_{e-1}(E)\right) \in H^{i+2(e-1)}(\mathbf{P}(E)).$$

Then $\pi_*\big(b \cap [\mathbf{P}(E)]\big) = a \cap [X]$, and hence $b \ne 0$ so long as $a \ne 0$. Moreover, by (7.5) the hypothesis on $a$ implies that

$$\xi \cdot b \;=\; \pm c_e(E) \cdot a \;=\; 0.$$

Now assume that $i \le n - e$, so that $i + 2(e-1) < \dim_{\mathbf{C}}(\mathbf{P}(E))$. Then since $\xi$ is the class of an ample line bundle, the classical hard Lefschetz theorem in the form of Corollary 3.1.40 implies that the mapping

$$L_\xi : H^{i+2(e-1)}(\mathbf{P}(E)) \longrightarrow H^{i+2e}(\mathbf{P}(E))$$

given by multiplication by $\xi$ is injective. Therefore $b = 0$, whence $a = 0$, and the first statement follows. The last assertion follows from this since the two groups $H^{n-e}(X)$ and $H^{n+e}(X)$, being Poincaré dual, have the same dimension. When $i \ge n - e$ write $i = n - e + d$ with $d \ge 0$; we may suppose that $d \le n - e$, the statement being trivial otherwise. Let $A$ be an ample line bundle on $X$, and consider the ample vector bundle

$$F = E \oplus A^{\oplus d}$$

of rank $e + d \leq n$. Then $c_{e+d}(F) = c_e(E)c_1(A)^d$, and by applying what we have already proved to $F$ we deduce that the composition

$$H^{n-e-d}(X) \xrightarrow{\cdot c_1(A)^d} H^{n-e+d}(X) \xrightarrow{\cdot c_e(E)} H^{n+e+d}(X)$$

is an isomorphism. Therefore the second map is surjective, as required. $\qquad\square$

For bundles of rank $e \geq n = \dim X$ one has

**Variant 7.1.11.** *In the setting of the theorem, suppose that $E$ has rank $e \geq n$. Then multiplication by $c_n(E) \in H^{2n}(X)$ determines an isomorphism $H^0(X) \longrightarrow H^{2n}(X)$.*

*Proof.* The issue is to show that $c_n(E) \neq 0$. Keeping the notation of the previous proof, set

$$b = \xi^{n-1} - c_1(E) \cdot \xi^{n-2} + \ldots + (-1)^{n-1} c_{n-1}.$$

Then $b \neq 0$, but the vanishing of $c_n(E)$ implies that $b \cdot \xi^{e-n+1} = 0$, which again contradicts 3.1.40. $\qquad\square$

As an application, we give a result of Sommese [551, (2.5)] generalizing parts of Barth's theorem (Theorem 3.2.1):

**Proposition 7.1.12. (Subvarieties with ample normal bundles).** *Let $P$ be a smooth projective variety and $X \subseteq P$ a smooth subvariety of dimension $n$ and codimension $e$. Assume that the normal bundle $N = N_{X/P}$ is ample. Then the restriction maps*

$$H^i(P; \mathbf{C}) \longrightarrow H^i(X; \mathbf{C})$$

*are surjective when $i \geq n + e$.*

In view of 6.3.11, this applies for instance to any smooth subvariety of a simple abelian variety.

**Remark 7.1.13.** By contrast to Barth's statement 3.2.1, which deals with cohomology in low degrees, the assertion here involves cohomology in high degrees. However, it is easily seen that when $P = \mathbf{P}^{n+e}$ is projective space, 7.1.12 is equivalent to Barth's theorem.

*Proof of Proposition 7.1.12.* The argument follows Hartshorne's proof of Theorem 3.2.1. In fact, the inclusion $j : X \hookrightarrow P$ determines a Gysin map

$$j_* : H^{i-2e}(X) \longrightarrow H^i(P),$$

which sits in a commutative diagram:

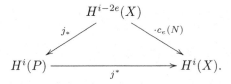

If $i \geq n + e$ then the diagonal map on the right is surjective thanks to 7.1.10, and the assertion follows. □

**Remark 7.1.14. (k-ample bundles).** The theorem of Bloch and Gieseker, and with it Proposition 7.1.12, extend to the $k$-ample setting (Remark 6.2.18). If $E$ is a $k$-ample bundle of rank $e \leq n$ on a smooth projective $n$-fold $X$, then the homomorphism

$$H^i(X; \mathbf{C}) \xrightarrow{\cdot c_e(E)} H^{i+2e}(X; \mathbf{C})$$

is injective if $i \leq n - e - k$ and surjective if $i \geq n - e + k$ [551, Proposition 1.17]. In the situation of 7.1.12, if the normal bundle $N = N_{X/P}$ is $k$-ample then $H^i(P) \longrightarrow H^i(X)$ is surjective for $i \geq n + e + k$ [551, Proposition 2.6]. □

### 7.1.C A Barth-Type Theorem for Branched Coverings

In this subsection we prove a Barth-type theorem of the author [387] for branched coverings of projective space, illustrating the principle that branched coverings of low degree share some of the properties of projective subvarieties of small codimension. The statement is deduced from a more general statement about subvarieties of the total space of an ample bundle. We follow the presentation of [387] quite closely. In the sequel, $H^*(\ )$ denotes cohomology with complex coefficients.

The first formulation of the result in question deals with coverings of $\mathbf{P}^n$:

**Theorem 7.1.15. (Barth-type theorem for coverings, I).** *Let $X$ be a smooth irreducible projective variety of dimension $n$ and $f : X \longrightarrow \mathbf{P}^n$ a finite surjective mapping of degree $d$. Then the induced homomorphisms*

$$f^* : H^i(\mathbf{P}^n) \longrightarrow H^i(X)$$

*are isomorphisms for $i \leq n + 1 - d$.*

If $d \geq n + 1$ there is no assertion. On the other hand, as $d$ becomes small with respect to $n$ one finds stronger and stronger topological obstructions to expressing a variety as a $d$-sheeted covering of projective space.

The connection between coverings and subvarieties flows from the construction of Section 6.3.D. Specifically, we proved there that a covering $f : X \longrightarrow \mathbf{P}^n$ of degree $d$ factors through an embedding of $X$ into the total space of an ample vector bundle of rank $d - 1$ on $\mathbf{P}^n$ (Proposition 6.3.49 and Theorem 6.3.55). Theorem 7.1.15 therefore follows from

**Theorem 7.1.16. (Barth-type theorem for coverings, II).** *Let $Y$ be a smooth irreducible projective variety of dimension $n$, and $E$ an ample vector bundle on $Y$ of rank $e$. Denote by $p : \mathbf{E} \longrightarrow Y$ the total space of $E$, and suppose that $X$ is a smooth projective variety of dimension $n$ embedded in $\mathbf{E}$:*

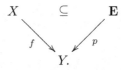

*Then the indicated composition $f$ induces isomorphisms*

$$f^* : H^i(Y) \xrightarrow{\cong} H^i(X) \quad \text{for } i \le n - e.$$

Note that $f$, being proper and affine, is finite.

**Example 7.1.17.** Theorem 7.1.16 includes Barth's theorem for embeddings $X \subseteq \mathbf{P}^{n+e}$. (Choose a linear space $L \subseteq \mathbf{P}^{n+e}$ of dimension $e - 1$ that is disjoint from $X$, and consider the projection $\mathbf{P}^{n+e} - L \longrightarrow \mathbf{P}^n$ to a complementary $\mathbf{P}^n$. The variety $\mathbf{P}^{n+e} - L$ is isomorphic to the total space of the direct sum $\mathcal{O}_{\mathbf{P}^n}(1)^{\oplus e}$ of $e$ copies of the hyperplane bundle, and 7.1.16 implies that $H^i(\mathbf{P}^n) \longrightarrow H^i(X)$ is bijective for $i \le n - e$. But this is equivalent to Barth's statement 3.2.1 (i).) □

We now give the proof of Theorem 7.1.16. The argument once again follows the ideas of Hartshorne's proof of Barth's theorem, as well as Sommese's Proposition 7.1.12.

*Proof of Theorem 7.1.16.* We start by compactifying $\mathbf{E}$ and working out some cohomology classes on this compactification. Specifically, consider the projective closure

$$\pi : \overline{\mathbf{E}} = \mathbf{P}_{\text{sub}}(E \oplus \mathcal{O}_Y) \longrightarrow Y$$

of $\mathbf{E}$, where $\mathbf{P}_{\text{sub}}(F)$ denotes the projective bundle of one-dimensional **sub**-bundles of a bundle $F$. One has a commutative diagram:

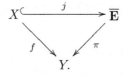

Let $\eta_X \in H^{2e}(\overline{\mathbf{E}})$ be the cohomology class of $X$, and put

$$\xi = c_1\big(\mathcal{O}_{\overline{\mathbf{E}}}(1)\big) \in H^2(\overline{\mathbf{E}}).$$

The class $\xi$ represents the divisor at infinity $\mathbf{P}_{\text{sub}}(E) \subseteq \mathbf{P}_{\text{sub}}(E \oplus \mathcal{O}_Y)$, and $X$ is disjoint from this divisor. Therefore $j^*\xi = 0$. On the other hand,

$$\eta_X = (\deg f) \cdot \left( \sum_{i=1}^{e} c_i(\pi^* E) \, \xi^{e-i} \right)$$

thanks to the fact that $[X]$ is homologous to the indicated multiple of the class of the zero-section in $\overline{\mathbf{E}}$. It follows that

$$j^*(\eta_X) = (\deg f) \cdot c_e(f^* E). \tag{7.6}$$

Note next that to prove the theorem it suffices to establish

$$f^* : H^{n+e+\ell}(Y) \longrightarrow H^{n+e+\ell}(X) \text{ is surjective for } \ell \geq 0. \tag{7.7}$$

In fact $H^*(Y)$ injects into $H^*(X)$, so (7.7) is Poincaré dual to the statement of 7.1.16. For (7.7) we consider the commutative diagram

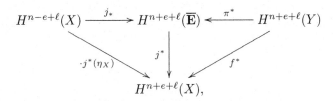

where as usual $j_* : H^{n-e+\ell}(X) \longrightarrow H^{n+e+\ell}(\overline{\mathbf{E}})$ is the Gysin map. Equation (7.6) shows that the diagonal map on the left is given by cup product with a multiple of the top Chern class of an ample vector bundle. It follows from 7.1.10 that the homomorphism in question is surjective for all $\ell \geq 0$. Therefore the vertical map in the middle is surjective. But $H^*(\overline{\mathbf{E}})$ is generated over $H^*(Y)$ by $\xi$, and $j^*$ kills $\xi$. The surjectivity of $j^*$ therefore implies the surjectivity of $f^*$, and we are done. $\qquad \square$

**Remark 7.1.18.** The homotopy analogue of Theorem 7.1.15 for coverings of projective space is established in [389, Theorem 3.1]. $\qquad \square$

**Remark 7.1.19. (Homogeneous varieties).** Some extensions of Theorem 7.1.15 for coverings of rational homogeneous varieties appear in [338], [336], [335], [417], and [337]. For example, Manivel [417], strengthening earlier work of Kim [335], shows that Theorem 7.1.15 holds for coverings of Grassmannians (with the dimension $n$ of projective space replaced by the dimension of the Grassmannian in question). Debarre [110] considers the analogous questions for branched coverings of an abelian variety $A$. He conjectures that if $f : X \longrightarrow A$ is a covering that does not factor through an isogeny of $A$, then the bundle $E_f$ associated to $f$ should be ample (compare Example 6.3.59). Then 7.1.16 would apply. $\qquad \square$

**Example 7.1.20. (Picard groups of coverings).** Let $f : X \longrightarrow \mathbf{P}^n$ be a branched covering of degree $d \leq n - 1$, with $X$ non-singular as usual. Then $f$ induces an isomorphism

$$\text{Pic}(\mathbf{P}^n) \xrightarrow{\cong} \text{Pic}(X).$$

(Arguing as in 3.2.3, it is enough to show that $f^* : H^i(\mathbf{P}^n; \mathbf{Z}) \longrightarrow H^i(X; \mathbf{Z})$ is an isomorphism for $i = 1, 2$. For this one can either invoke the result quoted in Remark 7.1.18, or else combine 7.1.15 with 3.4.10. See [387, Proposition 3.1].) This result is used in [387, Proposition 3.2] to show that if $n \geq 4$ then any degree three covering $f : X \longrightarrow \mathbf{P}^n$ factors through an embedding of $X$ into the total space of a line bundle on $\mathbf{P}^n$: see also [196].    □

## 7.2 Degeneracy Loci

We present in this section results of Fulton and the author from [212] and [389] concerning the non-emptiness and connectedness of rank-dropping loci associated to a homomorphism of vector bundles. The theme is that positivity hypotheses on a homomorphism of vector bundles can force the map in question to drop rank, and to do so on a connected locus. The results have applications to a number of concrete questions, including the loci of special divisors on a curve and projective embeddings of varieties uniformized by bounded domains.

### 7.2.A Statements and First Examples

Let $X$ be an irreducible projective variety of dimension $n$, and suppose given a homomorphism

$$u : E \longrightarrow F$$

between vector bundles on $X$ of ranks $e$ and $f$ respectively. For every $0 \leq k \leq \min\{e, f\}$ one can associate to $u$ its $k^{\text{th}}$ degeneracy locus:

$$D_k(u) = \{x \in X \mid \text{rank } u(x) \leq k \}. \tag{7.8}$$

This is a closed algebraic subset of $X$, locally defined by the vanishing of the determinants of the $(k + 1) \times (k + 1)$ minors of a matrix for $u$. The expected dimension of $D_k(u)$ is

$$\rho_k = \rho_k(e, f) =_{\text{def}} n - (e - k)(f - k),$$

but of course it may be empty or have strictly larger dimension. We refer to [15, Chapter 2] for a general discussion of such determinantal varieties, and to [208, Chapter 14] for their associated enumerative geometry. The reader interested in a more panoramic view might consult the book [215] of Fulton and Pragacz.

The result for which we are aiming, from [212], shows that under suitable positivity hypotheses on the bundles involved, all degeneracy loci are non-empty and even connected whenever it is dimensionally reasonable to expect them to be so.

**Theorem 7.2.1. (Non-emptiness and connectedness of degeneracy loci).** *Let* $u : E \longrightarrow F$ *be a homomorphism between bundles of ranks* $e$ *and* $f$ *on an irreducible projective variety* $X$ *of dimension* $n$, *and assume that the bundle*

$$\mathrm{Hom}(E, F) \; = \; E^* \otimes F$$

*on* $X$ *is ample.*

(i). *If* $n \geq (e - k)(f - k)$ *then* $D_k(u)$ *is non-empty.*

(ii). *If* $n > (e - k)(f - k)$ *then* $D_k(u)$ *is connected.*

The numerical hypotheses in (i) and (ii) are of course equivalent to asking that the expected dimension $\rho_k(e, f)$ of $D_k(u)$ be $\geq 0$ and $> 0$ respectively.

After a few examples to illustrate the result, we will give a relatively elementary proof of the first statement, from [389]: this will suffice for many of the applications. The (more involved) proof of the general case of Theorem 7.2.1 occupies the next subsection, while Section 7.2.C presents some applications. Some variants and extensions are sketched in 7.2.D.

**Example 7.2.2. (Rank obstructions).** Let $E$ be an ample vector bundle of rank $e$ on a projective variety $X$ of dimension $n$. Then $E$ cannot be a quotient of a trivial bundle of rank $< n + e$. More generally, if $E$ is a quotient of a bundle $F$ of rank $f$ whose dual $F^*$ is nef, then $f \geq n + e$. (Note that $F^* \otimes E$ is ample by virtue of 6.2.12 (iv).)    □

**Example 7.2.3. (Embedding obstructions).** Let $A$ be an abelian variety of dimension $r$, and let $X \subset A$ be a smooth subvariety of dimension $n$ and codimension $r - n < n$. Then the cotangent bundle $\Omega_X^1$ of $X$ cannot be ample. Similarly, if $X \subseteq \mathbf{P}^r$ is a smooth subvariety of dimension $n$ whose cotangent bundle $\Omega_X^1$ is nef, then $r \geq 2n$. (In either case, the cotangent bundle of the ambient variety surjects onto that of the submanifold.) See [523] for another approach to statements of this sort, and Corollary 7.2.18 for a generalization.    □

**Example 7.2.4. (Hypersurfaces containing a subvariety of small codimension).** Let $X \subseteq \mathbf{P}^r$ be a non-singular projective variety of dimension $n$ and codimension $e = r - n$, and denote by $N^* = N^*_{X/\mathbf{P}}$ the conormal bundle to $X$ in $\mathbf{P}^r$. We fix any natural number $d \in \mathbf{N}$ that is sufficiently positive so that the twisted conormal bundle $N^*(d)$ is ample: this will hold, for example, if $X \subseteq \mathbf{P}^r$ is cut out as a scheme by hypersurfaces of degrees $\leq d - 1$.

(i). Suppose that $S = S_d \supseteq X$ is a hypersurface of degree $d$ containing $X$. If $e \leq n$ then $X \cap \mathrm{Sing}\, S \neq \varnothing$, and if $e < n$, then $X \cap \mathrm{Sing}\, S$ is connected.

(ii). More generally, suppose that $S = S_{d_1, \dots, d_k} \subset \mathbf{P}^r$ is the complete intersection of hypersurfaces of degrees $d_1, \dots, d_k \geq d$ containing $X$. If $e \leq n + k - 1$ then $X \cap \mathrm{Sing}\, S \neq \varnothing$, and if $e < n + k - 1$, then $X \cap \mathrm{Sing}\, S$ is connected.

(For (i), observe that if $s \in \Gamma\big(\mathbf{P}^r, \mathcal{I}_X(d)\big)$ is the section defining $S$, then $s$ determines a section

$$ds \in \Gamma\big(\mathbf{P}^r, \mathcal{I}_X(d)/\mathcal{I}_X^2(d)\big) \;=\; \Gamma\big(X, N^*(d)\big),$$

which one may think of as giving the first-order terms of $s$ in directions normal to $X$. In particular, $X \cap \operatorname{Sing} S$ is given by the zero-locus of $s$. For (ii), $X \cap \operatorname{Sing} S$ arises in an analogous fashion as the degeneracy locus of a vector bundle map $\mathcal{O}_X(-d_1) \oplus \cdots \oplus \mathcal{O}_X(-d_k) \longrightarrow N^*$.) $\qquad\square$

We now turn to the first proof of statement (i) of Theorem 7.2.1. In this approach the non-emptiness is deduced from a stronger "stepwise" result:

**Proposition 7.2.5. (Stepwise degeneration).** *Let $Y$ be an irreducible projective variety of dimension $m$, and let*

$$v : E \longrightarrow F$$

*be a homomorphism of vector bundles on $Y$ of ranks $e$ and $f$ respectively that has rank $\le \ell$ at every point of $Y$. Assume that $E^* \otimes F$ is ample. If*

$$m \;\ge\; (e+f) - 2\ell + 1,$$

*then $D_{\ell-1}(v)$ is non-empty.*

Noting that $(e+f) - 2\ell + 1$ is the expected codimension of $D_{\ell-1}(v)$ in $D_\ell(v)$, the first statement of Theorem 7.2.1 follows immediately:

*First Proof of Theorem 7.2.1 (i).* Set $r = \min(e, f)$ and consider the chain

$$X \;=\; D_r(u) \;\supseteq\; D_{r-1}(u) \;\supseteq\; \ldots \;\supseteq\; D_k(u).$$

If $D_k(u) = \varnothing$, then we can find an index $\ell > k$ together with an irreducible component $Y \subseteq D_\ell(u)$ of dimension $\ge (e+f) - 2\ell + 1$ such that $D_{\ell-1}(u|Y) = \varnothing$. This contradicts the proposition. $\qquad\square$

*Proof of Proposition 7.2.5.* We assume that $v$ has rank exactly $\ell$ at every point of $Y$, and we will show that then $m \le (e+f) - 2\ell$. Let

$$N = \ker v \quad , \quad K = \operatorname{im} v.$$

Because the rank of $v$ is everywhere constant these are vector bundles on $Y$, of ranks $e - \ell$ and $\ell$ respectively, and we have an exact sequence

$$0 \longrightarrow K^* \longrightarrow E^* \longrightarrow N^* \longrightarrow 0.$$

Pass now to the projective bundle $\pi : \mathbf{P}(E^*) \longrightarrow Y$: we find there a diagram

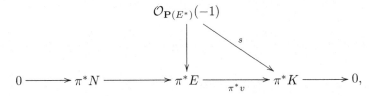

which defines a section

$$s \in \Gamma\Big( \mathbf{P}(E^*), \, \pi^* K \otimes \mathcal{O}_{\mathbf{P}(E^*)}(1) \Big).$$

The zero-locus $\mathrm{Zeroes}(s)$ of $s$ is exactly the subvariety $\mathbf{P}(N^*) \subseteq \mathbf{P}(E^*)$ determined by the quotient $E^* \twoheadrightarrow N^*$, and the idea is to study the complement $\mathbf{P}(E^*) - \mathbf{P}(N^*)$.

We would like to apply the vanishing (7.3) in 7.1.3 to this complement, but one does not expect the bundle $\pi^* K \otimes \mathcal{O}_{\mathbf{P}(E^*)}(1)$ of which $s$ is a section to be ample. However, consider the section $t \in \Gamma\big(\mathbf{P}(E^*), \pi^* F \otimes \mathcal{O}_{\mathbf{P}(E^*)}(1)\big)$ defined by the composition

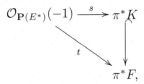

$K \hookrightarrow F$ being the natural inclusion. Then evidently $\mathrm{Zeroes}(t) = \mathrm{Zeroes}(s)$. On the other hand, we assert that

$$\text{the bundle } \pi^* F \otimes \mathcal{O}_{\mathbf{P}(E^*)}(1) \text{ is ample on } \mathbf{P}(E^*). \qquad (*)$$

Granting this for the moment, it follows that $\mathbf{P}(N^*) \subseteq \mathbf{P}(E^*)$ is the zero-locus of a section of an ample vector bundle of rank $f$. Therefore by Remark 7.1.3,

$$H_i\big( \mathbf{P}(E^*) - \mathbf{P}(N^*)\big) \ = \ 0 \ \text{ for } \ i \ \geq \ (m + e - 1) + f, \qquad (**)$$

where as above $m = \dim Y$. But linear projection centered along $\mathbf{P}(N^*) \subseteq \mathbf{P}(E^*)$ defines a morphism

$$\big(\mathbf{P}(E^*) - \mathbf{P}(N^*)\big) \longrightarrow \mathbf{P}(K^*)$$

over $Y$ realizing this complement as the total space of a $\mathbf{C}^{e-\ell}$-bundle over $\mathbf{P}(K^*)$. Consequently,

$$H_i\big( \mathbf{P}(E^*) - \mathbf{P}(N^*)\big) \ = \ H_i\big(\mathbf{P}(K^*)\big) \quad \text{ for all } \ i.$$

In particular, $H_i\big(\mathbf{P}(K^*)\big) = 0$ if $i \geq m + e + f - 1$ thanks to $(**)$. But $\mathbf{P}(K^*)$ is a compact variety of dimension $m + \ell - 1$, and hence $H_{2(m+\ell-1)}\big(\mathbf{P}(K^*)\big) \neq 0$. We conclude that

$$2(m + \ell - 1) \; < \; m + e + f - 1,$$

i.e. $m \leq e + f - 2\ell$, as required.

It remains to verify (*). Consider to this end the projectivization

$$\mathbf{P} = \mathbf{P}\big(\pi^* F \otimes \mathcal{O}_{\mathbf{P}(E^*)}(1)\big) \longrightarrow \mathbf{P}(E^*).$$

The issue is to show that the Serre line bundle $\mathcal{O}_{\mathbf{P}}(1)$ on $\mathbf{P}$ is ample. But $\mathbf{P}$ is isomorphic to the fibre product $\mathbf{P}(E^*) \times_X \mathbf{P}(F)$, and $\mathcal{O}_{\mathbf{P}}(1)$ is the restriction of the Serre line bundle $\mathcal{O}_{\mathbf{P}(E^* \otimes F)}(1)$ under the Segre embedding

$$\mathbf{P}(E^*) \times_X \mathbf{P}(F) \; \subseteq \; \mathbf{P}(E^* \otimes F).$$

Therefore $\mathcal{O}_{\mathbf{P}(E^* \otimes F)}(1)$ is ample since $E^* \otimes F$ is.      □

**Remark 7.2.6.** Ilic and Landsberg [306], and more recently Graham [231], [230], have extended this stepwise result to other sorts of vector bundle maps. See Section 7.2.D for an account.      □

### 7.2.B  Proof of the Connectedness Theorem for Degeneracy Loci

This subsection is devoted to the proof of the general case of Theorem 7.2.1, closely following [212].

We start by describing a generalization of the construction appearing in the proof of Theorem 7.1.1, which will play a central role:

**Construction 7.2.7.** Suppose given a homomorphism $h : A \longrightarrow B$ of vector bundles on a projective variety $Y$. We will want to analyze the zero-locus $Z = \mathrm{Zeroes}(h) \subseteq Y$, i.e. the scheme defined by the vanishing of $h$, considered as a global section of the bundle $A^* \otimes B$. Consider to this end the projective bundle

$$p : \mathbf{P} = \mathbf{P}\big(\mathrm{Hom}(A, B)\big) \; = \; \mathbf{P}(A^* \otimes B) \longrightarrow Y.$$

On $\mathbf{P}$ one has a tautological map $\phi : p^* B \longrightarrow p^* A \otimes \mathcal{O}_{\mathbf{P}}(1)$, and we form the composition $h^*$:

$$
\begin{array}{ccc}
p^* A & \xrightarrow{\;\;p^* h\;\;} & p^* B \\[2pt]
 & \searrow^{\,h^*} & \downarrow{\scriptstyle \phi} \\[4pt]
 & & p^* A \otimes \mathcal{O}_{\mathbf{P}}(1).
\end{array}
$$

Denote by $\mathrm{tr}(h^*) \in \Gamma\big(\mathbf{P}, \mathcal{O}_{\mathbf{P}}(1)\big)$ the trace of $h^*$, and by $Z^* = \mathrm{Zeroes}\big(\mathrm{tr}(h^*)\big)$ its divisor of zeroes. We assert that $p$ induces a morphism

$$p : \big(\mathbf{P} - Z^*\big) \longrightarrow \big(Y - Z\big)$$

realizing $\mathbf{P} - Z^*$ as an affine-space bundle over $Y - Z$. In fact, fixing a point $y \in Y$ we may view $p^{-1}(y) = \mathbf{P}\big(\mathrm{Hom}(A(y), B(y))\big)$ as the space of all linear maps $\phi : B(y) \longrightarrow A(y)$, $\phi$ being defined up to scalars.[1] Then the fibre of $Z^*$ over a given point $y \in Y$ consists of the subset

$$\{\, \phi \mid \mathrm{trace}\big((\phi \circ h)(y)\big) \,=\, 0 \,\}. \tag{*}$$

But (*) is a linear subspace of codimension one in $p^{-1}(y)$ if $h(y) \neq 0$, whereas it holds for every $\phi$ if $h(y) \neq 0$, and the assertion follows.    $\square$

We now turn to the

*Proof of Theorem 7.2.1.* To begin with, it suffices to prove the theorem under the assumption that $X$ is normal. In fact, if $\nu : X' \longrightarrow X$ is the normalization of $X$, then the degeneracy loci of $\nu^* u : \nu^* E \longrightarrow \nu^* F$ map onto those of $u$, and $\nu^* E^* \otimes \nu^* F$ is an ample bundle on $X'$. So we suppose henceforth that $X$ is normal. Furthermore, replacing $u$ if necessary by its transpose, we will assume in addition that $f \geq e$. To avoid trivialities we suppose also that $k \geq 1$: when $k = 0$ the statement is a special case of Example 7.1.4.

Let $\pi : \mathbf{G} = \mathbf{G}(E, k) \longrightarrow X$ be the Grassmannian of $k$-dimensional quotients of $E$. The pullback of $E$ to $\mathbf{G}$ acquires a tautological rank $(e - k)$ sub-bundle $S \subset \pi^* E$, and one has the diagram

defining $\tau$. Thus we may view $\tau$ as the map taking a variable $(e - k)$-plane in $S(x) \subseteq E(x)$ to its image in $F(x)$. In particular, $\tau$ vanishes at the point $(x, S(x)) \in \mathbf{G}$ if and only if $S(x) \subset \ker u(x)$. Therefore the zero-locus

$$Y =_{\mathrm{def}} \mathrm{Zeroes}(\tau) \subseteq \mathbf{G}$$

projects onto $D_k(u)$. So it is sufficient to show that $Y$ is non-empty and connected in the appropriate range of dimensions.

To this end we will study the cohomology $H^*(\mathbf{G} - Y) = H^*(\mathbf{G} - Y; \mathbf{Q})$ of the complement of $Y$. Specifically, letting $r = n + k(e - k)$ denote the dimension of $\mathbf{G}$, we will establish the

*Main Claim*: If $i \geq r + f(e - k)$, then $H^i(\mathbf{G} - Y) = 0$.

Let us check that the claim does indeed imply the theorem. In fact, if $n \geq (e - k)(f - k)$, then $r = n + k(e - k) \geq f(e - k)$, and hence $2r \geq r + f(e - k)$. Therefore

---

[1] Lest there be any confusion here, recall that we are using $\mathbf{P}(V)$ to denote the projective space of all one-dimensional quotients of a vector space (or vector bundle). Therefore we may view an element of $\mathbf{P}(V)$ as a one-dimensional subspace of the dual space (or bundle) $V^*$.

$H^{2r}(\mathbf{G} - Y) = 0$, which implies that $\mathbf{G} - Y$ cannot be compact, and hence that $Y$ is non-empty. Similarly, if $n > (e-k)(f-k)$, then $H^{2r-1}(\mathbf{G} - Y) = 0$. But this implies that $Y$ is connected: when $X$ is smooth this is a consequence of the duality $H^{2r-1}(\mathbf{G} - Y) = H_1(\mathbf{G}, Y)$, and if $X$ is merely normal the vanishing of $H_1(\mathbf{G}, Y)$ follows from the universal coefficient theorem and the Zeeman spectral sequence as in Example 7.1.4.

Turning to the proof of the Main Claim, the plan is to play off two applications of Construction 7.2.7. In fact, first apply 7.2.7 to the map

$$\tau : S \longrightarrow \pi^* F \quad \text{on } \mathbf{G}.$$

Thus we pass to the projectivization

$$p : \mathbf{P} = \mathbf{P}(\mathrm{Hom}(S, \pi^* F)) \longrightarrow \mathbf{G},$$

and form the corresponding affine-space bundle $(\mathbf{P} - Y^*) \longrightarrow (\mathbf{G} - Y)$, where $Y^* \subseteq \mathbf{P}$ is the divisor of zeroes of the trace $\mathrm{tr}(\tau^*) \in \Gamma(\mathbf{P}, \mathcal{O}_{\mathbf{P}}(1))$ of the map $p^* S \longrightarrow p^* S \otimes \mathcal{O}_{\mathbf{P}}(1)$. So we are reduced to proving the vanishing of $H^i(\mathbf{P} - Y^*)$ for $i \geq r + f(e - k)$.

The point of the argument is that we can also study $\mathbf{P} - Y^*$ via a different application of Construction 7.2.7. Specifically, consider the projective bundle

$$q : \mathbf{P}' = \mathbf{P}(\mathrm{Hom}(E, F)) \longrightarrow X.$$

We assert first that there is a mapping $h : \mathbf{P} \longrightarrow \mathbf{P}'$ sitting in a commutative diagram

$$\begin{array}{ccc} \mathbf{P} & \xrightarrow{\;p\;} & \mathbf{G} \\ {\scriptstyle h}\big\downarrow & & \big\downarrow{\scriptstyle \pi} \\ \mathbf{P}' & \xrightarrow{\;q\;} & X. \end{array}$$

In fact, specifying a point in $\mathbf{P}'$ is the same as giving a point $x \in X$, plus a linear transformation $\psi : F(x) \longrightarrow E(x)$ defined up to scalars: we designate this point by $(\psi, x) \in \mathbf{P}'$. In a similar fashion, a point $(\phi, S, x) \in \mathbf{P}$ over $x$ is determined by a subspace $S(x) \subseteq E(x)$ and a homomorphism $\phi : F(x) \longrightarrow S(x)$ (mod scalars). Then $h(\phi, S, x)$ is represented by the composition

$$F(x) \xrightarrow{\;\phi\;} S(x) \lhook\joinrel\longrightarrow E(x).$$

Now apply Construction 7.2.7 on $X$ starting from $u : E \longrightarrow F$ to form the trace $\mathrm{tr}(u^*) \in \Gamma(\mathbf{P}', \mathcal{O}_{\mathbf{P}'}(1))$ of the composition $q^* E \xrightarrow{q^* u} q^* F \longrightarrow q^* E \otimes \mathcal{O}_{\mathbf{P}'}(1)$. Setting $Y' = \mathrm{Zeroes}(\mathrm{tr}(u^*))$, we assert that

$$h^{-1}(Y') = Y^*. \tag{$*$}$$

Indeed, (*) follows from the fact that the trace of a linear transformation $E(x) \longrightarrow E(x)$ whose image is contained in a subspace $S(x) \subseteq E(x)$ coincides with the trace of the restriction of the transformation to $S(x)$. Thus $h$ restricts to a proper morphism

$$h : V = (\mathbf{P} - Y^*) \longrightarrow (\mathbf{P}' - Y') = U. \tag{**}$$

On the other hand, since by assumption $\operatorname{Hom}(E, F)$ is ample, $Y'$ is an ample divisor on $\mathbf{P}'$ and hence $U$ is an affine variety. The idea now is to use the Leray spectral sequence to deduce the Main Claim from the vanishing of cohomology on an affine variety.

To this end, we start by analyzing the fibres of $h$. Keeping notation as above, fix a point $(\psi, x) \in \mathbf{P}'$, $\psi$ being a homomorphism $\psi : F(x) \longrightarrow E(x)$ defined up to scalars. Then

$$h^{-1}(\psi, x) = \{ (\psi, S, x) \mid \operatorname{im} \psi(x) \subseteq S(x) \subseteq E(x) \}.$$

In particular, if rank $\psi = e - k - \ell$, then $h^{-1}(\psi, x)$ is isomorphic to the Grassmannian of $\ell$-dimensional subspaces of an $(\ell + k)$-dimensional vector space, and hence has dimension $\ell k$. Now define

$$\mathbf{P}'_\ell = \{ (\psi, x) \in \mathbf{P}(\operatorname{Hom}(E, F)) \mid \operatorname{rank}(\psi) \le e - k - \ell \},$$

and put

$$U_\ell = U \cap \mathbf{P}'_\ell.$$

Then each $U_\ell$ is an affine variety, with $U_{\ell+1} \subseteq U_\ell$, and $h(V) = U_0$. Thinking of the fibre of $q$ as the projective space of all $e \times f$ matrices, the dimension of $U_\ell$ is determined by a fibrewise computation, and we find that

$$\operatorname{codim}_{U_0} U_\ell = 2k\ell + \ell^2 + (f - e)\ell.$$

In particular, $\operatorname{codim}_{U_0} U_\ell \ge 2\ell k$, and so the following lemma applies to the mapping $h : V \longrightarrow U_0$ to yield the Main Claim, and complete the proof of the theorem. $\qquad\square$

**Lemma 7.2.8.** *Let $f : X \longrightarrow Y$ be a proper surjective mapping of irreducible varieties, with $Y$ affine. Assume that for each $d \ge 0$ the set*

$$Y_d = \{ y \in Y \mid \dim f^{-1}(y) \ge d \}$$

*has codimension $\ge 2d$ in $Y$. Then*

$$H^i(X; \mathbf{Z}) = 0 \quad \text{for } i \ge \dim X + 1.$$

*Proof.* Consider the Leray spectral sequence

$$E_2^{p,q} = H^p(Y, R^q f_* \mathbf{Z}) \Rightarrow H^{p+q}(X, \mathbf{Z}).$$

Since $f$ is proper, the sheaves $R^i f_* \mathbf{Z}$ compute the cohomology of the fibres, and consequently the direct images $R^{2d-1} f_* \mathbf{Z}$ and $R^{2d} f_* \mathbf{Z}$ are supported on $Y_d$. But these are constructible sheaves on the affine variety $Y_d$ (Remark 3.1.12), and hence their cohomology vanishes in degrees above the dimension of $Y_d$ thanks to the Artin–Grothendieck Theorem 3.1.13. Since $\dim Y_d + 2d \le \dim Y = \dim X$, one has $E_2^{p,q} = 0$ for $p + q \ge \dim X + 1$, and the lemma follows.                                                                  $\square$

**Remark 7.2.9.** The proof of the lemma draws on the somewhat non-trivial fact that if $f : X \longrightarrow Y$ is a proper map of algebraic varieties, then the direct images $R^q f_* \mathbf{Z}$ are constructible sheaves on $Y$. So it may be worth noting that in the particular setting of the theorem, the constructibility in question is evident. In fact, keeping notation as in the proof of Theorem 7.2.1, the mapping $h : V \longrightarrow U_0$ restricts to a Grassmannian bundle over each of the locally closed sets $(U_\ell - U_{\ell+1}) \subseteq U_0$. Therefore on each $(U_\ell - U_{\ell+1})$, $R^q f_* \mathbf{Z}$ is a local system.                                                                  $\square$

**Remark 7.2.10.** The dimensional hypothesis in the lemma is equivalent to asking that $f$ be "semismall" in the sense of Goresky and MacPherson [227, p. 151].                                                                  $\square$

**Remark 7.2.11. (k-ample bundles).** Tu [577, §6] has generalized Theorem 7.2.1 to the $k$-ample setting (Remark 6.2.18). Specifically, he shows in the situation of 7.2.1 that if $E^* \otimes F$ is $k$-ample, then $D_\ell(u)$ is non-empty provided that $n \ge (e - \ell)(f - \ell) + k$ and connected when $n > (e - \ell)(f - \ell) + k$. See also [571, §3b]                                                                  $\square$

### 7.2.C Some Applications

Now we present some applications of the non-emptiness and connectedness theorems for degeneracy loci.

**Special linear series.** The original motivation for Theorem 7.2.1 came from the theory of special divisors on a curve. Let $C$ be a smooth irreducible projective curve of genus $g$. A great deal of the geometry of $C$ is captured in the analysis of how $C$ can map to projective spaces of various dimensions. More formally, let $\mathrm{Pic}^d(C)$ be the Jacobian of $C$, viewed as parameterizing isomorphism classes of line bundles of fixed degree $d$ on $C$, and consider the loci

$$W_d^r(C) =_{\mathrm{def}} \{ A \in \mathrm{Pic}^d(C) \mid h^0(C, A) \ge r + 1 \}.$$

Thus $W_d^r(C)$ is the set of all linear series of degree $d$ and (projective) dimension $\ge r$ on $C$. It is an algebraic subset of $\mathrm{Pic}^d(C)$, and (as we shall see shortly) its expected dimension is given by the Brill–Noether number

$$\rho = \rho(g, r, d) =_{\mathrm{def}} g - (r + 1)(g - d + r).$$

We refer to [450], Lectures I and III, for a beautiful introduction to the geometry of special divisors, and to [15] for a very complete and careful account of the theory.

The application we have in mind asserts that these loci are non-empty and connected whenever their expected dimension would predict that they might be so:

**Theorem 7.2.12. (Non-emptiness and connectedness of loci of special divisors).** *If $\rho \geq 0$ then $W_d^r(C)$ is non-empty, and if $\rho > 0$ then $W_d^r(C)$ is connected.*

The first statement was originally proved by Kempf and Kleiman–Laksov [347], [348] by enumerative methods. The present approach, via considerations of positivity, is due to Fulton and the author [212]. Note that while the theorem here applies to every curve $C$, it is known that the varieties of special divisors are particularly well behaved on a generic curve (cf. [250], [225], [390], and [94]). In particular, the connectedness of $W_d^r(C)$ can be combined with [225] to show that the locus in question is actually irreducible when $\rho > 0$ provided that $C$ is sufficiently general (see [212, Corollary 2.4]).

*Proof of Theorem 7.2.12.* Fix a large integer $\lambda \geq 2g - 1$ and a finite set $Z = \{x_1, \ldots, x_w\}$ of $w = \lambda - d$ points on $C$. Given a line bundle $L$ of degree $\lambda$, consider the evaluation map

$$\sigma_{Z,L} : H^0(C, L) \longrightarrow H^0(C, L \otimes \mathcal{O}_Z).$$

Then evidently $\ker \sigma_{Z,L} = H^0(C, L(-\sum x_i))$, and every line bundle $A$ of degree $d$ is of the form $A = L(-\sum x_i)$ for some $L \in \mathrm{Pic}^\lambda(C)$. Therefore

$$W_d^r(C) \cong \{ L \in \mathrm{Pic}^\lambda(C) \mid \dim \ker(\sigma_{Z,L}) \geq r + 1 \}.$$

On the other hand, we observed in 6.3.C that the maps $\sigma_{Z,L}$ globalize to a homomorphism of vector bundles

$$\sigma_Z : E_\lambda \longrightarrow \Sigma_Z$$

on $\mathrm{Pic}^\lambda(C)$, with $\mathrm{rk}(E_\lambda) = \lambda + 1 - g$ and $\mathrm{rk}(\Sigma_Z) = w = \lambda - d$. In other words, $W_d^r(C) \cong D_k(\sigma_Z)$ with $k = \lambda - g - r$: in particular, the postulated dimension of $D_k(u)$ is exactly $\rho = g - (r+1)(g - d + r)$. But according to Theorem 6.3.48, $\mathrm{Hom}(E_\lambda, \Sigma_Z)$ is an ample vector bundle on $\mathrm{Pic}^\lambda(C)$. Therefore the non-emptiness and connectedness of $W_d^r(C)$ follow in the indicated range of the parameters from 7.2.1. □

**Example 7.2.13. (Ghione's theorem).** Ghione [221] found a generalization of Theorem 7.2.12 involving vector bundles of higher rank. Specifically, fix a vector bundle $V$ on $C$ of rank $v$ and degree $p$, so that

$$\chi = \chi(C, V) = p + v(1 - g).$$

Given an integer $r \geq 0$, set

$$W^r(C; V) = \{ A \in \operatorname{Pic}^0(C) \mid h^0(C, V \otimes A) \geq r+1 \}.$$

This is a determinantal locus having expected codimension $(r+1)(r+1-\chi)$ in $\operatorname{Pic}^0(C)$.[2]

**Proposition.** *If $g \geq (r+1)(r+1-\chi)$ then $W^r(C; V)$ is non-empty, and if $g > (r+1)(r+1-\chi)$ then $W^r(C; V)$ is connected.*

(Following [389] we sketch an argument along the lines of the proof of 7.2.12. Fixing a line bundle $M$ of degree $\lambda \gg 0$, we may construct an exact sequence

$$0 \longrightarrow V \longrightarrow M \otimes \mathcal{O}_C^v \longrightarrow M \otimes \mathcal{O}_Z \longrightarrow 0$$

with $Z$ a finite subscheme of $C$ of degree $v\lambda - p$. Given $A \in \operatorname{Pic}^0(C)$, this realizes $h^0(C, V \otimes A)$ and $h^1(C, V \otimes A)$ as the kernel and cokernel respectively of the resulting map

$$H^0(C, M \otimes A \otimes \mathcal{O}_C^v) \longrightarrow H^0(C, M \otimes A \otimes \mathcal{O}_Z).$$

Globalizing, $W^r(C; V)$ arises as a degeneracy locus of a bundle homomorphism

$$\bigoplus^v E_\lambda \longrightarrow \Sigma_Z,$$

and Theorem 7.2.1 applies.) □

**Example 7.2.14. (Theorem of Segre–Nagata).** Let $V$ be a vector bundle of rank $v$ and degree $p$ on $C$. Then $V$ contains a line sub-bundle $B \subseteq V$ with

$$\deg B \geq \frac{p}{v} - \frac{v-1}{v}(g-1).$$

(Use Ghione's theorem to compute the least degree $a$ such that

$$H^0(C, V \otimes A) \neq 0$$

for some $A \in \operatorname{Pic}^a(C)$, and then take $B$ to be the line sub-bundle spanned by the image of the resulting map $A^* \longrightarrow V$.) □

---

[2] Observe that working with the loci

$$W_d^r(C; V) =_{\text{def}} \{ A \in \operatorname{Pic}^d(C) \mid h^0(C, V \otimes A) \geq r+1 \}$$

actually involves no increase in generality, for if $L$ is a fixed line bundle of degree $d$ then $W_d^r(C; V) \cong W^r(C; V \otimes L)$.

**Example 7.2.15. (Analogue for irregular varieties of higher dimension).** The negativity of the Picard bundles established in Section 6.3.C leads to a generalization of Theorem 7.2.12 to higher dimensional irregular varieties. Specifically, suppose that $X$ is a smooth projective variety of dimension $n$ and irregularity $q(X) = h^1(X, \mathcal{O}_X)$. Fix first an algebraic equivalence class $\lambda$ that satisfies the positivity conditions (6.10) and (6.11) from 6.3.C and then choose a finite subscheme $Z \subseteq X$ of length $w$. As $t$ varies over $\mathrm{Pic}^\lambda(X)$, we can ask to what extent the points of $Z$ fail to impose independent conditions on the sections of $H^0(X, L_t)$, $L_t$ being the bundle on $X$ corresponding to the point $t$. Specifically, define

$$\mathrm{SP}^s_{\lambda, Z}(X) = \{\, t \in \mathrm{Pic}^\lambda(X) \mid h^0(X, L_t \otimes \mathcal{I}_Z) \geq s \,\}.$$

("SP" stands for Special Position: in effect we are looking for those line bundles $L_t$ with respect to which $Z$ is in special position by a specified amount.) Denoting by $\chi = \chi(\lambda) = \chi(X, L_t) = h^0(X, L_t)$ the common value of the Euler characteristic of the line bundles in the given class, one has the

> **Proposition.** *If $q(X) \geq s(w - \chi + s)$ then $\mathrm{SP}^s_{\lambda, Z}(X)$ is non-empty, and if $q(X) > s(w - \chi + s)$ then it is connected.*

(In fact, $\mathrm{SP}^s_{\lambda, Z}(X) = D_{\chi-s}(\sigma_Z)$, where $\sigma_Z : E_\lambda \longrightarrow \Sigma_Z$ is the bundle map on $\mathrm{Pic}^\lambda(X)$ globalizing the restrictions $H^0(X, L_t) \longrightarrow H^0(X, L_t \otimes \mathcal{O}_Z)$. Then Theorems 6.3.48 and 7.2.1 apply.)     □

**Example 7.2.16.** In the setting of the previous example, suppose in addition that for every $t \in \mathrm{Pic}^\lambda(X)$ the corresponding line bundle $L_t$ is globally generated. Then $\chi(\lambda) > q(X)$. Similarly if each $L_t$ is very ample then $\chi(\lambda) > q(X) + \dim(X) + 1$. (For the first statement, apply the result from 7.2.15 with $Z$ a single point; for the second, take $Z$ to be the first infinitesimal neighborhood of a point.)     □

**Projective embeddings of quotients of bounded domains.** We next give some applications to singularities of mappings to projective space. In particular, we establish a (nearly) exponential lower bound from [148] on the degree of any projective embedding of a variety whose cotangent bundle is nef.

Consider to begin with non-singular projective varieties $X$ and $Y$ of dimensions $n$ and $m$ respectively, with $n \leq m$, and let

$$f : X \longrightarrow Y$$

be a morphism. Given a positive integer $i$, the $i^{\mathrm{th}}$ *singularity locus* of $f$ is the set of points where the derivative of $f$ drops rank by $\geq i$:

$$S^i(f) = \{\, x \in X \mid \mathrm{rank}\, df_x \leq n - i \,\}.$$

In other words, $S^i(f)$ is the degeneracy locus $D_{n-i}(df)$ associated to the derivative $df : TX \longrightarrow f^*TY$. Thus $S^i(f)$ has expected codimension $i(m - n + i)$ in $X$. The following immediate consequence of Theorem 7.2.1 (i) was already noted in [212, p. 278].

**Proposition 7.2.17.** *With $f : X \longrightarrow Y$ as above, assume that the vector bundle*

$$\mathrm{Hom}(TX, f^*TY)$$

*is ample on $X$. If $n \geq i(m - n + i)$ then $S^i(f)$ is non-empty, and if $n > i(m - n + i)$ then $S^i(f)$ is connected.*                                   □

We propose to apply this to study projective embeddings of smooth varieties whose cotangent bundles are nef. Recall from Remark 6.3.41 that according to the theorem of Kratz [372], this includes in particular any projective manifold whose universal covering is a bounded domain. Likewise, any submanifold of an abelian variety has a nef cotangent bundle, and the class of all such varieties is closed under taking products and smooth subvarieties.

**Corollary 7.2.18.** *If the cotangent bundle $T^*X = \Omega^1_X$ is nef, and if $f : X \longrightarrow \mathbf{P}^m$ is any finite morphism, then $S^i(f) \neq \emptyset$ provided that $n \geq i(m - n + i)$.*

*Proof.* In fact, $f^*T\mathbf{P}^m$ is ample since $f$ is finite, and consequently $\Omega^1_X \otimes f^*T\mathbf{P}^m$, being the product of a nef and an ample bundle, is itself ample.     □

We now use the corollary to establish a lower bound on the degree of any projective embedding of a variety with non-negative cotangent bundle. Specifically, given a positive integer $n$, set

$$\delta(n) = 2^{[\sqrt{n}]},$$

where as usual $[x]$ denotes the integer part of a real number $x$.

**Theorem 7.2.19.** *Let $X$ be a smooth projective variety of dimension $n$ whose cotangent bundle $\Omega^1_X$ is nef, and let $X \subseteq \mathbf{P}$ be any projective embedding of $X$. Then the degree $\deg X$ of $X$ in $\mathbf{P}$ must be at least $\delta(n)$.*

*Proof.* By taking a suitable linear projection of $X$, we construct a branched covering $f : X \longrightarrow \mathbf{P}^n$ whose degree is $\deg X$. So it suffices to show that $\deg f \geq \delta(n)$. To this end, set $k = [\sqrt{n}]$. Then Corollary 7.2.18 implies that $S^k(f) \neq \emptyset$: fix $x_0 \in S^k(f)$. It follows from Lemma 7.2.20 below that the local degree $e_f(x_0)$ of $f$ at $x_0$ (Definition 3.4.7) satisfies the inequality $e_f(x_0) \geq 2^k = \delta(n)$. But if $y_0 = f(x_0)$ then

$$\deg f = \sum_{f(x)=y_0} e_f(x),$$

and consequently $\deg f \geq \delta(n)$, as claimed.                          □

**Lemma 7.2.20.** *Let* $f : X \longrightarrow Y$ *be a finite surjective mapping between irreducible non-singular varieties of dimension* $n$. *Fix a point* $x \in X$, *let* $y = f(x) \in Y$, *and denote by* $e_f(x)$ *the local degree of* $f$ *at* $x$. *Suppose that the derivative* $df_x : T_x X \longrightarrow T_y Y$ *of* $f$ *at* $x$ *has rank* $n - k$. *Then* $e_f(x) \geq 2^k$.

*Proof.* By assumption, the co-derivative $df_x^* : T_y^* Y \longrightarrow T_x^* X$ has a $k$-dimensional kernel. Therefore we can find a system $z_1, \ldots, z_n$ of local coordinates on $Y$ centered at $y$ such that

$$\mathrm{ord}_x(f^* z_1) \geq 2, \ldots, \mathrm{ord}_x(f^* z_k) \geq 2. \qquad (*)$$

By property (3.10) from Section 3.4.A, the local degree $e_f(x)$ is computed as the intersection multiplicity at $x$ of the (germs of the) divisors defined by the $f^* z_i$, i.e.

$$e_f(x) = \dim_{\mathbf{C}} \frac{\mathcal{O}_x X}{(f^* z_1, \ldots, f^* z_n)}.$$

On the other hand, it is well known that this multiplicity is at least the product of the multiplicities $\mathrm{ord}_x(f^* z_i)$ (cf. [208, 12.4]), and the lemma then follows from $(*)$. $\qquad \square$

### 7.2.D Variants and Extensions

In this subsection we summarize without proof some extensions and generalizations.

**Symmetric and skew degeneracy loci.** It is natural to ask for analogues of Theorem 7.2.1 and Proposition 7.2.5 for symmetric or skew symmetric maps. Specifically, let $E$ be a vector bundle of rank $e$ and $L$ a line bundle on an irreducible variety $X$, and suppose that

$$u : E^* \longrightarrow E \otimes L$$

is a homomorphism. The transpose of $u$ (twisted by $L$) gives rise to a second mapping $u^* : E^* \longrightarrow E \otimes L$, and naturally enough one says that $u$ is *symmetric* if $u^* = u$ and *skew-symmetric* if $u^* = -u$. Such a homomorphism is given locally by a symmetric or skew symmetric matrix of functions. If $X$ has dimension $n$, then the expected dimensions of the degeneracy loci $D_k(u)$ are given by

$$\rho_{k,\mathrm{symm}} = n - \binom{e - k + 1}{2} \quad \text{for } u \text{ symmetric,}$$

$$\rho_{k,\mathrm{skew}} = n - \binom{e - k}{2} \quad \text{for } u \text{ skew,}$$

where in the skew case we assume that $k$ is even. Supposing that $u$ is symmetric and that $S^2 E \otimes L$ is ample, it is natural to expect that $D_k(u)$ is non-empty if $\rho_{k,\text{symm}} \geq 0$ and connected if $\rho_{k,\text{symm}} > 0$, with an analogous statement for skew maps assuming the amplitude of $\Lambda^2 E \otimes L$. The non-emptiness was established by Fulton and the author in [214, §3b]: the argument appears below in Section 8.4.B. For skew maps and symmetric maps with $k$ even, Tu proved connectedness in [576]. The connectedness of $D_k(u)$ for symmetric maps and odd $k$ was established by Harris–Tu in [273] under slightly stronger dimensional hypotheses. Another approach to the connectedness, via vanishing theorems, is explored by Laytimi in [385].

The stepwise degeneration (Proposition 7.2.5) has been generalized to symmetric maps by Ilic and Landsberg [306]. Specifically, consider a symmetric mapping $u : E^* \longrightarrow E \otimes L$ on a simply connected non-singular projective variety $X$ of dimension $n$, with $S^2 E \otimes L$ ample, and suppose that $u$ has constant even rank $k$. Then $n \leq e - k$. Ilic and Landsberg apply this to the linear system of quadrics arising from the second fundamental form of a variety in projective space in order to study projective varieties with degenerate duals. Graham [231] has recently completed the result of Ilic–Landsberg by removing the hypothesis that $X$ is simply connected and that $r$ is even. He proves an analogous stepwise result for skew maps in [230].

**Orthogonal degeneracy loci and Prym special divisors.** Another sort of degeneracy locus arises in the study of Prym special divisors [449], [215, Appendix H]. Here, following Mumford, one starts with a vector bundle $V$ on a projective variety $X$ together with a non-degenerate quadratic form $S^2 V \longrightarrow L$ taking values in a line bundle $L$. One then considers two isotropic subbundles $F_1, F_2 \subseteq V$, and the loci in question are the sets

$$O_k = \left\{ x \in X \;\middle|\; \begin{array}{l} \dim \left( F_1(x) \cap F_2(x) \right) \geq k \\ \dim \left( F_1(x) \cap F_2(x) \right) \equiv k \pmod 2 \end{array} \right\}.$$

Here the natural expectation is that if $F_1^* \otimes F_2^* \otimes L$ is ample, then $O_k$ should be non-empty (resp. connected) provided that its expected dimension $\binom{k}{2}$ is $\geq 0$ (resp. $> 0$). The non-emptiness was established by Bertram [53] in the course of proving a conjecture of Welters about the existence of Prym-special divisors. Debarre proves the connectedness in [113, Proposition 6.1].

**Results of Lefschetz type.** In the spirit of the Lefschetz hyperplane theorem, it is natural to ask whether one can say anything about the higher Betti numbers of the loci $D_k(u)$. Some interesting results along these lines are given by Debarre [113]. For example, suppose in the setting of 7.2.1 that $X$ is smooth and that $D_{k-1}(u) = \varnothing$. Then Debarre is able to describe the integral cohomology of $D_k(u)$ up to degrees $\rho_k(e, f) - 1$. He obtains analogous assertions in smaller ranges of degrees without the assumption that $D_{k-1}(u)$ be empty. Debarre also treats symmetric, skew, and orthogonal loci. Related results on the homotopy level were established by Muñoz and Presas [454] under assumptions of Griffiths-positivity.

**Relative results.** Some interesting generalizations appear in the paper [557] of Steffen (see also [558]). She considers a proper surjective mapping $\phi : X \longrightarrow Y$ of irreducible complex varieties and a homomorphism $u : E \longrightarrow F$ of bundles of ranks $e$ and $f$ on $X$. Assume that $E^* \otimes F$ is ample for $\phi$ in the sense that the Serre line bundle $\mathcal{O}_{\mathbf{P}}(1)$ on $\mathbf{P}(E^* \otimes F)$ is ample for the evident composition $\mathbf{P}(E^* \otimes F) \longrightarrow Y$ (Section 1.7). Steffen shows that if $Z$ is any irreducible component of the image $\phi\big(D_k(u)\big)$ of the degeneracy locus $D_k(u)$, then

$$\dim Z \;\geq\; \rho_k \;=\; \dim X - (e-k)(f-k)$$

provided that $\dim Y \geq \rho_k$. In other words, the image of $D_k(u)$ cannot be any smaller than its expected dimension. Steffen's arguments are algebraic in nature; subsequently Geertsen [219] found an alternative more geometric proof. An illustrative special case is treated in [186, §3.6].

## 7.3 Vanishing Theorems

In this section we discuss Kodaira-type vanishing theorems for ample vector bundles. We will give the proofs of the most fundamental statements, and refer to the literature for more recent refinements.

Throughout the section $X$ is a smooth irreducible complex projective variety of dimension $n$, and $\omega_X = \mathcal{O}_X(K_X)$ denotes the canonical line bundle on $X$. We fix a vector bundle $E$ on $X$ of rank $e$, and write as usual $\pi : \mathbf{P}(E) \longrightarrow X$ for the corresponding projective bundle.

### 7.3.A Vanishing Theorems of Griffiths and Le Potier

The vanishing statements for vector bundles that we consider here will be deduced from vanishing theorems for line bundles on $\mathbf{P}(E)$. We will repeatedly use two basic exact sequences — the cotangent bundle sequence for $\pi$ and the relative Euler sequence — to pass from $\mathbf{P}(E)$ to $X$:

$$0 \longrightarrow \pi^* \Omega^1_X \longrightarrow \Omega^1_{\mathbf{P}(E)} \longrightarrow \Omega^1_{\mathbf{P}(E)/X} \longrightarrow 0, \qquad (7.9)$$

$$0 \longrightarrow \Omega^1_{\mathbf{P}(E)/X} \longrightarrow \pi^* E(-1) \longrightarrow \mathcal{O}_{\mathbf{P}(E)} \longrightarrow 0. \qquad (7.10)$$

In the relative Euler sequence (7.10), $\pi^* E(-1)$ denotes the tensor product $\pi^* E \otimes \mathcal{O}_{\mathbf{P}(E)}(-1)$; we will use analogous abbreviations in the sequel. Note that as a consequence of these sequences one obtains the familiar formula

$$\omega_{\mathbf{P}(E)} \;=\; \pi^* \big(\omega_X \otimes \det(E)\big) \otimes \mathcal{O}_{\mathbf{P}(E)}(-e)$$

for the canonical bundle of $\mathbf{P}(E)$.

One of the earliest vanishing theorems for bundles is due to Griffiths [247]:

**Theorem 7.3.1. (Griffiths vanishing theorem).** *If $E$ is ample, then*

$$H^i\big(X, \omega_X \otimes S^m E \otimes \det E\big) \;=\; 0 \quad \text{for all } i > 0 \,, \; m \geq 0.$$

*Proof.* This is a simple consequence of Kodaira vanishing on $\mathbf{P}(E)$. In fact, note that for any $m \geq 0$,

$$
\begin{aligned}
\pi_*\big(\omega_{\mathbf{P}(E)} \otimes \mathcal{O}_{\mathbf{P}(E)}(m + e)\big) &= \pi_*\big(\pi^*(\omega_X \otimes \det E) \otimes \mathcal{O}_{\mathbf{P}(E)}(m)\big) \\
&= \omega_X \otimes \det E \otimes \pi_*\big(\mathcal{O}_{\mathbf{P}(E)}(m)\big) \\
&= \omega_X \otimes \det E \otimes S^m E.
\end{aligned}
$$

Furthermore, when $m \geq 0$ the higher direct images of the sheaf in question vanish. Therefore

$$H^i\big(\mathbf{P}(E), \omega_{\mathbf{P}(E)} \otimes \mathcal{O}_{\mathbf{P}(E)}(m + e)\big) \;=\; H^i\big(X, \omega_X \otimes \det E \otimes S^m E\big).$$

But $\mathcal{O}_{\mathbf{P}(E)}(1)$ is an ample line bundle on $\mathbf{P}(E)$ since $E$ is an ample vector bundle, and so the theorem follows from Kodaira vanishing (Theorem 4.2.1). $\quad\square$

An essentially identical argument proves:

**Variant 7.3.2.** If $E$ is an ample vector bundle, and $L$ is any nef line bundle, then

$$H^i\big(X, \omega_X \otimes S^m E \otimes \det E \otimes L\big) \;=\; 0 \quad \text{for all } \; i > 0 \,, \; m \geq 0.$$

The same statement holds if $E$ is nef and $L$ is ample. $\quad\square$

**Example 7.3.3.** Appealing in the proof to vanishing for nef and big line bundles (Theorem 4.3.1) instead of the classical Kodaira vanishing theorem, it is enough for 7.3.1 to assume that $E$ is big and nef in the sense that the line bundle $\mathcal{O}_{\mathbf{P}(E)}(1)$ is so. Similarly, in the situation of Variant 7.3.2 it suffices that $E$ and $L$ are both nef, and that one or the other is big. $\quad\square$

**Example 7.3.4. (Global generation).** Let $X$ be a smooth projective variety of dimension $n$, let $E$ an ample vector bundle on $X$, and let $B$ be a globally generated ample line bundle on $X$. Then

$$E \otimes \det E \otimes \omega_X \otimes B^{\otimes k}$$

is globally generated for $k \geq n$. If $E$ is nef, the same holds when $k \geq n + 1$. (By 7.3.2, $E \otimes \det E \otimes \omega_X$ is $n$-regular with respect to $B$ in the sense of Castelnuovo–Mumford (Section 1.8.A) provided that $E$ is ample. When $E$ is nef, then the bundle in question is $(n + 1)$-regular.) $\quad\square$

In practice one often wants a statement that does not involve the factor of $\det E$. In this direction, the fundamental result was proved by Le Potier [397]:

**Theorem 7.3.5. (Le Potier vanishing theorem).** *Assume that $E$ is ample. Then*

$$H^i(X, \omega_X \otimes \Lambda^a E) = 0 \quad \text{for } a > 0 \text{ and } i > e - a. \tag{7.11}$$

*Furthermore,*

$$H^i(X, \Omega_X^p \otimes E) = 0 \quad \text{for } i + p \geq n + e. \tag{7.12}$$

**Remark 7.3.6.** By Serre duality, the statements are equivalent to the assertions:

$$H^j(X, \Lambda^a E^*) = 0 \quad \text{for } a > 0 \text{ and } j < n - e + a;$$
$$H^j(X, \Omega_X^q \otimes E^*) = 0 \quad \text{for } j + q \leq n - e.$$

Note also that if $E$ is a line bundle, so that $e = 1$, then Theorem 7.3.5 exactly reduces to Kodaira–Nakano vanishing.

Deferring the proof until later in this subsection, we start with some examples and applications.

**Example 7.3.7.** Le Potier's theorem differs from Griffiths' result in that the degrees in which one gets vanishing depend on the rank of the bundle. However, simple examples show that this is unavoidable. For example, in the setting of 7.3.6 consider the negative bundle $E^* = \Omega_{\mathbf{P}^n}^1$ of rank $n$ on $X = \mathbf{P}^n$. Then $H^1(\mathbf{P}^n, E^*) \neq 0$, and more generally $H^a(\mathbf{P}^n, \Lambda^a E^*) \neq 0$ for $a \leq n$.  □

**Remark 7.3.8. (Counter-examples).** Contrary to what one might hope from interpolating between the two statements in Le Potier's theorem, it is not the case in general that $H^i(X, \Omega_X^p \otimes \Lambda^a E) = 0$ for $E$ ample and $i + p > n + e - a$. Counter-examples are constructed by Demailly in [121, p. 205]. Similarly, Peternell, Le Potier and Schneider [502, §3] give examples to show that the groups $H^i(X, \Omega_X^p \otimes S^m E)$ need not vanish if $i + p \geq n + e$, answering in the negative a question that had been posed by Le Potier.  □

**Example 7.3.9.** Le Potier's theorem leads to another proof of the fact (Example 7.1.4) that if $E$ is an ample bundle of rank $e \leq n$, then any section $s \in \Gamma(X, E)$ must vanish at some point of $X$. (In fact, if $s$ is nowhere vanishing then the Koszul complex it determines takes the form of a long exact sequence

$$0 \longrightarrow \Lambda^e E^* \longrightarrow \ldots \longrightarrow \Lambda^2 E^* \longrightarrow E^* \longrightarrow \mathcal{O}_X \longrightarrow 0.$$

Le Potier's theorem gives the vanishings $H^{j-1}(X, \Lambda^j E^*) = 0$ for $j > 0$, and chasing through the sequence one finds that $H^0(X, \mathcal{O}_X) = 0$, which is absurd.) Similarly, if $e \leq n - 1$ and if the zero-scheme $Z = \text{Zeroes}(s)$ of $s$ has the expected dimension $n - e$, then $H^0(X, \mathcal{O}_Z) = \mathbf{C}$, and in particular $Z$ is connected. This line of attack is used to study higher and more general degeneracy loci in [385] and [417].  □

**Example 7.3.10. (Theorem of Evans and Griffith, II).** Following Ein [140], we use Le Potier's theorem to establish a result originally proved algebraically by Evans and Griffith [175] concerning vector bundles of small rank on projective space:

> **Theorem.** *Let $E$ be a vector bundle of rank $e \geq 2$ on $\mathbf{P}^n$, and assume that*
>
> $$H^i\big(\mathbf{P}^n, E(k)\big) = 0 \quad \text{for all } 1 \leq i \leq e-1 \text{ and all } k \in \mathbf{Z}.$$
>
> *Then $E$ is a direct sum of line bundles.*

Note that since $H^n\big(\mathbf{P}^n, E(k)\big) \neq 0$ when $k \ll 0$, the hypothesis implies that $e \leq n$. We refer to Section 3.2.B — in particular Remark 3.2.12 — for a discussion of the context and interest of the statement.

To establish the theorem, we suppose for a contradiction that $E$ is not the direct sum of a line bundle and a vector bundle of smaller rank. After replacing $E$ by a suitable twist, we may also assume that $E$ is $(-1)$-regular in the sense of Castelnuovo–Mumford but not $(-2)$-regular (Section 1.8). By the definition of $(-2)$-regularity, the latter hypothesis means that

$$H^i\big(\mathbf{P}^n, E(-2-i)\big) \neq 0 \quad \text{for some } i \geq 1. \tag{*}$$

On the other hand, thanks to Mumford's Theorem 1.8.3 the $(-1)$-regularity implies that $E$ is the quotient of a direct sum of copies of $\mathcal{O}_{\mathbf{P}^n}(1)$. In particular, $E$ is ample. Since $\omega_{\mathbf{P}^n} = \mathcal{O}_{\mathbf{P}^n}(-n-1)$, Le Potier's theorem (applied to a positive twist of $E$) therefore yields:

$$H^i\big(\mathbf{P}^n, E(-2-i)\big) = 0 \quad \text{for } e \leq i \leq n-1. \tag{**}$$

Comparing (*) with (**) and the vanishings appearing in the hypothesis, it follows that $H^n\big(\mathbf{P}^n, E(-n-2)\big) \neq 0$. By Serre duality, this implies that $H^0\big(\mathbf{P}^n, E^*(1)\big) \neq 0$, which in turn is equivalent to the existence of a nonzero morphism $u : E \longrightarrow \mathcal{O}_{\mathbf{P}^n}(1)$. But $E$ is a quotient of a direct sum of copies of $\mathcal{O}_{\mathbf{P}^n}(1)$, and so $u$ splits to realize $\mathcal{O}_{\mathbf{P}^n}(1)$ as a summand of $E$. This contradiction completes the proof. □

We now turn to the proof of Le Potier's theorem. The approach we follow, which is due to Schneider [522], considerably simplifies the original arguments in [396], [397]. The plan is to use some computations of direct images to reduce the question to the Nakano vanishing theorem on $\mathbf{P}(E)$. During the course of the proof we shall have occasion to study the cohomology of a sheaf via a resolution and a filtration. We refer the reader to Section B.1 in Appendix B for a summary of the elementary diagram-chasing lemmas that are (silently) at work here.

**Lemma 7.3.11.** *Fix an integer $0 \leq m \leq e-1$. Then:*

(i).  $\pi_*\big(\Omega^m_{\mathbf{P}(E)/X}(\ell)\big) = 0$  *for* $\ell \leq m$.

(ii).  $\pi_*\big(\Omega^m_{\mathbf{P}(E)/X}(m+1)\big) = \Lambda^{m+1}E$.

(iii).  $R^j\pi_*\big(\Omega^m_{\mathbf{P}(E)/X}(\ell)\big) = 0$  *for all* $j > 0$, $\ell > 0$.

*Proof.* The Koszul complex associated to the surjection $\pi^*E(-1) \to \mathcal{O}_{\mathbf{P}(E)}$ appearing in the relative Euler sequence (7.10) gives rise — upon truncation — to a long exact sequence

$$0 \longrightarrow \pi^*(\Lambda^e E)(-e) \longrightarrow \dots \longrightarrow \pi^*(\Lambda^{m+2}E)(-m-2) \longrightarrow$$
$$\longrightarrow \pi^*(\Lambda^{m+1}E)(-m-1) \longrightarrow \Omega^m_{\mathbf{P}(E)/X} \longrightarrow 0 \quad (7.13)$$

resolving $\Omega^m_{\mathbf{P}(E)/X}$. But recall that $R^j\pi_*\big(\mathcal{O}_{\mathbf{P}(E)}(t)\big) = 0$ for all $j > 0$ and $t > -e$, and similarly $\pi_*\big(\mathcal{O}_{\mathbf{P}(E)}(t)\big) = 0$ when $t < 0$. With this in mind the direct images in question can be read off (using the projection formula) from the complex obtained by twisting (7.13) by $\mathcal{O}_{\mathbf{P}(E)}(\ell)$. $\qquad\square$

*Proof of Theorem 7.3.5.* For (7.11), fix an integer $a > 0$. We will show that

$$R^j\pi_*\left(\Omega^{n+a-1}_{\mathbf{P}(E)}(a)\right) = \begin{cases} \Omega^n_X \otimes \Lambda^a E & \text{if } j = 0, \\ 0 & \text{if } j > 0. \end{cases} \quad (7.14)$$

Granting this, it follows that

$$H^i\big(\mathbf{P}(E), \Omega^{n+a-1}_{\mathbf{P}(E)}(a)\big) = H^i\big(X, \omega_X \otimes \Lambda^a E\big) \quad (7.15)$$

for all $i$. But $\mathcal{O}_{\mathbf{P}(E)}(a)$ is ample thanks to the amplitude of $E$, so by Nakano vanishing (Theorem 4.2.3) the group on the left in (7.15) vanishes when

$$i + (n + a - 1) > \dim \mathbf{P}(E) = n + e - 1,$$

as required.

Turning to (7.14), we use the cotangent bundle sequence (7.9) to study $\Omega^m_{\mathbf{P}(E)}$. Specifically, it follows from (7.9) that $\Omega^{n+a-1}_{\mathbf{P}(E)}(a)$ has a filtration whose graded pieces are the sheaves

$$\pi^*\Omega^\ell_X \otimes \Omega^m_{\mathbf{P}(E)/X}(a) \quad \text{for } \ell + m = n + a - 1. \quad (7.16)$$

Note that the bundles in question vanish unless $\ell \leq n$, in which case $m \geq a-1$. But Lemma 7.3.11 shows that these have vanishing higher direct images and only one non-zero direct image in the indicated range of $\ell$ and $m$, viz.

$$\pi_*\left(\pi^*\Omega^n_X \otimes \Omega^{a-1}_{\mathbf{P}(E)/X}(a)\right) = \Omega^n_X \otimes \Lambda^a E$$

coming from the lowest term in the filtration. The assertion (7.14) follows.

The proof of (ii) is similar but simpler. Arguing as above one shows that

$$R^j \pi_* \left( \Omega^p_{\mathbf{P}(E)}(1) \right) = \begin{cases} \Omega^p_X \otimes E & \text{if } j = 0, \\ 0 & \text{if } j > 0. \end{cases}$$

As before, the statement then follows from Nakano vanishing.        □

**Remark 7.3.12. (k-ample bundles).** Sommese [551, Proposition 1.14] shows that if $E$ is a $k$-ample bundle of rank $e$ (Remark 6.2.18) then

$$H^i\left(X, \omega_X \otimes \Lambda^a E\right) = 0 \quad \text{for } a > 0 \text{ and } i > e + k - a.$$

**Example 7.3.13. (Another proof of vanishing).** Bogomolov and De Oliveira [63] have given an interesting new proof of the vanishing of $H^1(X, E^*)$ for an ample vector bundle $E$ of rank $e < n$ on the smooth projective $n$-fold $X$, and we outline their argument here. The cohomology group in question parameterizes extensions of $E$ by $\mathcal{O}_X$, so supposing that $H^1(X, E^*) \neq 0$ there is a non-split extension

$$0 \longrightarrow \mathcal{O}_X \longrightarrow V \longrightarrow E \longrightarrow 0 \tag{*}$$

of bundles on $X$. This determines an embedding of $\mathbf{P}(E)$ into $\mathbf{P}(V)$ as a divisor with ample normal bundle $N_{\mathbf{P}(E)/\mathbf{P}(V)} = \mathcal{O}_{\mathbf{P}(E)}(1)$. Then a theorem of Hartshorne (Example 1.2.30) implies that there is a proper birational morphism $\overline{f} : \mathbf{P}(V) \longrightarrow \overline{U}$ that is an isomorphism except over a finite subset $S \subset \overline{U}$, and that embeds $\mathbf{P}(E)$ as an ample divisor in $\overline{U}$. Thus $U =_{\text{def}} \overline{U} - \mathbf{P}(E)$ is affine, and $\overline{f}$ restricts to a proper morphism

$$f : \mathbf{P}(V) - \mathbf{P}(E) \longrightarrow U.$$

Now suppose that $Z \subseteq \mathbf{P}(V) - \mathbf{P}(E)$ is any closed subset contracted to a point by $f$. Grant for the time being that

$$\dim Z \leq n - 1. \tag{**}$$

Then the higher direct images $R^j f_* \mathbf{C}$ are supported on a finite set, and thanks to (**) they vanish for $j > 2n - 2$. On the other hand, since $U$ is affine, the cohomology of the constructible sheaf $f_* \mathbf{C}$ on $U$ vanishes in degrees $> \dim U = n + e$ (Theorem 3.1.13). Observing that $2n - 1 \geq n + e$, it follows from the Leray spectral sequence that

$$H^i\left(\mathbf{P}(V) - \mathbf{P}(E), \mathbf{C}\right) = 0 \quad \text{for } i > 2n - 1.$$

On the other hand, $\mathbf{P}(V) - \mathbf{P}(E)$ is an affine space bundle over $X$, and hence

$$H^{2n}\left(\mathbf{P}(V) - \mathbf{P}(E), \mathbf{C}\right) = H^{2n}\left(X, \mathbf{C}\right) \neq 0,$$

a contradiction.

Turning to (**), note first that any complete subvariety $Z \subseteq \mathbf{P}(V) - \mathbf{P}(E)$, being a complete subset of an affine bundle, must map finitely to $X$. Therefore, $\dim Z = n$ if and only if some component of $Z$ dominates $X$. Assume then for a contradiction that $Z$ is an irreducible complete variety that projects onto $X$ under $p : \mathbf{P}(V) - \mathbf{P}(E) \longrightarrow X$. Now since $\mathbf{P}(E) \subseteq \mathbf{P}(V)$ is the divisor of a section of $\mathcal{O}_{\mathbf{P}(V)}(1)$, there is a natural trivialization of this line bundle on $\mathbf{P}(V) - \mathbf{P}(E)$, and hence $p^*V$ admits a trivial quotient on the complement of $\mathbf{P}(E)$. One sees using this that the pullback of (*) to $Z$ must split. But this can happen only if (*) splits already on $X$: in fact, the homomorphism $H^1(X, E^*) \longrightarrow H^1(Z, p^*E^*)$ is injective thanks to Lemma 4.1.14. Thus we have established (**), completing the proof. Bogomolov and De Oliveira also make some interesting suggestions about how this approach might fit into the circle of ideas surrounding the Shafarevich conjecture.                                  □

## 7.3.B Generalizations

We turn now to a survey — mostly without proofs — of some generalizations of the vanishing theorems of Griffiths and Le Potier. We continue to consider a vector bundle $E$ of rank $e$ on a complex projective manifold $X$ of dimension $n$.

**Associated bundles: theorems of Demailly and Manivel.** Given a representation $\rho : \mathrm{GL}(e, \mathbf{C}) \longrightarrow \mathrm{GL}(N, \mathbf{C})$ of algebraic groups, one can associate to $E$ a bundle $E_\rho$ of rank $N$ by applying $\rho$ to the transition matrices describing $E$. The symmetric, exterior, and tensor powers of $E$ are the most familiar examples of this construction. Demailly [121] studied the question of what sort of vanishing theorems hold for $E_\rho$ when $E$ is ample.

To fix notation, recall that the irreducible finite-dimensional representations of $\mathrm{GL}(e, \mathbf{C})$ are parametrized by non-increasing sequences $\lambda$ of non-negative integers:

$$\lambda = (\lambda_1, \ldots, \lambda_e) \quad , \quad \lambda_1 \geq \ldots \geq \lambda_e \geq 0.$$

(See [210, §15.5], [208, Chapter 8]). The height $h(\lambda)$ of $\lambda$ is the number of non-zero components ("parts") $\lambda_i$. Given $E$ as above, we denote by $\Gamma^\lambda E$ the bundle associated to the representation corresponding to $\lambda$. For example,

$$\Gamma^\lambda E = \begin{cases} S^m E & \text{when } \lambda = (m, 0, 0, \ldots, 0), \\ \Lambda^m E & \text{when } \lambda = 1^{\times m} =_{\text{def}} (1, \ldots, 1, 0, \ldots, 0) \;\; (m \text{ repetitions}). \end{cases}$$

Note that $h(\lambda) = 1$ when $\lambda = (m, 0, \ldots, 0)$, whereas $h(1^{\times m}) = m$.

Demailly [121] proves the following generalization of Griffiths' theorem:

**Theorem 7.3.14. (Demailly's theorem).** *If $E$ is ample, then*

$$H^i \left( X, \omega_X \otimes \Gamma^\lambda E \otimes (\det E)^{\otimes h(\lambda)} \right) = 0 \quad \text{for all } i > 0.$$

*More generally, if $E$ is ample and $L$ is a nef line bundle, or if $E$ is nef and $L$ is ample, then*

$$H^i\left( X\,,\, \omega_X \otimes \Gamma^\lambda E \otimes (\det E)^{\otimes h(\lambda)} \otimes L \right) \;=\; 0 \quad \text{for all } i > 0.$$

It follows for example that $H^i\big(X, \omega_X \otimes \Lambda^m E \otimes (\det E)^{\otimes m}\big) = 0$ for all $i > 0$ when $E$ is ample. Demailly's strategy was to apply vanishing for line bundles on a suitable flag variety constructed from $E$. Here the computations of direct images become much more involved than those appearing in the proof of Le Potier's theorem. However Manivel then found an extremely simple trick, which we shall explain momentarily, allowing one to deduce this statement directly from Griffiths vanishing.

Demailly also considered the more subtle question of vanishings for the Dolbeaut-type groups

$$H^i\left( X\,,\, \Omega_X^p \otimes \Gamma^\lambda E \otimes (\det E)^{\otimes \ell} \right)$$

for suitable ranges of the parameters. His results were given very nice extensions by Manivel in [417]. Manivel's general statement involves some further combinatorics that would take us too far afield, so following [417] we content ourselves here with quoting a special case:

**Theorem. (Manivel's theorem).** *Assume that $E$ is ample, and suppose given sequences of positive integers $k_1, \ldots, k_\ell$ and $a_1, \ldots, a_m$. Then*

$$H^i\left( X\,,\, \Omega_X^p \otimes S^{k_1} E \otimes \ldots \otimes S^{k_\ell} E \otimes \Lambda^{a_1} E \otimes \cdots \otimes \Lambda^{a_m} E \otimes (\det E)^{\otimes \ell + n - p} \right) \;=\; 0$$

$$\text{for } i + p \;>\; n + \sum_{s=1}^m (e - a_s).$$

For example, taking $p = n$, $\ell = 0$, and $m = 1$, one recovers the first statement in Le Potier's theorem. Similarly, if $\ell = 0$ and $p = n$, then the assertion is that

$$H^i\big(X, \omega_X \otimes \Lambda^{a_1} E \otimes \cdots \otimes \Lambda^{a_m} E\big) = 0$$

provided that $i > \sum(e - a_s)$: this had been established in [153] in order to study the Koszul cohomology of varieties embedded by certain "hyperadjoint" linear series (Theorem 1.8.60).

**Remark 7.3.15.** A further generalization appears in the paper [386] of Laytimi and Nahm. Here one assumes not the amplitude of $E$ itself, but rather the positivity of the tensor product of certain symmetric and exterior products of $E$. $\qquad\square$

**Manivel's trick.** Manivel discovered a device that effortlessly yields a plethora of vanishings for associated bundles. The idea is simply to apply the Griffiths or Le Potier vanishing theorems to a direct sum of ample bundles, and to decompose the resulting symmetric or exterior powers into a direct sum of associated bundles. We illustrate the method by giving the proof of Demailly's theorem:

*Proof of Theorem 7.3.14.* We focus on the first assertion. Given an ample vector bundle $E$, and an integer $h > 0$, consider the $h$-fold direct sum

$$F = E \oplus \cdots \oplus E \quad (h \text{ times}).$$

Then $F$ is ample, and $\det F = (\det E)^{\otimes h}$. Now apply Griffiths vanishing (Theorem 7.3.1) to $F$. Recalling that

$$S^m \big( E \oplus \ldots \oplus E \big) = \bigoplus_{m_1 + \ldots + m_h = m} S^{m_1} E \otimes \ldots \otimes S^{m_h} E,$$

one finds that

$$H^i \Big( X \, , \, \omega_X \otimes S^{m_1} E \otimes \ldots \otimes S^{m_h} E \otimes (\det E)^{\otimes h} \Big) = 0$$

for $i > 0$ and any $m_i \geq 0$. On the other hand, if $\lambda = (\lambda_1, \lambda_2, \ldots, \lambda_h, 0, \ldots, 0)$, then $\Gamma^\lambda(E)$ is a direct summand of $S^{\lambda_1} E \otimes \ldots \otimes S^{\lambda_h} E$ (cf. [207, Chapter 8, Corollary 2]). The theorem follows. □

**Example 7.3.16.** Let $E_1, \ldots, E_h$ be vector bundles on $X$, and $L$ a line bundle. Assume that either all the $E_i$ are ample and $L$ is nef, or that all the $E_i$ are nef and $L$ is ample. Then

$$H^i \Big( X \, , \, \omega_X \otimes S^{m_1} E_1 \otimes \ldots \otimes S^{m_h} E_h \otimes \det E_1 \otimes \ldots \otimes \det E_h \otimes L \Big) = 0$$

for $i > 0$ and $m_1, \ldots, m_h \geq 0$. □

**Example 7.3.17.** In the situation of the previous example, suppose that rank $E_i = e_i$. Then

$$H^i \Big( X \, , \, \omega_X \otimes \Lambda^{a_1} E_1 \otimes \ldots \otimes \Lambda^{a_h} E_h \otimes L \Big) = 0$$

for $i > \sum (e_j - a_j)$. □

**Vanishing for Nakano-positive bundles.** We have mentioned in Section 6.1.D the notion of Nakano positivity for vector bundles. Here one has vanishing theorems that are considerably stronger than those holding for arbitrary ample bundles.

**Theorem. (Vanishing for Nakano-positive bundles).** *Let $X$ be a compact Kähler manifold of dimension $n$, and let $E$ be a Hermitian vector bundle on $X$ that is Nakano-semipositive at every point of $X$, and Nakano-positive in the neighborhood of at least one point of $X$. Then*

$$H^i\big(X, \omega_X \otimes E\big) = 0 \quad \text{for all} \quad i > 0.$$

We refer for instance to [533, Chapter VI] for the proof and further discussion.

**Example 7.3.18. (Tangent bundle of projective space).** The tangent bundle $T\mathbf{P}^n$ of projective $n$-space does not admit a Nakano-positive metric. (In fact,

$$H^{n-1}(\mathbf{P}^n, \omega_{\mathbf{P}^n} \otimes T\mathbf{P}^n) = H^{n-1}(\mathbf{P}^n, \Omega_{\mathbf{P}^n}^{n-1}) \neq 0,$$

which would contradict the theorem just stated.) This gives a sense of how strong a condition Nakano-positivity is.     □

**Remark 7.3.19.** Demailly and Skoda [134] show that if $E$ is a Griffiths-positive bundle, then $E \otimes \det E$ is Nakano-positive. One can see this as "explaining" the special case of Griffiths vanishing that $H^i(X, \omega_X \otimes E \otimes \det E) = 0$ for $i > 0$.

**Remark 7.3.20.** Kim and Manivel [337] give an interesting application of Nakano vanishing to establish the amplitude of vector bundles associated to branched coverings of certain rational homogeneous spaces. (See Sections 6.3.D and 7.1.C.)     □

**Vanishing theorems on special varieties.** Manivel [416] has shown that particularly strong vanishings hold on special varieties. For example:

**Theorem.** *Let $X$ be a non-singular projective toric variety of dimension $n$ (for example $X = \mathbf{P}^n$), and let $E$ be an ample vector bundle of rank $e$ on $X$. Then*

$$H^i(X, \Omega_X^p \otimes \Lambda^a E) = 0 \quad \text{for } i > e - a;$$
$$H^i(X, \Omega_X^p \otimes S^m E) = 0 \quad \text{for } i \geq e.$$

We have noted in Remark 7.3.8 that even weaker statements fail in general. We refer to [416] for related results and proofs. The idea very roughly speaking is that for the varieties in question, the tangent bundle $TX$ "can be expressed in terms of nef line bundles."

# Notes

Theorem 7.1.1 was established in [548, Appendix] and [551, (1.6)], although it is stated there under some dimensional and transversality hypotheses that Fulton observed to be extraneous. Griffiths [247, Theorem H] had previously proven the same result under stronger positivity and transversality assumptions via a direct Morse-theoretic argument. Bloch and Gieseker established Theorem 7.1.10 on the way to proving the positivity of the Chern classes of an ample vector bundle (Section 8.2.A). Once again Griffiths [247, (0.5)] had proved a statement along these lines under stronger assumptions.

The proof of Theorem 7.2.1 starts by passing to a Grassmannian and noting that the degeneracy locus in question is dominated by the zero-locus of a vector bundle map. This is a very well-known construction, which leads, for example, to Porteus' formula as in [208, §14.3]. The difficulties spring from the fact that going to the Grassmannian involves a loss of positivity, and one has to argue that in the end it doesn't matter. The determinantal description of the varieties $W_d^r(C)$ in Section 7.2.C is due to Kempf and Kleiman–Laksov [347]. Lemma 7.2.20 was suggested by some constructions of Flenner and Ran in connection with [511].

Le Potier's theorem had been conjectured in [247, (0.6)] under the (potentially) stronger hypothesis of Griffiths-positivity. As indicated in the text, we follow Schneider's approach [522] to this result. The book [533] of Shiffman and Sommese gives a nice introduction to vanishing theorems for bundles.

# 8

# Numerical Properties of Ample Bundles

This chapter deals with the numerical properties of ample vector bundles. The theme is that amplitude implies the positivity of various intersection and Chern numbers.

The first section is devoted to some intersection-theoretic preliminaries. The basic positivity theorems appear in Section 8.2: we prove a result of Bloch and Gieseker [60] to the effect that the Chern classes of a nef bundle are non-negative, and we deduce from this a theorem of Fulton and the author [214] concerning positivity of cone classes. This leads in Section 8.3 to the determination of all positive polynomials in the Chern classes of ample vector bundles. Some applications appear in Section 8.4.

The central results of this chapter draw on, and to some extent were motivated by, the Fulton–MacPherson approach to intersection theory. In dealing with this theory, we attempt to steer a middle course between developing all the requisite machinery ab initio on the one hand, and assuming expertise on the other. Specifically, we explain the basic constructions and present heuristic justifications of the facts we use, but refer to [208] or occasionally [214] for details and careful proofs.

## 8.1 Preliminaries from Intersection Theory

This section is devoted to some intersection-theoretic facts that we will require. We start in Section 8.1.A by setting up a formalism of Chern classes for $\mathbf{Q}$-twisted bundles. The most important material appears in 8.1.B, where we recall the construction and basic properties of the classes obtained by intersecting a cone sitting in the total space of a bundle with the zero-section of that bundle. Finally, these so-called cone classes are extended to the $\mathbf{Q}$-twisted setting in 8.1.C.

Throughout this section $X$ denotes an irreducible projective variety or scheme of dimension $n$, and $E$ is a vector bundle of rank $e$ on $X$. In this section and the next we work with the projective bundle $\mathbf{P}_{\mathrm{sub}}(E)$ of one-dimensional subspaces of $E$.

### 8.1.A Chern Classes for Q-Twisted Bundles

Our first object is to establish a formalism of Chern classes for the **Q**-twisted bundles introduced in Section 6.2.A. The plan is to use the singular cohomology groups $H^*(X; \mathbf{Q})$ to receive these classes, and we start with some remarks intended to show that it makes sense to do so.

Specifically, recall that if $P$ is a numerically trivial line bundle on $X$, then some power of $P$ is a deformation of the trivial bundle (Corollary 1.4.38). In particular, $c_1(P) \in H^2(X; \mathbf{Z})$ is a torsion class. Therefore there is a natural map

$$N^1(X)_{\mathbf{Q}} \longrightarrow H^2(X; \mathbf{Q}),$$

i.e. a numerical equivalence class of divisors determines a well-defined element in rational cohomology. By abuse of notation, given $\delta \in N^1(X)_{\mathbf{Q}}$ we write also $\delta \in H^2(X; \mathbf{Q})$ for the corresponding cohomology class. As usual, the cup product of classes $\alpha, \beta \in H^*(X; \mathbf{Q})$ is denoted by $\alpha \cdot \beta$ or simply $\alpha\beta$ when no confusion seems likely.

Recall next that if $E$ is a vector bundle of rank $e$ on $X$ and $L$ is any line bundle, then

$$c_i(E \otimes L) = \sum_{k=0}^{i} \binom{e-k}{i-k} c_k(E)\, c_1(L)^{i-k}.$$

We are led (in fact, forced) to use this expression to define the Chern classes of a **Q**-twisted bundle:

**Definition 8.1.1. (Chern and Segre classes of Q-twisted bundles).** Let $\delta \in N^1(X)_{\mathbf{Q}}$ be a **Q**-divisor class on $X$. The Chern classes of the **Q**-twisted bundle $E{<}\delta{>}$ are defined by

$$c_i\big(E{<}\delta{>}\big) = \sum_{k=0}^{i} \binom{e-k}{i-k} c_k(E) \cdot \delta^{i-k} \in H^{2i}(X; \mathbf{Q}). \tag{8.1}$$

In the usual way, we also introduce the Chern polynomial

$$c_t\big(E{<}\delta{>}\big) = \sum c_i\big(E{<}\delta{>}\big) t^i.$$

The Segre classes (or inverse Chern classes) $s_i(E{<}\delta{>})$ of $E{<}\delta{>}$ are determined via

$$s_t\big(E{<}\delta{>}\big) = \sum s_i\big(E{<}\delta{>}\big) t^i =_{\mathrm{def}} \frac{1}{c_t\big(E{<}\delta{>}\big)}. \qquad \square$$

By construction, this definition is compatible with the identification of $E<D>$ and $E \otimes \mathcal{O}_X(D)$ when $D$ is integral, i.e. it respects the relation of $\mathbf{Q}$-isomorphism (Definition 6.2.2). It also leads to the expected formal properties:

**Lemma 8.1.2.** (i).   *Let $f : Y \longrightarrow X$ be a morphism of projective varieties. Then*

$$c_i\big( f^*(E<\delta>) \big) = f^*\big( c_i(E<\delta>) \big).$$

(ii).   *Given an exact sequence $0 \longrightarrow E' \longrightarrow E \longrightarrow E'' \longrightarrow 0$ of bundles on $X$, together with a $\mathbf{Q}$-divisor class $\delta \in N^1(X)_\mathbf{Q}$, one has the Whitney product formula*

$$c_t\big( E<\delta> \big) = c_t\big( E'<\delta> \big) \cdot c_t\big( E''<\delta> \big).$$

(iii).   *Let $\pi : \mathbf{P}_{\mathrm{sub}}(E) \longrightarrow X$ denote the projectivization of $E$ and put $\xi_E = c_1(\mathcal{O}_{\mathbf{P}_{\mathrm{sub}}(E)}(1))$, so that the Grothendieck relation*

$$\xi_E^e + c_1(E) \cdot \xi_E^{e-1} + \ldots + c_e(E) = 0 \tag{8.2}$$

*holds on $\mathbf{P}_{\mathrm{sub}}(E)$, where by abuse of notation we write $c_i(E)$ to denote the pullback of this class via $\pi$. Given a $\mathbf{Q}$-divisor class $\delta$ on $X$, set*

$$\xi_{E<\delta>} = \xi_E - \pi^*\delta \in H^2\big( \mathbf{P}_{\mathrm{sub}}(E); \mathbf{Q} \big). \tag{8.3}$$

*Then*

$$\xi_{E<\delta>}^e + c_1(E<\delta>) \cdot \xi_{E<\delta>}^{e-1} + \ldots + c_e(E<\delta>) = 0 \tag{8.4}$$

*in $H^*\big( \mathbf{P}_{\mathrm{sub}}(E); \mathbf{Q} \big)$.*

(iv).   *If $E<\delta>$ is a $\mathbf{Q}$-twisted bundle and $\varepsilon \in N^1(X)_\mathbf{Q}$ is a $\mathbf{Q}$-divisor class, then*

$$c_i\big( E<\delta + \varepsilon> \big) = \sum_{k=0}^{i} \binom{e-k}{i-k} c_k(E<\delta>) \cdot \varepsilon^{i-k}.$$

**Remark 8.1.3.** Concerning the signs in (8.3), note that if $A$ is a line bundle, then

$$\mathbf{P}_{\mathrm{sub}}(E \otimes A) \cong \mathbf{P}_{\mathrm{sub}}(E)$$

by an isomorphism taking $\mathcal{O}_{\mathbf{P}_{\mathrm{sub}}(E \otimes A)}(1)$ to $\mathcal{O}_{\mathbf{P}_{\mathrm{sub}}(E)}(1) \otimes \pi^*A^*$.   $\square$

*Proof of Lemma 8.1.2.* The first statement follows from the functoriality of (usual) Chern classes and intersection products. For (ii), we may assume by the splitting principle that the Chern polynomials of the bundles in question factor into a product of linear terms. In this case note that if $c_t(E) = (1 + a_1 t) \cdot \ldots \cdot (1 + a_e t)$, then

$$c_t\big( E<\delta> \big) = \big(1 + (a_1 + \delta)t\big) \cdot \ldots \cdot \big(1 + (a_e + \delta)t\big),$$

with analogous expressions for $c_t\big(E'{<}\delta{>}\big)$ and $c_t\big(E''{<}\delta{>}\big)$. The assertion is then a consequence of the fact that the Chern roots of $E$ are the concatenation of the Chern roots of $E'$ and of $E''$. For (iii) we may again assume that the Chern roots of $E$ exist in $H^*(X; \mathbf{Q})$. In this case (8.2) can be written in the form

$$\big(\xi_E + \alpha_1\big) \cdot \ldots \cdot \big(\xi_E + \alpha_e\big) \;=\; 0,$$

where the $\alpha_i$ are the pullbacks under $\pi$ of the Chern roots $a_i$ of $E$. Then (8.4) follows from the identity

$$\big((\xi_E - \pi^*\delta) + (\alpha_1 + \pi^*\delta)\big) \cdot \ldots \cdot \big((\xi_E - \pi^*\delta) + (\alpha_e + \pi^*\delta)\big) \;=\; 0.$$

We leave (iv) to the reader.                                                    $\square$

**Remark 8.1.4. (Polynomials).** Given $t \in \mathbf{Q}$ we can form the twisted bundle $E{<}t \cdot \delta{>}$. Assuming that $\dim X = n$, the degree

$$P(t) \;=_{\mathrm{def}}\; \int_X c_n\big(E{<}t \cdot \delta{>}\big)$$

is a rational number. By using the definition of $c_n(E{<}t \cdot \delta{>})$ to expand out the integrand, we can view $P(t)$ as a polynomial in $t$. Provided that the top self-intersection number $\big(\delta^n\big)$ is non-zero, one has $\deg P(t) = n$. We can make a similar construction by replacing $c_n$ with any homogeneous polynomial of weighted degree $n$ in the Chern classes of $E$. Several of the arguments below will involve the analysis of such polynomials.                                          $\square$

### 8.1.B  Cone Classes

One of the essential ideas of the Fulton–MacPherson approach to intersection theory is that many questions ultimately reduce to a rather simple geometric construction. Namely, one starts with a vector bundle $E$ on a variety $X$, and considers a cone $C \subseteq \mathbf{E}$ sitting in the total space of $E$. One can then intersect $C$ with the zero-section $\mathbf{0_E} \subseteq \mathbf{E}$ of $E$ to get a well-defined class $z(C, E)$ on $X$. The main positivity result (Theorem 8.2.4 in the next section) will be that these classes are positive when $E$ is ample. In this subsection we recall the construction of these cone classes and sketch some of their properties.

To begin with, we require a homology theory for possibly non-compact varieties $V$ — analogous to the Chow groups $A_*(V)$ — in which every closed subvariety determines a fundamental class. In keeping with our policy of using topologically defined groups, we follow [208, Chapter 19] and use Borel–Moore homology. However one could just as well use the group of rational cycles modulo numerical equivalence in the sense of [208, Definition 19.1].

Given a complex variety or scheme $V$, denote by $H_*(V) = H_*^{\mathrm{BM}}(V; \mathbf{Q})$ the Borel–Moore homology of $V$ with rational coefficients.[1] The basic features of these groups are indicated in [208, Chapter 19, §1]. For us the most important are the following:

**Lemma 8.1.5.** (i). *The groups $H_*(\ )$ are covariant for proper morphisms.*

(ii). *If $V$ is an irreducible variety of dimension $n$, then $V$ has a fundamental class $[V] \in H_{2n}^{\mathrm{BM}}(V; \mathbf{Z})$ that freely generates the group in question.*

(iii). *There is a cycle map $\mathrm{cl}_V : Z_d(V) \longrightarrow H_{2d}(V)$ on the group of algebraic d-cycles on $V$, determined by taking a closed subvariety $W \subseteq V$ of dimension $d$ to the image of its fundamental class. This map factors through the Chow group $A_d(V)$ of rational equivalence classes of d-cycles.*

We refer to [208, Chapter 19, §1] for details and references. Given a cycle $\alpha \in Z_d(V)$, we often continue to write $\alpha \in H_{2d}(V)$ for the corresponding homology class in place of the more correct notation $\mathrm{cl}_V(\alpha)$.

Of particular importance are the various products one can define:

**Proposition 8.1.6. (Products).** *As above let $V$ be a complex variety or scheme, and continue to write $H_*(\ ) = H_*^{\mathrm{BM}}(\ ; \mathbf{Q})$.*

(i). *Given a closed subvariety $W \subseteq V$ there are cap-products*

$$H^j(V, V - W; \mathbf{Q}) \otimes H_k(V) \longrightarrow H_{k-j}(W)$$

(ii). *Let $\iota : W \longrightarrow V$ be a regular embedding of codimension $e$, i.e. assume that $\iota$ is the inclusion of a local complete intersection subscheme $W \subseteq V$ of the stated codimension. Then $\iota$ canonically determines an orientation class*

$$u_\iota = u_{W,V} \in H^{2e}(V, V - W; \mathbf{Q}).$$

*By (i) this class gives rise to refined Gysin mappings*

$$\iota^! : H_k(V) \longrightarrow H_{k-2e}(W).$$

We refer to [208, Chapter 19.2] for the proof and other properties. In (ii), one thinks of $\iota^!$ as obtained by intersecting a class $\alpha$ on $V$ with $W$. These homomorphisms are compatible with the corresponding maps on Chow groups constructed in [208, Chapter 6]: in particular, many of the references to [208] in the following paragraphs actually point to results about Chow groups.

We now specialize this machine to cones sitting in vector bundles. As before, let $X$ be an irreducible projective variety or scheme of dimension $n$, and

---

[1] We remark that everything we do in the present subsection works with integer coefficients. It is only in the next subsection, when we extend the theory to $\mathbf{Q}$-twists, that rational coefficients enter the picture.

fix a vector bundle $E$ on $X$ having rank $e$. Denote by $\mathbf{E} = \mathrm{Spec}_{\mathcal{O}_X} \mathrm{Sym}(E^*)$ the total space of $E$, with $p : \mathbf{E} \longrightarrow X$ the projection. We write

$$o_E : X = \mathbf{0_E} \hookrightarrow \mathbf{E}$$

for inclusion of the zero-section, so that $o_E$ is a regular embedding of codimension $e$.

By definition, a *cone* in $E$ is a subscheme $C \subseteq \mathbf{E}$ stable under the natural $\mathbf{C}^*$ action. Equivalently, $C$ is defined by a graded sheaf of ideals in the graded $\mathcal{O}_X$ algebra $\mathrm{Sym}(E^*)$. Assume that $C$ has pure dimension $\ell$: then as in [208, Chapter 1.5], [214], or Section 1.1.C, $C$ determines an $\ell$-cycle

$$[C] = \sum m_i[C_i] \in Z_\ell(\mathbf{E}) \tag{8.5}$$

(the $C_i$ being the irreducible components of $C$) and hence also a class $[C] = \sum m_i [C_i] \in H_{2\ell}(\mathbf{E})$.

**Definition 8.1.7. (Cone classes).** The *cone class* of $C$ in $E$ is the class

$$z(C, E) = o_E^!([C]) \in H_{2(\ell-e)}(X). \quad \square$$

Intuitively, $z(C, E)$ is obtained by intersecting $C$ with the zero-section in $\mathbf{E}$. By construction cone classes are linear with respect to the cycle decomposition (8.5):

$$z(C, E) = \sum m_i \, z(C_i, E). \tag{8.6}$$

Observe also that since $X$ is complete, the group $H_{2(\ell-e)}^{\mathrm{BM}}(X; \mathbf{Q})$ receiving cone classes is identified with usual singular homology.

**Example 8.1.8. (Chern classes).** One can realize all the Chern classes $c_i(E)$ of $E$ as cone classes. For a heuristic argument, suppose that $E$ is globally generated, and choose $e-i+1$ general sections $s_1, \ldots, s_{e-i+1} \in \Gamma(X, E)$. These determine a mapping

$$s : \mathcal{O}_X^{e-i+1} \longrightarrow E,$$

and as is well known, $c_i(E) \cap [X] \in A_{n-i}(X)$ is represented by the degeneracy cycle $[D_{e-i}(s)]$ where $s$ drops rank.[2] To say the same thing somewhat differently, fix a vector space $V$ of dimension $e - i + 1$, form the trivial bundle $V_X$ modeled on $V$, and consider the bundle $H = \mathrm{Hom}(V_X, E)$ of homomorphisms from $V_X$ to $E$. Inside the total space $\mathbf{H}$ of $H$ there is an irreducible cone $\Omega \subseteq \mathbf{H}$ of codimension $i$ whose fibre over $x \in X$ consists of all homomorphisms $s(x) : V_X(x) \longrightarrow E(x)$ of rank $\leq e - i$. Now $s$ determines a section $\sigma : X \hookrightarrow \mathbf{H}$, and (at least set-theoretically) $\sigma^{-1}(\Omega) = D_{e-i}(s)$. This suggests the formula

---

[2] If $X$ is not locally Cohen–Macaulay, then this cycle is to be understood in the refined sense of [208, Chapter 14, §4].

$$c_i(E) \cap [X] = z(\Omega, H),$$

which in fact holds whether or not $E$ is globally generated. (For the proof, see [208, Chapter 14, §3, especially Remark 14.3].) A more general formula appears in equation (8.12) below.     □

**Example 8.1.9.** Flat pullback determines an isomorphism

$$p^* : H_{2(\ell-e)}(X) \longrightarrow H_{2\ell}(\mathbf{E}),$$

and $z(C, E)$ is characterized as the unique class in $H_{2(\ell-e)}(X)$ such that $p^* z(C, E) = [C] \in H_{2\ell}(\mathbf{E})$. (See [214, §1].)     □

**Example 8.1.10.** The formation of cone classes commutes with flat pullback. More precisely, suppose that $f : Y \longrightarrow X$ is a flat morphism. A cone $C$ in $E$ gives rise to a cone $f^*C$ sitting in the total space of $f^*E$, and one has

$$f^* z(C, E) = z(f^*C, f^*E).$$

(See [208, Theorem 6.2 (b)].)     □

We will require two additional geometric constructions.

**Definition 8.1.11. (Projectivized cones).** Let $C \subseteq \mathbf{E}$ be a cone. Then $C$ determines in a natural way a subscheme

$$\mathbf{P}_{\mathrm{sub}}(C) \subseteq \mathbf{P}_{\mathrm{sub}}(E) :$$

if $C$ is defined in $\mathbf{E} = \mathrm{Spec}_{\mathcal{O}_X} \mathrm{Sym} E^*$ by a sheaf $\mathcal{J}$ of graded ideals, then $\mathbf{P}_{\mathrm{sub}}(C)$ is the subscheme of $\mathbf{P}_{\mathrm{sub}}(E) = \mathrm{Proj}_{\mathcal{O}_X} \mathrm{Sym} E^*$ defined by the sheaf on $\mathbf{P}_{\mathrm{sub}}(E)$ determined by $\mathcal{J}$.     □

**Definition 8.1.12. (Direct sum of cones).** Given cones $C_1 \subseteq \mathbf{E}_1, C_2 \subseteq \mathbf{E}_2$ lying in the total spaces of vector bundles $E_1$ and $E_2$ on $X$, the *direct sum* of $C_1$ and $C_2$ is the cone

$$C_1 \oplus C_2 =_{\mathrm{def}} C_1 \times_X C_2 \subseteq \mathbf{E}_1 \times_X \mathbf{E}_2 = \mathbf{E}_1 \oplus \mathbf{E}_2$$

sitting in the total space of the direct sum $E_1 \oplus E_2$.     □

For computational purposes, it is useful to compactify. Given a cone $C \subseteq \mathbf{E}$, consider the projective closures

$$\overline{C} =_{\mathrm{def}} \mathbf{P}_{\mathrm{sub}}(C \oplus \mathcal{O}_X) \subseteq \mathbf{P}_{\mathrm{sub}}(E \oplus \mathcal{O}_X) =_{\mathrm{def}} \overline{\mathbf{E}}, \tag{8.7}$$

with projection $\pi : \overline{\mathbf{E}} \longrightarrow X$. Geometrically, $\overline{\mathbf{E}}$ is obtained from $\mathbf{E}$ by adding a hyperplane at infinity to each fibre of $\mathbf{E}$, and $\overline{C}$ is then the closure of $C$ in $\overline{\mathbf{E}}$. Note that the zero-section $\mathbf{0}_{\mathbf{E}} \subseteq \mathbf{E} \subseteq \overline{\mathbf{E}}$ is identified with the inclusion $\mathbf{P}_{\mathrm{sub}}(\mathcal{O}_X) \subseteq \mathbf{P}_{\mathrm{sub}}(E \oplus \mathcal{O}_X)$. Denote by $Q_{\overline{\mathbf{E}}} = Q_{\mathbf{P}_{\mathrm{sub}}(E \oplus \mathcal{O}_X)}$ the universal quotient bundle of rank $e$ on $\overline{\mathbf{E}}$ sitting in the exact sequence

$$0 \longrightarrow \mathcal{O}_{\overline{\mathbf{E}}}(-1) \longrightarrow \pi^*(E \oplus \mathcal{O}_X) \longrightarrow Q_{\overline{\mathbf{E}}} \longrightarrow 0. \tag{8.8}$$

**Proposition 8.1.13. (Formula for cone classes, I).** *Assume that $C \subseteq \mathbf{E}$ has pure dimension $\ell \geq e$. Then the cone class $z(C, E)$ is given by*

$$z(C, E) = \pi_* \Big( c_e(Q_{\overline{\mathbf{E}}}) \cap [\,\overline{C}\,] \Big) \in H_{2(\ell - e)}(X).$$

*Idea of Proof.* This follows from [208, Proposition 3.3]. Geometrically, the key observation is that the composition

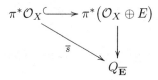

determines a section $\overline{s} \in \Gamma(\overline{\mathbf{E}}, Q_{\overline{\mathbf{E}}})$ that vanishes precisely on $\mathbf{0_E}$. So one expects the corresponding Gysin map determined by $o_E$ to be given by cap product with $c_e(Q_{\overline{\mathbf{E}}})$. □

Under an additional hypothesis, there is an even simpler formula. Specifically, consider the "small" compactification $\mathbf{P}_{\mathrm{sub}}(C) \subseteq \mathbf{P}_{\mathrm{sub}}(E)$, with

$$\rho : \mathbf{P}_{\mathrm{sub}}(E) \longrightarrow X$$

the bundle map. Denote by

$$Q_{\mathbf{P}_{\mathrm{sub}}(E)} = \rho^* E / \mathcal{O}_{\mathbf{P}_{\mathrm{sub}}(E)}(-1)$$

the rank $(e - 1)$ tautological quotient bundle on $\mathbf{P}_{\mathrm{sub}}(E)$.

**Corollary 8.1.14. (Formula for cone classes, II).** *Assume that every component of $C$ maps to its support in $X$ with positive-dimensional fibres, i.e. assume that no component of $C$ lies in the zero-section of $\mathbf{E}$. Then*

$$z(C, E) = \rho_* \Big( c_{e-1}(Q_{\mathbf{P}_{\mathrm{sub}}(E)}) \cap [\mathbf{P}_{\mathrm{sub}}(C)] \Big).$$

**Remark 8.1.15. (Cones in the zero-section).** If $C \subseteq \mathbf{E}$ is contained in the zero-section, then one may view $C$ as a complete subscheme of $X$, and

$$z(C, E) = c_e(E) \cap [C] \in H_{2(\ell - e)}(X). \quad \square$$

*Sketch of Proof of Corollary 8.1.14.* This follows formally from Proposition 8.1.13. In brief, the exact sequence (8.8) gives rise to the identity

$$c_e(Q_{\overline{\mathbf{E}}}) = \pi^* c_e(E) + c_{e-1}(Q_{\overline{\mathbf{E}}}) \cdot c_1(\mathcal{O}_{\overline{\mathbf{E}}}(1))$$

of classes on $\overline{\mathbf{E}}$. On the other hand, the hypothesis on $C$ implies that $\pi_*[\overline{C}] = 0$, whence $\pi_* \big( \pi^* c_e(E) \cap [\overline{C}] \big) = 0$. Therefore

$$z(C, E) = \pi_* \Big( c_e(Q_{\overline{\mathbf{E}}}) \cap [\,\overline{C}\,] \Big)$$
$$= \pi_* \Big( c_{e-1}(Q_{\overline{\mathbf{E}}}) \cdot c_1(\mathcal{O}_{\overline{\mathbf{E}}}(1)) \cap [\,\overline{C}\,] \Big).$$

But $\mathbf{P}_{\mathrm{sub}}(E) \subseteq \mathbf{P}_{\mathrm{sub}}(E \oplus \mathcal{O}_X) = \overline{\mathbf{E}}$ is the divisor of a section of $\mathcal{O}_{\overline{\mathbf{E}}}(1)$, so that $c_1(\mathcal{O}_{\overline{\mathbf{E}}}(1)) \cap [\overline{C}] = [\mathbf{P}_{\mathrm{sub}}(C)]$, and moreover

$$Q_{\overline{\mathbf{E}}} \mid \mathbf{P}_{\mathrm{sub}}(E) = Q_{\mathbf{P}_{\mathrm{sub}}(E)} \oplus \mathcal{O}_{\mathbf{P}_{\mathrm{sub}}(E)}.$$

The stated formula follows.    $\square$

We next discuss twisting cones by line bundles: this will guide the extension to $\mathbf{Q}$-twists in the next subsection. Let $C$ be a cone in a bundle $E$, and let $L$ be a line bundle on $X$. Then there is a naturally defined cone $C \otimes L$ sitting in the total space of $E \otimes L$. Specifically, if $C \subseteq \mathbf{E}$ is defined by the graded sheaf of ideals

$$\mathcal{J} = \oplus \mathcal{J}_m \subseteq \oplus S^m E^*,$$

then $C \otimes L$ is defined by the sheaf of ideals

$$\oplus \Big( \mathcal{J}_m \otimes L^{\otimes -m} \Big) \subseteq \oplus \Big( S^m E^* \otimes L^{\otimes -m} \Big) = \mathrm{Sym}\big( E^* \otimes L^* \big).$$

**Proposition 8.1.16. (Twisted cone classes).** *Keep the notation and assumptions of Corollary 8.1.14. Then*

$$z\big( C \otimes L, E \otimes L \big) = \rho_* \Big( c_{e-1}(Q_{\mathbf{P}_{\mathrm{sub}}(E)} \otimes \rho^* L) \cap [\mathbf{P}_{\mathrm{sub}}(C)] \Big).$$

*Proof.* There is a natural isomorphism of pairs

$$\Big( \mathbf{P}_{\mathrm{sub}}(E \otimes L), \mathbf{P}_{\mathrm{sub}}(C \otimes L) \Big) \cong \Big( \mathbf{P}_{\mathrm{sub}}(E), \mathbf{P}_{\mathrm{sub}}(C) \Big)$$

under which the bundle $Q_{\mathbf{P}_{\mathrm{sub}}(E \otimes L)}$ on $\mathbf{P}_{\mathrm{sub}}(E \otimes L)$ goes to the bundle $Q_{\mathbf{P}_{\mathrm{sub}}(E)} \otimes \rho^* L$ on $\mathbf{P}_{\mathrm{sub}}(E)$. The assertion then follows from 8.1.14.    $\square$

**Remark 8.1.17.** If $C \subseteq \mathbf{0_E}$ lies in the zero-section of $E$, then

$$z\big( C \otimes L, E \otimes L \big) = c_e\big( E \otimes L \big) \cap [C].  \ \square$$

We conclude with some additional properties and examples of cone classes.

**Example 8.1.18.** (i).   Let $E$ be a bundle of rank $e$ on $X$, and $F \subseteq E$ a sub-bundle of rank $f$. Viewing the total space of $F$ as a cone in $E$, one has

$$z(F, E) = c_{e-f}(E/F) \cap [\, X \,].$$

(ii).   If $C$ is a cone in $E$, and $F$ is any bundle, then $C \oplus F$ is a cone in $E \oplus F$, and

$$z(C \oplus F, E \oplus F) = z(C, E).$$

(iii). If $F$ is a bundle of rank $f$, and if $C$ is identified with the cone $C \oplus 0$ in $E \oplus F$, then
$$z(C, E \oplus F) = c_f(F) \cap [z(C, E)].$$

(See for example [214, §1].) □

**Remark 8.1.19. (Cone cohomology classes).** Continue to assume that $E$ is a bundle of rank $e$ on an irreducible variety $X$ of dimension $n$. In general one cannot make sense of the product of two cone classes: this reflects the fact that in general the $z(C, E)$ exist as homology rather than cohomology classes. However, there is one important case when they lift to cohomology, giving classes that can then be multiplied. Specifically, suppose that $C \subseteq \mathbf{E}$ is *flat* over $X$, of pure codimension $i$ in $\mathbf{E}$. Then $C$ determines a class

$$\mathrm{cl}(C, E) \in A^i(X)$$

in the operational Chow cohomology ring of $X$.[3] This class lifts $z(C, E)$ in the sense that
$$\mathrm{cl}(C, E) \cap [X] = z(C, E).$$

Given bundles $E_1, E_2$ on $X$ together with flat cones $C_1 \subseteq \mathbf{E}_1$, $C_2 \subseteq \mathbf{E}_2$ of codimensions $i$ and $j$, one has the basic formula :

$$\mathrm{cl}(C_1, E_1) \cdot \mathrm{cl}(C_2, E_2) = \mathrm{cl}(C_1 \oplus C_2, E_1 \oplus E_2). \tag{8.9}$$

We refer to [214, §3c] for details. An application appears in Example 8.3.12. □

### 8.1.C Cone Classes for Q-Twists

Finally, we extend the definition of cone classes to the **Q**-twisted setting. The quickest approach is to make the definitions in the first instance for irreducible cones, using linearity to extend to the general situation.

Let $E$ be a bundle of rank $e$, and $C \subseteq \mathbf{E}$ a cone of pure dimension $\ell \geq e$. Fixing a class $\delta \in N^1(X)_\mathbf{Q}$, we will define cone classes

$$z\big(C{<}\delta{>}, E{<}\delta{>}\big) \in H_{2(\ell-e)}(X).$$

**Definition 8.1.20. (Q-twisted cone classes).** Assume first that $C$ is irreducible and reduced.

---

[3] $A^*(X)$ is a cohomology-like ring that admits cup and cap products satisfying the usual formalisms (cf. [208, Chapter 17]): in particular, there is a cap product map $A^i(X) \xrightarrow{\cap[X]} H_{2(n-i)}(X)$.

(i).   If $C$ is not contained in the zero-section of $E$, define

$$z\big(C{<}\delta{>}, E{<}\delta{>}\big) \;=\; \rho_*\Big(c_{e-1}\big(Q_{\mathbf{P}_{\mathrm{sub}}(E)}{<}\rho^*\delta{>}\big) \,\cap\, [\mathbf{P}_{\mathrm{sub}}(C)]\Big),$$

where $\rho : \mathbf{P}_{\mathrm{sub}}(E) \longrightarrow X$ denotes as in Corollary 8.1.14 the bundle projection, and $Q_{\mathbf{P}_{\mathrm{sub}}(E)}$ is the tautological quotient bundle of rank $e-1$.

(ii).  If $C \subseteq \mathbf{0_E}$ is contained in the zero-section of $E$, then define

$$z\big(C{<}\delta{>}, E{<}\delta{>}\big) \;=\; c_e\big(E{<}\delta{>}\big) \cap [C].$$

(iii). If $C$ is an arbitrary (possibly reducible or non-reduced) cone of pure dimension $\ell$, use the cycle decomposition (8.5) to reduce the definition to the case of irreducible cones:

$$z\big(C{<}\delta{>}, E{<}\delta{>}\big) \;=\; \sum m_i\, z\big(C_i{<}\delta{>}, E{<}\delta{>}\big). \quad \square$$

The Chern classes appearing in (i) and (ii) are of course those associated in Section 8.1.A to $\mathbf{Q}$-twisted bundles. Observe that we are not trying to attach any formal meaning to the symbol $C{<}\delta{>}$, although of course one has 8.1.16 in mind.

It follows from the results of the previous subsection, in particular 8.1.16 and 8.1.17, that this definition is "correct" if $\delta$ is the class of an integral divisor:

**Proposition 8.1.21.** (i).   *If $D$ is an integral divisor on $X$, then*

$$z\big(C \otimes \mathcal{O}_X(D), E \otimes \mathcal{O}_X(D)\big) \;=\; z\big(C{<}D{>}, E{<}D{>}\big).$$

(ii).  *The formation of these classes commutes with flat pullback.*    $\square$

## 8.2 Positivity Theorems

In this section we prove the basic positivity theorems for ample bundles. We begin with positivity of Chern classes and then deduce the positivity of cone classes. The original arguments of [60] and [214] were reconsidered from an analytic perspective by Demailly, Peternell, and Schneider in [133], and we incorporate here several of the simplifications they introduced.

### 8.2.A Positivity of Chern Classes

We start with a theorem of Bloch and Gieseker [60] concerning the Chern classes of an ample bundle. The essential statement deals with nef bundles:

**Theorem 8.2.1. (Non-negativity of Chern classes).** *Let $X$ be an irreducible projective variety or scheme of dimension $n$, and let $E$ be a nef vector bundle on $X$. Then*

$$\int_X c_n(E) \;\geq\; 0.$$

*The same statement holds if $E$ is replaced by a nef $\mathbf{Q}$-twisted bundle $E\langle\delta\rangle$.*

**Corollary 8.2.2.** *Suppose that $E$ is ample and that $e = \operatorname{rank}(E) \geq n$. Then*

$$\int_X c_n(E) \;>\; 0,$$

*and again the analogous statement for $\mathbf{Q}$-twists is valid.*

*Proof of Corollary.* We will give the proof for a $\mathbf{Q}$-twisted bundle $E\langle\delta\rangle$. We can assume without loss of generality that $X$ is reduced, and we argue by induction on $n = \dim X$, the case $n = 0$ being trivial since $c_0(E\langle\delta\rangle) = 1$. Assuming $X$ has dimension $n \geq 1$, fix a very ample divisor $H$ on $X$ and consider for $t \in \mathbf{Q}$ the $\mathbf{Q}$-twisted bundle $E\langle\delta + tH\rangle$. Put

$$
\begin{aligned}
P(t) \;&=\; \int_X c_n(E\langle\delta + tH\rangle) \\
&=\; \int_X c_n(E\langle\delta\rangle) + t \cdot (e - n + 1)\left(\int_X c_{n-1}(E\langle\delta\rangle) \cdot H\right) \\
&\quad + (\text{higher-order terms in } t)\,.
\end{aligned}
$$

Here we are using Lemma 8.1.2 (iv) to expand $c_n(E\langle\delta + tH\rangle)$, and we view $P(t)$ in the evident way as a polynomial in $t$ with rational coefficients. Now $E\langle\delta + tH\rangle$ remains ample for small negative $t$, so by the theorem we can fix $\varepsilon > 0$ such that $P(t) \geq 0$ for $t > -\varepsilon$. On the other hand, by the induction hypothesis

$$\int_X c_{n-1}(E\langle\delta\rangle) \cdot H \;=\; \int_H c_{n-1}(E\langle\delta\rangle \,|\, H) \;>\; 0.$$

Consequently $P'(0) > 0$, i.e. $P(t)$ is increasing near $0$. Being non-negative in a neighborhood of $0$, it follows that $P(0) > 0$, as asserted.    □

We now turn to the

*Proof of Theorem 8.2.1.* The cleanest and most natural argument, following [133], makes implicit use of $\mathbf{R}$-twisted bundles, the formalism of which we leave to the interested reader. For present purposes, it is simplest to work directly with (what would be) the definitions.

Suppose then that $E\langle\delta\rangle$ is a nef $\mathbf{Q}$-twisted bundle on $X$. Aiming for a contradiction, we assume that $\int_X c_n(E\langle\delta\rangle) < 0$. Note that this is possible only if $e = \operatorname{rank}(E) \geq n$.

Observe first that we can suppose without loss of generality that $X$ is non-singular. In fact, consider a resolution (or alteration) of singularities $f : Y \longrightarrow X$.[4] Then $f^* E{<}\delta{>}$ is a nef $\mathbf{Q}$-twisted bundle on $Y$, and

$$\int_Y c_n(f^* E{<}\delta{>}) = (\deg f) \int_X c_n(E{<}\delta{>}).$$

Therefore it suffices to get a contradiction on $Y$. Replacing $X$ by $Y$, we may — and do — assume henceforth that $X$ is non-singular.

Fix next a very ample divisor $H$ on $X$. Then $\int_X c_n(E{<}\delta + tH{>}) < 0$ for small positive $t$. So there is no loss in generality in assuming — which we also do — that $E{<}\delta{>}$ is actually ample. Consider as above the polynomial

$$P(t) = \int_X c_n(E{<}\delta + tH{>}).$$

Then $P(t)$ has positive leading coefficient $(H^n)$, and $P(0) < 0$. Thus $P(t)$ has a positive real root, i.e. there is a real number $t_0 > 0$ such that

$$P(t_0) = \int_X c_n(E{<}\delta + t_0 H{>}) = 0. \tag{*}$$

As indicated at the start of the argument, the path of expediency is not to attempt to attach any meaning to $E {<} \delta + t_0 H {>}$ itself, but simply to use equation (8.1) of Definition 8.1.1 to define in the evident manner the class

$$\gamma_n = c_n(E{<}\delta + t_0 H{>}) \in H^{2n}(X; \mathbf{R}) = \mathbf{R}$$

appearing as the integrand in (*) (which we of course view as being the $n^{\text{th}}$ Chern class of the ample $\mathbf{R}$-twisted bundle in question). Now suppose that we knew that the hard Lefschetz theorem (Variant 7.1.11) of Bloch and Gieseker extends to the setting of such real Chern classes. Then multiplication by $\gamma_n$ would determine an isomorphism $H^0(X, \mathbf{C}) \longrightarrow H^{2n}(X, \mathbf{C})$, which contradicts (*). So to complete the proof, we simply repeat the argument leading to 7.1.11.

We henceforth pass to singular cohomology with complex coefficients, and assume for a contradiction that $\gamma_n = 0$. We also revert — for this paragraph only — to the projective bundle $\mathbf{P}(E)$ of one-dimensional quotients of $E$. Consider then on $\mathbf{P}(E)$ the class

$$\xi_0 = \xi_E + \pi^*(\delta + t_0 H),$$

where as usual $\xi_E = c_1(\mathcal{O}_{\mathbf{P}(E)}(1))$. As in Lemma 8.1.2 (iv), using the formula in Definition 8.1.1 to define the classes $\gamma_i = c_i(E{<}\delta + t_0 H{>})$, the Grothendieck relation

---

[4] Recall following [108] that an alteration of singularities is a generically finite proper surjective map $f : Y \longrightarrow X$ with $Y$ non-singular. De Jong shows that these exist in all characteristics.

$$\xi_0^e - \gamma_1 \cdot \xi_0^{e-1} + \ldots + (-1)^n \gamma_n \cdot \xi_0^{e-n} = 0 \qquad (**)$$

continues to hold in $H^*(\mathbf{P}(E), \mathbf{C})$ (where we recall from the beginning of the proof that $e \geq n$). Consider the non-zero cohomology class

$$\beta = \xi_0^{n-1} - \gamma_1 \cdot \xi_0^{n-2} + \ldots + (-1)^{n-1} \gamma_{n-1} \in H^{2(n-1)}(\mathbf{P}(E); \mathbf{C}).$$

It follows from the assumption $\gamma_n = 0$ and $(**)$ that

$$\beta \cdot \xi_0^{e-n+1} = 0 \in H^{2e}(\mathbf{P}(E); \mathbf{C}). \qquad (+)$$

On the other hand, $\xi_{E<\delta>} = \xi_E + \pi^* \delta$ is an ample class since $E<\delta>$ is ample, and hence $\xi_0 = \xi_{E<\delta>} + \pi^*(t_0 H)$ lies in (the interior of) the ample cone in $N^1(\mathbf{P}_{\mathrm{sub}}(E))_{\mathbf{R}}$, and hence determines a Kähler class in $H^2(\mathbf{P}(E); \mathbf{R})$. But the hard Lefschetz theorem holds for such classes (Theorem 3.1.39), and therefore the map

$$L^{e-n+1} : H^{2n-2}(\mathbf{P}(E), \mathbf{C}) \longrightarrow H^{2e}(\mathbf{P}(E), \mathbf{C})$$

given by cup product with $\xi_0^{e-n+1}$ is injective. This contradicts $(+)$.     $\square$

**Remark 8.2.3.** The reduction to non-singular varieties in the proof just completed is needed in order to apply the hard Lefschetz theorem of Bloch and Gieseker. As in [214] one can avoid this step by following the argument of [60] and appealing to the hard Lefschetz theorem for the intersection cohomology of possibly singular varieties. This approach also works in positive characteristics. It remains a tantalizing open problem to find a proof of the non-negativity of the Chern classes of a nef bundle that bypasses hard Lefschetz altogether.     $\square$

## 8.2.B Positivity of Cone Classes

The main result of the present section appeared in the paper [214] of Fulton and the author:

**Theorem 8.2.4. (Positivity of cone classes).** *Let $E$ be an ample vector bundle of rank $e$ on an irreducible projective variety or scheme $X$ of dimension $n$, and let $C \subseteq \mathbf{E}$ be a cone of pure dimension $\ell$. If $\ell = e$, then the cone class $z(C, E)$ has strictly positive degree:*

$$\int_X z(C, E) > 0.$$

*The analogous statement holds also for $\mathbf{Q}$-twists: if $E<\delta>$ is ample, then*

$$\int_X z(C<\delta>, E<\delta>) > 0.$$

**Corollary 8.2.5.** *In the situation of the theorem, assume that $\ell \geq e$, and let $L$ be an ample line bundle on $X$. Then*

$$\deg_L z(C, E) > 0,$$

*where the $L$-degree of a cycle class $\alpha$ of dimension $k$ is $\deg_L \alpha = \int \alpha \cdot c_1(L)^k$. Again the corresponding statement for **Q**-twists is valid.*

*Proof.* In fact, let $F$ be the direct sum of $\ell - e$ copies of $L$. Then

$$\deg_L z(C, E) = \int z(C, E \oplus F)$$

thanks to 8.1.18 (iii), and then the assertion follows from the theorem. This argument does not directly apply in the setting of a **Q**-twisted bundle $E{<}\delta{>}$ because we do not have a notion of direct sums. Instead, take a finite flat map $f : Y \longrightarrow X$ under which $\delta$ pulls back to an integral class (Theorem 4.1.10), and use Example 8.1.10 to reduce to the case just treated.     □

*Proof of Theorem 8.2.4.* We consider an ample **Q**-twisted bundle $E{<}\delta{>}$. We may — and do — assume that $C$ is irreducible and reduced, for the general case follows from this by linearity. If $C$ is contained in the zero section of $E$, then in view of Definition 8.1.20 (ii), the positivity of $z(C{<}\delta{>}, E{<}\delta{>})$ follows from the theorem of Bloch and Gieseker. So we may suppose that we are in the setting of 8.1.20 (i), and replacing $X$ by the support of $C$, we may assume also that $C$ surjects onto $X$: so in particular, $n = \dim X \leq \dim \mathbf{P}_{\mathrm{sub}}(C) = e - 1$. As before, we write $\rho : \mathbf{P}_{\mathrm{sub}}(E) \longrightarrow X$ for the bundle map.

Consider for the moment any **Q**-twist $E{<}\delta_1{>}$ of $E$. Fix a very ample divisor $H$ on $X$, and set

$$P_1(t) = \int_X z\big(C{<}\delta_1 + tH{>}, \; E{<}\delta_1 + tH{>}\big)$$

$$= \int_{[\mathbf{P}_{\mathrm{sub}}(C)]} c_{e-1}\big(Q_{\mathbf{P}_{\mathrm{sub}}(E)}{<}\rho^*(\delta_1 + tH){>}\big),$$

which we view in the usual way as a polynomial in $t$. The essential point is

> *Main Claim:* The polynomial $P_1(t)$ is not identically zero, and if $E{<}\delta_1{>}$ is nef then all of its coefficients are non-negative.

In fact, use Lemma 8.1.2 (iv) to expand out $P_1(t)$:

$$P_1(t) = \sum_{i=0}^{e-1} \left( \int_{[\mathbf{P}_{\mathrm{sub}}(C)]} c_{e-1-i}\big(Q_{\mathbf{P}_{\mathrm{sub}}(E)}{<}\rho^*\delta_1{>}\big) \cdot (\rho^* H)^i \right) t^i$$

$$= \sum_{i=0}^{e-1} \left( \int_{[W_i]} c_{e-1-i}\big(Q_{\mathbf{P}_{\mathrm{sub}}(E)}{<}\rho^*\delta_1{>}\big) \right) t^i, \tag{8.10}$$

where $W_i$ is an effective cycle on $\mathbf{P}_{\mathrm{sub}}(C)$ representing $\big((\rho^*H)^i\big)$. If $E{<}\delta_1{>}$ is nef, then $Q_{\mathbf{P}_{\mathrm{sub}}(E)}{<}\rho^*\delta_1{>}$, being a quotient of $\rho^*\big(E{<}\delta_1{>}\big)$, is also nef. It follows from Theorem 8.2.1 that in this case each of the coefficients appearing in (8.10) is non-negative, which proves the second assertion of the claim. To see that $P_1(t)$ is in any event non-vanishing consider the term corresponding to $i = n = \dim X$ in (8.10). In this case we can take $[W_i]$ to be a positive linear combination of fibres of $\rho \mid \mathbf{P}_{\mathrm{sub}}(C) : \mathbf{P}_{\mathrm{sub}}(C) \longrightarrow X$. But on a general fibre $F$ of this map one has

$$\int_F c_{e-1-n}\big(Q_{\mathbf{P}_{\mathrm{sub}}(E)}{<}\rho^*\delta_1{>}\big) \;=\; \int_F c_{e-1-n}\big(Q_{\mathbf{P}_{\mathrm{sub}}(E)}\big)$$
$$> 0$$

thanks to the fact that on a fixed projective space $\mathbf{P}_{\mathrm{sub}}(W)$, the Chern classes of the tautological quotient bundle $Q_{\mathbf{P}_{\mathrm{sub}}(W)}$ are represented by linear spaces.

Returning to the ample $\mathbf{Q}$-twisted bundle $E{<}\delta{>}$, the theorem now follows at once. In fact, choose $\varepsilon > 0$ such that $E{<}\delta - \varepsilon H{>}$ remains ample. Applying the claim with $\delta_1 = \delta - \varepsilon H$ we conclude that the polynomial

$$P(t) \;=_{\mathrm{def}}\; \int_X z\big(C{<}\delta + (t - \varepsilon)H{>}\,,\; E{<}\delta + (t - \varepsilon)H{>}\big)$$

has non-negative coefficients and does not vanish identically. But this implies that $P(\varepsilon) > 0$. $\qquad\square$

**Example 8.2.6. (Nef cone classes).** Theorem 8.2.4 and Corollary 8.2.5 extend in the natural way to nef bundles. Specifically, let $C \subseteq \mathbf{E}$ be a cone of pure dimension $\ell \geq e$, and assume that $E{<}\delta{>}$ is nef. Then

$$\deg_L z\big(C{<}\delta{>}\,,\; E{<}\delta{>}\big) \;\geq\; 0$$

for any ample $L$. (Fix an ample class $h \in N^1(X)$ and consider the polynomial

$$P(t) \;=\; \deg_L z\big(C{<}\delta + t \cdot h{>}\,,\; E{<}\delta + t \cdot h{>}\big).$$

It follows from 8.2.5 that $P(t) > 0$ for $t > 0$ and hence $P(0) \geq 0$, as claimed. When $\ell = e$ one can alternatively argue directly from Theorem 8.2.1 by observing (as we did during the proof of 8.2.4) that $Q_{\mathbf{P}_{\mathrm{sub}}(E)}{<}\rho^*\delta{>}$, being a quotient of $\rho^*E{<}\delta{>}$, is nef.) $\qquad\square$

**Example 8.2.7. (A bound for cone classes).** With notation as before, let $C \subseteq \mathbf{E}$ be an irreducible cone of pure dimension $\ell \geq e$, and denote by $\Gamma \subseteq X$ the support of $C$ (considered as a reduced subscheme of $X$). Let $H$ be an ample divisor on $X$, and $\delta > 0$ a rational number such that $E{<}-\delta \cdot H{>}$ is nef. Then

$$\deg_H z(C, E) \;\geq\; \delta^{\dim \Gamma - (\ell - e)} \cdot \deg_H \Gamma. \tag{8.11}$$

Here as in 8.2.5 $\deg_H z(C, E)$ denotes the $H$-degree of the indicated cone class, and $\deg_H \Gamma$ denotes the degree $\big(H^{\dim \Gamma} \cdot \Gamma\big)$ of $\Gamma$ with respect to $L$. The inequality (8.11) holds in particular if $\delta = \delta(X, E, H)$ is the Barton invariant (Example 6.2.14) of $E$ with respect to $H$. (Consider first the case $\ell = e$. It is enough to prove that if $\delta_1 \in N^1(X)_{\mathbf{Q}}$ is any class such that $E<\delta_1>$ is nef, then

$$\int_X z\Big(C<\delta_1 + tH>, E<\delta_1 + tH>\Big) \geq t^{\dim \Gamma} \cdot \deg_H \Gamma \qquad (*)$$

when $t \geq 0$: taking $\delta_1 = -\delta H$ and $t = \delta$ yields (8.11). For (*) one can suppose that $X = \Gamma$, and then the argument used to prove 8.2.4 — in particular the analysis of the polynomial $P_1(t)$ via (8.10) — applies. When $\ell \geq e$ one can assume after passing to a covering that $\delta H \equiv_{\text{num}} D$ for some integral divisor $D$. Then as in the proof of 8.2.5 one applies the case already treated to $C$ considered as a cone in $E \oplus \mathcal{O}_X(D)^{\oplus (\ell - e)}$.)                     $\square$

## 8.3 Positive Polynomials for Ample Bundles

Still following the paper [214] of Fulton and the author, we now apply the positivity theorem for cone classes to determine all positive polynomials for ample vector bundles.

Consider the ring $\mathbf{Q}[c_1, \ldots, c_e]$, graded by setting $\deg c_i = i$, and fix a weighted homogeneous polynomial $P \in \mathbf{Q}[c_1, \ldots, c_e]$ of degree $n$. If $X$ is an irreducible projective variety of dimension $n$, and $E$ is a vector bundle of rank $e$ on $X$, then $P$ determines a Chern number

$$\int_X P\big(c(E)\big) =_{\text{def}} \int_X P\big(c_1(E), \ldots, c_e(E)\big) \in \mathbf{Q}.$$

We are interested in those Chern numbers that are universally positive for ample vector bundles:

**Definition 8.3.1. (Positive polynomials).** A weighted homogeneous polynomial $P \in \mathbf{Q}[c_1, \ldots, c_e]$ as above is *numerically positive for ample vector bundles of rank $e$* if

$$\int_X P(c(E)) > 0$$

for every ample vector bundle $E$ of rank $e$ on every irreducible projective variety $X$ of dimension $n$.                     $\square$

**Example 8.3.2.** The determinant of an ample vector bundle is an ample line bundle, and therefore the polynomial $P = c_1^n$ is numerically positive for ample bundles of any rank $e \geq 1$.                     $\square$

**Example 8.3.3. (Q-twists).** Note that if $E<\delta>$ is a $\mathbf{Q}$-twisted bundle of rank $e$, then a polynomial $P$ as above again determines a Chern number $\int_X P\big(c(E<\delta>)\big) \in \mathbf{Q}$. Allowing $\mathbf{Q}$-twisted bundles does not change the class of positive polynomials. Specifically, $P$ is numerically positive in the sense of the previous definition if and only if $\int_X P\big(c(E{<}\delta{>})\big) > 0$ for all ample $\mathbf{Q}$-twisted bundles $E{<}\delta{>}$. (Given $E{<}\delta{>}$ pass to a covering to reduce to the case when $\delta$ is integral.) $\qquad\square$

Our main object in this section is to determine the cone of all such positive polynomials. In order to state the answer, we recall some combinatorial definitions. To begin with, denote by $\Lambda(n,e)$ the set of all partitions of $n$ by non-negative integers $\le e$. So an element $\lambda \in \Lambda(n,e)$ is specified by a sequence

$$e \ge \lambda_1 \ge \lambda_2 \ge \ldots \ge \lambda_n \ge 0, \quad \text{with} \quad \sum \lambda_i = n.$$

Associated to each such $\lambda \in \Lambda(n,e)$ one has the *Schur polynomial* $s_\lambda \in \mathbf{Q}[c_1,\ldots,c_e]$ of weighted degree $n$. This is defined as the determinant of the $n \times n$ matrix

$$s_\lambda = \begin{vmatrix} c_{\lambda_1} & c_{\lambda_1+1} & \cdots & c_{\lambda_1+n-1} \\ c_{\lambda_2-1} & c_{\lambda_2} & \cdots & c_{\lambda_2+n-2} \\ \cdots & \cdots & \cdots & \cdots \\ c_{\lambda_n-n+1} & c_{\lambda_n-n+2} & \cdots & c_{\lambda_n} \end{vmatrix},$$

where by convention $c_0 = 1$ and $c_i = 0$ if $i \notin [0,e]$. These polynomials form a basis for the vector space of all homogeneous polynomials of weighted degree $n$ in $e$ variables. See [210, Appendix A], or [208] for an overview of some of the basic facts.

**Example 8.3.4.** In low degrees, the Schur polynomials are

$$s_{(1)} = c_1$$

$$s_{(2,0)} = c_2 \;\;,\;\; s_{(1,1)} = c_1^2 - c_2$$

$$s_{(3,0,0)} = c_3 \;\;,\;\; s_{(2,1,0)} = c_1 c_2 - c_3 \;\;,\;\; s_{(1,1,1)} = c_1^3 - 2c_1 c_2 + c_3. \quad \square$$

**Example 8.3.5.** If $\lambda = 1^k = (1,\ldots,1)$ ($k$ repetitions of 1), then

$$s_\lambda(E) \;=\; s_k(E^*)$$

is the $k^{\text{th}}$ Segre (or inverse Chern) class of $E^*$. $\qquad\square$

**Remark 8.3.6. (Schur polynomials and Grassmannians).** As explained e.g. in [248, Chapter I, §5], [208, Chapter 14, §6] or [207, Chapter 9], Schur polynomials arise geometrically in describing the cohomology or Chow rings of Grassmannians. Let $V$ be a vector space of dimension $m \ge e + n$, let $\mathbf{G} = \mathbf{G}(V,e)$ be the Grassmannian of $e$-dimensional quotients of $V$, and denote by $Q$ the tautological rank-$e$ quotient bundle on $\mathbf{G}$. Then the classes $s_\lambda(c(Q))$ ($\lambda \in \lambda(n,e)$) are linearly independent and span the cone of effective codimension-$n$ cycles on $\mathbf{G}$. More specifically, fix a complete flag

$$0 = V_0 \subseteq V_1 \subseteq V_2 \subseteq \ldots \subseteq V_m = V,$$

and view elements of $\mathbf{G}$ as $(m-e)$-dimensional subspaces $W \subset V$ of $V$. Given a partition $\lambda \in \Lambda(n,e)$, consider the Schubert variety $\Omega_\lambda \subset \mathbf{G}$ defined by

$$\Omega_\lambda = \big\{ W_{m-e} \subset V \mid \dim \big( W \cap V_{e+i-\lambda_i} \big) \geq i \quad \text{for all } i \big\}.$$

Then $\Omega_\lambda$ is an irreducible subvariety of codimension $n$ whose class is represented by $s_\lambda(Q)$. $\qquad \square$

**Example 8.3.7. (Schur polynomials and degeneracy loci).** A variant of Remark 8.3.6 allows one to realize Schur polynomials as degeneracy loci. Let $E$ be a globally generated bundle of rank $e$ on an irreducible projective variety $X$ of dimension $n$. Fix a vector space $V \subset H^0(X, E)$ of sections spanning $E$, with $\dim V \geq n + e$, and let

$$u : V_X \twoheadrightarrow E$$

be the corresponding evaluation map. Given a partition $\lambda \in \Lambda(n, e)$, fix a general flag

$$A_\bullet : \quad 0 \subseteq A_1 \subseteq A_2 \subseteq \ldots \subseteq A_n \subseteq V$$

with $\dim A_i = e + i - \lambda_i$. We associate to these data the degeneracy locus

$$D_\lambda(u) = D_{A_\bullet}(u) = \big\{ x \in X \mid \dim \big( \ker u(x) \cap A_i \big) \geq i \quad \text{for all } i \big\}.$$

Then

$$[D_{A_\bullet}(u)] = s_\lambda \big( c(E) \big) \cap [X]$$

(where as usual the left-hand side in general is to be understood as a suitable positive cycle supported on $D_{A_\bullet}(u)$ as per [208, Chapter 14, §3] or [215, Appendix A]). $\qquad \square$

**Remark 8.3.8.** The connection between degeneracy loci and symmetric polynomials is explored in detail in the notes [215] of Fulton and Pragacz. In particular, Section 1.1 of that book contains a very nice sketch of the early history of these connections. $\qquad \square$

Suppose now that $P \in \mathbf{Q}[c_1, \ldots, c_e]$ is any homogeneous polynomial of weighted degree $n$. Since the Schur polynomials span the space of all such, we can uniquely write $P$ as a linear combination of the $s_\lambda$:

$$P = \sum_{\lambda \in \Lambda(n,e)} a_\lambda(P) \cdot s_\lambda.$$

The main result is:

**Theorem 8.3.9. (Cone of positive polynomials).** *A weighted homogeneous polynomial $P$ is numerically positive for ample vector bundles of rank $e$ if and only if*

$$P \neq 0 \quad and \quad a_\lambda(P) \geq 0 \quad for \ all \ \lambda \in \Lambda(n, e).$$

*Equivalently, the Schur polynomials $s_\lambda$ ($\lambda \in \Lambda(n, e)$) span the cone of numerically positive polynomials.*

*Proof.* The first step is to establish the numerical positivity of the $s_\lambda$. We do this by realizing them as cone classes, generalizing the approach of Example 8.1.8. Specifically, suppose given a partition $\lambda \in \Lambda(n, e)$ and an ample vector bundle $E$ of rank $e$ on an $n$-dimensional projective variety $X$. Fix first a vector space $V$ of dimension $n + e$, and a flag $A_\bullet$ of subspaces:

$$A_\bullet : \quad 0 \subseteq A_1 \subseteq A_2 \subseteq \ldots \subseteq A_n \subseteq V$$

with $\dim A_i = e + i - \lambda_i$. Now consider the total space

$$p : \mathbf{H} = \mathbf{Hom}(V_X, E) \longrightarrow X$$

of the bundle $H = \mathrm{Hom}(V_X, E)$ of homomorphisms from the trivial bundle modeled on $V$ to $E$. Inside $\mathbf{H}$ there is a naturally defined cone $\Omega_\lambda = \Omega_{A_\bullet} \subset \mathbf{H}$ whose fibre over a point $x \in X$ consists of all homomorphisms $u \in \mathrm{Hom}(V, E(x))$ such that

$$\dim \big( \ker(u) \cap A_i \big) \geq i$$

for every $i$. The basic point is that

$$z(\Omega_{A_\bullet}, H) = s_\lambda\big(c(E)\big). \tag{8.12}$$

This follows informally in the spirit of 8.1.8 from Example 8.3.7; a formal proof appears in [208, 14.3]. But $H$ is ample if $E$ is, so the positivity of $\int_X s_\lambda\big(c(E)\big)$ follows from the positivity of cone classes (Theorem 8.2.4).

Conversely, suppose given a weighted homogeneous polynomial

$$P = \sum a_\lambda s_\lambda$$

of degree $n$ with $a_\mu < 0$ for some $\mu \in \Lambda(n, e)$. We need to produce an irreducible projective variety $X$ of dimension $n$, carrying an ample bundle $E$ of rank $e$, such that $\int_X P\big(c(E)\big) < 0$. As noted in Example 8.3.3, it suffices in fact to produce an ample **Q**-twisted bundle with the analogous property, for then one can pass to a covering to obtain an example involving a "genuine" bundle $E$.

To this end, start with the Grassmannian $\mathbf{G} = G(V, e)$ of $e$-dimensional quotients of a vector space $V$ of dimension $n + e$. Letting $Q$ denote the rank-$e$ tautological quotient bundle on $\mathbf{G}$, the classes $s_\lambda\big(c(Q)\big)$ are represented by irreducible Schubert varieties $\Omega_\lambda \subset \mathbf{G}$ of codimension $n$ determined as in Remark 8.3.6. Fixing $\mu$, recall that one can construct another Schubert variety dual to $\Omega_\mu$, i.e. there is an irreducible Schubert variety $X \subset \mathbf{G}$ of dimension $n$ having the property that

$$[X] \cdot [\Omega_\lambda] = \delta_{\mu\lambda}$$

for every $\lambda$. Setting $E = Q \,|\, X$ one therefore has

$$\int_X P\big(c(E)\big) = a_\mu < 0.$$

This does not quite give the required example because $E$ — although globally generated, and hence nef — may not be ample. However, if $H$ is an ample divisor on $X$, then for a sufficiently small rational number $0 < \varepsilon \ll 1$, the $\mathbf{Q}$-twisted ample bundle $E\!<\!\varepsilon H\!>$ will still satisfy $\int_X P\big(c(E\!<\!\varepsilon H\!>)\big) < 0$, which shows that $P$ is not numerically positive. Finally, note that by passing to a resolution or alteration of $X$, there is no difficulty in constructing an example with $X$ non-singular.    □

**Example 8.3.10. (Non-negative polynomials).** A homogeneous polynomial $P \in \mathbf{Q}[c_1, \ldots, c_e]$ of weighted degree $n$ is non-negative for nef bundles (in the evident sense) if and only if $P = \sum a_\lambda s_\lambda$ with all $a_\lambda \geq 0$.    □

**Example 8.3.11. ($\mathbf{c_1^2 - 2c_2}$).** The weight two polynomial

$$P(c_1, c_2) = c_1^2 - 2c_2$$

appeared as an example several times in the early literature on the subject. Note that $P$ is not a positive combination of the two Schur polynomials $s_{(2,0)} = c_2$ and $s_{(1,1)} = c_1^2 - c_2$, and so $P$ does not lie in the positive cone. In fact, $P$ is negative already for the nef bundle $T\mathbf{P}^2(-1)$ on $\mathbf{P}^2$. (Compare [248, p. 418].)    □

**Example 8.3.12. (Products of positive polynomials).** Note that the cones $\Omega_\lambda$ used to establish the positivity of the Schur polynomials are flat over $X$. Therefore we are in the situation of Remark 8.1.19, and can apply the formula (8.9) appearing there. One deduces, for instance, that if $E_1$ and $E_2$ are possibly different ample bundles, and if $P_1$ and $P_2$ are positive polynomials of degrees adding to $n = \dim X$, then

$$\int_X P_1\big(c(E_1)\big) \cdot P_2\big(c(E_2)\big) > 0. \quad □$$

**Example 8.3.13. (A theorem of Pragacz).** Let $E$ be a bundle of rank $e$ on an $n$-dimensional projective variety $X$, and $\mu$ an $e$-rowed Young tableau indexing an irreducible representation of $\mathrm{GL}(e, \mathbf{C})$, say of dimension $d(\mu)$. Denote by $\Gamma^\mu E$ the corresponding associated bundle built from $E$. If $\nu \in \Lambda(n, d(\mu))$ then one can express universally the Chern number $s_\nu\big(c(\Gamma^\mu E)\big)$ in terms of the Schur polynomials for $E$, i.e.

$$s_\nu\big(c(\Gamma^\mu E)\big) = \sum a_\lambda \cdot s_\lambda(c(E)), \qquad (*)$$

where the $a_\lambda$ depend only on $\mu$ and $\nu$. Pragacz [506, Corollary 7.2] shows that all of the $a_\lambda$ are non-negative. (In fact, since $\Gamma^\mu E$ is ample if $E$ is, the left-hand side of (*) is a positive polynomial.)    □

**Remark 8.3.14. (Numerical criteria for amplitude).** In the early days of the theory, it was hoped that one could find a numerical criterion for amplitude analogous to the Nakai criteria for line bundles or Hartshorne's result (Theorem 6.4.15) characterizing ample bundles on curves. More precisely, the hope was that one could find collections $\mathcal{P} = \mathcal{P}(n, e)$ of Chern polynomials so that the amplitude of a bundle $E$ of rank $e$ on a projective $n$-fold $X$ would be characterized by the following property:

For every subvariety $V \subseteq X$ of dimension $d > 0$, and every quotient $Q$ of $E \,|\, V$ of rank $f$,

$$\int_V P(c(Q)) \; > \; 0 \quad \text{for all } \; P \in \mathcal{P}(d, f). \tag{$*$}$$

However, Fulton [201] gave an example to show that no such criterion can exist. Specifically, he constructs a non-ample vector bundle $E$ on $\mathbf{P}^2$ for which ($*$) holds with $\mathcal{P}$ the collection of all positive polynomials in the appropriate rank and dimension. In his example, there is a singular curve $C \subseteq \mathbf{P}^2$ for which the restriction $E \,|\, C$ is a counterexample to Hartshorne's criterion for amplitude of bundles on a smooth curve (Example 6.4.18). $\quad\square$

**Example 8.3.15. (Positive polynomials for very ample bundles).** Motivated by some results and conjectures of Ballico ([23], [24]), Beltrametti, Schneider, and Sommese [49] observed that one can obtain much stronger inequalities for *very* ample bundles. Specifically, suppose that $E$ is a vector bundle of rank $e$ on an irreducible $n$-dimensional projective variety $X$ with the property that $\mathcal{O}_{\mathbf{P}(E)}(1)$ is very ample on the projectivized bundle $\mathbf{P}(E)$, and let $P \in \mathbf{Q}[c_1, \ldots, c_r]$ be any homogeneous positive polynomial of weighted degree $n$. Then

$$\int_X P(c_1(E), \ldots, c_e(E)) \; \geq \; P\left(\binom{e}{1}, \ldots, \binom{e}{n}\right). \tag{8.13}$$

For example, if $0 < i_1, \ldots, i_k \leq e$ are positive integers such that $i_1 + \cdots + i_k = n$, then

$$\int_X c_{i_1}(E) \cdot \ldots \cdot c_{i_k}(E) \; \geq \; \binom{e}{i_1} \cdot \ldots \cdot \binom{e}{i_k}.$$

Similarly, denoting by $s_n(E)$ the top Segre class of $E^*$, one has

$$\int_X s_n(E^*) \; \geq \; \binom{e + n - 1}{n}$$

(cf. [49, p. 103]). (It is enough to prove (8.13) when $P = s_\lambda$ is a Schur polynomial. To this end, fix any smooth point $x \in X$, and let $\mu : Y = \mathrm{Bl}_x(X) \longrightarrow X$ be the blowing up of $X$ at $x$, with exceptional divisor $f \subseteq Y$. Set $E' = \mu^* E$. The hypothesis on $\mathcal{O}_{\mathbf{P}(E)}(1)$ implies that $E'(-f)$ is globally generated, and hence

$$\int_Y s_\lambda\Big(c_1(E'(-f)),\ldots,c_e(E'(-f))\Big) \; \geq \; 0.$$

But for $i, j > 0$,

$$\big(c_i(E') \cdot f^j\big) \; = \; \big(\mu^* c_i(E) \cdot f^j\big) \; = \; 0, \tag{8.14}$$

and therefore

$$c_k\big(E'(-f)\big) \; = \; c_k(E') + (-1)^k \binom{e}{k}(f^k)$$

for every $k \geq 0$. The required inequality (8.13) follows by inserting this into the determinant defining $s_\lambda$ and using again (8.14).) Note that this argument shows that (8.13) holds as soon as there is a single smooth point $x \in X$ such that $E \otimes \mathcal{I}_x^2$ is globally generated, $\mathcal{I}_x \subset \mathcal{O}_X$ denoting the ideal sheaf of $x \in X$. □

**Example 8.3.16. (Counter-examples).** The inequalities of Beltrametti, Schneider, and Sommese discussed in the previous example can fail if $E$ is merely ample or even ample and globally generated. For instance, let $C$ be a smooth curve of genus $g > 2$, and for $d \geq 2g - 1$ denote by $E = E_d$ the rank $d + 1 - g$ Picard bundle on $\mathrm{Pic}^d(C)$ discussed in Section 6.3.C. Thus $E$ is negative, i.e. $E^*$ is ample, but

$$\int_{\mathrm{Pic}^d(C)} c_g(E^*) = 1.$$

(In fact, it follows from the proof of Theorem 7.2.12 that $c_g(E) = [\, W_0(C)\, ]$.) It was conjectured in [49] that the inequalities in question are valid as soon as $E$ is ample and globally generated, but Hacon constructed counter-examples in [260]. □

**Remark 8.3.17. (Positive polynomials for filtered bundles).** In his paper [206], Fulton generalized Theorem 8.3.9 to the setting of filtered bundles. Fix a positive integer $e > 0$ and a sequence $\mathbf{r}$ of natural numbers

$$0 = r_0 \; < \; r_1 \; < \; \ldots \; < \; r_k = e.$$

An $\mathbf{r}$-filtered bundle on a variety $X$ is by definition a vector bundle $E$ of rank $e$ on $X$ together with a filtration by subbundles

$$E = E_0 \supseteq E_1 \supseteq \ldots \supseteq E_{k-1} \supseteq E_k = 0$$

such that each of the quotients $F_j = E/E_j$ has rank $r_j$. In this situation, one is interested in polynomials that can be expressed in terms of the Chern classes of the graded pieces $E_j/E_{j+1}$. To this end, denote by $\mathcal{P}(r, n)$ the rational vector space of all homogeneous polynomials of degree $n$ in variables $x_1, \ldots, x_e$ that are symmetric in each of the sets of variables $\{x_1, \ldots, x_{r_1}\}$, $\{x_{r_1+1}, \ldots, x_{r_2}\}, \ldots, \{x_{r_{k-1}+1}, \ldots, x_e\}$. By substituting the Chern roots of $E$

for the $x_i$, labeled in the natural way so that $x_{r_j+1}, \ldots, x_{r_{j+1}}$ correspond to the Chern roots of $E_j/E_{j+1}$, a polynomial $P \in \mathcal{P}(r,n)$ determines a class

$$P(E) \in H^{2n}(X, \mathbf{Q}).$$

The space $\mathcal{P}(r,n)$ has a natural basis consisting of so-called Schubert polynomials $S_w$ indexed by certain sequences $w = \{w(1), \ldots, w(n)\}$.

One says that $P \in \mathcal{P}(r,n)$ is positive for ample filtered bundles if $\int_X P(E) > 0$ whenever $E$ is an ample bundle with an **r**-filtration on an $n$-dimensional projective variety $X$. The main result of [206] is that a non-zero polynomial $P \in \mathcal{P}(r,n)$ is positive if and only if $P \neq 0$ and

$$P = \sum a_w \cdot S_w \quad \text{with all} \quad a_w \geq 0.$$

The proof is similar to that given above, the main additional ingredient being a formula from [204] interpreting the Schubert polynomials via degeneracy loci. □

**Example 8.3.18. (Vector bundles of small rank on special varieties).** A. Holme has observed that in certain situations one can use Theorem 8.3.9 to deduce the positivity of polynomials that do not in general lie in the positive cone. For example, suppose that $E$ is an ample vector bundle of rank 2 on $\mathbf{P}^n$ for some $n \geq 3$. We view the Chern classes $c_i = c_i(E)$ of $E$ as integers in the natural way. Since $c_3 = 0$, the positivity of the Segre class $s_3(E^*) = s_{(1,1,1)}(E)$ yields $c_1(c_1^2 - 2c_2) > 0$, and since $c_1 > 0$ this implies that

$$c_1^2 - 2c_2 > 0.$$

However as observed in 8.3.11, $c_1^2 - 2c_2$ is not a positive linear combination of Schur polynomials. Note that this argument works on any variety $X$ of dimension $\geq 3$ with the property that the Hodge groups $H^{p,p}(X, \mathbf{Q}) = H^{2p}(X, \mathbf{Q}) \cap H^{p,p}(X, \mathbf{C})$ are one-dimensional for $p \leq 3$, so that all the Chern classes in question are multiples of a generator of $H^{1,1}(X, \mathbf{Q})$ and its powers. □

**Remark 8.3.19. (Positive polynomials for k-ample bundles).** Tin [571] has extended Theorem 8.3.9 to the $k$-ample setting. Specifically, denote by $\Pi_{e,k,n}(m)$ the cone of all weighted homogeneous polynomials $P \in \mathbf{Q}[c_1, \ldots, c_e]$ of degree $m$ such that

$$\int_X c_1(L)^{\dim X - m} P(c(E)) > 0$$

for every ample line bundle $L$ and $k$-ample vector bundle $E$ of rank $e$ on every irreducible projective variety of dimension $n$. Tin shows that $\Pi_{e,k,n}(m) = \varnothing$ when $n < m + k$, whereas $\Pi_{e,k,n}(m)$ is spanned by the Schur polynomials $s_\lambda$ with $\lambda \in \Lambda(m, e)$ when $n \geq m + k$. □

## 8.4 Some Applications

This section is devoted to some applications of the positivity theorems. We start by establishing the positivity of intersection numbers in the presence of ample normal bundles. In 8.4.B we return to degeneracy loci, and extend the non-emptiness statements in Theorem 7.2.1 to symmetric and skew-symmetric homomorphisms. Finally, in the third subsection we derive some inequalities on the number of singular points that can appear on a hypersurface containing a smooth curve.

### 8.4.A Positivity of Intersection Products

Closely following the note [213], of Fulton and the author, we apply the positivity theorem for cone classes to study intersection products.

Let $X$ be a projective variety sitting as a local complete intersection of codimension $e$ inside some variety $M$ (which need not be smooth or complete), and let $Y \subseteq M$ be a subvariety (or subscheme) of pure dimension $\ell \geq e$. Then one has the Cartesian diagram

$$
\begin{array}{ccc}
X \cap Y & \subseteq & Y \\
\cap & & \cap \\
X & \subseteq & M \; .
\end{array}
\qquad (8.15)
$$

In this situation, the intersection cycle

$$
\big( X \cdot Y \big) \;\in\; H_{2(\ell-e)}(X)
$$

is defined.[5] If $X$ and $Y$ meet properly in $M$ — i.e. if $X \cap Y$ has pure dimension $(\ell - e)$ — then $\big( X \cdot Y \big)$ is represented by a non-negative linear combination of the irreducible components of $X \cap Y$. In particular,

$$
\deg_H \big( X \cdot Y \big) \;\geq\; 0
\qquad (*)
$$

for any ample line divisor $H$ on $X$. On the other hand, if $X$ and $Y$ meet in larger than expected dimension then of course (*) can fail.

The main result of the present section shows that hypotheses of amplitude on the embedding of $X$ in $M$ imply the positivity of intersection classes even in the case of excess intersection.

**Theorem 8.4.1. (Positivity of intersection products).** *Let $N = N_{X/M}$ be the normal bundle to $X$ in $M$, and fix an ample divisor $H$ on $X$.*

---

[5] If $\iota : X \longrightarrow M$ is the inclusion, then $\big( X \cdot Y \big) = \iota^!([Y])$, where

$$
\iota^! : H_{2\ell}(Y) \longrightarrow H_{2(\ell-e)}(X)
$$

is the Gysin map determined by $\iota$ (Proposition 8.1.6).

(i). *If $N$ is nef, then $\deg_H (X \cdot Y) \geq 0$.*

(ii). *If $N$ is ample and $X \cap Y$ is non-empty, then $\deg_H (X \cdot Y) > 0$. In fact, if $\delta \leq 1$ is any positive rational number such that $N < -\delta \cdot H >$ is nef, then*

$$\deg_H (X \cdot Y) \geq \delta^{\dim X - (\ell - e)} \cdot \deg_H (X \cap Y). \tag{8.16}$$

*Here $\deg_H (X \cap Y)$ denotes the sum of the $H$-degrees of the irreducible components of $X \cap Y$, taken with their reduced scheme structures.*

**Remark 8.4.2.** The hypothesis $\delta \leq 1$ in (ii) is not seriously restrictive. For suppose that $N < -\delta_1 H_1 >$ is nef for some ample divisor $H_1$ and rational number $\delta_1 > 1$. Then choose $p \geq \delta_1$ and apply the theorem with $H = pH_1$ and $\delta = \frac{\delta_1}{p}$.     □

We will prove the theorem shortly, but first we record some consequences.

**Corollary 8.4.3. (Numerically effective subvarieties).** *Given $X \subseteq M$ as in (8.15), assume that $N = N_{X/M}$ is nef. Then $X$ is numerically effective in the sense that*

$$\int_X (X \cdot Y) \geq 0$$

*for every subvariety $Y \subseteq M$ of pure dimension $e$ (= $\operatorname{codim} X$). If moreover $N$ is ample then strict inequality holds provided that $[Y]$ is homologous to an effective algebraic cycle whose support meets $X$.*     □

**Corollary 8.4.4. (Manifolds with nef tangent bundles).** *Let $X$ be a smooth projective variety whose tangent bundle $TX$ is nef. Then*

$$\int_X (Z \cdot W) \geq 0$$

*for any two subschemes of complementary dimension. In particular, any effective divisor $D$ on $X$ is nef.*

The nefness of any effective divisor on $X$ was proved by Demailly [122] by a different argument.

*Proof of Corollary 8.4.4.* For the first statement apply 8.4.3 with $X \subseteq X \times X = M$ the diagonal embedding and $Y = Z \times W$. As a special case, if $D$ is an effective divisor on $X$, then $(D \cdot C) \geq 0$ for every curve $C \subseteq X$.     □

**Example 8.4.5. (Intersections in projective space).** Let $V_1, \ldots, V_r \subseteq \mathbf{P}^n$ be subvarieties of degrees $d_1, \ldots, d_r$. Then

$$\deg (V_1 \cap \ldots \cap V_r) \leq d_1 \cdot \ldots \cdot d_r.$$

(By passing to a larger projective space and cones over the $V_i$ one can suppose that $\sum \dim V_i \geq (r-1)n$. Then consider as above the diagonal embedding

$$\mathbf{P}^n \subseteq \mathbf{P}^n \times \ldots \times \mathbf{P}^n \quad (r \text{ times})$$

and apply 8.4.1 (ii) with $H$ a hyperplane and $\delta = 1$.) This statement was originally established by Fulton and MacPherson by a slightly different argument.    □

**Example 8.4.6. (Subvarieties of homogeneous spaces).** With $X \subseteq M$ as in (8.15), suppose that $M$ is acted on transitively by a connected algebraic group. Then $X$ meets any subvariety $Y \subseteq X$ of dimension $\geq \operatorname{codim} X$. (The homogeneity of $M$ implies that $Y$ is algebraically equivalent to a subvariety $Y'$ that meets $X$, and the second statement of 8.4.3 applies.) This simplifies and extends somewhat a result of Lübke [409].    □

It remains only to give the

*Proof of Theorem 8.4.1.* The theorem follows directly from the results of Section 8.2.B together with the intersection theory of Fulton–MacPherson. Specifically, consider the normal cone $C = C_{(X \cap Y)/Y}$ of $X \cap Y$ in $Y$. Then $C$ has pure dimension $\ell$, and it embeds naturally into the total space $\mathbf{N}$ of the normal bundle $N$. The basic fact ([208, Chapter 6.1]) is that

$$(X \cdot Y) = z(C, N).$$

Statement (i) and the first assertion in (ii) then follow immediately from Example 8.2.6 and Corollary 8.2.5. For (8.16), consider the $\ell$-cycle

$$[C] = \sum m_i [C_i]$$

determined by the irreducible components of $C$. The linearity of cone classes (8.6) yields

$$(X \cdot Y) = \sum m_i \cdot z(C_i, N).$$

On the other hand, denote by $\Gamma_i \subseteq X$ the support of $C_i$ (with its reduced scheme structure). Then 8.2.7 gives

$$\deg_H z(C_i, E) \geq \delta^{\dim \Gamma_i - (\ell - e)} \cdot \deg_H \Gamma_i \geq \delta^{\dim X - (\ell - e)} \cdot \deg_H \Gamma_i,$$

the second inequality coming from the hypothesis that $\delta \leq 1$. The theorem then follows from the observation that each irreducible component of $X \cap Y$ occurs as one of the $\Gamma_i$.    □

### 8.4.B Non-Emptiness of Degeneracy Loci

We next revisit from Section 7.2 the degeneracy loci of vector bundle maps, and use the positivity theorem 8.2.4 to prove the non-emptiness of these loci under suitable positivity conditions. Our discussion closely follows [214, §3b].

Let $X$ be an irreducible projective variety of dimension $n$, and let $E$ and $F$ be vector bundles on $X$ of ranks $e$ and $f$ respectively. There are three natural sorts of bundle morphisms to consider:

(A). $u : E \longrightarrow F$, $u$ arbitrary;

(B). $u : E^* \longrightarrow E \otimes L$, $L$ a line bundle, $u$ symmetric;

(C). $u : E^* \longrightarrow E \otimes L$, $L$ a line bundle, $u$ skew.

Recall that in each case one is concerned with the degeneracy loci

$$D_k(u) = \{\, x \in X \mid \operatorname{rank} u(x) \le k \,\}.$$

(Since a skew map always has even rank, we limit ourselves in (C) to even values of $k$.) In each case, the expected codimensions $c_k$ in $X$ of these degeneracy loci are

In setting (A): $c_k = (e - k)(f - k)$;

In setting (B): $c_k = \binom{e-k+1}{2}$;

In setting (C): $c_k = \binom{e-k}{2}$ (for $k$ even).

**Theorem 8.4.7.** *Given a bundle homomorphism of type* (A), (B), *or* (C), *assume that* $\dim X \ge c_k$ *(for the appropriate choice of* $c_k$*). Assume in addition that the bundles in question satisfy*

*In setting* (A): $\operatorname{Hom}(E, F)$ *is ample;*

*In setting* (B): $\operatorname{Sym}^2(E) \otimes L$ *is ample;*

*In setting* (C): $\Lambda^2(E) \otimes L$ *is ample.*

*Then* $D_k(u)$ *is non-empty.*

*Proof.* We will give the argument in setting (B), the other cases being virtually identical. Consider the vector bundle $S = \operatorname{Sym}^2(E) \otimes L$. As in the proof of Theorem 8.3.9, symmetric homomorphisms of rank $\le k$ form a cone $\Omega_k \subseteq \mathbf{S}$ of codimension $c_k$ sitting inside the total space of $S$. On the other hand, the mapping $u$ gives rise to a section $s : X \longrightarrow \mathbf{S}$, and the issue is to show that $s^{-1}(\Omega_k) \ne \varnothing$. But $s^{-1}(\Omega_k)$ supports a cycle representing the cone class $z(\Omega_k, S)$, and so by the positivity of cone classes it cannot be empty.    □

As discussed in Sections 7.2.A and 7.2.D it is expected — and known in settings (A) and (C), and under slightly stronger numerical hypotheses in setting (B) — that if $\dim X > c_k$ then the degeneracy locus $D_k(u)$ is actually connected. This would follow in a uniform way from a general conjecture which remains open from [214]:

**Conjecture 8.4.8. (Connectedness of intersections).** Let $X$ be an irreducible projective variety of dimension $n$, and $E$ an ample vector bundle on $X$ of rank $e$. Suppose given an irreducible subvariety $V \subseteq \mathbf{E}$ in the total space of $E$. If $\dim V > n$, then the intersection of $V$ with the zero-section $X = \mathbf{0_E} \subseteq \mathbf{E}$ is connected.

**Example 8.4.9. (Non-empty intersections).** One can at least prove the non-emptiness statement that naturally accompanies 8.4.8. Specifically, let $X$ be an irreducible projective variety. In the setting of the conjecture, let $V \subseteq \mathbf{E}$ be an irreducible subvariety. If $\dim V \geq n$, then $V$ meets the zero-section $X = \mathbf{0_E} \subseteq \mathbf{E}$. (If $V$ is a cone this follows from Theorem 8.2.4. In general, consider for $t \neq 0$ the mapping $\mu_t : \mathbf{E} \longrightarrow \mathbf{E}$ given by multiplication by $t$. Then $V^* = \lim_{t \to 0} \mu_t(V)$ is a cone, and the emptiness of $V \cap \mathbf{0_E}$ would imply $z(V^*, E) = 0$. See [208, Theorem 12.1.(d)].)     □

## 8.4.C Singularities of Hypersurfaces Along a Curve

In the course of his work [462] on boundedness of Fano varieties, Nadel proved an interesting observation about divisors on a family of projective lines:

**Nadel's product lemma.** *Let $S$ be a smooth variety, and let*

$$D \subseteq \mathbf{P}^1 \times S$$

*be a divisor having degree $d$ on the typical fibre $\mathbf{P}^1 = \mathbf{P}^1 \times \{t\}$ of the projection $\mathrm{pr}_2$. Suppose that $p = (x, s) \in \mathbf{P}^1 \times S$ is a point at which $D$ has multiplicity $m$. Then*

$$\mathrm{mult}_{(y,s)}(D) \geq m - d \quad \text{for all } y \in \mathbf{P}^1, \tag{8.17}$$

*i.e. $D$ has multiplicity $\geq m - d$ along the curve $\mathbf{P}^1 \times \{s\}$.*

(See Example 8.4.11 for an elementary proof. The title Nadel gave to his result stems from the analogy with Faltings' product theorem in [178].) In this subsection we use the positivity theorem for cone classes to prove a geometric result that includes Nadel's lemma as a special case.

The result for which we are aiming concerns the singularities of a divisor along a smooth curve in an arbitrary non-singular variety:

**Theorem 8.4.10.** *Let $X$ be a smooth variety and $D \subseteq X$ an effective divisor. Consider a smooth irreducible projective curve $\Gamma \subseteq X$, denote by $N = N_{\Gamma/X}$ the normal bundle to $\Gamma$ in $X$, and fix a point $p \in \Gamma$. If $\delta \in \mathbf{Q}$ is a rational number such that $N_{<-\delta \cdot p>}$ is nef, then*

$$\mathrm{mult}_p(D) \leq (D \cdot \Gamma) + (1 - \delta) \cdot \mathrm{mult}_\Gamma(D). \tag{8.18}$$

*More generally, given any $r$ distinct points $p_1, \ldots, p_r \in \Gamma$ one has*

$$\sum \mathrm{mult}_{p_i}(D) \leq (D \cdot \Gamma) + (r - \delta) \cdot \mathrm{mult}_\Gamma(D). \tag{8.19}$$

Here $\mathrm{mult}_\Gamma(D)$ denotes the multiplicity of $D$ at a general point of $\Gamma$ (Definition 5.2.10); note that we do not assume that $\delta \geq 0$. If $X = \mathbf{P}^1 \times S$ and $\Gamma = \mathbf{P}^1 \times \{s\}$ then $N$ is trivial, so we can take $\delta = 0$ in (8.18) to recover Nadel's result.

**Example 8.4.11. (Proof of Nadel's lemma).** For an elementary proof of (8.17), let $t$ be an affine linear coordinate on $\mathbf{P}^1$ centered at $y$ and let $x = (x_1, \ldots, x_n)$ be local analytic coordinates on $S$ centered at $s$. Then over a neighborhood $U$ of $s$ the equation defining $D$ is written in the form

$$f(x, t) = a_0(x)t^d + a_1(x)t^{d-1} + \ldots + a_d(x) \tag{*}$$

for suitable holomorphic functions $a_i(x)$. Since $D$ has multiplicity $m$ at $p$, every term in (*) must vanish to order $\geq m$ at the origin, and hence

$$\mathrm{ord}_0 \, a_i(x) \geq m - d + i.$$

In particular, each of the coefficients $a_i(x)$ in (*) vanishes to order $\geq m - d$ at $x = 0$, and (8.17) follows.     □

**Example 8.4.12. (Hypersurfaces containing a line).** Let $D \subseteq \mathbf{P}^n$ be a hypersurface of degree $d$ containing a line $L$. In this case the Theorem applies with $\delta = 1$. So for instance if $D$ is generically smooth along $L$, then

$$\sum_{p \in L} \big(\mathrm{mult}_p(D) - 1\big) \leq d - 1 \tag{*}$$

Observe that equality holds if $D$ is the cone over a smooth hypersurface of degree $d$ in $\mathbf{P}^{n-1}$ and $L$ is a ruling of $D$. (One can also see (*) directly by considering the intersection of $L$ with the zeroes of a generic first derivative of an equation defining $D$.)     □

Turning to the proof of Theorem 8.4.10, assume for concreteness that $\Gamma \subseteq D$ (the result being evident otherwise), and set $n = \dim X$. The plan is to use the theory of Fulton–MacPherson to compute the intersection number $(D \cdot \Gamma)$ as a cone class. Specifically, let $C = C_{\Gamma/D}$ be the normal cone to $\Gamma$ in $D$. Thus $C$ is a cone of pure dimension $n - 1$ sitting as a divisor in the total space $\mathbf{N}$ of $N = N_{\Gamma/X}$, and

$$(D \cdot \Gamma) = \deg z(C, N) \tag{8.20}$$

([208, Chapter 6.1]).

The first point to observe is that the whole fibre of $\mathbf{N}$ over $p \in \Gamma$ appears as a component of $C$ whenever $\mathrm{mult}_p D > \mathrm{mult}_\Gamma D$:

**Lemma 8.4.13.** *Write* $m = \mathrm{mult}_p(D)$ *and* $k = \mathrm{mult}_\Gamma(D)$, *and denote by* $F \subseteq \mathbf{N}$ *the fibre of* $\mathbf{N}$ *over* $p$ (*considered as a reduced divisor*). *If* $m > k$ *then* $F$ *occurs in* $C$ *with multiplicity* $\geq m - k$, *i.e. we can write*

$$C = (m - k)F + C'$$

*where* $C'$ *is an effective divisor on* $\mathbf{N}$.

*Proof.* This may be verified by an explicit local calculation. Specifically, choose local analytic coordinates $t, x_1, \ldots, x_{n-1}$ on $X$ centered at $p$ in such a way that $\Gamma$ is locally defined by $x_1 = \ldots = x_{n-1} = 0$. Since $\mathrm{mult}_\Gamma(D) = k$, $D$ is locally defined by an equation of the form

$$f(x, t) = \sum_{|I|=k} a_I(x, t) \cdot x^I,$$

the sum being over monomials of degree $k$ in $x = (x_1, \ldots, x_{n-1})$, and moreover there is at least one index $I$ for which $a_I(0, t)$ does not vanish identically. The normal bundle $\mathbf{N}$ is locally a product $\Gamma \times \mathbf{C}^{n-1}$, and the coordinates $x_1, \ldots, x_{n-1}$ determine in the natural way "fibre coordinates" $x_1^\#, \ldots, x_{n-1}^\#$ on $\mathbf{N}$. In terms of these coordinates, $F$ is cut out by $t = 0$ and the normal cone $C \subseteq \mathbf{N}$ is defined by the equation

$$f^\#(x^\#, t) = \sum_{|I|=k} a_I(0, t) \cdot (x^\#)^I.$$

On the other hand, since $f$ vanishes to order $m$ at the origin, each of the coefficients $a_I(x, t)$ must vanish there to order $\geq m - k$. In particular

$$\mathrm{ord}_t\big(a_I(0, t)\big) \geq m - k$$

for every $I$, and the Lemma follows.    $\square$

For the next Lemma, decompose the divisor $C \subseteq \mathbf{N}$ as a sum

$$C = V + C_1$$

where every component of $V$ is "vertical" — i.e. supported on a fibre of $\mathbf{N} \longrightarrow \Gamma$ — while every component of $C_1$ dominates $\Gamma$.

**Lemma 8.4.14.** *For any $\mathbf{Q}$-divisor class $\varepsilon$ on $\Gamma$ one has*

$$\deg z(C_1{<}\varepsilon{>}, \mathbf{N}{<}\varepsilon{>}) = \deg z(C_1, \mathbf{N}) + k \cdot \deg(\varepsilon),$$

*where as above $k = \mathrm{mult}_\Gamma(D)$.*

Granting this for the moment, the Theorem follows at once:

*Proof of Theorem 8.4.10.* For each $1 \leq i \leq r$, write $m_i = \mathrm{mult}_{p_i}(D)$, and set $k = \mathrm{mult}_\Gamma(D)$. Noting that $z(F, \mathbf{N}) = 1$, the Lemmas imply:

$$
\begin{aligned}
(D \cdot \Gamma) &= \deg z(C, \mathbf{N}) \\
&= \deg z(V, \mathbf{N}) + \deg z(C_1, \mathbf{N}) \\
&\geq \sum_{i=1}^r (m_i - k) + \deg z(C{<}{-}\delta \cdot p{>}, \mathbf{N}{<}{-}\delta \cdot p{>}) + k\delta.
\end{aligned}
$$

But $\deg z(C<-\delta \cdot p>, N<-\delta \cdot p>) \geq 0$ thanks to 8.2.6 since $N<-\delta \cdot p>$ is nef. Therefore

$$(D \cdot \Gamma) \geq \sum_{i=1}^{r}(m_i - k) + k\delta,$$

as asserted.  □

*Proof of Lemma 8.4.14.* We use Definition 8.1.20 to compute the class in question. Specifically, write $e = \operatorname{rank} N = \dim X - 1$. Then 8.1.20 (i) gives

$$\deg z(C_1<\varepsilon>, N<\varepsilon>)$$
$$= \deg \Big( c_{e-1}(Q_{\mathbf{P}_{\mathrm{sub}}(N)}<\rho^*\varepsilon>) \cap [\mathbf{P}_{\mathrm{sub}}(C_1)] \Big)$$
$$= \deg z(C_1, N) + \deg \Big( \big(c_{e-2}(Q_{\mathbf{P}_{\mathrm{sub}}(N)}) \cdot \rho^*\varepsilon\big) \cap [\mathbf{P}_{\mathrm{sub}}(C_1)] \Big),$$

$\rho : \mathbf{P}_{\mathrm{sub}}(N) \longrightarrow \Gamma$ being the bundle map. But $\mathbf{P}_{\mathrm{sub}}(C_1)$ restricts to a hypersurface of degree $k$ on the generic fibre of $\rho$, and $(c_{e-2}(Q_{\mathbf{P}_{\mathrm{sub}}(N)}) \cdot \rho^*\varepsilon)$ is represented by $\deg(\varepsilon)$ copies of a line in this fibre. The assertion follows.  □

## Notes

The project of determining the numerically positive polynomials for ample bundles was initiated by Griffiths. In [247] he gave an essentially analytic description of a collection of classes which he proved to be positive for Griffiths-positive bundles, and he showed that these include the Chern and inverse Segre classes. Fulton and the author established in [214, Appendix A] that in fact these Griffiths-positive polynomials exactly coincide with those specified in Theorem 8.3.9. However as Fulton and Pragacz [215, p. 98] remark, the example of $c_1^2 - 2c_2$ — which through a miscalculation was thought to lie in the Griffiths cone — illustrates the difficulty of working with the analytic definition.

Kleiman [342] determined the cone of positive polynomials on a surface: here the essential point is to establish the positivity of $c_2$. This work in part inspired the result (Corollary 8.2.2) of Bloch and Gieseker in Section 8.2.A giving the positivity of all $c_n$. Usui and Tango [580] established the positivity of the Schur polynomials $s_\lambda$ for globally generated ample bundles. This gave an important hint for the shape of Theorem 8.3.9.

As we indicated, Section 8.4.A closely follows [213]. The material in 8.4.C is new, although J.-M. Hwang has independently obtained some similar results.

# Part Three

# Multiplier Ideals and Their Applications

# Introduction to Part Three

These last three chapters aim to give a systematic development of the theory of multiplier ideals from an algebro-geometric perspective.

We begin with some history. It was discovered by Kawamata and Viehweg [316], [589] in the early 1980s that the classical vanishing theorems can be extended to statements for $\mathbf{Q}$-divisors. The idea is that positivity hypotheses on a $\mathbf{Q}$-divisor $D$ imply a Kodaira-type vanishing statement for an integral divisor "close to" $D$. The theorem is actually rather elementary, and at first blush the resulting statement may not appear very geometric in nature. However a fundamental insight coming from higher-dimensional geometry is that this apparently technical improvement of Kodaira vanishing leads in fact to a quite dramatic and startling increase in the power of the result. Roughly speaking, the reason for this is that applying the classical theorem to study (say) a given linear series $|L|$ requires geometric hypotheses on $|L|$ that can be hard or impossible to verify in practice. By contrast, the new technology allows one to make instead asymptotic constructions involving a high multiple $|kL|$ of the linear series at hand, and then to "divide by $k$" to deduce geometric consequences for $|L|$ itself. As we shall see, this possibility of working asymptotically opens the door to fundamentally new sorts of applications.

The theorem of Kawamata–Viehweg contains a normal crossing hypothesis that in practice is never satisfied directly. In the early applications of the theorem — due to Kawamata, Shokurov, Reid, and others (see [368]) — one passed to an embedded resolution of singularities, and applied vanishing on this new space. Unfortunately this becomes quite cumbersome, and it obscures the underlying geometric ideas.

Two approaches emerged during the 1990s to overcome this difficulty. First, the theorem of Kawamata–Viehweg — along with some related ideas coming from the minimal model program — leads to various notions of singularities for a pair $(X, D)$ consisting of a normal (or even smooth) variety $X$ and a $\mathbf{Q}$-divisor $D$ on $X$. Many of the applications of vanishing can be phrased naturally in this language, thereby reintroducing geometry into the picture. Kollár's survey [364] gives an excellent overview.

An alternative strategy arose at around the same time in the complex-analytic side of the field. Nadel [461], Demailly [124], Siu [537], and others introduced the concept of a multiplier ideal associated to a singular metric on a line bundle, and proved a Kodaira-type vanishing theorem involving these ideals.[6] Working with multiplier ideals obviates the need to pass to a new space every time one wants to apply vanishing. This machine soon led to interesting applications to linear series and related questions, most notably in work of Demailly and Siu (cf. [124], [537], [10]). In the following years these methods achieved further successes, including for example Siu's proof of the deformation invariance of plurigenera [539] and the extensions of this work by Kawamata [323], [324]. These last papers, and ideas of Tsuji [574], [575], brought into focus the power of asymptotic constructions involving multiplier ideals.

It became clear quite soon that one could define multiplier ideals in a purely algebro-geometric setting, and indeed many of the basic facts had already been worked out in passing by Esnault and Viehweg (see [174, Chapter 7]). The idea in effect is to pass to an embedded resolution, and "push down" vanishing theorems from the blow-up. While in principle the analytic constructions are more general, in practice most of the actual applications have been essentially algebro-geometric in nature and translate without too much difficulty into algebraic language. The algebraic multiplier ideals quickly began finding applications in their own right (cf. [154], [324], [323], [155], [130], [159]).

It seems safe to predict that multiplier ideals and their variants are destined to become fundamental tools in algebraic geometry. At the same time, the subject is somewhat in flux: much of the material discussed here is of quite recent origin, and many of the questions that we take up remain the focus of intense current activity. But while aspects of the story are undoubtedly not yet in their definitive form, our sense is that the basic foundations have achieved some stability even if one can anticipate many further developments and applications. Our hope is that these chapters can contribute to this evolution by making the theory available to a wider audience of algebraic geometers. We stress, however, that one should see the machinery developed here as complementing — rather than supplanting — parallel ideas involving singularities of pairs or analytic methods. In any given situation, the appropriate tool will certainly depend on the nature of the problem and the tastes of the researcher.

After proving the Kawamata–Viehweg vanishing theorem for **Q**-divisors, we develop in Chapter 9 the basic definitions, examples, and properties of

---

[6] Lipman [403] simultaneously and independently made some closely related constructions in commutative algebra.

multiplier ideals. A number of applications of the theory are presented in Chapter 10. The final Chapter 11 revolves around asymptotic multiplier ideals that can be defined in various settings.

# 9

---

# Multiplier Ideal Sheaves

This chapter is devoted to the definition and basic properties of multiplier ideals.

Let $X$ be a smooth complex variety and $D$ an effective $\mathbf{Q}$-divisor on $X$. The multiplier ideal $\mathcal{J}(D) \subseteq \mathcal{O}_X$ of $D$ is a coherent sheaf of ideals on $X$. One can think of it as reflecting in a somewhat subtle way the singularities of $D$, and it has many good formal properties. When $X$ is projective and $L$ is an integral divisor such that $L - D$ is ample, the Kawamata–Viehweg–Nadel vanishing theorem states that

$$H^i\big(X, \mathcal{O}_X(K_X + L) \otimes \mathcal{J}(D)\big) = 0 \quad \text{for } i > 0. \tag{*}$$

If $D$ is integral then $\mathcal{J}(D) = \mathcal{O}_X(-D)$ and this reduces to the classical Kodaira vanishing. On the other hand, if $A \in |kL|$ for some $k \gg 0$ and $D = \frac{1}{k+1}A$, then $\mathcal{J}(D)$ will be cosupported on the locus at which $A$ is very singular. In this case, (*) leads to interesting connections between singularities of divisors and the geometry of adjoint linear series. From another point of view, the theorem opens the door to proving Kodaira-type vanishing statements for ideal sheaves: when we are able to realize a given ideal sheaf $\mathcal{I}$ (or something related) as a multiplier ideal $\mathcal{J}(D)$, one can expect that (*) will have strong geometric consequences.

We attach multiplier ideals to objects of three sorts: $\mathbf{Q}$-divisors, ideal sheaves, and linear series. Although these different contexts lead to essentially equivalent theories, they vary somewhat in flavor. Our usual policy is to give statements in each setting but to write proofs only in one. In all cases multiplier ideals will be defined by passing to suitable log resolutions, and an essential point is to check that they are actually independent of the resolutions used to construct them.[1]

---

[1] In the analytic approach, which is summarized in Section 9.3, they arise as sheaves of "multipliers" (hence the name).

We start in Section 9.1 with some preliminary material, most notably the Kawamata–Viehweg vanishing theorem for **Q**-divisors. Multiplier ideals are introduced in Section 9.2. Section 9.3 focuses on examples and illustrations, while the basic vanishing theorems appear in 9.4. These vanishings are used in the extended Section 9.5 to deduce local geometric properties of multiplier ideals. We conclude in Section 9.6 with an account of Skoda's theorem. Numerous applications of the theory are collected in the next chapter.

## 9.1 Preliminaries

In this section, after some preliminaries on **Q**-divisors and log resolutions, we state and prove the Kawamata–Viehweg vanishing theorem.

**Convention 9.1.1. (Divisors).** In the present Part Three of this book, by *divisor* we will understand unless otherwise stated a Weil divisor, i.e. a formal sum of codimension one subvarieties.    □

Note that this differs from Convention 1.1.2 in Parts One and Two. We institute the change because the critical rounding operations we are about to discuss make sense only for Weil divisors (Remark 9.1.3). In fact, for the most part we work on smooth ambient varieties, where of course Weil and Cartier divisors coincide.

### 9.1.A Q-Divisors

We start by recalling some definitions surrounding **Q**-divisors. Let $X$ be an irreducible variety, which for the moment is required only to be normal. A **Q**-divisor $D$ on $X$ is a finite formal linear combination

$$D = \sum a_i D_i \qquad (*)$$

of codimension-one irreducible subvarieties $D_i \subseteq X$ with rational coefficients $a_i \in \mathbf{Q}$. Unless otherwise stated, in writing expressions such as $(*)$ we will always assume that the $D_i$ are distinct prime divisors, so that the coefficients $a_i$ are uniquely defined. As usual, one says that $D$ is *effective* if every $a_i \geq 0$, and we write $D_1 \preccurlyeq D_2$ if $D_2 - D_1$ is effective. A **Q**-divisor $D$ is *integral* if in fact $a_i \in \mathbf{Z}$ for each $i$. If $D$ is a **Q**-divisor, we say that an integer $m$ *clears the denominators of $D$* if $mD$ is integral.

Starting with a **Q**-divisor, one can pass to integral divisors by rounding its coefficients up or down.

**Definition 9.1.2. (Rounds).** Let $D = \sum a_i D_i$ be a **Q**-divisor on $X$. The *round-up* $\lceil D \rceil$ and *round-down* or *integral part* $\lfloor D \rfloor = [D]$ of $D$ are the integral divisors

$$\ulcorner D \urcorner \;=\; \sum \ulcorner a_i \urcorner D_i,$$

$$\llcorner D \lrcorner \;=\; [D] \;=\; \sum [a_i] \, D_i,$$

where as usual for $x \in \mathbf{Q}$ one denotes by $\ulcorner x \urcorner$ the least integer $\geq x$, and by $\llcorner x \lrcorner = [x]$ the greatest integer $\leq x$. The *fractional part* $\{D\}$ of $D$ is

$$\{D\} \;=\; D - [\,D\,]. \quad \square$$

**Remark 9.1.3.** Note that rounds are defined strictly at the cycle level. In particular, 9.1.2 would not make sense if we worked with Cartier divisors as in Chapter 1. This explains why we deal here with Weil divisors. However, if $X$ is not smooth we then need extra hypotheses in order to guarantee that the usual functorial operations are defined. $\hspace{2cm}\square$

As in Remark 1.3.8 we say that a $\mathbf{Q}$-divisor $D$ is $\mathbf{Q}$-*Cartier* if some multiple of $D$ is an (integral) Cartier divisor. Note that when $X$ is smooth — which will be our main focus — any $\mathbf{Q}$-divisor is $\mathbf{Q}$-Cartier. It follows from the theory developed in Section 1.3 that all of the usual operations and equivalences defined for Cartier divisors extend naturally to $\mathbf{Q}$-Cartier divisors. For the convenience of the reader, we review quickly the most important of these.

To begin with, when $X$ is complete, there is a $\mathbf{Q}$-valued intersection theory for $\mathbf{Q}$-Cartier $\mathbf{Q}$-divisors obtained from the classical theory by clearing denominators. Explicitly, if $D = \sum a_i D_i$ is a $\mathbf{Q}$-divisor such that $D' = rD$ is integral, and if $V \subseteq X$ has dimension $k$, then

$$\left(D^k \cdot V\right) \;=\; \frac{1}{r^k} \int_V c_1\big(\mathcal{O}_X(D')\big)^k.$$

This gives rise to a notion of numerical equivalence for $\mathbf{Q}$-Cartier $\mathbf{Q}$-divisors, which we denote by $\equiv_{\mathrm{num}}$. Specifically,

$$D \equiv_{\mathrm{num}} 0 \;\; \text{iff} \;\; \left(D \cdot C\right) = 0$$

for every reduced and irreducible curve $C$ on $X$. Similarly, if $D$ is a $\mathbf{Q}$-Cartier $\mathbf{Q}$-divisor on $X$ and if $f : Y \longrightarrow X$ is any morphism of varieties such that no component of $Y$ maps into the support of $D$, then the pullback $f^*D$ is determined as a $\mathbf{Q}$-divisor by clearing denominators. For an arbitrary morphism, one can always define $f^*D$ up to numerical equivalence by passing in the usual fashion to $\mathrm{Pic}(X) \otimes \mathbf{Q}$.

Finally, assuming again that $X$ is complete, the usual notions of positivity apply to a $\mathbf{Q}$-Cartier $\mathbf{Q}$-divisor $D$. Specifically, $D$ is *nef* if $(D \cdot C) \geq 0$ for every irreducible curve $C \subset X$, and $D$ is *ample* if it satisfies the conclusion of Nakai's criterion, i.e. if $(D^k \cdot V) > 0$ for every irreducible subvariety $V \subseteq X$ of positive dimension. Equivalently, $D$ is nef or ample if $mD$ is so for some (or any) positive integer $m$ clearing the denominators of $D$. Similarly we define $D$ to be big if $mD$ is. For nef divisors, bigness is equivalent to the positivity of the top self-intersection number. We refer to Section 1.4 for a fuller discussion of these matters.

**Remark 9.1.4. (Warning on rounds).** It is critical to note that *pullback and numerical equivalence do not commute with rounding.* To give a simple example, let $X = \mathbf{C}^2$ with coordinates $x, y$, and let $A \subset X$ be the parabola $\{y = x^2\}$. Let $Y \subset X$ denote the $x$-axis and $P$ the origin, so that $Y$ is tangent to $A$ at $P$. Consider the **Q**-divisor $D = \frac{1}{2}A$ on $X$. Then $[D] = 0$, so $[D] \,|\, Y = 0$, but

$$[D \,|\, Y] = [\tfrac{1}{2} \cdot 2P] = P.$$

The situation is illustrated in Figure 9.1.

Similarly, let $L, C \subset \mathbf{P}^2$ denote respectively a line and an irreducible conic. Then $L \equiv_{\mathrm{num}} \frac{1}{2}C$, but

$$[\tfrac{1}{2}C] = 0 \quad \text{whereas} \quad [L] = L. \quad \square$$

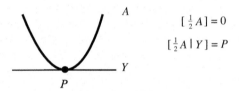

$$[\tfrac{1}{2}A] = 0$$
$$[\tfrac{1}{2}A \,|\, Y] = P$$

**Figure 9.1.** Rounding does not commute with restriction

**Remark 9.1.5. (Warning on fractional parts).** The theorems to be discussed below involve using positivity hypotheses on a **Q**-divisor $D$ to deduce vanishings for rounds-up or down of $D$, and as we have just seen rounding does not commute with numerical equivalence. So it is imperative that the fractional part of $D$ be defined as an actual divisor, and not merely as a numerical equivalence class. In fact, many applications depend on exploiting this non-commutativity through the construction of **Q**-divisors with geometrically interesting fractional parts.    $\square$

**Remark 9.1.6. (R-divisors).** All of these definitions and constructions extend in the natural way to divisors with real coefficients. If $X$ is singular (but normal) the appropriate Cartier condition on an **R**-divisor $D$ is that one can write $D = \sum b_j E_j$ where $E_j$ is an integral Cartier divisor and $b_j \in \mathbf{R}$. We leave the remaining definitions and statements to the reader.    $\square$

### 9.1.B Normal Crossing Divisors and Log Resolutions

Assume now that $X$ is a non-singular variety of dimension $n$. We start by recalling some definitions from Section 4.1.

**Definition 9.1.7.** A divisor $D = \sum D_i$ has *simple normal crossings* (and $D$ is an *SNC divisor*) if each $D_i$ is smooth, and if $D$ is defined in a neighborhood of any point by an equation in local analytic coordinates of the type

$$z_1 \cdot \ldots \cdot z_k \;=\; 0$$

for some $k \le n$. A **Q**-divisor $\sum a_i D_i$ has *simple normal crossing support* if $\sum D_i$ is an SNC divisor. $\qquad\square$

In other words, intersections among the components of $D$ should occur "as transversely as possible," i.e. the singularities of $D$ should look no worse than those of a union of coordinate hyperplanes.

We will eventually see that vanishing theorems for **Q**-divisors $D$ are closely connected with singularities of the fractional part of $D$. The theorems will be proved in the first instance for divisors whose fractional parts have simple normal crossing support, where the local picture can be completely analyzed.

**Example 9.1.8. (Restrictions of normal crossing divisors).** The sort of problem flagged in Remark 9.1.4 can be avoided by normal crossing hypotheses. Specifically, let $D = \sum a_i D_i$ be a **Q**-divisor on a smooth variety $X$, and let $E \subset X$ be a non-singular hypersurface. Assume that $E + \sum D_i$ is an SNC divisor. Then rounding commutes with restriction to $E$:

$$[D] \,|\, E \;=\; [D \,|\, E]. \quad \square$$

Free linear series satisfy a useful transversality property with respect to a normal crossing divisor:

**Lemma 9.1.9. (Free linear series and SNC divisors).** *Let $X$ be a smooth variety, and let $|V|$ be a finite dimensional free linear series on $X$. Suppose given a simple normal crossing divisor $\sum E_i$ on $X$. If $A \in |V|$ is a general divisor, then $A + \sum E_i$ again has simple normal crossings.*

*Proof.* It suffices to show that a general element $A \in |V|$ is smooth, and that it meets each of the intersections $E_{i_1} \cap \cdots \cap E_{i_t}$ either transversely or not at all. Since $\sum E_i$ has normal crossings we are reduced to proving that given any finite collection of smooth subvarieties $Z_1, \ldots, Z_s \subset X$, a general divisor $A \in |V|$ meets each $Z_j$ transversely or not at all. But this follows from the fact that a general member of a free linear series — in particular, the restriction of $|V|$ to each $Z_j$ — is non-singular. $\qquad\square$

We will define multiplier ideals in three settings: for **Q**-divisors, for linear series, and for ideal sheaves. In each case, the first step is to form an appropriate log resolution. We now specify the properties that will be required of these resolutions.

**Definition 9.1.10. (Log resolutions of Q-divisors).** Let $D = \sum a_i D_i$ be a **Q**-divisor on $X$. A *log resolution* of $D$ (or of the pair $(X, D)$) is a projective birational mapping

$$\mu : X' \longrightarrow X,$$

with $X'$ non-singular, such that the divisor $\mu^* D + \text{except}(\mu)$ has simple normal crossing support. Here $\text{except}(\mu)$ denotes the sum of the exceptional divisors of $\mu$.    □

A log resolution of a linear series is defined similarly. Specifically, let $L$ be an integral divisor on $X$, and let $V \subseteq H^0(X, \mathcal{O}_X(L))$ be a non-zero finite-dimensional space of sections of the line bundle determined by $L$. As usual write $|V|$ for the corresponding linear series of divisors in the linear series $|L|$.

**Definition 9.1.11. (Log resolution of a linear series).** A *log resolution* of $|V|$ is a projective birational map $\mu : X' \longrightarrow X$ as above, again with $X'$ non-singular, such that

$$\mu^* |V| = |W| + F,$$

where $F + \text{except}(\mu)$ is a divisor with SNC support, and

$$W \subseteq H^0(X', \mathcal{O}_{X'}(\mu^* L - F))$$

defines a free linear series.    □

In other words, we are separating $\mu^* |V|$ into a fixed part $F$ and a free part $|W|$. If $V$ is a one-dimensional subspace, defining an integral divisor $D$ on $X$, then a log resolution of $D$ also resolves the linear series $|V|$ (in this case $|W|$ will be empty). In general it follows from Lemma 9.1.9, applied to the free linear series $|W|$, that a log resolution of $|V|$ gives a log resolution of a general divisor $D \in |V|$.

In the same spirit, we define resolutions of ideal sheaves:

**Definition 9.1.12. (Log resolution of an ideal).** Let $\mathfrak{a} \subseteq \mathcal{O}_X$ be a non-zero ideal sheaf on $X$. A *log resolution* of $\mathfrak{a}$ is a projective birational map $\mu : X' \longrightarrow X$ as before — so that in particular $X'$ is non-singular — such that

$$\mu^{-1} \mathfrak{a} =_{\text{def}} \mathfrak{a} \cdot \mathcal{O}_{X'} = \mathcal{O}_{X'}(-F),$$

where $F$ is an effective divisor on $X$ such that $F + \text{except}(\mu)$ has simple normal crossing support.    □

**Example 9.1.13.** Consider the ideal

$$\mathfrak{a} = (x^3, y^2) \subset \mathbf{C}[x, y]$$

(which we may view as an ideal sheaf on $X = \mathbf{C}^2$). Then a log resolution $\mu : X' \longrightarrow X$ of $\mathfrak{a}$ is obtained by the familiar sequence of three blowings up leading to an embedded resolution of the cuspidal cubic $\{y^2 = x^3\}$. The

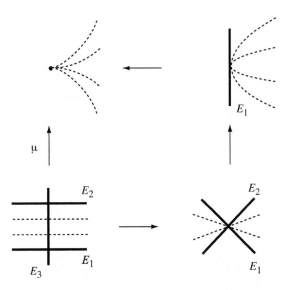

**Figure 9.2.** Log resolution of $\mathfrak{a} = (x^3, y^2)$

resolution is illustrated in Figure 9.2, where by way of reference we have indicated the proper transforms of the divisors of typical elements in the ideal.[2] One finds that

$$\mathfrak{a} \cdot \mathcal{O}_{X'} = \mathcal{O}_{X'}\left(-2E_1 - 3E_2 - 6E_3\right). \quad \Box$$

**Example 9.1.14. (Base ideal of a linear series).** Let $|V| \subseteq |L|$ be a non-empty linear series. Recall (Definition 1.1.8) that the *base ideal*

$$\mathfrak{b} = \mathfrak{b}\big(|V|\big) \subseteq \mathcal{O}_X$$

of $|V|$ is defined to be the image of the map $V \otimes_{\mathbf{C}} \mathcal{O}_X(-L) \longrightarrow \mathcal{O}_X$ determined by evaluation. In other words, $\mathfrak{b}\big(|V|\big)$ is the ideal sheaf generated by the sections in $V$. Then a log resolution of $|V|$ is the same thing as a log resolution of $\mathfrak{b}\big(|V|\big)$. $\quad \Box$

As we noted in Section 4.1.A, the existence of such resolutions was established by Hironaka [288]. In fact, Hironaka proves that one can construct log resolutions by a sequence of blowings-up along smooth centers: in the case of ideal sheaves, for instance, this is the content of [288, Main Theorem II]. Streamlined proofs have recently been given by Bierstone–Milman [56] and by Encinas and Villamayor [166]. An extremely simple approach to the existence of resolutions of singularities, growing out of ideas of De Jong, appears in [2],

---

[2] Note that by a slight abuse of notation, the exceptional divisors pictured at each stage of the resolution are actually the proper transforms of the divisors having the same label in the previous stage.

[64], and [494], although the arguments don't give very precise control over the resolution process.[3]

Given any such resolution $\mu : X' \longrightarrow X$, we denote by

$$K_{X'/X} = K_{X'} - \mu^* K_X \tag{9.1}$$

the relative canonical divisor of $X'$ over $X$. Note that this is naturally defined as an effective divisor supported on the exceptional locus of $\mu$, and not merely as a linear equivalence class: the determinant of the derivative $\det(d\mu)$ gives a local equation. Recalling that $\mu_* \mathcal{O}_{X'}(K_{X'/X}) = \mathcal{O}_X$, one sees that if $N$ is an effective (integral) divisor on $X'$, then

$$\mu_* \mathcal{O}_{X'}(K_{X'/X} - N) \subseteq \mathcal{O}_X \tag{9.2}$$

is naturally an ideal sheaf of $\mathcal{O}_X$. Multiplier ideals will arise in this manner.

**Example 9.1.15.** Let $\mu : X' \longrightarrow X$ be the log resolution of $\mathfrak{a} = (x^3, y^2)$ constructed in Example 9.1.13 and Figure 9.2. Then

$$K_{X'/X} = E_1 + 2E_2 + 4E_3. \quad \square$$

**Example 9.1.16. (Comparing log resolutions).** Let $D$ be a $\mathbf{Q}$-divisor on $X$, and let

$$\mu_1 : X_1 \longrightarrow X \quad , \quad \mu_2 : X_2 \longrightarrow X$$

be two log resolutions of $(X, D)$. Then there is a log resolution $\mu : X' \longrightarrow X$ of $(X, D)$ that dominates both $X_1$ and $X_2$: in particular $X'$ is a common log resolution of $\mu_1^*(D) + \mathrm{except}(\mu_1)$ and $\mu_2^*(D) + \mathrm{except}(\mu_2)$. (Start by constructing a smooth variety $X_{12}$ sitting in a commutative diagram:

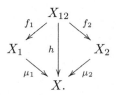

Then replace $X_{12}$ by a log resolution of $h^*(D) + \mathrm{except}(h)$.) A similar statement holds for log resolutions of ideals and linear series. $\square$

**Generalization 9.1.17. (Relative linear series).** Definition 9.1.11 extends to a relative setting. Specifically, let $L$ be an integral divisor on a variety $X$, and suppose given a proper surjective map $f : X \longrightarrow T$. Then there is a naturally defined homomorphism

---

[3] The published versions of these three papers do not mention explicitly the possibility of constructing log resolutions in the specified sense. However Paranjape has posted a revised version [494] in which this is worked out in detail: see `math.AG/9806084` in the math arXiv.

$$\rho : f^* f_* \mathcal{O}_X(L) \longrightarrow \mathcal{O}_X(L).$$

Assume that the sheaf on the left is non-zero, or equivalently that the restriction of $L$ to a general fibre of $f$ has a non-vanishing section. Then a log resolution of $\rho$, or of $|L|$ with respect to $f$, is defined to be a proper and birational mapping $\mu : X' \longrightarrow X$, with $X'$ non-singular, having the property that the image of the induced homomorphism

$$\mu^* \rho : \mu^* f^* f_* \mathcal{O}_X(L) \longrightarrow \mu^* \mathcal{O}_X(L)$$

of sheaves on $X'$ is the subsheaf $\mathcal{O}_{X'}(\mu^* L - F)$ of $\mathcal{O}_{X'}(\mu^* L)$, $F$ being an effective divisor on $X'$ such that $F + \text{except}(\mu)$ has simple normal crossing support. If $\mathfrak{b} = \mathfrak{b}(f, |L|) \subseteq \mathcal{O}_X$ is the *base ideal* of $\rho$ — i.e. the image of the map $f^* f_* \mathcal{O}_X(L) \otimes \mathcal{O}_X(-L) \longrightarrow \mathcal{O}_X$ determined by $\rho$ — then a log resolution of $\rho$ is the same thing as a log resolution of $\mathfrak{b}$.

One can think of $\rho : f^* f_* \mathcal{O}_X(L) \longrightarrow \mathcal{O}_X(L)$ as giving the relative analogue of the complete linear series associated to $L$. A possibly incomplete linear series would then be given by a coherent subsheaf $\mathcal{V} \subseteq f_* \mathcal{O}_X(L)$, together with the homomorphism $\rho_{\mathcal{V}} : f^* \mathcal{V} \longrightarrow \mathcal{O}_X(L)$ arising from $\rho$. We leave it to the reader to formulate the natural notion of a log resolution in this setting.   □

### 9.1.C The Kawamata–Viehweg Vanishing Theorem

We now come to the statement and proof of the basic vanishing theorem for **Q**-divisors.

**Theorem 9.1.18. (Kawamata–Viehweg vanishing).** *Let $X$ be a non-singular projective variety of dimension $n$, and let $N$ be an integral divisor on $X$. Assume that*

$$N \equiv_{\text{num}} B + \Delta \tag{9.3}$$

*where $B$ is a nef and big **Q**-divisor, and $\Delta = \sum a_i \Delta_i$ is a **Q**-divisor with simple normal crossing support and fractional coefficients:*

$$0 \le a_i < 1 \quad \text{for all } i.$$

*Then*

$$H^i \big( X, \mathcal{O}_X(K_X + N) \big) = 0 \quad \text{for } i > 0.$$

*Equivalently,*

$$H^j \big( X, \mathcal{O}_X(-N) \big) = 0 \quad \text{for } j < n.$$

**Remark 9.1.19.** In other words, the hypothesis on $\Delta$ is that it is an effective divisor whose support $\sum \Delta_i$ has simple normal crossings, with $[\Delta] = 0$. A divisor satisfying these conditions is sometimes called a *boundary divisor*.

The next statements are sometimes easier to remember:

**Corollary 9.1.20.** *As in the theorem, assume that $X$ is a smooth projective variety. Let $B$ be a big and nef $\mathbf{Q}$-divisor whose fractional part has simple normal crossing support. Then*

$$H^i\big(X, \mathcal{O}_X(K_X + \ulcorner B \urcorner)\big) \;=\; 0 \quad \text{for } i > 0. \quad \square$$

**Corollary 9.1.21.** *In the setting of Corollary 9.1.20, let $L$ be an integer divisor and $D$ a $\mathbf{Q}$-divisor on $X$. Assume that $L - D$ is big and nef and that $D$ has SNC support. Then*

$$H^i\big(X, \mathcal{O}_X(K_X + L - [D])\big) \;=\; 0 \quad \text{for } i > 0. \quad \square$$

There are now several different approaches to proving 9.1.18 that simultaneously establish the classical Kodaira vanishing theorem. Specifically, Esnault and Viehweg [170] have given an argument based on Hodge theory for logarithmic forms, and Kollár [362] presents a related topological argument. The $L^2$-methods leading to the Nadel vanishing theorem also apply (see [126]), and there is even a characteristic $p$ proof [174] based on the approach of Deligne and Illusie.

However the original strategy of Kawamata was simply to use covering constructions to reduce the statement to vanishing for integral divisors.[4] We will take this path here, following an argument given by Kollár and Mori [368] that simplifies the details of Kawamata's proof in [316]. The covering constructions that we use appear in Section 4.1.

*Proof of Theorem 9.1.18.* For the duration of the proof we allow the divisors $\Delta_i$ to have several connected components (each non-singular and reduced), and argue by induction on the number of fractional terms appearing on the right hand side of (9.3). If $N \equiv_{\text{num}} B$, then the statement reduces to vanishing for big and nef divisors (Theorem 4.3.1) and there is nothing further to prove. So we may assume that $a_1 = \frac{c}{d} \neq 0$ for positive integers $0 < c < d$. We focus on the vanishing of $H^j\big(X, \mathcal{O}_X(-N)\big)$ for $j < n$. Recall from Lemma 4.1.14 that it suffices to exhibit a finite surjective map

$$p : X' \longrightarrow X,$$

with $X'$ non-singular, such that

$$H^j\big(X', p^*(\mathcal{O}_X(-N))\big) \;=\; 0 \quad \text{for } j < n. \tag{9.4}$$

The vanishing (9.4) follows from a slight elaboration of Proposition 4.1.12, which is easiest to prove ab initio. Specifically, start by taking a Bloch–Gieseker covering (Theorem 4.1.10) to produce a non-singular branched covering

---

[4] Viehweg took a somewhat different tack in [589]. Specifically, he used cyclic covering constructions to deduce Theorem 9.1.18 from a vanishing result of Bogomolov (Remark 4.2.5).

$$p : X' \longrightarrow X$$

such that $\Delta'_i =_{\text{def}} p^* \Delta_i$ is smooth for all $i$, $\sum \Delta'_i$ is a simple normal crossing divisor, and

$$\Delta'_1 \equiv_{\text{lin}} dA' \tag{9.5}$$

for some (possibly ineffective) divisor $A'$ on $X'$. For ease of notation, set $N' = p^* N$ and $B' = p^* B$.

Now $\Delta'_1$ is divisible by $d$ in $\text{Pic}(X')$ thanks to (9.5), and so by 4.1.6 we can form next the $d$-fold cyclic covering

$$q : X'' \longrightarrow X'$$

branched along $\Delta'_1$. Writing $\Delta''_i = q^* \Delta'_i$, $A'' = q^* A'$, etc., we then have

$$N'' = q^* N' \equiv_{\text{lin}} B'' + cA'' + \sum_{i \geq 2} a_i \Delta''_i.$$

Then $N'' - cA'' \equiv_{\text{num}} B'' - \sum_{i \geq 2} a_i \Delta''_i$ and we have reduced the number of fractional divisors in play. So by induction we can assume that

$$H^j \left( X'', \mathcal{O}_{X''}(-N'' + cA'') \right) = 0$$

for $j < n$, and hence also that

$$H^j \left( X', q_* \left( \mathcal{O}_{X''}(-N'' + cA'') \right) \right) = 0. \tag{9.6}$$

But

$$q_* \mathcal{O}_{X''} = \mathcal{O}_{X'} \oplus \mathcal{O}_{X'}(-A') \oplus \dots \oplus \mathcal{O}_{X'}(-(d-1)A')$$

by Remark 4.1.7, and consequently

$$\begin{aligned} q_* \left( \mathcal{O}_{X''}(-N'' + cA'') \right) \\ = (q_* \mathcal{O}_{X''}) \left( -N' + cA' \right) \\ = \mathcal{O}_{X'}(-N' + cA') \oplus \dots \oplus \mathcal{O}_{X'}(-N' + (c - d + 1)A'). \end{aligned}$$

Since $c \leq d - 1$, $\mathcal{O}_{X'}(-N')$ occurs as one of the summands on the right, and so (9.4) follows from (9.6). $\qquad \square$

**Generalization 9.1.22. (Vanishing for a projective morphism).** On a few occasions we will require an extension of Theorem 9.1.18 to a relative setting. Specifically, suppose that $f : X \longrightarrow Y$ is a surjective projective (or merely proper) morphism of quasi-projective varieties, with $X$ non-singular. Recall that a **Q**-divisor $B$ on $X$ is said to be *f-nef* if its restriction to every fibre of $f$ is nef, and $D$ is *f-big* if its restriction to a general fibre of $f$ is big. The result in question is the following:

**Theorem.** *With $f : X \longrightarrow Y$ as above, suppose that $B$ is a $\mathbf{Q}$-divisor on $X$ that is $f$-nef and $f$-big, and whose fractional part has simple normal crossing support. Then*

$$R^j f_* \mathcal{O}_X \big( K_X + \lceil B \rceil \big) \; = \; 0 \quad \text{for} \;\; j > 0.$$

Roughly speaking, the idea is to construct a "compactification" $\bar{f} : \overline{X} \longrightarrow \overline{Y}$ of the given data, with $\overline{X}, \overline{Y}$ projective, and apply the theorem of Kawamata and Viehweg on $\overline{X}$ to deduce the stated result. We will carry out the argument in a special case during the course of the proof of the local vanishing theorem for multiplier ideals (Theorem 9.4.1). However in general the compactification process is somewhat technical, and we do not give details here. The reader may consult [326, (1.2.3)] for a sketch. □

**Remark 9.1.23. (Vanishing for R-divisors).** The theorem of Kawamata and Viehweg remains valid for **R**-divisors. Specifically:

Let $X$ be a smooth projective variety, $L$ an integer divisor, and $D$ an effective **R**-divisor on $X$ having SNC support. Assume that $L - D$ is big and nef. Then

$$H^i \big( X, \mathcal{O}_X(K_X + L - [D]) \big) \; = \; 0 \quad \text{for} \;\; i > 0. \qquad (*)$$

In fact, if $L - D$ is *ample* then by perturbing slightly the coefficients of $D$ one can find a $\mathbf{Q}$-divisor $D'$ such that $[D] = [D']$ and $L - D'$ is ample. So when $L - D$ is ample one can directly apply Corollary 9.1.21. The general case reduces to this one by passing to a suitable blow-up. In outline, using the argument of [368, Proposition 2.61] and Example 2.2.23, one constructs a modification $\mu : X' \longrightarrow X$ of $X$ on which there is an effective **R**-divisor $N'$ having the properties that $\mu^* D + N' + \operatorname{except}(\mu)$ has SNC support and $\mu^*(L - D) - \frac{1}{k} N'$ is ample for all $k \gg 0$. Now if $k$ is sufficiently large then

$$\big[ \mu^* D + \tfrac{1}{k} N' \big] \; = \; \big[ \mu^* D \big],$$

so by the case of ample divisors

$$K_{X'} + \mu^* L - [\mu^* D + \tfrac{1}{k} N'] \; = \; K_{X'/X} + \mu^*(K_X + L) - [\mu^* D]$$

has vanishing higher cohomology on $X'$. On the other hand,

$$\mu_* \mathcal{O}_{X'} \big( K_{X'/X} - [\mu^* D] \big) \; = \; \mathcal{O}_X \big( -[D] \big)$$

thanks to Lemma 9.2.19 and Remark 9.2.20. Moreover, as in the proof of Theorem 9.4.1, Lemma 4.3.10 shows that the corresponding higher direct images vanish, and $(*)$ follows. □

**Remark 9.1.24. (Injectivity theorem).** Just as the classical Kodaira vanishing is a special case of Kollár's injectivity theorem (Remark 4.3.8), so too is the theorem of Kawamata and Viehweg a special case of a more general statement due to Kollár [356] and Esnault–Viehweg [171]. Specifically:

> **Theorem.** *Let* $f : X \longrightarrow Y$ *be a surjective morphism of irreducible projective varieties, with* $X$ *smooth and* $Y$ *normal. Consider integral divisors* $N$ *and* $D$ *on* $X$, *with* $D$ *effective and* $f(D) \neq Y$. *Suppose that*
>
> $$N \equiv_{\mathrm{num}} f^*B + \Delta,$$
>
> *where* $B$ *is a nef and big* **Q**-*divisor on* $Y$ *and* $\Delta$ *is a* **Q**-*divisor on* $X$ *with SNC support that rounds down to zero:* $[\Delta] = 0$. *Then for every* $i > 0$ *the natural homomorphism*
>
> $$H^i\big(X, \mathcal{O}_X(K_X + N)\big) \xrightarrow{\;\cdot D\;} H^i\big(X, \mathcal{O}_X(K_X + N + D)\big)$$
>
> *determined by* $D$ *is injective.*

(It is actually enough that $\Delta$ be an effective **Q**-divisor such that the pair $(X, \Delta)$ is KLT (Definition 9.3.9). When $Y = X$ and $f$ is the identity, one recovers the theorem of Kawamata and Viehweg by taking $D$ to be a suitably large multiple of an ample divisor and applying Serre vanishing. See [174] and [362, Chapter 10] for further discussion. The stated result appears as Theorem 10.13 in [362]. □

# 9.2 Definition and First Properties of Multiplier Ideal Sheaves

In this section we introduce multiplier ideals. Working as usual on a smooth irreducible variety $X$, they will be defined in three contexts:

- For an effective **Q**-divisor $D$ on $X$, its multiplier ideal $\mathcal{J}(D) \subseteq \mathcal{O}_X$ reflects the extent to which the fractional part of $D$ fails to have normal crossing support;

- For an ideal sheaf $\mathfrak{a} \subseteq \mathcal{O}_X$ and a rational number $c > 0$, the multiplier ideal $\mathcal{J}(\mathfrak{a}^c) \subseteq \mathcal{O}_X$ measures the singularities of divisors of elements $f \in \mathfrak{a}$, with "nastier" singularities giving rise to "deeper" ideals;

- When $|V| \subseteq |L|$ is a linear series and $c > 0$ is a rational number, $\mathcal{J}(c \cdot |V|) \subseteq \mathcal{O}_X$ is controlled by the geometry of the divisors appearing in $|V|$.

One can pass back and forth between these settings, but in a given application there is often a conceptual preference for one over the other. Generally speaking we state results in each of the three contexts, but provide proofs only once.

We start with the definitions and some simple examples, and establish independence of the resolution appearing in the construction. Then we spell out the relations between the different sorts of multiplier ideals and present some first properties.

### 9.2.A Definition of Multiplier Ideals

We begin with **Q**-divisors.

**Definition 9.2.1. (Multiplier ideal of an effective Q-divisor).** Let $D$ be an effective **Q**-divisor on a smooth complex variety $X$, and fix a log resolution $\mu : X' \longrightarrow X$ of $D$. Then the *multiplier ideal sheaf*

$$\mathcal{J}(D) = \mathcal{J}(X, D) \subseteq \mathcal{O}_X$$

associated to $D$ is defined to be

$$\mathcal{J}(D) = \mu_* \mathcal{O}_{X'} \left( K_{X'/X} - [\mu^* D] \right). \tag{9.7}$$

(It follows from (9.2) that $\mathcal{J}(D) \subseteq \mathcal{O}_X$ is indeed a sheaf of ideals).     □

We will check shortly (Theorem 9.2.18) that $\mathcal{J}(D)$ is independent of the log resolution $\mu$. Remark 9.2.11 presents some preliminary motivation for this definition. The analytic construction of multiplier ideals is indicated in Section 9.3.D.

**Remark 9.2.2. (Ineffective divisors).** In case $D$ is possibly ineffective one can use the same equation (9.7) to define $\mathcal{J}(D)$, but then it is generally not a submodule of $\mathcal{O}_X$. However, we may view $\mathcal{J}(D)$ as a fractional ideal sheaf, i.e. a rank-one $\mathcal{O}_X$-submodule of the field of rational functions $\mathbf{C}(X)$ on $X$, and we will still refer to $\mathcal{J}(D)$ as a multiplier ideal sheaf.     □

We next consider the case of ideal sheaves:

**Definition 9.2.3. (Multiplier ideal associated to an ideal sheaf).** Let $\mathfrak{a} \subseteq \mathcal{O}_X$ be a non-zero ideal sheaf, and $c > 0$ a rational number. Fix a log resolution $\mu : X' \longrightarrow X$ of $\mathfrak{a}$ as in 9.1.12, with $\mathfrak{a} \cdot \mathcal{O}_{X'} = \mathcal{O}_{X'}(-F)$. The *multiplier ideal* $\mathcal{J}(c \cdot \mathfrak{a}) = \mathcal{J}(\mathfrak{a}^c)$ associated to $c$ and $\mathfrak{a}$ is defined as

$$\mathcal{J}(\mathfrak{a}^c) = \mathcal{J}(X, \mathfrak{a}^c) = \mu_* \mathcal{O}_{X'} \left( K_{X'/X} - [c \cdot F] \right).$$

Observe that we sometimes write this ideal "additively" as $\mathcal{J}(c \cdot \mathfrak{a})$, while on other occasions we prefer the "exponential" notation $\mathcal{J}(\mathfrak{a}^c)$. When $c = 1$ we write simply $\mathcal{J}(\mathfrak{a})$.     □

**Remark 9.2.4.** If $\mathfrak{a} \subseteq \mathcal{O}_X$ is an ideal, and $m \in \mathbf{N}$ is a whole number, then it follows at once from the definition that

$$\mathcal{J}(m \cdot \mathfrak{a}) = \mathcal{J}(\mathfrak{a}^m),$$

where $\mathfrak{a}^m$ denotes the usual $m^{\text{th}}$ power of $\mathfrak{a}$. This justifies our use of "exponential notation" $\mathcal{J}(\mathfrak{a}^c)$ for arbitrary rational $c > 0$.    □

**Remark 9.2.5.** Note that we are not trying to attach any actual meaning to the symbols $c \cdot \mathfrak{a}$ or $\mathfrak{a}^c$, although we will see shortly that the resulting multiplier ideals are related to the multiplier ideals of the $\mathbf{Q}$-divisors $c \cdot \operatorname{div}(f)$ for $f \in \mathfrak{a}$ (Proposition 9.2.26).    □

**Convention 9.2.6.** When $\mathfrak{a} = (0)$ is the zero ideal, it is convenient to decree that $\mathcal{J}(c \cdot \mathfrak{a}) = (0)$ for every $c > 0$.    □

**Remark 9.2.7. (Ideals in affine coordinate rings).** Assume that $X$ is affine, with coordinate ring $A = \mathbf{C}[X]$. Then any ideal $\mathfrak{a} \subseteq A$ determines a sheaf on $X$. Therefore given $c > 0$, its multiplier ideal $\mathcal{J}(c \cdot \mathfrak{a})$ is defined, and by the same token we can view $\mathcal{J}(c \cdot \mathfrak{a})$ as an ideal in $A$.    □

On a number of occasions it will be useful to have an analogous notion for several ideals:

**Generalization 9.2.8. ("Mixed" multiplier ideals).** Suppose given non-zero ideals $\mathfrak{a}_1, \ldots, \mathfrak{a}_t \subseteq \mathcal{O}_X$ and rational numbers $c_1, \ldots, c_t > 0$. Then

$$\mathcal{J}\left(X, \mathfrak{a}_1^{c_1} \cdot \ldots \cdot \mathfrak{a}_t^{c_t}\right) \subseteq \mathcal{O}_X$$

is defined as follows. Take a common log resolution $\mu : X' \longrightarrow X$ of the $\mathfrak{a}_i$, with $\mathfrak{a}_i \cdot \mathcal{O}_{X'} = \mathcal{O}_{X'}(-F_i)$ and $\sum F_i + \operatorname{except}(\mu)$ having SNC support, and put

$$\mathcal{J}\left(X, \mathfrak{a}_1^{c_1} \cdot \ldots \cdot \mathfrak{a}_t^{c_t}\right) = \mu_* \mathcal{O}_{X'}\left(K_{X'/X} - [c_1 F_1 + \ldots + c_t F_t]\right).$$

In "additive" notation (which is certainly less preferable here!) one might denote this by

$$\mathcal{J}\left((c_1 \cdot \mathfrak{a}_1) \cdot \ldots \cdot (c_t \cdot \mathfrak{a}_t)\right). \quad \square$$

**Example 9.2.9.** In the situation of 9.2.8 one can find a single ideal $\mathfrak{b} \subseteq \mathcal{O}_X$ and exponent $d \in \mathbf{Q}$ such that $\mathcal{J}(\mathfrak{a}_1^{c_1} \cdot \ldots \cdot \mathfrak{a}_t^{c_t}) = \mathcal{J}(\mathfrak{b}^d)$. (Choose integers $C_i, N > 0$ such that $c_i = \frac{C_i}{N}$. Then take $\mathfrak{b} = \mathfrak{a}_1^{C_1} \cdot \ldots \cdot \mathfrak{a}_t^{C_t}$ and $d = \frac{1}{N}$.)    □

Finally, we consider the case of linear series:

**Definition 9.2.10. (Multiplier ideal of a linear series).** Let $|V| \subseteq |L|$ be a non-empty linear series on $X$, and let $\mu : X' \longrightarrow X$ be a log resolution of $|V|$ as in 9.1.11, with

$$\mu^* |V| = |W| + F.$$

Given a rational number $c > 0$, the *multiplier ideal* $\mathcal{J}(c \cdot |V|)$ corresponding to $c$ and $|V|$ is

$$\mathcal{J}(c \cdot |V|) = \mathcal{J}(X, c \cdot |V|) = \mu_* \mathcal{O}_{X'}(K_{X'/X} - [c \cdot F]). \quad \square$$

As before, we do not pretend to assign any independent meaning to the expression $c \cdot |V|$. In the spirit of 9.2.6 we adopt the convention that if $|V| = \varnothing$, then $\mathcal{J}(c \cdot |V|) = 0$ for every $c > 0$.

**Remark 9.2.11. (Preview of Nadel vanishing).** It might be helpful to give at once a few words of motivation for these definitions. Assume then that $X$ is projective, and suppose given an integral divisor $L$ and effective **Q**-divisor $D$ on $X$ with $L - D$ big and nef. We attempt to apply the theorem of Kawamata and Viehweg to prove a vanishing involving $L - D$. To this end, let $\mu : X' \longrightarrow X$ be a log resolution of $(X, D)$. Then $\mu^*(L - D)$ is nef and big on $X'$, with simple normal crossing support, so we can apply Corollary 9.1.21 to conclude:

$$H^i(X', \mathcal{O}_{X'}(K_{X'} + \mu^* L - [\mu^* D])) = 0 \qquad (*)$$

for $i > 0$. The plan is to take direct images to deduce a statement on $X$. For this note that

$$K_{X'} + \mu^* L - [\mu^* D] = \mu^*(K_X + L) + (K_{X'/X} - [\mu^* D]),$$

and consequently

$$\begin{aligned} \mu_* \mathcal{O}_{X'}(K_{X'} + \mu^* L - [\mu^* D]) &= \mathcal{O}_X(K_X + L) \otimes \mu_* \mathcal{O}_X(K_{X'/X} - [\mu^* D]) \\ &= \mathcal{O}_X(K_X + L) \otimes \mathcal{J}(D) \end{aligned}$$

thanks to the projection formula. Once one knows that the corresponding higher direct images disappear, $(*)$ will imply the vanishing of the higher cohomology of $\mathcal{O}_X(K_X + L) \otimes \mathcal{J}(D)$. This is the Nadel vanishing theorem, which we shall (re)prove in Section 9.4.B. For the moment, the point we wish to emphasize is that the multiplier ideal $\mathcal{J}(D)$ arises naturally when one tries to "push down" vanishings from a log resolution of $D$. $\qquad \square$

We next give a few simple examples. The last two hint at the fact that the multiplier ideal of a **Q**-divisor reflects the singularities of its fractional part.

**Example 9.2.12.** Let $A$ be an integral divisor on $X$. Then $\mathcal{J}(A) = \mathcal{O}_X(-A)$. (Observe that $[\mu^* A] = \mu^* A$, and apply the projection formula.) $\qquad \square$

**Example 9.2.13.** If $D$ is any **Q**-divisor on $X$ with normal crossing support, then $\mathcal{J}(D) = \mathcal{O}_X(-[D])$. $\qquad \square$

**Example 9.2.14.** Suppose that $\dim X = n$, and let $\mathfrak{m} \subseteq \mathcal{O}_X$ denote the maximal ideal of a point $x \in X$. Then

$$\mathcal{J}(\mathfrak{m}^c) = \mathfrak{m}^{(\lceil c \rceil + 1 - n)}$$

provided that $c \geq n$ and $\mathcal{J}(\mathfrak{m}^c) = \mathcal{O}_X$ otherwise. (A log resolution of $\mathfrak{m}$ is obtained by simply blowing up $x$.) □

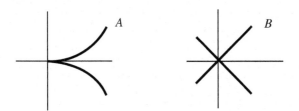

**Figure 9.3.** Curves with double points

**Example 9.2.15.** Let $X = \mathbf{C}^2$ with coordinates $x$ and $y$, and let $A \subseteq \mathbf{C}^2$ be the cuspidal curve $A = \{y^2 = x^3\}$ (Figure 9.3). Then

$$\mathcal{J}(\tfrac{5}{6}A) = (x, y),$$

while $\mathcal{J}(cA) = \mathcal{O}_X$ for $0 < c < \tfrac{5}{6}$. (Use the standard log resolution of $A$ (cf. Example 9.1.13 and Figure 9.2) to make an explicit calculation. A more general result appears in Theorem 9.3.27.) By contrast, if $B \subseteq \mathbf{C}^2$ is the curve $\{y^2 = x^2\}$ having an ordinary double point, then

$$\mathcal{J}(cB) = \mathcal{O}_X \quad \text{for } 0 < c < 1.$$

In other words, multiplier ideals recognize the curve $A$ as having a "worse" singularity than does $B$. □

**Example 9.2.16.** Take $X = \mathbf{C}^n$, and let $A \subseteq X$ be a divisor that is non-singular except for an ordinary $(n + 1)$-fold point at the origin $0$ (i.e. assume that the origin is an isolated singular point of $X$ at which its projectivized tangent cone is a non-singular hypersurface of degree $n + 1$). Then

$$\mathcal{J}\left(\mathbf{C}^n, \tfrac{n}{n+1}A\right) = \mathfrak{m},$$

where $\mathfrak{m} \subseteq \mathcal{O}_X$ denotes the maximal ideal of the origin. (The hypotheses imply that a log resolution of $(X, A)$ may be constructed by a single blow-up of $\mathbf{C}^n$ at the origin.) See Proposition 9.3.2 for a more general statement. □

**Remark 9.2.17. (R-divisors).** Definition 9.2.1 makes perfectly good sense when $D$ is an **R**-divisor, and by the same token one can work with real coefficients in 9.2.3, 9.2.8, and 9.2.10. □

We refer the reader to Section 9.3 for further examples.

The first point to verify is that these constructions yield well-defined ideal sheaves:

**Theorem 9.2.18. (Esnault–Viehweg, cf. [174, 7.3]).** *In the situations of Definitions 9.2.1, 9.2.10, and 9.2.3, the multiplier ideal sheaves $\mathcal{J}(D)$, $\mathcal{J}(c \cdot |V|)$ and $\mathcal{J}(c \cdot \mathfrak{a})$ are independent of the log resolutions used to construct them.*

*Proof.* We will focus on the ideals $\mathcal{J}(D)$. Consider first a sequence of maps

$$V \xrightarrow{g} X' \xrightarrow{\mu} X,$$

where $\mu$ is a log resolution of $D$, and $g$ is a log resolution of $\mu^*D + \mathrm{except}(\mu)$. Assume for the moment that we are able to prove

$$g_* \mathcal{O}_V \left( K_{V/X'} - [g^*\mu^*D] \right) = \mathcal{O}_{X'} \left( -[\mu^*D] \right). \tag{9.8}$$

Then, setting $h = \mu \circ g$, one finds using the projection formula:

$$
\begin{aligned}
\mu_* \mathcal{O}_{X'} \left( K_{X'/X} - [\mu^*D] \right) &= \mu_* \left( \mathcal{O}_{X'}(K_{X'/X}) \otimes g_* \mathcal{O}_V \left( K_{V/X'} - [g^*\mu^*D] \right) \right) \\
&= \mu_* g_* \left( g^* \mathcal{O}_{X'}(K_{X'/X}) \otimes \mathcal{O}_V \left( K_{V/X'} - [g^*\mu^*D] \right) \right) \\
&= h_* \mathcal{O}_V \left( K_{V/X} - [h^*D] \right).
\end{aligned}
$$

In other words, we get the same multiplier ideal working from $h$ as working from $\mu$. But any two resolutions can be dominated by a third (Example 9.1.16), and so the theorem follows. Thus we are reduced to checking that if the underlying divisor — $\mu^*D$ in (9.8) — has normal crossing support, then nothing is changed by passing to a further resolution. This is the content of the next lemma.  $\square$

**Lemma 9.2.19.** *Let $X$ be a smooth variety of dimension $n$, and $D$ any **Q**-divisor on $X$ with simple normal crossing support. Suppose that $\mu : X' \longrightarrow X$ is a log resolution of $D$. Then*

$$\mu_* \mathcal{O}_{X'} \left( K_{X'/X} - [\mu^*D] \right) = \mathcal{O}_X \left( -[D] \right). \tag{9.9}$$

*Proof.* It follows from the projection formula that if (9.9) holds for a given **Q**-divisor $D$, then it holds also for $D + A$ whenever $A$ is integral. Therefore it suffices to prove the lemma under the additional hypothesis that $[D] = 0$, in which case we need to show that $[\mu^*D] \preccurlyeq K_{X'/X}$.

We start by fixing some notation. Let $E \subset X'$ be a prime divisor. We are required to establish that

$$\mathrm{ord}_E \left( [\mu^*D] \right) \leq \mathrm{ord}_E (K_{X'/X}). \tag{9.10}$$

In view of the hypothesis that $[D] = 0$ this is clear if $E$ is non-exceptional. So we may suppose that $E$ appears in the exceptional locus of $\mu$. Let

$$Z = \mu(E) \subseteq X \quad , \quad e = \mathrm{codim}_X Z.$$

Choose a general point $y \in E$ and put $x = \mu(y) \in Z$. Denote by $D_1, \ldots, D_\ell$ the prime components of $D$ that pass through $Z$. Then $\ell \le e$, and since only these components contribute to the left-hand side of (9.10) we can suppose moreover that

$$D = \sum_{i=1}^{\ell} a_i D_i \quad , \quad 0 \le a_i < 1.$$

We next choose local analytic coordinates $z_1, \ldots, z_n$ on $X'$ centered at $y$ and $w_1, \ldots, w_n$ on $X$ centered at $x$ such that $E$ is locally defined by ($z_1 = 0$). Since $D$ has SNC support we can assume that

$$D_i =_{\text{locally}} \operatorname{div}(w_i) \ (1 \le i \le \ell).$$

Let $c_i = \operatorname{ord}_E(\mu^* D_i)$. Thus

$$\operatorname{ord}_E(\mu^* D) = \sum_{i=1}^{\ell} a_i c_i < \sum_{i=1}^{\ell} c_i ,$$

and there exist functions $b_i \in \mathbf{C}\{z\}$ such that

$$w_i = z_1^{c_i} \cdot b_i \ (1 \le i \le \ell).$$

Then $\mu^* dw_i = c_i z_1^{c_i-1} b_i \cdot dz_1 + z_1^{c_i} \cdot db_i$ for $1 \le i \le \ell$ and consequently

$$\mu^*\big(dw_1 \wedge \ldots \wedge dw_n\big) = z_1^{\gamma-1} g \cdot dz_1 \wedge \ldots \wedge dz_n$$

for some $g \in \mathbf{C}\{z\}$, where $\gamma = \sum_{i=1}^{\ell} c_i$. So

$$\operatorname{ord}_E\big(K_{X'/X}\big) \ge \Big(\sum_{i=1}^{\ell} c_i\Big) - 1$$
$$> \operatorname{ord}_E\big(\mu^* D\big) - 1,$$

and (9.10) follows. $\qquad\qquad\qquad\qquad\qquad\qquad\qquad\qquad\qquad\square$

**Remark 9.2.20. (R-divisors).** The argument just completed does not use that $D$ has rational coefficients, and so the lemma remains true for an **R**-divisor $D$ with SNC support. Therefore Theorem 9.2.18 extends to **R**-divisors or multiplier ideals with real coefficients. $\qquad\qquad\qquad\qquad\qquad\square$

**Generalization 9.2.21. (Relative linear series).** Let $f : X \longrightarrow T$ be a projective (or merely proper) mapping, let $L$ be an integral divisor on $X$ whose restriction to a general fibre of $f$ has a non-zero section, and let $\rho : f^* f_* \mathcal{O}_X(L) \longrightarrow \mathcal{O}_X(L)$ be the canonical homomorphism. Fix a log resolution $\mu : X' \longrightarrow X$ of $\rho$ in the sense of 9.1.17, so that $\operatorname{im}\big(\mu^* \rho\big) = \mathcal{O}_X(\mu^* L - F)$. Given $c > 0$ we define

$$\mathcal{J}(f, c \cdot \rho) \;=\; \mathcal{J}\big(f, c \cdot |L|\big) \;=\; \mu_* \mathcal{O}_{X'}\big(K_{X'/X} - [cF]\big).$$

The notation is intended to emphasize the viewpoint that these are relative variants of the the multiplier ideals associated to a complete linear series. We leave it to the reader to check that this relative multiplier ideal is independent of the choice of resolution. We also leave to the reader the natural extension of this definition to the case of "incomplete linear series" given by a coherent subsheaf $V \subseteq f_* \mathcal{O}_X(L)$. □

### 9.2.B First Properties

Having established that multiplier ideals are well-defined, we now spell out some first properties. Numerous examples appear in the next section. As usual, $X$ denotes a non-singular variety.

We begin by discussing the relations between the various sorts of multiplier ideals we have defined. The first statement shows that the multiplier ideals attached to a linear series coincide with those arising from the base ideal of the series in question.

**Proposition 9.2.22. (Linear series and base ideals, I).** *Let $L$ be an integral divisor on $X$. Given a linear series $|V| \subseteq |L|$ on $X$, denote by $\mathfrak{b} = \mathfrak{b}\big(|V|\big)$ its base ideal. Then*

$$\mathcal{J}\big(c \cdot |V|\big) \;=\; \mathcal{J}\big(c \cdot \mathfrak{b}\big)$$

*for every $c > 0$.*

*Proof.* Take a log resolution $\mu : X' \longrightarrow X$ of $|V|$ as in 9.1.11 with $f^*|V| = |W| + F$. Then $\mu$ is also a log resolution of $\mathfrak{b}$ with $\mathfrak{b} \cdot \mathcal{O}_{X'} = \mathcal{O}_{X'}(-F)$, and the assertion follows. □

**Example 9.2.23. (Linear series and base ideals, II).** One can also pass from ideal sheaves to linear series. Specifically, let $\mathfrak{a} \subseteq \mathcal{O}_X$ be an ideal sheaf and $L$ an integral divisor on $X$ such that $\mathfrak{a} \otimes \mathcal{O}_X(L)$ is globally generated. Choose a finite-dimensional space of sections

$$V \;\subseteq\; H^0\big(X, \mathfrak{a} \otimes \mathcal{O}_X(L)\big)$$

that generates the sheaf in question, giving rise to a linear series $|V| \subseteq |L|$. Then for every $c > 0$,

$$\mathcal{J}\big(c \cdot \mathfrak{a}\big) \;=\; \mathcal{J}\big(c \cdot |V|\big).$$

(The hypotheses imply that $\mathfrak{a}$ is the base ideal of $|V|$.) □

**Example 9.2.24. (Linear series and base ideals, III).** In the situation of 9.2.23, suppose more generally that $\mathfrak{a}^p \otimes \mathcal{O}_X(pL)$ is globally generated for some integer $p \geq 1$. If $W \subseteq H^0\big(X, \mathfrak{a}^p \otimes \mathcal{O}_X(pL)\big)$ is a finite-dimensional subspace that generates, then

$$\mathcal{J}(c \cdot \mathfrak{a}) \;=\; \mathcal{J}\big(\frac{c}{p} \cdot |W|\big). \quad \square$$

**Example 9.2.25.** In the setting of Generalization 9.2.21, let $\mathfrak{b} \subseteq \mathcal{O}_X$ be the base ideal of $\rho : f^* f_* \mathcal{O}_X(L) \longrightarrow \mathcal{O}_X(L)$. Then $\mathcal{J}\big(f; c \cdot \rho\big) = \mathcal{J}\big(c \cdot \mathfrak{b}\big)$. $\quad \square$

We next show that the multiplier ideals coming from linear series or ideal sheaves can be computed in terms of the **Q**-divisors arising from sufficiently general elements of the series or ideal in question.

**Proposition 9.2.26. (Linear series and Q-divisors).** *Let $X$ be a smooth variety, and $|V| \subseteq |L|$ a non-empty linear series on $X$. Fix a rational number $c > 0$, choose $k > c$ general divisors $A_1, \ldots, A_k \in |V|$, and set*

$$D \;=\; \frac{1}{k} \cdot (A_1 + \cdots + A_k).$$

*Then $\mathcal{J}\big(c \cdot |V|\big) = \mathcal{J}\big(c \cdot D\big)$.*

*Proof.* Write $\mu^* |V| = |W| + F$ as in Definition 9.1.11. Then

$$\mu^* A_i \;=\; A_i' + F$$

for general divisors $A_i' \in |W|$. Since $|W|$ is free, Lemma 9.1.9 implies that $F + \sum A_i' + \text{except}(\mu)$ has SNC support. So $X'$ is a log resolution of $(X, D)$, and we can compute the multiplier ideal $\mathcal{J}(cD)$ on $X'$. But since $k > c$:

$$K_{X'/X} - \big[\mu^*(cD)\big] \;=\; K_{X'/X} - \Big[k \cdot \frac{c}{k}F + \sum \frac{c}{k} \cdot A_i'\Big]$$
$$\;=\; K_{X'/X} - \big[cF\big].$$

The assertion then follows from the definition of $\mathcal{J}\big(c \cdot |V|\big)$. $\quad \square$

Turning to the analogue of Proposition 9.2.26 for ideal sheaves, it will be helpful to be able to speak of general elements in an ideal.

**Definition 9.2.27. (General elements of an ideal).** Assuming that $X$ is affine, let $\mathfrak{a} \subseteq \mathbf{C}[X]$ be a non-zero ideal and choose generators $g_1, \ldots, g_p \in \mathfrak{a}$. A *general element* of $\mathfrak{a}$ (with respect to the given generators) is a general **C**-linear combination of the $g_i$. $\quad \square$

Then one has

**Proposition 9.2.28.** *Assume that $X$ is affine, let $\mathfrak{a} \subseteq \mathbf{C}[X]$ be an ideal, and $c > 0$ a rational number. Fix $k > c$ general elements*

$$f_1, \ldots, f_k \in \mathfrak{a}$$

*(with respect to any set of generators), and let $A_i = \operatorname{div} f_i$ be the divisor of $f_i$. Put $D = \frac{1}{k} \sum A_i$, so that $D$ is an effective **Q**-divisor on $X$. Then*

$$\mathcal{J}(\mathfrak{a}^c) \;=\; \mathcal{J}(X, c \cdot D). \quad \square$$

The next examples deal with perturbations of a given divisor.

**Example 9.2.29. (Kollár–Bertini theorem, I).** Let $D$ be an effective **Q**-divisor on a smooth variety $X$, and let $|V|$ be a free linear series on $X$. If $A \in |V|$ is a general divisor, then

$$\mathcal{J}(D + cA) = \mathcal{J}(D)$$

for every $0 < c < 1$. (Argue as in the proof of Proposition 9.2.26.) This is the generalization to multiplier ideals of some Bertini-type statements given by Kollár [364, Theorem 4.8] for singularities of pairs. See 9.3.50 for a related result.     □

**Example 9.2.30. (Small perturbations).** Let $D$ and $D'$ be **Q**-divisors on a non-singular variety $X$, with $D'$ effective. Then

$$\mathcal{J}(X, D) = \mathcal{J}(X, D + \varepsilon D')$$

for all $0 < \varepsilon \ll 1$. Similarly, if $\mathfrak{a}, \mathfrak{b} \subseteq \mathcal{O}_X$ are ideals, then given $c > 0$,

$$\mathcal{J}(\mathfrak{a}^c) = \mathcal{J}(\mathfrak{a}^c \cdot \mathfrak{b}^d)$$

for sufficiently small $d > 0$. (For the first assertion, let $\mu : X' \longrightarrow X$ be a log resolution of $D + D'$. If $\varepsilon > 0$ is sufficiently small, then

$$\mathrm{ord}_E\big(K_{X'/X} - [\mu^* D]\big) = \mathrm{ord}_E\big(K_{X'/X} - [\mu^*(D + \varepsilon D')]\big)$$

for every prime divisor $E$ on $X'$, and the first statement follows. The proof of the second is similar.)     □

We next study the effect of adding integral divisors, and show that multiplier ideals satisfy various natural containment relations.

**Proposition 9.2.31. (Adding integral divisors).** *Let $X$ be a smooth variety, $A$ an integral divisor, and $D$ any **Q**-divisor on $X$ (effective or not). Then*

$$\mathcal{J}(D + A) = \mathcal{J}(D) \otimes \mathcal{O}_X(-A).$$

*Proof.* Let $\mu : X' \longrightarrow X$ be a log resolution of $(X, D)$. If $A$ is integral, then $[\mu^* D + \mu^* A] = [\mu^* D] + \mu^* A$ for any **Q**-divisor $D$. Therefore

$$
\begin{aligned}
\mu_* \mathcal{O}_{X'}\big(K_{X'/X} - [\mu^*(D + A)]\big) &= \mu_* \big(\mathcal{O}_{X'}(K_{X'/X} - [\mu^* D]) \otimes \mathcal{O}_{X'}(-\mu^* A)\big) \\
&= \mu_* \big(\mathcal{O}_{X'}(K_{X'/X} - [\mu^* D])\big) \otimes \mathcal{O}_X(-A) \\
&= \mathcal{J}(D) \otimes \mathcal{O}_X(-A),
\end{aligned}
$$

as required.     □

**Proposition 9.2.32. (Containments).** *Let $X$ be a smooth variety.*

(i).  If $D_2 \preccurlyeq D_1$ are **Q**-divisors on $X$, then $\mathcal{J}(D_1) \subseteq \mathcal{J}(D_2)$.

(ii).  If $|V_1| \subseteq |V_2| \subseteq |L|$ are two linear series on $X$, then $\mathcal{J}(c \cdot |V_1|) \subseteq \mathcal{J}(c \cdot |V_2|)$ for any rational number $c > 0$.

(iii).  If $\mathfrak{a}_1 \subseteq \mathfrak{a}_2 \subseteq \mathcal{O}_X$ are ideal sheaves, then for any rational number $c > 0$,

$$\mathcal{J}(\mathfrak{a}_1^c) \subseteq \mathcal{J}(\mathfrak{a}_2^c).$$

(iv).  Given any ideal $\mathfrak{a} \subseteq \mathcal{O}_X$, one has the inclusion

$$\mathfrak{a} \subseteq \mathcal{J}(\mathfrak{a}).$$

(v).  If $|V| \subseteq |L|$ is a non-empty linear series, with base ideal $\mathfrak{b} = \mathfrak{b}(|V|)$, then

$$\mathfrak{b} \subseteq \mathcal{J}(|V|).$$

*Proof.* Statement (i) is clear from the definition. For (ii), let $\mu : X' \longrightarrow X$ be a common log resolution of $|V_1|$ and $|V_2|$, with

$$\mu^* |V_1| = |W_1| + F_1 \ , \ \mu^* |V_2| = |W_2| + F_2.$$

Then $|W_1| \subseteq |W_2|$, and consequently their fixed divisors satisfy the reverse inclusion $F_2 \preccurlyeq F_1$. This being said, (ii) follows from the definition, and (iii) is similar. Assertion (iv) is a consequence of the fact that the relative canonical bundle $K_{X'/X}$ of a log resolution of $\mathfrak{a}$ is effective, and (v) is a restatement of (iv).  $\qquad\square$

Finally, we record an elementary but important fact about the behavior of multiplier ideals under birational morphisms. It will be the key to comparing the ideals defined here with their analytic counterparts.

**Theorem 9.2.33. (Birational transformation rule).** *Let $f : Y \longrightarrow X$ be a projective (or proper) birational map of non-singular varieties, and let $D$ be any **Q**-divisor on $X$ (not necessarily effective). Then*

$$\mathcal{J}(X, D) = f_*\Big(\mathcal{J}(Y, f^*D) \otimes \mathcal{O}_Y(K_{Y/X})\Big).$$

*More symmetrically, the assertion is that*

$$\mathcal{O}_X(K_X) \otimes \mathcal{J}(X, D) = f_*\Big(\mathcal{O}_Y(K_Y) \otimes \mathcal{J}(Y, f^*D)\Big). \qquad (9.11)$$

Of course there are analogous statements for ideal sheaves and linear series which we leave to the reader to formulate.

*Proof of Theorem 9.2.33.* This is a consequence of the projection formula, as in the proof of Proposition 9.2.18. Specifically, let $\nu : V \longrightarrow Y$ be a log resolution of $f^*D + \text{except}(f)$, and denote by $h : V \longrightarrow X$ the composition $h = f \circ \nu$. Then

$$
\begin{aligned}
\mathcal{J}(X, D) &= h_* \left( \mathcal{O}_V(K_{V/X} - [h^*D]) \right) \\
&= f_* \nu_* \left( \mathcal{O}_V(K_{V/Y} - [\nu^* f^* D]) \otimes \nu^* \mathcal{O}_Y(K_{Y/X}) \right) \\
&= f_* \left( \nu_* \mathcal{O}_V(K_{V/Y} - [\nu^* f^* D]) \otimes \mathcal{O}_Y(K_{Y/X}) \right) \\
&= f_* \left( \mathcal{J}(Y, f^*D) \otimes \mathcal{O}_Y(K_{Y/X}) \right),
\end{aligned}
$$

as asserted.                                                                    □

**Remark 9.2.34. (R coefficients).** As before, all of these results remain valid for **R**-divisors and for real values of the coefficient $c$.                    □

**Remark 9.2.35. (Realizing ideals as multiplier ideals).** Assuming for simplicity that $X$ is affine with coordinate ring $\mathbf{C}[X]$, it is very interesting to ask which ideals $\mathfrak{j} \subseteq \mathbf{C}[X]$ arise as the multiplier ideal of some **Q**-divisor $D$. Other than integral closure (Corollary 9.6.13), there are no known local obstructions to an ideal being a multiplier ideal. In fact, Lipman–Watanabe [405] and Favre–Jonsson [184] have shown that on a smooth surface, any integrally closed ideal is locally a multiplier ideal. It is unclear what to expect in higher dimensions.                                                            □

## 9.3 Examples and Complements

This section is devoted to examples and further constructions involving multiplier ideals.

We start by discussing the effect of multiplicity on multiplier ideals, and then survey some interesting invariants of singularities defined by means of these ideals. The third subsection presents Howald's computation [293], [292], [294] of the multiplier ideals attached to monomial ideals, and in 9.3.D we give a brief survey of the analytic approach to the theory. After discussing adjoint ideals, and the relationship between multiplier and Jacobian ideals, we conclude in 9.3.G with the situation on singular varieties. Unless otherwise stated, $X$ denotes as usual a smooth algebraic variety.

### 9.3.A Multiplier Ideals and Multiplicity

We investigate in this subsection interactions between the multiplicity of a **Q**-divisor and its multiplier ideals.

**Definition 9.3.1. (Multiplicity of a Q-divisor).** Let $D = \sum a_i D_i$ be an effective **Q**-divisor on the smooth variety $X$, and let $x \in X$ be a fixed point. The *multiplicity* $\text{mult}_x D$ is the rational number

$$\mathrm{mult}_x D \;=\; \sum a_i \cdot \mathrm{mult}_x D_i.$$

More generally, given any irreducible subvariety $Z \subseteq X$,

$$\mathrm{mult}_Z D \;=\; \sum a_i \cdot \mathrm{mult}_Z D_i,$$

where as in Section 5.2.B $\mathrm{mult}_Z D_i$ denotes the multiplicity of $D_i$ at a generic point of $Z$. □

A basic fact is that points of high multiplicity force the non-triviality of multiplier ideals:

**Proposition 9.3.2.** *Assume that $X$ has dimension $n$, and let $D$ be an effective **Q**-divisor on $X$. If $\mathrm{mult}_x D \ge n$ at some point $x \in X$, then $\mathcal{J}(D)$ is non-trivial at $x$, i.e. $\mathcal{J}(D) \subseteq \mathfrak{m}_x$, where $\mathfrak{m}_x$ is the maximal ideal of $x$. More generally, if*

$$\mathrm{mult}_x D \;\ge\; n + p - 1$$

*for some integer $p \ge 1$, then $\mathcal{J}(D) \subseteq \mathfrak{m}_x^p$.*

*Proof.* Let $\mu : X' \longrightarrow X$ be a log resolution of $D$ constructed by first blowing up $X$ at $x$. The exceptional divisor of this initial blow-up determines a prime divisor $E \subseteq X'$ such that

$$\mathrm{ord}_E(K_{X'/X}) \;=\; n - 1 \;\;,\;\; \mathrm{ord}_E(\mu^* D) \;=\; \mathrm{mult}_x D.$$

The divisor $E$ corresponds to the valuation given by order of vanishing at $x$, and in particular $\mu_* \mathcal{O}_{X'}(-mE) = \mathfrak{m}_x^m$. Therefore if $\mathrm{mult}_x D \ge n + p - 1$, then

$$\mathrm{ord}_E(K_{X'/X} - [\mu^* D]) \;\le\; -p,$$

and consequently $\mathcal{J}(D) \subseteq \mu_* \mathcal{O}_{X'}(-pE) = \mathfrak{m}_x^p$, as required. □

**Remark 9.3.3. (Intermediate multiplicities).** We will prove later that if $\mathrm{mult}_x D < 1$, then $\mathcal{J}(X, D)$ is trivial at $x$ (Proposition 9.5.13). However, if

$$1 \;\le\; \mathrm{mult}_x D \;<\; n,$$

then $\mathcal{J}(X, D)$ may or may not be trivial near $x$. In Example 9.5.14 we construct effective (fractional) **Q**-divisors $D$ with multiplicities arbitrarily close to 1 whose multiplier ideals are non-trivial.[5] □

For the analogous statements involving the multiplicity of a **Q**-divisor along a subvariety, it will be helpful to recall a definition that appeared in Example 2.4.16 (iv):

---

[5] Of course if $D$ is integral and effective, then $\mathcal{J}(D) = \mathcal{O}_X(-D)$ is non-trivial at any point $x \in X$ at which $\mathrm{mult}_x D \ge 1$ (i.e. at any point $x \in D$).

**Definition 9.3.4. (Symbolic powers).** Let $Z \subseteq X$ be an irreducible subvariety, with ideal sheaf $\mathcal{I}_Z \subseteq \mathcal{O}_X$. The $p^{\text{th}}$ *symbolic power* of $\mathcal{I}_Z$ is the ideal sheaf consisting of germs of functions that have multiplicity $\geq p$ at a general point of $Z$:

$$\mathcal{I}_Z^{<p>} = \left\{ f \in \mathcal{O}_X \mid \operatorname{ord}_x(f) \geq p \text{ for a general point } x \in Z \right\}. \quad \square$$

By definition, to say that $\operatorname{ord}_x(f) \geq p$ means simply that all the partials of $f$ of order $< p$ vanish at $x$. Note that once the inequality $\operatorname{ord}_x(f) \geq p$ holds at a general point of $Z$, then by semicontinuity it holds at every point of $Z$. If $Z$ is reduced but possibly reducible, $\mathcal{I}_Z^{<p>}$ is defined in a similar fashion by working at a general point of each irreducible component.

**Example 9.3.5. (Effect of multiplicity along a subvariety).** Let $Z \subseteq X$ be an irreducible subvariety of codimension $e$, and let $D$ be an effective **Q**-divisor on $X$. If $\operatorname{mult}_Z D \geq e + p - 1$ then

$$\mathcal{J}(D) \subseteq \mathcal{I}_Z^{<p>}.$$

(Much as in Proposition 9.3.2, consider a log resolution $\mu : X' \longrightarrow X$ obtained by first blowing up $X$ along $Z$. Since $Z$ is generically smooth, there is a unique irreducible component of the exceptional divisor of $\operatorname{Bl}_Z(X)$ mapping onto $Z$, and as before one can compute the discrepancy of the resulting divisor $E \subset X'$. Finally, observe that $\mu_* \mathcal{O}_{X'}(-pE) = \mathcal{I}_Z^{<p>}$ by virtue of the fact that membership in $\mathcal{I}_Z^{<p>}$ is tested at a generic point of $Z$.) $\quad \square$

There are similar statements for ideal sheaves:

**Definition 9.3.6. (Order of vanishing of an ideal).** Let $\mathfrak{a} \subseteq \mathcal{O}_X$ be an ideal sheaf, and $Z \subseteq X$ an irreducible subvariety. The *order of vanishing* $\operatorname{ord}_Z(\mathfrak{a})$ of $\mathfrak{a}$ along $Z$ is the largest integer $m$ such that $\mathfrak{a} \subseteq \mathcal{I}_Z^{<m>}$. Given $c > 0$ rational, we set $\operatorname{ord}_Z(c \cdot \mathfrak{a}) = c \cdot \operatorname{ord}_Z(\mathfrak{a})$. $\quad \square$

Equivalently, working locally around a general point $x \in Z$, one could pick generators

$$g_1, \ldots, g_p \in \mathfrak{a}.$$

Then $\operatorname{ord}_Z(\mathfrak{a})$ is the order along $Z$ of a general **C**-linear combination of the $g_i$.

Proposition 9.3.2 and the previous example then admit the following generalization.

**Example 9.3.7.** Suppose that $Z$ has codimension $e$ in $X$, and that

$$\operatorname{ord}_Z(c \cdot \mathfrak{a}) \geq e + p - 1.$$

Then $\mathcal{J}(c \cdot \mathfrak{a}) \subseteq \mathcal{I}_Z^{<p>}$. (Argue along the lines of 9.3.5.) $\quad \square$

**Remark 9.3.8.** We leave it to the reader to state the analogous results for linear series. $\quad \square$

### 9.3.B Invariants Arising from Multiplier Ideals

We have noted on several occasions that one can view multiplier ideals as measuring singularities. In particular, they give rise to various interesting conditions and invariants which we survey briefly here.

To begin with, there are several important classes of singularities that can be defined in terms of the triviality of the multiplier ideal associated to a **Q**-divisor.

**Definition 9.3.9. (Singularities of pairs).** Let $D$ be an effective **Q**-divisor on a smooth variety $X$.

(i). The pair $(X, D)$ is *Kawamata log terminal* (KLT) if $\mathcal{J}(X, D) = \mathcal{O}_X$.

(ii). The pair $(X, D)$ is *log canonical* if

$$\mathcal{J}\big(X, (1 - \varepsilon) \cdot D\big) = \mathcal{O}_X \quad \text{for all } 0 < \varepsilon < 1.$$

By abuse of notation we will sometimes say simply that $D$ is KLT or log canonical if the appropriate condition holds.[6] $\qquad\square$

These and related concepts are discussed at length in Kollár's excellent survey [364]. The equivalence of 9.3.9 with the definitions appearing e.g. in [364, §3] follows from the observation that $\mathcal{J}(X, D) = \mathcal{O}_X$ if and only if

$$\mathrm{ord}_E\big(K_{X'/X} - \mu^* D\big) > -1$$

for every divisor $E$ appearing in a log resolution $\mu : X' \longrightarrow X$.

**Example 9.3.10. (Singularities of pairs and multiplicity loci).** Given an effective **Q**-divisor $D$ on $X$, consider for $k \geq 0$ the multiplicity loci

$$\Sigma_k(D) = \big\{ x \in X \mid \mathrm{mult}_x(D) \geq k \big\}.$$

If $(X, D)$ is log-canonical then every component of $\Sigma_k(D)$ has codimension $\geq k$ in $X$, while if $(X, D)$ is KLT then every component has codimension $> k$. (Use 9.3.5.) Note however that the converse statements can fail. For instance if $A \subseteq \mathbf{C}^2$ is the cuspidal cubic $(y^2 = x^3)$ and $\frac{5}{6} < c < 1$, then $D = cA$ has multiplicity $< 2$ at the origin but $D$ is not log canonical. $\qquad\square$

**Remark 9.3.11. (Local analogues).** It is sometimes convenient to have corresponding local notions. Given an effective **Q**-divisor $D$ on $X$ and a point $x \in X$, we say that $(X, D)$ is *log-canonical at* $x$ if $\mathcal{J}(X, (1 - \varepsilon) \cdot D)$ is trivial in a neighborhood of $X$ for every $0 < \varepsilon < 1$. The pair in question is KLT *at* $x$ if $\mathcal{J}(X, D)_x = \mathcal{O}_x X$. More generally, given a Zariski-closed subset $Z \subseteq X$, the pair $(X, D)$ is KLT or *log-canonical along* $Z$ if it is KLT or log-canonical at every point $x \in Z$. $\qquad\square$

---

[6] One can define analogous notions also when $X$ is allowed to be singular, in which case it is more important to view $(X, D)$ as a pair. See [364] or [368].

Suppose that $D$ is an effective $\mathbf{Q}$-divisor on $X$, and fix a point $x \in X$ lying in the support of $D$. Then $cD$ will be log-canonical at $x$ for sufficiently small $c$, whereas it cannot be so when $c$ is very large. The critical value of $c$ for this property is called the log-canonical threshold of $D$ at $x$:

**Definition 9.3.12. (Log-canonical threshold of a Q-divisor).** The *log-canonical threshold* of an effective $\mathbf{Q}$-divisor $D$ at $x \in X$ is

$$\mathrm{lct}(D;x) \;=\; \inf\left\{\, c \in \mathbf{Q} \mid \mathcal{J}\big(X\,,\, c \cdot D\big)_x \;\subseteq\; \mathfrak{m}_x \,\right\}, \qquad (9.12)$$

$\mathfrak{m}_x \subseteq \mathcal{O}_X$ being the maximal ideal sheaf of $x$. (When $x$ is not contained in the support of $D$ one can take $\mathrm{lct}(D;x) = \infty$.) More generally, given a closed subset $Z \subseteq X$, the *log-canonical threshold of $D$ along $Z$* is the infimum of all rational $c > 0$ for which $\mathcal{J}\big(X, c \cdot D\big)_x \subseteq \mathfrak{m}_x$ for some $x \in Z$. When $Z = X$ we write $\mathrm{lct}(D)$ for the threshold $\mathrm{lct}(D;X)$. $\qquad \square$

**Example 9.3.13.** Let $A \subseteq \mathbf{C}^2 = X$ denote the cuspidal cubic $(y^2 - x^3 = 0)$. Then the computation of Example 9.2.15 shows that $\mathrm{lct}(A;0) = \frac{5}{6}$. $\qquad \square$

The log-canonical threshold of an ideal sheaf is defined similarly:

**Definition 9.3.14. (Log-canonical threshold of an ideal sheaf).** Given an ideal sheaf $\mathfrak{a} \subseteq \mathcal{O}_X$, the *log-canonical threshold of $\mathfrak{a}$ at $x \in X$* is

$$\mathrm{lct}(\mathfrak{a};x) \;=\; \inf\left\{\, c \in \mathbf{Q} \mid \mathcal{J}\big(X\,,\, \mathfrak{a}^c\big)_x \;\subseteq\; \mathfrak{m}_x \,\right\}. \qquad (9.13)$$

The threshold $\mathrm{lct}(\mathfrak{a};Z)$ is the infimum of the numbers $\mathrm{lct}(\mathfrak{a};x)$ for $x \in Z$, and we write $\mathrm{lct}(\mathfrak{a}) = \mathrm{lct}(\mathfrak{a};X)$. $\qquad \square$

As above we take $\mathrm{lct}(\mathfrak{a};x) = \infty$ if $x \notin \mathrm{Zeroes}(\mathfrak{a})$. We leave it to the reader to define $\mathrm{lct}(|V|;x)$ and $\mathrm{lct}(|V|;Z)$ for a linear series $|V|$ on $X$.

The intuition is that smaller values of $\mathrm{lct}(D;x)$ correspond to "nastier" singularities of $D$ at $x$. Similarly, the log-canonical threshold $\mathrm{lct}(\mathfrak{a};x)$ reflects the singularities at $x$ of divisors of functions $f \in \mathfrak{a}$. We refer to [364, §8–§10] and [131] for much more information about these and related invariants.

**Example 9.3.15. (Diagonal ideals).** Working on $\mathbf{C}^n$ with coordinates $x_1, \dots, x_n$, fix positive integers $d_1, \dots, d_n > 0$ and let $\mathfrak{a} \subseteq \mathbf{C}[x_1, \dots, x_n]$ be the ideal

$$\mathfrak{a} \;=\; \big(x_1^{d_1}, \dots, x_n^{d_n}\big).$$

Then we will see in Example 9.3.31 that

$$\mathrm{lct}(\,\mathfrak{a}\,;\,0\,) \;=\; \sum_{i=1}^{n} \frac{1}{d_i}. \quad \square$$

**Example 9.3.16. (Rationality of log-canonical thresholds).** The log-canonical thresholds $\mathrm{lct}(D;x)$ and $\mathrm{lct}(\mathfrak{a};x)$ are rational numbers, and the infima appearing in equations (9.12) and (9.13) are actually minima. (In the situation of 9.3.12, let $\mu : X' \longrightarrow X$ be a log resolution of $D$, and write

$$\mu^* D = \sum r_i E_i \quad , \quad K_{X'/X} = \sum b_i E_i$$

for suitable $r_i \in \mathbf{Q}$, $b_i \in \mathbf{N}$ and prime divisors $E_i$ on $X'$. Then

$$\mathrm{lct}(D; x) = \min \left\{ \frac{b_j + 1}{r_j} \right\},$$

where the minimum is taken over all indices $j$ for which $\mu(E_j)$ passes through $x$.[7] We remark that we will later use asymptotic multiplier ideals to define log-canonical thresholds in more general settings in which the corresponding rationality statement can fail (Example 11.1.22).  □

**Example 9.3.17. (Semicontinuity of log-canonical threshold).** For a fixed effective $\mathbf{Q}$-divisor $D$ or ideal sheaf $\mathfrak{a} \subseteq \mathcal{O}_X$, the functions $x \mapsto \mathrm{lct}(D; x)$ and $x \mapsto \mathrm{lct}(\mathfrak{a}; x)$ are lower semicontinuous on $X$ for the Zariski topology. (This follows from the combinatorial description of these thresholds appearing in the previous example.)  □

**Example 9.3.18.** Let $D$ be an effective $\mathbf{Q}$-divisor on $X$, and let $c = \mathrm{lct}(D)$ be the (global) log-canonical threshold of $D$. Then $\mathcal{J}(X, cD) \subseteq \mathcal{O}_X$ is a radical ideal on $X$. (Arguing as in 9.3.16 observe that $\mathcal{J}(X, cD) = \mu_* \mathcal{O}_X(-E)$ for a reduced effective divisor $E$ on $X'$.)  □

**Remark 9.3.19. (Inequalities).** Suppose that $\mathfrak{a} \subseteq \mathcal{O}_X$ vanishes at a single point $x \in X$, so that $\mathfrak{a}$ is $\mathfrak{m}$-primary, $\mathfrak{m} = \mathfrak{m}_x$ being the maximal ideal of $x$. Then the Samuel multiplicity $e(\mathfrak{a})$ of $\mathfrak{a}$ is defined (cf. Section 1.6.B). De Fernex, Ein and Mustaţă [106] establish an interesting inequality relating the multiplicity and the log-canonical threshold of $\mathfrak{a}$. Specifically, they prove that if $\dim X = n$, then

$$e(\mathfrak{a}) \geq \left( \frac{n}{\mathrm{lct}(\mathfrak{a}; x)} \right)^n. \tag{*}$$

When $\mathfrak{a}$ is the "diagonal ideal" of Example 9.3.15, (*) comes down to the inequality between the arithmetic and geometric means: the general case is reduced to this via a degeneration to initial ideals. The inequality (*) is generalized and applied to questions of birational rigidity in [107].  □

**Remark 9.3.20. (Log-canonical thresholds via jet schemes).** Mustaţă [455], [458] discovered a beautiful and surprising characterization of log-canonical thresholds in terms of arc spaces. Specifically, let $X$ be a smooth variety and $\mathfrak{a} \subseteq \mathcal{O}_X$ an ideal sheaf. An $m^{\mathrm{th}}$-order arc along $\mathfrak{a}$ is by definition a map of schemes

$$\psi : \mathrm{Spec}\, \frac{\mathbf{C}[t]}{(t^{m+1})} \longrightarrow \mathrm{Zeroes}(\mathfrak{a});$$

when $X$ is affine, this is equivalent to a homomorphism

---

[7] The stated expression is determined by requiring that each $E_j$ appear with coefficient $\geq -1$ in $K_{X'/X} - [\mu^*(cD)]$ and that at least one $E_j$ have coefficient $= -1$.

$$\mathbf{C}[X]/\mathfrak{a} \longrightarrow \mathbf{C}[t]/(t^{m+1})$$

of algebras. The set of all such carries in the natural way the structure of an algebraic scheme, which we denote by $\mathrm{Arc}_m(\mathfrak{a})$. Mustață's theorem is that the log-canonical threshold of $\mathfrak{a}$ is computed by the rate of growth of these arc spaces:

**Theorem.** *With $\mathfrak{a} \subseteq \mathcal{O}_X$ as above, one has*

$$\mathrm{lct}(\mathfrak{a}; X) \;=\; \dim X \;-\; \sup_{m \geq 0} \frac{\dim \mathrm{Arc}_m(\mathfrak{a})}{m+1}.$$

Mustață's proof is based on the theory of motivic integration [136], [408]; the crucial point is that motivic integrals satisfy a change of variable formula closely parallel to the birational transformation rule 9.2.33. A related but somewhat simpler argument, which avoids explicit use of motivic integrals, appears in the paper [156] of Ein, Mustață, and the author.                  □

The log-canonical threshold depends only on the singularities of a pair. By taking into account the richer structure of multiplier ideals one can define a whole sequence of invariants that generalize the log-canonical threshold in a natural way. The idea is that the multiplier ideals $\mathcal{J}(X, cD)$ decrease as $c$ grows, and one singles out those values of the parameter $c$ at which these ideals jump.

The formal definition begins with a lemma characterizing the intervals on which $\mathcal{J}(c \cdot D)$ is constant:

**Lemma 9.3.21.** *Let $D$ be an effective $\mathbf{Q}$-divisor on $X$, and $x \in X$ a fixed point contained in the support of $D$. There is an increasing sequence*

$$0 = \xi_0(D; x) \;<\; \xi_1(D; x) \;<\; \xi_2(D; x) < \;\ldots$$

*of rational numbers $\xi_i = \xi_i(D; x)$ characterized by the properties that*

$$\mathcal{J}(X, c \cdot D)_x \;=\; \mathcal{J}(X, \xi_i \cdot D)_x \quad for \quad c \in [\xi_i, \xi_{i+1}),$$

*while $\mathcal{J}(X, \xi_{i+1} \cdot D)_x \subsetneqq \mathcal{J}(X, \xi_i \cdot D)_x$ for every $i$.*

Here we agree by convention that $\mathcal{J}(X, 0 \cdot D) = \mathcal{O}_X$. Thus $\xi_1(D; x) = \mathrm{lct}(D; x)$ is the log-canonical threshold of $D$ at $x$.

**Definition 9.3.22. (Jumping numbers).** The rational numbers $\xi_i(D; x)$ are the *jumping coefficients* or *jumping numbers* of $D$ at $x$. We say that $\xi$ is a *jumping coefficient* of $D$ on a closed subset $Z \subseteq X$ if it is a jumping number of $D$ at some point $x \in Z$.                  □

Given an ideal sheaf $\mathfrak{a} \subseteq \mathcal{O}_X$, the jumping numbers $\xi_i(\mathfrak{a}; x)$ and $\xi_i(\mathfrak{a}; Z)$ are defined similarly using the multiplier ideals $\mathcal{J}(X, \mathfrak{a}^c)$. Jumping coefficients are also attached in the evident manner to linear series.

*Proof of Lemma 9.3.21.* As in 9.3.16, fix a log resolution $\mu : X' \longrightarrow X$ of $D$ and write

$$\mu^* D = \sum r_i E_i \quad , \quad K_{X'/X} = \sum b_i E_i,$$

so that

$$\mathcal{J}(X, c \cdot D) = \mu_* \mathcal{O}_{X'}\left(\sum (b_j - [cr_j]) \cdot E_j\right).$$

Starting with a given positive rational number $c$, each of the coefficients appearing on the right remains constant if we increase $c$ slightly. Therefore the corresponding multiplier ideals are constant on intervals of the indicated shape. Moreover, the endpoints of these intervals occur among the numbers

$$\left\{ \frac{b_j + m}{r_j} \right\}$$

for some index $j$ and $m \geq 1$. Therefore the $\xi_i$ are indeed rational. □

These jumping numbers seem to have first appeared (at least implicitly) in the papers [399], [406], and [582] of Libgober and Loeser–Vaquié. They are studied more systematically by Ein, Smith, Varolin, and the author in [161], from which the following examples and remarks are taken.

**Example 9.3.23.** Returning to the setting of Example 9.3.15, we will see in 9.3.32 that the jumping numbers on $X = \mathbf{C}^n$ of the ideal $\mathfrak{a} = \left(x_1^{d_1}, \ldots, x_n^{d_n}\right) \subseteq \mathbf{C}[x_1, \ldots, x_n]$ consist of all rational numbers of the form $\frac{e_1 + 1}{d_1} + \ldots + \frac{e_n + 1}{d_n}$ for $e_1, \ldots, e_n \in \mathbf{N}$. □

**Example 9.3.24. (Periodicity of jumping numbers).** Let $A$ be an effective integer divisor on $X$ passing through a point $x \in X$. Then $\xi$ is a jumping coefficient of $A$ at $x$ if and only if $\xi + 1$ is likewise a jumping number. So all the jumping coefficients of $A$ are determined by the finitely many lying in the unit interval $[0, 1]$. (It follows from 9.2.31 that $\mathcal{J}((1 + \xi) \cdot A) = \mathcal{J}(\xi \cdot A)(-A)$.) Similarly, Skoda's Theorem 9.6.21 — together with the fact that multiplier ideals are integrally closed (Example 9.6.13) — implies that if $\xi > (\dim X - 1)$, then $\xi$ is a jumping coefficient of an ideal sheaf $\mathfrak{a}$ if and only if $\xi + 1$ is. □

**Remark 9.3.25. (Jumping coefficients and Bernstein–Sato polynomials).** Let $X = \mathbf{C}^n$ and consider a polynomial $f \in \mathbf{C}[X]$: we will speak of the log-canonical threshold and jumping numbers of $f$ for the corresponding invariants of the divisor $\mathrm{div} f$. One can also associate to $f$ its *Bernstein–Sato polynomial* $b_f(s) \in \mathbf{C}[s]$. This is the monic polynomial of minimal degree in a new variable $s$ characterized by the existence of a linear differential operator

$$P = \sum h_{I,j} s^j \frac{\partial^I}{\partial z^I}$$

with polynomial coefficients that satisfies the identity

$$b(s) \cdot f^s = P f^{s+1}.$$

These polynomials are very interesting and delicate invariants of the singularities of $\{f = 0\}$ (cf. [413], [414], [313]). Yano [615], Lichtin [400] and Kollár [364] show that if $\mathrm{lct}(f)$ is the log-canonical threshold of $f$ on $\mathbf{C}^n$, then $-\mathrm{lct}(f)$ is the largest root of $b_f(s)$. This extends in a natural way to higher jumping numbers. Specifically, Kollár's proof from [364, §10] is generalized in [161] to show that if $\xi$ is a jumping coefficient of $f$ on $\mathbf{C}^n$ lying in the interval $(0, 1]$, then $b_f(-\xi) = 0$. Another proof has recently been given by Budur and Saito [75]. When $f$ has an isolated singularity at the origin, a stronger result was established previously by Loeser and Vaquié [406] based on results of Varchenko [583]. □

**Remark 9.3.26. (Uniform Artin–Rees numbers).** Assume that $X$ is affine, and fix a non-zero polynomial $f \in \mathbf{C}[X]$. Then the classical Artin–Rees lemma (see [164, Lemma 5.1]) states that given any ideal $\mathfrak{b} \subseteq \mathbf{C}[X]$, there exists an integer $k \geq 0$ such that

$$\mathfrak{b}^m \cap (f) \subseteq \mathfrak{b}^{m-k} \cdot (f)$$

for all $m \geq k$. Classically $k$ is allowed to vary with both $\mathfrak{b}$ and $f$, but Huneke [296] proved the surprising result that one can find a single $k$ that works for all $\mathfrak{b}$.[8] One says in this case that $k$ is a uniform Artin–Rees number for $f$, and it is interesting to ask what geometric information it depends on. The paper [161] contains a bound involving jumping numbers:

> **Theorem.** Let $\ell = \ell(f)$ be the number of jumping coefficients of $f$ on $X$ that are $\leq 1$. Then $k = \ell \cdot \dim X$ is a uniform Artin–Rees number for $(f)$.

Combining the proof of the theorem with some ideas of [296], one deduces that if $(f = 0)$ has an isolated singularity at a point $x \in X$ but otherwise is smooth, then $\tau(f, x) + \dim X$ is a uniform Artin–Rees number, where $\tau(f, x)$ denotes the Tyurina number $\tau(f, x) = \dim_{\mathbf{C}} \left( \mathcal{O}_x X / \left( f, \frac{\partial f}{\partial z_1}, \dots, \frac{\partial f}{\partial z_n} \right) \right)$ of $f$ at the point $x$. □

### 9.3.C Monomial Ideals

We now present a theorem of Howald [293] that computes the multiplier ideal associated to an arbitrary monomial ideal. This provides a rich source of nontrivial examples.

Let $X = \mathbf{C}^n$, with coordinates $x_1, \dots, x_n$, and put $\mathbf{C}[X] = \mathbf{C}[x_1, \dots, x_n]$. We will be concerned with *monomial ideals* $\mathfrak{a} \subseteq \mathbf{C}[X]$, i.e. ideals generated by monomials in the $x_i$. Any such monomial is specified by its *vector of exponents*

---

[8] Both the Artin–Rees lemma and Huneke's theorem hold in much more general algebraic settings: see [296].

$v \in \mathbf{N}^n$, and given $v$ we write $x^v$ for the corresponding monomial. Viewing $\mathbf{N}^n$ as a subset of $\mathbf{R}^n$ in the natural way, we define the *Newton polyhedron*

$$P = P(\mathfrak{a}) \subseteq \mathbf{R}^n$$

of $\mathfrak{a}$ to be the convex hull in $\mathbf{R}^n$ of the set of all exponent vectors of monomials in $\mathfrak{a}$. Thus $P(\mathfrak{a})$ is an unbounded closed set lying in the first orthant. By way of example, the Newton polyhedron associated to the monomial ideal

$$\mathfrak{a} = (y^6, xy^2, x^3 y, x^7)$$

in two variables is pictured in Figure 9.4: generators of $\mathfrak{a}$ are indicated by open circles, and $P(\mathfrak{a})$ is the shaded region.

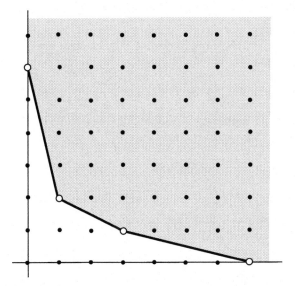

**Figure 9.4.** Newton polyhedron of $\mathfrak{a} = (y^6, xy^2, x^3 y, x^7)$

Given $c > 0$ we denote by $c \cdot P(\mathfrak{a})$ the convex region obtained upon scaling $P(\mathfrak{a})$ by $c$ (so that $P(\mathfrak{a}) \subseteq c \cdot P(\mathfrak{a})$ when $c \leq 1$), and we declare that $P(c \cdot \mathfrak{a}) = c \cdot P(\mathfrak{a})$. The *interior* $\mathrm{int}(P(c \cdot \mathfrak{a}))$ is the topological interior of $P(c \cdot \mathfrak{a})$ with respect to the standard topology on $\mathbf{R}^n$. Finally, we introduce the vector

$$\mathbf{1} = (1, \ldots, 1) \in \mathbf{N}^n,$$

corresponding to the monomial $x^{\mathbf{1}} = x_1 x_2 \cdot \ldots \cdot x_n$, which will play a special role.

A monomial ideal $\mathfrak{a} \subseteq \mathbf{C}[X]$ determines a sheaf on $X$, and so $\mathcal{J}(c \cdot \mathfrak{a}) \subseteq \mathbf{C}[X]$ is defined for every rational $c > 0$. Howald's result is the following:

**Theorem 9.3.27. (Howald's theorem).** *The multiplier ideal* $\mathcal{J}(c \cdot \mathfrak{a})$ *is the monomial ideal generated by all monomials* $x^v$ *whose exponent vectors satisfy the condition that*

$$v + 1 \ \in \ \text{int}\Big( P(c \cdot \mathfrak{a}) \Big).$$

Before turning to the proof, we give some illustrations and applications.

**Example 9.3.28.** Consider the ideal $\mathfrak{a} = (y^6, xy^2, x^3y, x^7)$ illustrated in Figure 9.4. Then it follows from the theorem (and Figure 9.4) that

$$\mathcal{J}(\mathfrak{a}) \ = \ (y^2, xy, x^3).$$

Note that although $(0, 1) + 1 \in P(\mathfrak{a})$, the vector in question does not lie in the interior of $P(\mathfrak{a})$. Therefore the monomial $y$ does not belong to $\mathcal{J}(\mathfrak{a})$. On the other hand, $y \in \mathcal{J}(c \cdot \mathfrak{a})$ — and similarly $x^2 \in \mathcal{J}(c \cdot \mathfrak{a})$ — as soon as $c < 1$. $\qquad\square$

**Example 9.3.29. ("Diagonal" ideals).** Fix positive integers $d_1, \ldots, d_n \geq 1$, and consider the ideal

$$\mathfrak{a} \ = \ \big( x_1^{d_1}, \ldots, x_n^{d_n} \big).$$

Then for $c > 0$ the multiplier ideal $\mathcal{J}(c \cdot \mathfrak{a})$ is spanned by all monomials $x_1^{e_1} \cdot \ldots \cdot x_n^{e_n}$ satisfying the inequality

$$\sum_{i=1}^{n} \frac{e_i + 1}{d_i} \ > \ c.$$

In particular, this ideal is trivial if and only if $c < \sum \frac{1}{d_i}$ (so that $x^0 \in \mathcal{J}(c \cdot \mathfrak{a})$). This example is analyzed analytically in [126, (5.10)]. $\qquad\square$

**Example 9.3.30. (Multiplier ideals of sparse polynomials).** Given a polynomial $f \in \mathbf{C}[X]$, let $\mathfrak{a} = \mathfrak{a}_f \subseteq \mathbf{C}[X]$ be the ideal spanned by the monomials appearing in $f$ (the so-called *term ideal* of $f$). Then

$$\mathcal{J}\big( c \cdot \text{div} f \big) \subseteq \mathcal{J}\big( c \cdot \mathfrak{a} \big) \quad \text{for all } 0 < c < 1.$$

(Observe that the multiplier ideal on the left coincides with $\mathcal{J}(c \cdot (f))$.) If $f^* \in \mathbf{C}[X]$ is the polynomial obtained from $f$ by taking the coefficient of each monomial in $f$ to have a sufficiently general value, then $\mathcal{J}(c \cdot \text{div} f^*) = \mathcal{J}(c \cdot \mathfrak{a})$ for every $0 < c < 1$. (Use (9.2.26) and (9.2.22).) See Theorem 9.3.37 for a more precise statement. $\qquad\square$

**Example 9.3.31. (Log canonical thresholds).** Let $\mathfrak{a} \subseteq \mathbf{C}[X]$ be a monomial ideal. Then Theorem 9.3.27 gives a simple way of computing the log-canonical threshold $\text{lct}(\mathfrak{a})$ of $\mathfrak{a}$. In fact, let $\zeta_1, \ldots, \zeta_n$ be the natural coordinates on $\mathbf{R}^n = \mathbf{Z}^n \otimes \mathbf{R}$ adapted to the standard basis of $\mathbf{Z}^n$. The Newton polyhedron $P(\mathfrak{a})$ is cut out within the first orthant $\{\zeta_1 \geq 0, \ldots, \zeta_n \geq 0\}$ by finitely many inequalities of the form

$$f_\alpha(\zeta_1, \ldots, \zeta_n) \geq 1,$$

where the $f_\alpha$ are linear forms with non-negative rational coefficients. Then

$$\mathrm{lct}(\mathfrak{a}) = \min_\alpha \{ f_\alpha(1, \ldots, 1) \}. \tag{*}$$

For instance, if $\mathfrak{a} = (x_1^{d_1}, \ldots, x_n^{d_n})$ then $P = \{ \frac{\zeta_1}{d_1} + \ldots + \frac{\zeta_n}{d_n} \geq 1 \}$, so that $\mathrm{lct}(\mathfrak{a}) = \sum \frac{1}{d_i}$ (as we saw in Example 9.3.29). (The right-hand side of (*) computes the least rational $c > 0$ such that the origin $0 \in \mathbf{Z}^n$ satisfies $0 + 1 \in c \cdot P(\mathfrak{a})$.) This computation appears in a somewhat different context in [18, §6.2.3, 13.1.7].    □

**Example 9.3.32. (Jumping numbers of monomial ideals).** Keeping the set-up and notation of the previous example, one can also compute the jumping numbers of the monomial ideal $\mathfrak{a}$. In fact, every exponent vector $v \in \mathbf{N}^n$ determines a jumping coefficient $\xi_v$ characterized as the smallest rational $c > 0$ such that $x^v \notin \mathcal{J}(c \cdot \mathfrak{a})$, and every jumping number is of this form.[9] As in 9.3.31, Howald's theorem implies that

$$\xi_v = \min_\alpha \{ f_\alpha(v + 1) \}.$$

For example, the jumping numbers of the ideal $\mathfrak{a} = (x_1^{d_1}, \ldots, x_n^{d_n})$ consist precisely of the rational numbers

$$\frac{e_1 + 1}{d_1} + \ldots + \frac{e_n + 1}{d_n}$$

as $(e_1, \ldots, e_n)$ ranges over all vectors in $\mathbf{N}^n$.    □

We now give a

*Sketch of Proof of Theorem 9.3.27.* The fact that $\mathcal{J}(c \cdot \mathfrak{a})$ is a monomial ideal follows from the remark that it is invariant under the natural torus action on $\mathbf{C}^n$. So the issue is to determine which monomials this multiplier ideal contains. For this, fix a log resolution $\mu : X' \longrightarrow X = \mathbf{C}^n$ of $\mathfrak{a}$. We assume for simplicity that $\mu$ is obtained by a sequence of blowings up along intersections of coordinate hyperplanes and the components of their total transforms.[10] Write $\mathbf{1}_X$ for the divisor of $x^1$ on $X = \mathbf{C}^n$, so that $\mathbf{1}_X$ is the sum of the coordinate hyperplanes each with coefficient one. Similarly, denote by $\mathbf{1}_{X'}$ the

---

[9] Note however that different $v \in \mathbf{N}^n$ might give rise to the same coefficient $\xi$.

[10] The existence of such resolutions is established directly by Goward in [229]. However with a little more care one can make do with the assumption that $\mu$ is any toric resolution: see [292].

divisor on $X'$ consisting of the sum of the exceptional divisors of $\mu$ (each with coefficient one) and the proper transform of the coordinate axes in $\mathbf{C}^n$. The next point to observe is that

$$\mu^*(\mathbf{1}_X) \;=\; \mathbf{1}_{X'} + K_{X'/X}. \tag{9.14}$$

This can be understood torically (cf. [292]), but it is also elementary to argue inductively by verifying the assertion for a single blow-up of an intersection of components of $\mathbf{1}_{X''}$ for an intermediate blow-up $X''$. Set $\mathfrak{a} \cdot \mathcal{O}_{X'} = \mathcal{O}_{X'}(-F)$, so that $F$ is supported on the components of $\mathbf{1}_{X'}$.

We will show that if $v + \mathbf{1} \in \text{int}\big(P(c \cdot \mathfrak{a})\big)$, then $x^v \in \mathcal{J}(c \cdot \mathfrak{a})$. To this end, observe to begin with that $v + \mathbf{1}$ lies in the interior of $P(c \cdot \mathfrak{a})$ if and only if

$$m \cdot (v + \mathbf{1}) \;\in\; \mathbf{1} + m \cdot P(c \cdot \mathfrak{a}) \quad \text{for all } m \gg 0. \tag{9.15}$$

So it is enough to verify that $x^v \in \mathcal{J}(c \cdot \mathfrak{a})$ provided that $v$ satisfies (9.15). Suppose then that (9.15) holds. Then

$$m \cdot \mu^*\big(\text{div}(x^v) + \mathbf{1}_X\big) \;\succcurlyeq\; m \cdot cF + \mu^*(\mathbf{1}_X).$$

Using (9.14) and dividing by $m$, one sees that this is equivalent to the condition:

$$\mu^*\big(\text{div}(x^v)\big) + \big(\mathbf{1}_{X'} - \tfrac{1}{m}\mu^*(\mathbf{1}_X)\big) + K_{X'/X} - cF \;\succcurlyeq\; 0. \tag{*}$$

But every component of $\mathbf{1}_{X'}$ appears with coefficient one, and $\mu^*\mathbf{1}_X$ has the same support as $\mathbf{1}_{X'}$. Consequently

$$\Big[\mathbf{1}_{X'} - \tfrac{1}{m}\mu^*(\mathbf{1}_X) - cF\Big] \;=\; -\big[cF\big]$$

provided that $m \gg 0$. Therefore (*) implies that

$$\mu^*\big(\text{div}(x^v)\big) + K_{X'/X} - \big[cF\big] \;\succcurlyeq\; 0,$$

i.e. that $x^v \in \mu_*\mathcal{O}_{X'}\big(K_{X'/X} - [cF]\big) = \mathcal{J}(c \cdot \mathfrak{a})$, as required. The reverse inclusion is left to the reader.  $\square$

We conclude this subsection by describing without proof another result of Howald [294] giving an explicit condition under which the multiplier ideals of a polynomial can be computed via Theorem 9.3.27.

Continuing to work on $X = \mathbf{C}^n$, let

$$f \in \mathbf{C}[x_1, \dots, x_n] \;=\; \mathbf{C}[X]$$

be a polynomial, and write $\mathcal{J}(c \cdot f)$ for the multiplier ideal associated to $\text{div} f$. Consider the *term ideal* of $f$, i.e. the ideal $\mathfrak{a}_f \subseteq \mathbf{C}[X]$ generated by the monomials appearing in $f$. Then

$$\mathcal{J}(c \cdot f) \subseteq \mathcal{J}(c \cdot \mathfrak{a}_f)$$

provided that $0 < c < 1$, and if the coefficients of $f$ are "sufficiently general" then equality holds (Example 9.3.30). One wishes to have an effective and concrete test for genericity in this sense. To this end, we recall some definitions from [18]:

**Definition 9.3.33. (Non-degenerate polynomials).** Let $f \in \mathbf{C}[X]$ be a polynomial with term ideal $\mathfrak{a}_f$, and consider its Newton polyhedron $P(f) = P(\mathfrak{a}_f)$. Given any face $\sigma$ of $P(f)$ — including $\sigma = P(f)$ — denote by $f_\sigma \in \mathbf{C}[X]$ the sum of those terms of $f$ corresponding to points lying on $\sigma$. One says that $f$ is *non-degenerate along* $\sigma$ if the 1-form $df_\sigma$ is nowhere vanishing on the torus $(\mathbf{C}^*)^n$. One says that $f$ is *non-degenerate* if $f$ is non-degenerate along each face of $P(f)$. Finally, we say that $f$ has *non-degenerate principal part* if this condition holds for every *compact* face $\sigma$ of $P(f)$.[11]  □

**Remark 9.3.34.** If $f$ has a non-zero constant term, then $P(f)$ contains the origin $\{0\}$ as a face. In this case we agree by convention that $f$ is non-degenerate along $\{0\}$ even though $f_{\{0\}}$ is constant and $df_{\{0\}}$ vanishes.  □

**Example 9.3.35.** For any choices of natural numbers $d_1, \ldots, d_n \geq 1$, the polynomial

$$f = x_1^{d_1} + \ldots + x_n^{d_n}$$

is non-degenerate.  □

**Example 9.3.36. (Some degenerate polynomials).** Here are some examples, taken from [292], of polynomials in two variables for which the condition in 9.3.33 fails.

(i).  Let $f = y^2 - y(x-1)^2$. Then $\mathfrak{a}_f = (y)$, and one of the faces of $\mathfrak{a}_f$ is the half-line $\sigma$ generated by the monomials $x^i y$ ($i \geq 0$). Then $f_\sigma = y(x-1)^2$ and $df_\sigma(1,1) = 0$. Note however that $df$ itself is nowhere vanishing on the torus.

(ii).  Let $f = (x+y)^2 - (x-y)^d$ with $d \gg 0$. Then the Newton polyhedron has a face $\sigma$ with $f_\sigma = (x+y)^2$, for which 9.3.33 fails. This example becomes monomial after a linear change of variables, but there exist $c < 1$ for which the multiplier ideal $\mathcal{J}(c \cdot f)$ is not itself monomial. In particular, $\mathcal{J}(c \cdot f) \neq \mathcal{J}(c \cdot \mathfrak{a}_f)$.  □

Howald's result is that if $f \in \mathbf{C}[X]$ is non-degenerate, then the multiplier ideals of $f$ are computed by those of its term ideal:

**Theorem 9.3.37.** *Assume that $f \in \mathbf{C}[X]$ is non-degenerate. Then*

---

[11] In fact only the latter notion appears in [18]: it essentially governs the local geometry of $f$ near the origin. The stronger condition of non-degeneracy rules out (among other things) the presence of any singular points away from the coordinate hyperplanes.

$$\mathcal{J}(c \cdot f) = \mathcal{J}(c \cdot \mathfrak{a}_f)$$

*for every $0 < c < 1$. If $f$ has non-degenerate principal part, then the same statement holds in a neighborhood of the origin.*

Starting as above with a monomial log resolution $\mu : X' \longrightarrow X$ of $\mathfrak{a}_f$, with $\mathfrak{a}_f \cdot \mathcal{O}_{X'} = \mathcal{O}_{X'}(-F)$, write $\mu^*(\mathrm{div} f) = D' + F$ where $D'$ is the proper transform of $\mathrm{div} f$. The idea of the proof is to show that the non-degeneracy condition implies that $F + D'$ has SNC support: then the statement follows immediately as in 9.2.26. We refer to [292], [294] for the details.

**Remark 9.3.38.** In the setting of the theorem, if $c > 1$ then the corresponding multiplier ideal is determined by the relation

$$\mathcal{J}(c \cdot f) = (f) \cdot \mathcal{J}((c-1) \cdot f)$$

coming from 9.2.31.    □


### 9.3.D  Analytic Construction of Multiplier Ideals

We describe very briefly — without proofs — the analytic interpretation of the multiplier ideals we have defined. We refer to [126] for a detailed account. At the end of the subsection we also say a word about algebraic interpretations of multiplier ideals.

In the analytic approach, multiplier ideals are attached to an arbitrary plurisubharmonic (PSH) function:

**Definition 9.3.39. (Analytic multiplier ideal of a PSH function).** Let $X$ be a complex manifold and $\varphi$ a PSH function on $X$. The *analytic multiplier ideal sheaf*

$$\mathcal{J}(\varphi) \subseteq \mathcal{O}_X$$

associated to $\varphi$ is the sheaf of germs of holomorphic functions $f$ such that $|f|^2 e^{-2\varphi}$ is locally integrable with respect to Lebesgue measure in some local coordinate system.    □

It is known that this is a coherent analytic sheaf of ideals on $X$ ([126, Proposition 5.7]).

The algebraic multiplier sheaves defined above arise by taking suitable weight functions $\varphi$.

**Example 9.3.40. (Weight function locally associated to a Q-divisor).** Let $X$ be a smooth algebraic variety (viewed as a complex manifold), and consider an effective **Q**-divisor $D = \sum_{i=1}^t a_i D_i$ on $X$. Fixing an open set $U \subseteq X$ on which each $D_i$ is principal, let $g_i \in \Gamma(U, \mathcal{O}_U)$ be a holomorphic function locally defining $D_i$ in $U$. Then

$$\varphi_D \;=\; \sum_{i=1}^{t} a_i \log |g_i|$$

is plurisubharmonic on $U$ and

$$\mathcal{J}(\varphi_D) \;=_{\text{locally}} \left\{ f \in \mathcal{O}_X \;\Big|\; \frac{|f|^2}{\prod |g_i|^{2a_i}} \in \mathrm{L}^1_{\text{loc}} \right\}.$$

Although $\varphi_D$ depends on the local defining equations $g_i$, the resulting integrability condition is independent of this choice. So $\mathcal{J}(\varphi_D)$ itself is well-defined.  $\square$

**Example 9.3.41. (Weight function associated to an ideal).** Assume that $X$ is affine and consider an ideal $\mathfrak{a} \subseteq \mathbf{C}[X]$. Fix generators $g_1, \dots, g_p \in \mathfrak{a}$, which one views as holomorphic functions on $X$. Given $c > 0$ set

$$\varphi_{c \cdot \mathfrak{a}} \;=\; \frac{1}{2} \cdot \log \Big( \sum |g_i|^{2c} \Big).$$

The resulting ideal $\mathcal{J}(\varphi_{c \cdot \mathfrak{a}})$ is independent of the choice of generators of $\mathfrak{a}$, and so is defined for an ideal sheaf $\mathfrak{a} \subseteq \mathcal{O}_X$ on an arbitrary smooth variety $X$. The case of linear series reduces to 9.3.41 by virtue of 9.2.22.  $\square$

A basic fact is that these are the analytic ideals corresponding to the multiplier ideals defined above:

**Theorem 9.3.42.** *In the settings of 9.3.40 and 9.3.41, the multiplier ideal associated to the specified weight function is the analytic sheaf determined by the corresponding algebraic multiplier ideal. In other words,*

$$\mathcal{J}(\varphi_D) \;=\; \mathcal{J}(D)^{an} \quad \textit{and} \quad \mathcal{J}(\varphi_{c \cdot \mathfrak{a}}) \;=\; \mathcal{J}(c \cdot \mathfrak{a})^{an}.$$

The proof proceeds in two steps. The first point is that analytic multiplier ideals satisfy the same birational transformation rules as do the algebraic ideals:

**Proposition 9.3.43. (Change of variables formula for analytic multiplier ideals).** *Let $f : Y \longrightarrow X$ be a modification, i.e. a projective (or merely proper) bimeromorphic morphism of complex manifolds, and let $\varphi$ be a PSH function on $X$. Then*

$$f_* \big( \mathcal{O}_Y(K_Y) \otimes \mathcal{J}(f^* \varphi) \big) \;=\; \mathcal{O}_X(K_X) \otimes \mathcal{J}(\varphi).$$

In brief, one can interpret the right-hand side as the subsheaf of $\Omega_X^n$ ($n = \dim X$) given by

$$\Big\{ \eta \in \Omega_X^n \;\Big|\; (e^{-\varphi}\eta) \wedge \overline{(e^{-\varphi}\eta)} \text{ is locally integrable} \Big\}.$$

Since $f$ is proper one can test integrability after pulling back to $Y$, and the assertion follows. See [126, Proposition 5.8] for details.

Focusing on the case of a **Q**-divisor $D$, one is then reduced by virtue of (9.11) to the simple normal crossing case. Here one has:

**Lemma 9.3.44.** *Let* $D = \sum a_i D_i$ *be an effective* **Q**-*divisor on* $X$ *having simple normal crossing support. Then*

$$\mathcal{J}(\varphi_D) = \mathcal{O}_X(-[D])^{an}.$$

Choosing local analytic coordinates $z_1, \ldots, z_n$ adapted to $D$, this amounts to the assertion that for a holomorphic function germ $f$,

$$\frac{|f|^2}{\prod |z_i|^{2a_i}}$$

is locally integrable at the origin if and only if

$$\left( z_1^{[a_1]} z_2^{[a_2]} \cdot \ldots \cdot z_n^{[a_n]} \right) \mid f.$$

We can assume here that $f$ is a monomial. Then the variables separate and one reduces to the elementary

**Lemma 9.3.45.** *In a single complex variable* $z$, *the function* $\frac{1}{|z|^{2c}}$ *is integrable near* 0 *iff* $c < 1$. $\qquad\square$

**Remark 9.3.46. (Algebraic interpretation of multiplier ideals).** In connection with his work on the Briançon–Skoda theorem, Lipman [403] defined and studied the multiplier ideal associated to any ideal $\mathfrak{a}$ in a regular Noetherian domain.[12] Lipman's definition is rather close to the one appearing here. The Briançon–Skoda theorem, and its connection to multiplier ideals, appears in Section 9.6 below.

As we will see, multiplier ideals can be used to prove several theorems of a purely algebraic nature: besides Briançon–Skoda one could mention the comparison theorem of [159] for symbolic powers (Section 11.3). Experience showed that such results could also be established — in more general algebraic settings — via the Hochster–Huneke theory of tight closure (see [297]). However an actual connection between these two theories has emerged only very recently in work of Hara–Yoshida [271] and Takagi [563], building on earlier results of Hara [270] and Smith [544]. In brief, Hara–Yoshida and Takagi start with **Q**-divisor $D$ or ideal $\mathfrak{a}$ on a suitable variety $X$ in characteristic $p > 0$. They use tight closure ideas involving the Frobenius to define "test ideals"

$$\tau(D) , \ \tau(\mathfrak{a}^c) \subseteq \mathcal{O}_X.$$

When the characteristic $p$ data lift to characteristic zero, they show that this test ideal is in fact the reduction (mod $p$) of the corresponding multiplier ideal provided that $p \gg 0$. Interestingly, these authors are able to establish directly that these test ideals satisfy many of the basic local properties of multiplier ideals discussed in Section 9.5. $\qquad\square$

---

[12] Lipman called this the "adjoint" of $\mathfrak{a}$: however we use this term for a somewhat different construction appearing in the next subsection.

### 9.3.E  Adjoint Ideals

Example 9.2.12 shows that the multiplier ideal associated to an integral divisor $D$ is uninteresting. However, when $D$ is reduced, one can define a variant of a multiplier ideal — the adjoint ideal of $D$ in $X$ — which does carry significant information about the singularities of $D$. These adjoint ideals seem first to have appeared in [582]. They were rediscovered in [154], where they were used to study the singularities of theta divisors (see Theorem 10.1.8).

Suppose then that $D$ is a reduced integral divisor on a smooth variety $X$, which we view in the natural way as a (possibly reducible) subvariety $D \subseteq X$ of $X$. Fix a log resolution $\mu : X' \longrightarrow X$ of $(X, D)$ with the property that the proper transform $D' \subset X'$ of $D$ is non-singular (but possibly disconnected), and write $\mu^* D = D' + F$, where $F$ is a $\mu$-exceptional effective integral divisor on $X'$.

**Definition 9.3.47. (Adjoint ideals).** Keeping the notation just introduced, the *adjoint ideal* $\mathrm{adj}(D) = \mathrm{adj}(X, D) \subseteq \mathcal{O}_X$ of $D$ in $X$ is defined to be

$$\mathrm{adj}(D) \; = \; \mu_* \mathcal{O}_{X'} (K_{X'/X} - F). \quad \square$$

One can verify in the spirit of Proposition 9.2.18 that this is independent of the choice of resolution.

The basic property of adjoint ideals is given by:

**Proposition 9.3.48.** *Assume as above that $D \subseteq X$ is a reduced divisor on the smooth variety $X$, and let $\nu : D' \longrightarrow D$ be any resolution of singularities of $D$.*

(i). *One has an exact sequence*

$$0 \longrightarrow \mathcal{O}_X(K_X) \xrightarrow{\cdot D} \mathcal{O}_X(K_X + D) \otimes \mathrm{adj}(D) \longrightarrow \nu_* \mathcal{O}_{D'}(K_{D'}) \longrightarrow 0.$$
$$(9.16)$$

(ii). *The adjoint ideal $\mathrm{adj}(D)$ is trivial — i.e. $\mathrm{adj}(D) = \mathcal{O}_X$ — if and only if $D$ is normal and has at worst rational singularities.*

*Proof.* Note to begin with that the sheaf $\nu_* \mathcal{O}_{D'}(K_{D'})$ is independent of the choice of resolution $\nu$, so we are free to work with the one coming from the log resolution $\mu : X' \longrightarrow X$ chosen above: thus we take $\nu = \mu | D'$. Recalling that $\mu^* D = D' + F$ one has the identity

$$K_{X'} + D' \; = \; \mu^*(K_X + D) + (K_{X'/X} - F).$$

Therefore

$$\mu_* \mathcal{O}_{X'}(K_{X'} + D') \; = \; \mathcal{O}_X(K_X + D) \otimes \mathrm{adj}(D),$$

and keeping in mind that $R^1\mu_*\mathcal{O}_{X'}(K_{X'}) = 0$ thanks e.g. to Grauert-Riemenschneider, (9.16) arises as the pushforward under $\mu$ of the exact sequence

$$0 \longrightarrow \mathcal{O}_{X'}(K_{X'}) \longrightarrow \mathcal{O}_{X'}(K_{X'} + D') \longrightarrow \mathcal{O}_{D'}(K_{D'}) \longrightarrow 0.$$

This proves (i). It follows from (9.16) that $\mathrm{adj}(D) = \mathcal{O}_X$ if and only if

$$\nu_*\mathcal{O}_{D'}(K_{D'}) = \omega_D, \tag{9.17}$$

where $\omega_D = \mathcal{O}_D(K_X + D)$ is the dualizing line bundle on $D$. Since $\nu$ factors through the normalization of $D$, and since $D$ is non-normal if and only if it is singular in codimension one, this equality can hold only if $D$ is normal. And since $D$ is in any event Cohen-Macaulay, (9.17) is then known to be equivalent to the condition that $D$ has at worst rational singularities (cf. [364, (11.10)]).    □

**Example 9.3.49. (Adjoint vs multiplier ideals).** Let $X$ be a smooth affine variety, and let $\mathfrak{a} \subseteq \mathbf{C}[X]$ be an ideal in the coordinate ring of $X$. Assume that the zeroes of $\mathfrak{a}$ have codimension $\geq 2$ in $X$. Let $f \in \mathfrak{a}$ be a general element, i.e. a general $\mathbf{C}$-linear combination of a set of generators of $\mathfrak{a}$ (Definition 9.2.27). Then $\mathrm{div} f$ is reduced, and

$$\mathcal{J}(X, \mathfrak{a}) = \mathrm{adj}(\mathrm{div}\, f).$$

(Argue as in the proof of Proposition 9.2.26.)    □

**Example 9.3.50. (Kollár–Bertini theorem, II).** Keeping the notation and assumptions of the previous example, suppose that for each $x \in \mathrm{Zeroes}(\mathfrak{a})$ there is an element $g_x \in \mathfrak{a}$ such that $\mathrm{div}(g_x)$ has only rational singularities near $x$. Then the divisor of a general element $f \in \mathfrak{a}$ has only rational singularities at every point of $X$. (Applying 9.3.48 and 9.3.49 locally near a given point, the hypothesis implies that $\mathcal{J}(\mathfrak{a}) = \mathcal{O}_X$. The assertion then follows upon applying the same statements globally.) This is one of several Bertini-type statements due to Kollár ([364], §4): the present argument is essentially his, transposed into somewhat different language.    □

**Example 9.3.51. (Adjoint vs. conductor ideal).** Let $X$ be a smooth surface and $D \subseteq X$ a reduced curve. Writing $\nu : D' \longrightarrow D$ for the normalization, the *conductor* of $D$ is defined to be the ideal sheaf $\mathfrak{c} = \mathrm{Ann}(\nu_*\mathcal{O}_{D'}/\mathcal{O}_D) \subseteq \mathcal{O}_D$. Then

$$\mathrm{adj}(D) \cdot \mathcal{O}_D = \mathfrak{c}.$$

(In fact, one has $\mathfrak{c} \otimes \omega_D = \nu_*\mathcal{O}_{D'}(K_{D'})$ [562, p. 26], where $\omega_D = \mathcal{O}_D(K_X + D)$ is the dualizing sheaf on $D$. The assertion then follows from 9.3.48.)    □

**Example 9.3.52. (Adjoint and Jacobian ideals).** Consider as above a reduced integral divisor $D \subseteq X$ on a smooth variety $X$. One can associate to $D$ its *Jacobian ideal* Jacobian$(D) \subseteq \mathcal{O}_X$ whose zeroes define the singular locus of $D$: if locally

$$D = \{\, f(z_1, \ldots, z_n) = 0 \,\},$$

$z_1, \ldots, z_n$ being local coordinates on $X$, then by definition Jacobian$(D)$ is the ideal sheaf locally generated by $f$ and its partials:[13]

$$\text{Jacobian}(D) =_{\text{locally}} \left( f, \frac{\partial f}{\partial z_1}, \ldots, \frac{\partial f}{\partial z_n} \right) \subseteq \mathcal{O}_X.$$

One then has the inclusion

$$\text{Jacobian}(D) \subseteq \text{adj}(X, D).$$

(This can be verified by an explicit computation in the spirit of the proof of Lemma 9.2.19.) A related result is stated in the next subsection.    □

### 9.3.F Multiplier and Jacobian Ideals

We state here without proof a result of Ein, Smith, Varolin, and the author from [161] relating Jacobian and multiplier ideals.

Let $X$ be a smooth variety of dimension $n$, and $\mathfrak{a} \subseteq \mathcal{O}_X$ an ideal. Write $Z = \text{Zeroes}(\mathfrak{a})$ for the corresponding subscheme of $X$.

**Definition 9.3.53.** Given $m \geq 1$, the $m^{\text{th}}$ *Jacobian ideal*

$$\text{Jacobian}_m(\mathfrak{a}) \subseteq \mathcal{O}_X$$

of $\mathfrak{a}$ is the $m^{\text{th}}$ Fitting ideal of the module of differentials $\Omega_Z^1$ of $Z$ ($\Omega_Z^1$ being considered in the natural way as an $\mathcal{O}_X$-module).    □

Very concretely, suppose that $\mathfrak{a}$ is locally generated by $t$ regular functions $f_1, \ldots, f_t \in \mathcal{O}_X$ in a neighborhood of $x$, and let $z_1, \ldots, z_n$ be local coordinates at $x \in X$ (so that $dz_1, \ldots, dz_n$ are a local basis for $\Omega_X^1$ near $x$). Consider the $n \times 2t$ matrix

$$A = \begin{pmatrix} f_1 \cdots f_t & \frac{\partial f_1}{\partial z_1} & \cdots & \frac{\partial f_t}{\partial z_1} \\ \vdots \;\; \vdots \;\; \vdots & \vdots & \vdots & \vdots \\ f_1 \cdots f_t & \frac{\partial f_1}{\partial z_n} & \cdots & \frac{\partial f_t}{\partial z_n} \end{pmatrix}.$$

Then $A$ is a presentation matrix for $\Omega_Z^1$, and Jacobian$_m(\mathfrak{a})$ is the ideal locally generated by the $m \times m$ minors of $A$. In particular, when $\mathfrak{a} = (f)$ is the principal ideal defined by a single function $f$, Jacobian$_1(\mathfrak{a}) = $ Jacobian$(f)$ is the Jacobian ideal considered in Example 9.3.52.

It is established in [161] that these Jacobian ideals are at least as deep as the multiplier ideal of a suitable power of $\mathfrak{a}$:

---

[13] This differs slightly from the possibly more usual definition of Jacobian$(D)$ as the ideal generated only by the partials of $f$. The present definition has the advantage of being intrinsic, and we trust that no confusion will result.

**Theorem 9.3.54.** *Let* $\mathfrak{a} \subseteq \mathcal{O}_X$ *be any ideal sheaf on* $X$, *and fix a natural number* $m$.

(i). *Assume that the multiplier ideal* $\mathcal{J}(\mathfrak{a}^m)$ *cuts out a scheme of codimension* $\geq m + 1$ *in* $X$. *Then*

$$\mathrm{Jacobian}_m(\mathfrak{a}) \subseteq \mathcal{J}(\mathfrak{a}^m).$$

(ii). *Assume that the multiplier ideal* $\mathcal{J}(\mathfrak{a}^m)$ *cuts out a scheme of codimension exactly* $m$ *in* $X$. *Then*

$$\mathrm{Jacobian}_m(\mathfrak{a}) \subseteq \mathcal{J}(\mathfrak{a}^{(1-\varepsilon)m}) \quad \text{for all } 0 < \varepsilon \leq 1.$$

In brief, let $\mu : X' \longrightarrow X$ be a log resolution of $\mathfrak{a}$, with $\mathfrak{a} \cdot \mathcal{O}_{X'} = \mathcal{O}_{X'}(-F)$. Fix a prime divisor $E \subseteq X'$, and write $a = \mathrm{ord}_E(F)$, $b = \mathrm{ord}_E(K_{X'/X})$. For (i), the issue is to show that if $\delta$ is one of the minors of $A$ locally generating $\mathrm{Jacobian}_m(\mathfrak{a})$, then

$$\mathrm{ord}_E(\mu^*\delta) \geq ma - b.$$

The argument proceeds in the spirit of the proof of Lemma 9.2.19, but the computations are now somewhat more involved. We refer to [161, §4] for details.

### 9.3.G Multiplier Ideals on Singular Varieties

In this subsection we discuss briefly how one can extend the definition of multiplier ideals to singular ambient spaces. We follow the approach of Ein in [145] (with some small modifications).

Let $X$ be an irreducible normal variety. The definition of multiplier ideals in Section 9.2 does not directly make sense on $X$ because the canonical bundle $K_X$ exists only as a Weil divisor (class), and so one cannot discuss its pullback to a resolution. However there is a standard way to deal with this difficulty. Namely, one perturbs the canonical divisor $K_X$ by adding a "boundary divisor" $\Delta$ in order to render it $\mathbf{Q}$-Cartier. More precisely, for the purpose of this discussion we make the following

**Definition 9.3.55. (Pairs).** A *pair* $(X, \Delta)$ consists of a normal variety $X$, together with a Weil $\mathbf{Q}$-divisor $\Delta = \sum d_i \Delta_i$ on $X$, such that the $\mathbf{Q}$-divisor $K_X + \Delta$ is $\mathbf{Q}$-Cartier on $X$. (Observe that we do not actually require the coefficients $d_i$ appearing in $\Delta$ to be small or non-negative.) □

We refer to [368, §2.3] for a careful account of the constructions that can be made in this setting. Here we quickly recall how one defines the discrepancies of $(X, \Delta)$. Let $\mu : X' \longrightarrow X$ be a log resolution of the pair in question. Then there are canonically defined rational numbers

$$a(E) = a(E, X, \Delta) \in \mathbf{Q}$$

attached to each prime divisor $E$ on $X'$ having the property that

$$K_{X'} \equiv_{\text{num}} \mu^*(K_X + \Delta) + \sum a(E) \cdot E, \tag{9.18}$$

the sum running over all prime divisors of $X'$.[14] These numbers are called the *discrepancies* of $E$, and are independent of $\mu$ when one views $E$ as being a valuation of the function field $\mathbf{C}(X)$ of $X$. Of course all but finitely many are zero.

Suppose now that $D$ is a $\mathbf{Q}$-Cartier $\mathbf{Q}$-divisor on $X$, and let $\mu : X' \longrightarrow X$ be a log resolution of $(X, D + \Delta)$. Then there are unique rational numbers $b(E)$ such that

$$\mu^*(-D) = \sum b(E) \cdot E,$$

the sum again running over all prime divisors of $X'$. Thus

$$K_{X'} - \mu^*(K_X + \Delta + D) \equiv_{\text{num}} \sum (a(E) + b(E)) \cdot E. \tag{9.19}$$

We will be concerned with the round-up of the divisor on the right, and (somewhat abusively) we write

$$K_{X'} - [\mu^*(K_X + \Delta + D)] = \sum \ulcorner a(E) + b(E) \urcorner \cdot E. \tag{9.20}$$

**Definition 9.3.56. (Multiplier ideal of a Q-divisor on a pair).** The multiplier ideal of the $\mathbf{Q}$-Cartier divisor $D$ for the pair $(X, \Delta)$ is the sheaf

$$\mathcal{J}\big((X, \Delta); D\big) = \mu_* \mathcal{O}_{X'}(K_{X'} - [\mu^*(K_X + \Delta + D)]).$$

If no confusion seems possible, we denote this simply by $\mathcal{J}(D)$. □

Note that for typographical clarity, we modify slightly the punctuation associated to multiplier ideals when dealing with pairs. We trust that this will not lead to any confusion.

If $\Delta$ and $D$ are effective, then $\mathcal{J}\big((X, \Delta); D\big) \subseteq \mathcal{O}_X$ is actually an ideal sheaf. In the general case, we consider $\mathcal{J}(D)$ to be a sheaf of fractional ideals. The proof of Theorem 9.2.18 goes through with little change to show that $\mathcal{J}\big((X, \Delta); D\big)$ is independent of the resolution used to construct it.

**Example 9.3.57.** (i). If $X$ is non-singular, then

$$\mathcal{J}\big((X, \Delta); D\big) = \mathcal{J}(X, \Delta + D).$$

(ii). Suppose that $(X, \Delta)$ is a pair as in 9.3.55, and $D$ is any effective $\mathbf{Q}$-Cartier divisor on $X$. Then $(X, \Delta + D)$ again satisfies 9.3.55, and

$$\mathcal{J}\big((X, \Delta); D\big) = \mathcal{J}\big((X, \Delta + D); 0\big). \quad \Box$$

---

[14] In brief, (9.18) determines $a(E)$ when $E$ is exceptional for $\mu$; for the proper transform $\Delta_i'$ of $\Delta_i$ one puts $a(\Delta_i') = -d_i$, while all other $a(E)$ vanish: see [368, Definition 2.5].

**Remark 9.3.58.** The second statement in 9.3.57 shows that there is some redundancy built into the definition, and that we could have worked directly with $J(X, D)$ for any divisor $D$ such that $K_X + D$ is **Q**-Cartier. However particularly when we define below the multiplier ideals $J((X, \Delta); c \cdot \mathfrak{a})$ associated to ideal sheaves, it seems preferable to have arranged things so that one can assign a multiplier ideal to an arbitrary Cartier divisor.     □

The main new feature when the ambient space is singular is that these multiplier ideals may be non-trivial even when $D$ (and $\Delta$) vanish.

**Example 9.3.59. (Ordinary d-fold points).** Let $X \subseteq \mathbf{C}^{n+1}$ be an $n$-dimensional hypersurface that is smooth except for an ordinary $d$-fold point at $P \in X$: thus $\text{mult}_P X = d$, and the blowing-up $\mu : X' = \text{Bl}_P(X) \longrightarrow X$ of $P$ is a log resolution of $(X, 0)$. Then $X$ is Gorenstein (i.e. $K_X$ is already Cartier), and

$$K_{X'} - \mu^* K_X \equiv_{\text{num}} (n - d) E,$$

$E$ being the exceptional divisor of $X'$. Writing $\mathfrak{m} \subseteq \mathcal{O}_X$ for the maximal ideal at $P$, it follows that if $d \geq n$, then

$$J((X, 0), 0) = \mathfrak{m}^{d-n}. \quad □$$

One can proceed similarly for ideal sheaves:

**Definition 9.3.60. (Multiplier ideals associated to ideal sheaves on a pair).** In the situation of 9.3.55, let $\mathfrak{a} \subseteq \mathcal{O}_X$ be an ideal sheaf, and $c > 0$ a rational number. Fix a log resolution

$$\mu : X' \longrightarrow X$$

of $\mathfrak{a}$ that also resolves the pair $(X, \Delta)$. Suppose that $\mathfrak{a} \cdot \mathcal{O}_{X'} = \mathcal{O}_{X'}(-F)$, where

$$-F = \sum b(E) \cdot E$$

(the sum being taken over all divisors in $X'$), and as in (9.20) write

$$K_{X'} - [\mu^*(K_X + \Delta) + cF] = \sum \lceil a(E) + c b(E) \rceil \cdot E.$$

Then define

$$J((X, \Delta); \mathfrak{a}^c) = \mu_* \mathcal{O}_{X'} \Big( K_{X'} - [\mu^*(K_X + \Delta) + cF)] \Big). \quad □$$

**Example 9.3.61.** As in Example 9.3.59, let $X \subseteq \mathbf{C}^{n+1}$ be a hypersurface that is non-singular except for an ordinary $d$-fold point at $p \in X$, and let $\mathfrak{m} \subseteq \mathcal{O}_X$ be the maximal ideal of $p$. Then

$$J((X, 0); \mathfrak{m}^c) = \mathfrak{m}^{[c]+d-n}$$

(provided that the exponent is non-negative).     □

One can likewise attach multiplier ideals to linear series on a pair $(X, \Delta)$: we leave it to the reader to formulate the definitions.

Most of the results in Section 9.2 go through in this more general setting. Specifically, Propositions 9.2.28, 9.2.31, and 9.2.32 (i)–(iii), as well as the analogous statements 9.2.22 and 9.2.26 for linear series, remain valid as stated for a pair $(X, \Delta)$. However, the previous example shows that the analogues of 9.2.32 (iv) and (v) can fail even when $\Delta = 0$.

The birational transformation rule (Theorem 9.2.33) is also still true, but it needs to be reformulated slightly as follows:

**Proposition 9.3.62. (Birational transformation rule for pairs).** *Let*

$$(X, \Delta_X) \quad , \quad (Y, \Delta_Y)$$

*be pairs satisfying the conditions of 9.3.55, and let $f : Y \longrightarrow X$ be a birational morphism. Assume that*

$$K_Y + \Delta_Y \equiv_{\mathrm{num}} f^*(K_X + \Delta_X) \quad \text{and} \quad f_* \Delta_Y = \Delta_X.$$

*If $D$ is any $\mathbf{Q}$-Cartier $\mathbf{Q}$-divisor on $X$, then*

$$\mathcal{J}\big((X, \Delta_X); D\big) = f_*\Big( \mathcal{J}\big((Y, \Delta_Y); f^*D\big) \Big).$$

When $X$ and $Y$ are non-singular one recovers 9.2.33 by taking $\Delta_X = 0$ and $\Delta_Y = -K_{Y/X}$.

*Indication of Proof of 9.3.62.* For any divisor $E$ over $X$, one has the equality of discrepancies

$$a(E, X, \Delta_X) = a(E, Y, \Delta_Y)$$

[368, Lemma 2.30]. If $\nu : V \longrightarrow Y$ is a log resolution of the data at hand and $h = f \circ \nu$, this implies that

$$K_V - \big[h^*(K_X + \Delta_X + D)\big] = K_V - \big[\nu^*(K_Y + \Delta_Y + f^*D)\big].$$

The assertion follows by taking direct images. $\square$

Similar statements — whose formulation we leave to the reader — hold for the multiplier ideals associated to ideal sheaves and linear series.

## 9.4 Vanishing Theorems for Multiplier Ideals

We now deduce the basic vanishings for multiplier ideals from the theorem of Kawamata and Viehweg. There are two essential statements: a local vanishing theorem for the higher direct images of the sheaves computing multiplier ideals, and the corresponding global statement (originally due in the analytic setting to Nadel). As a first application, we derive useful criteria for the non-vanishing and global generation of suitable twists of multiplier ideals.

### 9.4.A Local Vanishing for Multiplier Ideals

We begin with a basic "acyclicity" property of the divisors computing multiplier ideals. Most of the fundamental local properties of multiplier ideals discussed in the next section ultimately reduce to this result.

**Theorem 9.4.1. (Local vanishing).** *Let $D$ be any **Q**-divisor on a smooth variety $X$, and let $\mu : X' \longrightarrow X$ be a log resolution of $(X, D)$. Then*

$$R^j \mu_* \mathcal{O}_{X'} \big( K_{X'/X} - [\mu^* D] \big) = 0 \quad \text{for } j > 0.$$

*Proof.* Suppose first that $X$ and $X'$ are projective. Let $A$ be any ample divisor on $X$ that is sufficiently positive so that $A - D$ is itself ample. Then $\mu^*(A - D)$ is big and nef, and so Kawamata–Viehweg vanishing (in the form of Corollary 9.1.21) implies that

$$H^j \big( X', \mathcal{O}_{X'} (K_{X'} + \mu^* A - [\mu^* D]) \big) = 0 \quad \text{for } j > 0.$$

But this holds for any sufficiently positive divisor $A$, so it follows from Lemma 4.3.10 that

$$
\begin{aligned}
R^j \mu_* \mathcal{O}_{X'} &\big( K_{X'} + \mu^* A - [\mu^* D] \big) \\
&= R^j \mu_* \mathcal{O}_{X'} \big( K_{X'/X} - [\mu^* D] \big) \otimes \mathcal{O}_X (K_X + A) \\
&= 0 \quad \text{for } j > 0.
\end{aligned}
$$

This proves the theorem under the stated projectivity hypotheses.

Turning to the general situation, the plan is to reduce to the case just treated. The statement being local on $X$, we suppose that $X$ is affine. Then we can construct a fibre square

$$
\begin{array}{ccc}
X' & \hookrightarrow & \overline{X}' \\
\mu \downarrow & & \downarrow \overline{\mu} \\
X & \hookrightarrow & \overline{X},
\end{array}
$$

together with a **Q**-divisor $\overline{D}$ on $\overline{X}$, such that $\overline{X}$ and $\overline{X}'$ are non-singular and projective and $\overline{D} \,|\, X = D$. However it requires more precise information about resolution of singularities to arrange that $\overline{X}'$ is a log resolution of $(\overline{X}, \overline{D})$, and we wish to avoid this (see Remark 9.4.2). So let $\overline{\nu} : \overline{Y} \longrightarrow \overline{X}'$ be a log resolution of $\overline{\mu}^* D + \text{except}(\overline{\mu})$, and set $\overline{h} = \overline{\mu} \circ \overline{\nu}$:

$$
\begin{array}{ccc}
 & \overset{\overline{h}}{\overgroup{\hspace{3cm}}} & \\
\overline{Y} & \overset{\overline{\nu}}{\longrightarrow} \ \overline{X}' \ \overset{\overline{\mu}}{\longrightarrow} & \overline{X}.
\end{array}
$$

The projective case already treated applies to $\overline{\nu}$ and to $\overline{h}$. This gives first of all the vanishing

$$R^k \overline{\nu}_* \mathcal{O}_{\overline{Y}}\big(K_{\overline{Y}/\overline{X}'} - [\overline{\nu}^* \overline{\mu}^* \overline{D}]\big) \; = \; 0 \quad \text{for } k > 0.$$

Using the Grothendieck–Leray spectral sequence we also find that if $j > 0$, then

$$R^j \overline{\mu}_* \Big( \overline{\nu}_* \, \mathcal{O}_{\overline{Y}}\big(K_{\overline{Y}/\overline{X}'} - [\overline{\nu}^* \overline{\mu}^* \overline{D}]\big) \otimes \mathcal{O}_{\overline{X}'}\big(K_{\overline{X}'/\overline{X}}\big)\Big)$$
$$= \; R^j \overline{h}_* \Big( \mathcal{O}_{\overline{Y}}\big(K_{\overline{Y}/\overline{X}} - [\overline{h}^* \overline{D}]\big) \Big) \qquad (*)$$
$$= \; 0.$$

On the other hand, by hypothesis $\overline{\mu}^* \overline{D}$ has SNC support on the open subset $X' \subseteq \overline{X}'$. Therefore

$$\overline{\nu}_* \, \mathcal{O}_{\overline{Y}}\big(K_{\overline{Y}/\overline{X}'} - [\overline{\nu}^* \overline{\mu}^* \overline{D}]\big) \mid X' \; = \; \mathcal{O}_{\overline{X}'}\big(-[\mu^* D]\big) \mid X'$$

thanks to Lemma 9.2.19, and the required vanishing follows from $(*)$. □

**Remark 9.4.2.** One could streamline the argument just completed by showing that one can take the variety $\overline{X}'$ constructed there to be a log resolution of $(\overline{X}, \overline{D})$. For this one would invoke the fact that a log resolution of an effective divisor in a smooth variety can be obtained by a succession of blowings up along smooth centers contained in the locus along which the given divisor fails to have SNC support. See [70] (and [56, §12]). □

**Remark 9.4.3.** Theorem 9.4.1 is a special case of the vanishing theorem for a proper mapping (Remark 9.1.22), and the proof just completed is a special case of the argument used to establish the more general result. This more general result shows also that the theorem remains true if one starts with a proper (but possibly non-projective) log resolution. □

The evident analogues for ideal sheaves and linear series remain valid.

**Variant 9.4.4. (Local vanishing for ideals).** Let $\mathfrak{a} \subseteq \mathcal{O}_X$ be an ideal sheaf, and let $\mu : X' \longrightarrow X$ be a log resolution of $\mathfrak{a}$ with $\mathfrak{a} \cdot \mathcal{O}_{X'} = \mathcal{O}_{X'}(-F)$. Then for any rational $c > 0$,

$$R^j \mu_* \mathcal{O}_{X'}\big(K_{X'/X} - [c \cdot F]\big) = 0 \quad \text{for } j > 0.$$

(Either imitate the proof of Theorem 9.4.1 or use (the proof of) Proposition 9.2.26 to reduce to that result.) □

**Variant 9.4.5. (Local vanishing for linear series).** Let $\mu : X' \longrightarrow X$ be a log resolution of a linear series $|V| \subseteq |D|$, with $\mu^*|V| = |W| + F$, and let $c > 0$ be any rational number. Then

$$R^j \mu_* \mathcal{O}_{X'}\big(K_{X'/X} - [c \cdot F]\big) = 0 \quad \text{for } j > 0. \quad \square$$

**Example 9.4.6. (Local vanishing for a birational morphism).** Let $f :$ $Y \longrightarrow X$ be a projective birational morphism of smooth varieties, and let $D$ be a **Q**-divisor on $X$. Then

$$R^j f_* \Big( \mathcal{J}\big(Y, f^*D\big) \otimes \mathcal{O}_Y(K_{Y/X}) \Big) = 0 \quad \text{for } j > 0.$$

(Let $\nu : X' \longrightarrow Y$ be a log resolution of $f^*D$. One has vanishings for $\nu$ and for $f \circ \nu$, and the statement follows from the Grothendieck–Leray spectral sequence as in the proof of Theorem 9.4.1.) The analogous assertions for ideal sheaves and linear series — whose statements we leave to the reader — are also true. $\qquad \square$

**Remark 9.4.7. (R-divisors).** The fact that Kawamata–Viehweg vanishing remains true for **R**-divisors (Remark 9.1.23) shows that local vanishing holds for such divisors, and that one can allow the coefficient $c$ in Variants 9.4.4 and 9.4.5 to assume real values. $\qquad \square$

### 9.4.B The Nadel Vanishing Theorem

We now come to the basic global vanishing theorem for multiplier ideals:

**Theorem 9.4.8. (Nadel vanishing theorem).** *Let $X$ be a smooth complex projective variety, let $D$ be any **Q**-divisor on $X$, and let $L$ be any integral divisor such that $L - D$ is nef and big. Then*

$$H^i\big(X, \mathcal{O}_X(K_X + L) \otimes \mathcal{J}(D)\big) = 0 \quad \text{for } i > 0.$$

**Remark 9.4.9.** The actual analytic version of Nadel's theorem is more general, dealing as it does with the multiplier ideal associated to an arbitrary plurisubharmonic weight function. Nadel's original statement is briefly discussed below in Section 9.4.D. $\qquad \square$

**Remark 9.4.10. (Sociology).** In the algebraic setting treated here, Theorem 9.4.8 actually predates Nadel's work. For instance, an equivalent statement occurs as Corollary 4.6 in the paper [169] of Esnault and Viehweg. In fact, we will see momentarily that 9.4.8 is essentially just a restatement of Kawamata–Viehweg vanishing. However, the explicit study of multiplier ideals as stand-alone objects, and the systematic exploitation of the vanishing theorems they satisfy, originally occurred in the analytic side of the field starting with the work [460], [461] of Nadel. Therefore the result in question has generally come to be known as Nadel vanishing. But one should bear in mind that in the present context the title of 9.4.8 reflects as much sociological as purely mathematical developments. $\qquad \square$

*Proof of Theorem 9.4.8.* We argue much as in the proof of the local vanishing theorem. Specifically, let $\mu : X' \longrightarrow X$ be a log resolution of $(X, D)$. Then it follows from Kawamata–Viehweg vanishing that

$$H^i\Big(X', \mathcal{O}_{X'}\big(K_{X'/X} - [\mu^*D]\big) \otimes \mu^*\mathcal{O}_X(K_X + L)\Big)$$
$$= H^i\big(X', \mathcal{O}_{X'}(K_{X'} + \mu^*L - [\mu^*D])\big)$$
$$= 0 \quad \text{for} \quad i > 0.$$

But by the previous Theorem 9.4.1,

$$R^j\mu_*\Big(\mathcal{O}_{X'}\big(K_{X'/X} - [\mu^*D]\big) \otimes \mu^*\mathcal{O}_X(K_X + L)\Big)$$
$$= R^j\mu_*\big(\mathcal{O}_{X'}(K_{X'/X} - [\mu^*D])\big) \otimes \mathcal{O}_X(K_X + L)$$
$$= 0 \quad \text{for} \quad j > 0,$$

whereas

$$\mu_*\Big(\mathcal{O}_Y\big(K_{X'/X} - [\mu^*D]\big) \otimes \mu^*\mathcal{O}_X(K_X + L)\Big)$$
$$= \mu_*\big(\mathcal{O}_{X'}(K_{X'/X} - [\mu^*D])\big) \otimes \mathcal{O}_X(K_X + L)$$
$$= \mathcal{O}_X(K_X + L) \otimes \mathcal{J}(D).$$

The theorem then follows via the Leray spectral sequence.     □

A reformulation of Theorem 9.4.8 shows that one may view the presence of the multiplier ideal as a "correction term" appearing in the variant (9.1.20) of Kawamata–Viehweg vanishing in case the **Q**-divisor in question fails to have normal crossing support:

**Corollary 9.4.11.** *Let $B$ be any big and nef **Q**-divisor on a smooth projective variety $X$, and let $\Delta = \lceil B \rceil - B$ be the "upper fractional part" of $B$. Then*

$$H^i\big(X, \mathcal{O}_X(K_X + \lceil B \rceil) \otimes \mathcal{J}(\Delta)\big) = 0 \quad \text{for} \quad i > 0.$$

*Proof.* In fact, $\lceil B \rceil - \Delta = B$ is big and nef, so Nadel vanishing applies.     □

Note that if $\Delta$ has SNC support, then $\mathcal{J}(\Delta) = \mathcal{O}_X(-[\Delta]) = \mathcal{O}_X$, so that one does indeed recover (9.1.20).

**Example 9.4.12. (Sakai's lemma for surfaces).** If $X$ is a smooth projective surface, then Kawamata–Viehweg vanishing holds without any normal crossing hypotheses. In other words, if $B$ is any big and nef **Q**-divisor on $X$, then

$$H^i\big(X, \mathcal{O}_X(K_X + \lceil B \rceil)\big) = 0 \quad \text{for} \quad i > 0. \qquad (*)$$

(In fact, set $\Delta = \lceil B \rceil - B$. Then $\mathcal{J}(\Delta)$ is co-supported on a finite set of points, and hence the vanishing of $H^i\big(X, \mathcal{O}_X(K_X + \lceil B \rceil) \otimes \mathcal{J}(\Delta)\big)$ for $i > 0$ implies the vanishings $(*)$.) This example explains why vanishing theorems for **Q**-divisors are particularly easy to work with on surfaces (as for example in [394]).     □

**Example 9.4.13. (Miyaoka's vanishing theorem).** Let $D$ be an integral big divisor on a smooth projective surface $X$, with Zariski decomposition $D = P + N$ (Section 2.3.E). Then

$$H^i\big(X, \mathcal{O}_X(K_X + D - [N])\big) = 0 \quad \text{for} \quad i > 0.$$

(Use Sakai's lemma.) This statement, which is due to Miyaoka [427], was an important precursor of the theorem of Kawamata and Viehweg.    □

**Remark 9.4.14. (R-divisors).** Thanks to 9.4.7 and 9.1.23, Nadel vanishing and its proof remain valid for **R**-divisors.

There is an analogue of Theorem 9.4.8 for linear series and ideal sheaves:

**Corollary 9.4.15. (Nadel vanishing for linear series and ideals).** *Let $X$ be a smooth projective variety, let $c > 0$ be a rational number, and let $L$ and $A$ be integral divisors on $X$ such that $L - c \cdot A$ is big and nef.*

(i). *If $|V| \subseteq |A|$ is any linear series, then*

$$H^i\big(X, \mathcal{O}_X(K_X + L) \otimes \mathcal{J}(c \cdot |V|)\big) = 0 \quad \text{for} \quad i > 0.$$

*In particular, $H^i\big(X, \mathcal{O}_X(K_X + L) \otimes \mathcal{J}(c \cdot |A|)\big) = 0$ for $i > 0$.*

(ii). *Let $\mathfrak{a} \subseteq \mathcal{O}_X$ be an ideal sheaf such that $\mathfrak{a} \otimes \mathcal{O}_X(A)$ is globally generated. Then*

$$H^i\big(X, \mathcal{O}_X(K_X + L) \otimes \mathcal{J}(\mathfrak{a}^c)\big) = 0 \quad \text{for} \quad i > 0.$$

(iii). *The statement of (ii) holds more generally assuming that $\mathfrak{a}^p \otimes \mathcal{O}_X(pA)$ is globally generated for any integer $p > 0$.*

*Proof.* The statements follow from Nadel vanishing using respectively Proposition 9.2.26, Example 9.2.22, and Example 9.2.24.    □

**Generalization 9.4.16. (Nadel vanishing in the relative setting).** Let $X$ be a smooth variety and $f : X \longrightarrow T$ a surjective projective morphism. Keeping notation as in Generalization 9.2.21, Nadel's theorem generalizes to a relative statement as follows:

**Theorem.** *Let $L$ and $D$ be integral divisors on $X$, and $c > 0$ a positive rational number. Assume that $L - c \cdot D$ is nef and big for $f$. Then*

$$R^i f_*\big(\mathcal{O}_X(K_X + L) \otimes \mathcal{J}(f, c \cdot |D|)\big) = 0 \quad \text{for} \quad i > 0.$$

This follows from vanishing for a morphism (Remark 9.1.22) in essentially the same way that Theorem 9.4.8 is deduced from Kawamata–Viehweg vanishing. Very briefly, let $\mu : X' \longrightarrow X$ be a log resolution of $\rho : f^* f_* \mathcal{O}_X(D) \longrightarrow \mathcal{O}_X(D)$, with $\operatorname{im}(\mu^* \rho) = \mathcal{O}_{X'}(\mu^* D - F)$, and set $g = f \circ \mu$. Then

$$\mu^* \Big( L - c \cdot D \Big) + c \cdot \Big( \mu^* D - F \Big) \; = \; \mu^* L - c \cdot F$$

is nef and big for $g$, and consequently vanishing for $g$ implies that

$$R^i g_* \mathcal{O}_{X'} \Big( \mu^*(K_X + L) + (K_{X'/X} - [cF]) \Big) \; = \; 0 \quad \text{for } i > 0.$$

But $R^i g_* = R^i f_* \circ \mu_*$ for the sheaf in question thanks to Theorem 9.4.1 and the Grothendieck–Leray spectral sequence for $f \circ \mu$. The assertion of the theorem then follows from the projection formula.    □

### 9.4.C Vanishing on Singular Varieties

With small modifications, the previous results extend to the case of singular ambient spaces. Specifically:

**Theorem 9.4.17. (Nadel vanishing on singular varieties).** *Let $(X, \Delta)$ be a pair satisfying the conditions of 9.3.55, and let $D$ be a **Q**-Cartier **Q**-divisor on $X$.*

(i). *If $\mu : X' \longrightarrow X$ is a log resolution of $(X, \Delta + D)$, then*

$$R^j \mu_* \mathcal{O}_{X'} \Big( K_{X'} - [\mu^*(K_X + \Delta + D)] \Big) \; = \; 0 \quad \text{for } j > 0.$$

(ii). *Let $N$ be an integral Cartier divisor on $X$ having the property that*

$$N - \Big( K_X + \Delta + D \Big)$$

*is nef and big. Then*

$$H^i \Big( X \, , \, \mathcal{O}_X(N) \otimes \mathcal{J} \big( (X, \Delta) ; D \big) \Big) \; = \; 0 \quad \text{for } i > 0. \quad □$$

When $X$ is non-singular, one recovers Theorem 9.4.8 by taking $\Delta = 0$ and $N = K_X + L$. The argument is essentially the same as the proof in the smooth case, and we do not write it out separately.

There is also an analogue of Corollary 9.4.15:

**Proposition 9.4.18.** *With $(X, \Delta)$ as above, let $N$ and $A$ be integral Cartier divisors on $X$ such that $N - (K_X + \Delta + cA)$ is nef and big for some $c > 0$. Suppose that $\mathfrak{a} \subseteq \mathcal{O}_X$ is an ideal sheaf on $X$ with $\mathcal{O}_X(A) \otimes \mathfrak{a}$ globally generated. Then*

$$H^i \Big( X \, , \, \mathcal{O}_X(N) \otimes \mathcal{J} \big( (X, \Delta) ; \mathfrak{a}^c \big) \Big) \; = \; 0 \quad \text{for } i > 0. \quad □$$

We leave it to the reader to formulate the corresponding statements for linear series.

## 9.4.D Nadel's Theorem in the Analytic Setting

In this subsection, following [126, §3, §5] and [362, Chapter 10], we briefly survey the Nadel vanishing theorem in its native analytic setting.

Let $X$ be a complex manifold, and $L$ a holomorphic line bundle on $X$.

**Definition 9.4.19. (Singular Hermitian metric).** A *singular Hermitian metric* on $L$ is a Hermitian metric $h$ on $L$, defined over the complement of a set of measure zero, having the property that if

$$\theta \ : \ L \mid U \longrightarrow U \times \mathbf{C}$$

is any local trivialization of $L$, and if $\xi \in \Gamma(U, L)$ is a local generating section, then the length function $\| \ \|_h$ determined by $h$ satisfies

$$\| \xi \|_h \ = \ | \theta(\xi) | \cdot e^{-\varphi},$$

where $\varphi \in \mathrm{L}^1_{\mathrm{loc}}(U)$ is a locally integrable function on $U$, called the *local weight* of $h$ with respect to $\theta$.[15]                                      □

If $\theta' : U' \longrightarrow U' \times \mathbf{C}$ is a second local trivialization, giving rise to a transition function $g \in \mathcal{O}^*_X(U \cap U')$, then the corresponding local weight $\varphi'$ satisfies

$$\varphi' \ = \ \varphi + \log |g|. \tag{*}$$

This (and the integrability of $\varphi$) guarantees that

$$\Theta_h(L) \ =_{\mathrm{def}} \ 2 \partial \bar{\partial} \varphi$$

exists as a well-defined closed $(1,1)$-current on $X$. Moreover, $\frac{i}{2\pi}\Theta_h(L)$ represents $c_1(L)$ in $H^2(X, \mathbf{C})$.

**Example 9.4.20. (Singular metric determined by a Q-divisor).** Let $A$ be an effective integral divisor on a smooth variety $X$, and set $L = \mathcal{O}_X(A)$. Once we choose a local defining equation $g$ for $A$, $L$ acquires a canonical local generator $\xi$, characterized by the property that $g\xi \in \mathcal{O}_X(A)$ is the image of $1 \in \mathcal{O}_X$ under the natural inclusion $\mathcal{O}_X \longrightarrow \mathcal{O}_X(A)$. Then $L$ has a singular metric $h$ with $\|\xi\|_h = e^{-\log|g|}$. If $B \in |mA|$ has local defining equation $g_B$, then $L = \mathcal{O}_X(A)$ has a singular metric $h$ with local weight $\frac{1}{m}\log|g_B|$. Note that this coincides with the weight function $\varphi_D$ associated in Example 9.1.23 to the **Q**-divisor $D = \frac{1}{m}B$.                              □

Now suppose that the local weight functions $\varphi$ corresponding to $h$ are PSH. Then $\varphi$ determines a multiplier ideal $\mathcal{J}(\varphi) \subseteq \mathcal{O}_X$ (Definition 9.3.39). Moreover, it follows from (*) that this ideal is actually independent of the local weight function $\varphi$. Therefore $h$ gives rise to a globally defined multiplier ideal sheaf

$$\mathcal{J}(h) \ \subseteq \ \mathcal{O}_X.$$

---

[15] In practice, $\varphi$ will always be plurisubharmonic.

**Theorem 9.4.21. (Analytic Nadel vanishing theorem).** *Let $(X, \omega)$ be a compact Kähler manifold, and $L$ a holomorphic line bundle on $X$ equipped with a singular Hermitian metric $h$ as above. Assume that*

$$\sqrt{-1} \cdot \Theta_h(L) \geq \varepsilon \cdot \omega$$

*for some $\varepsilon > 0$. Then*

$$H^i\big(X, \mathcal{O}_X(K_X + L) \otimes \mathcal{J}(h)\big) = 0 \quad \text{for} \quad i > 0.$$

We refer to [126, §5] for the proof. This implies Theorem 9.4.8 upon taking $h$ to be the singular metric associated to the $\mathbf{Q}$-divisor $D$ (Example 9.4.20).

**Remark 9.4.22.** De Cataldo [103] has developed an analogous theory of singular Hermitian metrics on vector bundles of higher rank.  $\square$

### 9.4.E Non-Vanishing and Global Generation

We conclude with some statements (due to Esnault–Viehweg and Siu) asserting the non-vanishing and global generation of suitable twists of a multiplier ideal. These results — although elementary consequences of Nadel vanishing — will turn out to be quite important for applications (especially Siu's results [537], [539]). The essential point is the uniform nature of the statements.

We start with non-vanishing:

**Proposition 9.4.23. (Non-vanishing for multiplier ideals).** *Let $X$ be a smooth projective variety of dimension $n$. Let $D$ be an effective $\mathbf{Q}$-divisor and $L$ an integral divisor on $X$ such that $L - D$ is big and nef, and let $H$ be any ample (or nef and big) divisor. Then there is an integer $0 \leq k \leq n$ such that*

$$H^0\Big( X, \mathcal{O}_X(K_X + L + kH) \otimes \mathcal{J}(D)\Big) \neq 0.$$

*If $H$ is effective, then the same statement holds for every $k \geq n$.*

*Proof.* In fact,

$$H^i\big(X, \mathcal{O}_X(K_X + L + tH) \otimes \mathcal{J}(D)\big) = 0 \quad \text{for all} \quad i > 0 \text{ and } t \geq 0$$

thanks to Nadel vanishing. Hence for $t \geq 0$,

$$h^0\big(X, \mathcal{O}_X(K_X + L + tH) \otimes \mathcal{J}(D)\big) = \chi\big(X, \mathcal{O}_X(K_X + L + tH) \otimes \mathcal{J}(D)\big).$$

But the Euler characteristic on the right is given by a polynomial in $t$ of degree $n$. Therefore it must be non-zero for at least one integer $t \in [0, n]$. If $H$ is effective, then multiplication by the defining equation of $H$ determines for every $k \in \mathbf{Z}$ an injective map

$$\mathcal{O}_X(kH) \otimes \mathcal{J}(D) \longrightarrow \mathcal{O}_X((k+1)H) \otimes \mathcal{J}(D)$$

of coherent sheaves,[16] and the last statement follows.  $\square$

---

[16] The injectivity is automatic since $\mathcal{J}(D)$ is torsion-free.

The proposition implies a striking non-vanishing theorem due to Siu for certain adjoint bundles associated to big (or merely rationally effective) divisors. This result lies at the heart of Siu's proof — which appears in Section 10.2 — of an effective version of Matsusaka's Theorem.

**Corollary 9.4.24. (Non-vanishing for rationally effective divisors).** *Let $L$ be a divisor on a smooth projective variety of dimension $n$ having non-negative Iitaka dimension (Definition 2.1.3), i.e. satisfying the property that*

$$H^0(X, \mathcal{O}_X(mL)) \neq 0$$

*for some $m > 0$. Let $H$ be an ample (or nef and big) divisor on $X$. Then there exists at least one integer $k \in [1, n+1]$ such that*

$$H^0(X, \mathcal{O}_X(K_X + L + kH)) \neq 0.$$

*Proof.* In fact, choose $A \in |mL|$, and set $D = \frac{1}{m}A$. Then $D \equiv_{\mathrm{num}} L$, and so Proposition 9.4.23 (applied to $L + H$) implies that

$$H^0(X, \mathcal{O}_X(K_X + L + kH) \otimes \mathcal{J}(D)) \neq 0$$

for some $k$ in the indicated range. The assertion follows since this group is a subspace of $H^0(X, \mathcal{O}_X(K_X + L + kH))$. $\qquad\square$

**Remark 9.4.25.** When $D$ is big, one can take $k \in [0, n]$: see Example 11.2.14. $\qquad\square$

Turning to global generation, recall (Theorem 1.8.5) that if $B$ is a very ample divisor on a projective variety $X$, and if $\mathcal{F}$ is any coherent sheaf on $X$ satisfying the vanishing

$$H^i(X, \mathcal{F} \otimes \mathcal{O}_X(-iB)) = 0 \quad \text{for } i > 0,$$

then $\mathcal{F}$ is globally generated. Combining this with Nadel vanishing one immediately deduces:

**Proposition 9.4.26. (Global generation of multiplier ideals).** *Let $X$ be a smooth projective variety of dimension $n$, and fix a very ample divisor $B$ on $X$. Let $D$ be an effective $\mathbf{Q}$-divisor and $L$ an integral divisor on $X$ such that $L - D$ is big and nef. Then*

$$\mathcal{O}_X(K_X + L + mB) \otimes \mathcal{J}(D)$$

*is globally generated as soon as $m \geq n$.* $\qquad\square$

**Remark 9.4.27.** Using Theorem 9.4.17 in place of Nadel vanishing, the last two statements extend in a natural way to a pair $(X, \Delta)$ satisfying 9.3.55. We leave the statements to the interested reader. $\qquad\square$

## 9.5 Geometric Properties of Multiplier Ideals

This section is devoted to some geometric properties of multiplier ideals. We start by establishing the restriction theorem, which is undoubtably the most important local property of multiplier ideals. This is used to prove two statements (the subadditivity and summation theorems) of an algebraic nature. We conclude in Sections 9.5.D and 9.5.E by analyzing the behavior of multiplier ideals under specialization and generically finite coverings.

The reader who wishes to see multiplier ideals in action might want to pass directly to the next chapter. Several of the applications appearing there require only a couple of facts from the present section.

### 9.5.A Restrictions of Multiplier Ideals

Let $X$ be a non-singular variety, let $D$ be an effective $\mathbf{Q}$-divisor on $X$, and let $H \subseteq X$ be a smooth irreducible divisor that does not appear in the support of $D$. Then there are two ways to produce an ideal sheaf on $H$. In the first place, the restriction

$$D_H =_{\mathrm{def}} D \mid H$$

is an effective $\mathbf{Q}$-divisor on $H$, so we can form the multiplier ideal $\mathcal{J}(H, D_H) \subseteq \mathcal{O}_H$. On the other hand, we can take the multiplier ideal $\mathcal{J}(X, D)$ of $D$ on $X$ and restrict it to $H$ to get an ideal

$$\mathcal{J}(X,D)_H =_{\mathrm{def}} \mathcal{J}(X,D) \cdot \mathcal{O}_H$$
$$= \mathrm{im}\Big(\mathcal{J}(D) \hookrightarrow \mathcal{O}_X \twoheadrightarrow \mathcal{O}_H\Big).$$

The basic result, due in the algebro-geometric setting to Esnault–Viehweg [174], is a comparison between these two ideal sheaves on $H$:

**Theorem 9.5.1. (Restriction theorem).** *Keep the notation just introduced, so that $X$ is a smooth variety, $D$ is an effective $\mathbf{Q}$-divisor on $X$, and $H \subseteq X$ is a nonsingular irreducible hypersurface that is not contained in the support of $D$. Then there is an inclusion*

$$\mathcal{J}(H, D_H) \subseteq \mathcal{J}(X,D)_H.$$

One may think of this as reflecting the principle that "singularities can only get worse" upon restriction of a divisor to a hypersurface. See Example 9.5.3 for a strengthening due to Takagi.

**Example 9.5.2. (Strict inclusion).** It can easily happen that the inclusion in the theorem is strict. This occurs, for example, in the situation described in Example 9.1.4 and Figure 9.1. Specifically, take $X = \mathbf{C}^2$ with coordinates $s$ and $t$. Let $H$ be the $t$-axis, let $A = \{s - t^2 = 0\}$ be a parabola tangent to $H$, and put $D = \frac{1}{2}A$. Then $\mathcal{J}(X, D) = \mathcal{O}_X$, but $\mathcal{J}(H, D_H) = (t)$ is the ideal of the origin in $H$. $\square$

*Proof of Theorem 9.5.1.* The question being local, we can suppose that $X$ is quasi-projective. Let $\mu : X' \longrightarrow X$ be a log resolution of $(X, D + H)$. Write

$$\mu^* H \;=\; H' + \sum a_j E_j \;\; (a_j \geq 0),$$

where $H' \subseteq X'$ denotes the proper transform of $H$, and the $E_j \subseteq X'$ are exceptional divisors for $\mu$. Then

$$\mu_H \;=_{\text{def}}\; \mu \,|\, H' : H' \longrightarrow H$$

is a log resolution of $(H, D_H)$. The situation is summarized in the commutative diagram

$$\begin{array}{ccc}
H' & \lhook\joinrel\longrightarrow & X' \\
{\scriptstyle \mu_H} \downarrow & & \downarrow {\scriptstyle \mu} \\
H & \lhook\joinrel\longrightarrow & X.
\end{array}$$

By definition,

$$\begin{aligned}
\mathcal{J}(X, D) &\;=\; \mu_* \mathcal{O}_{X'}\big(K_{X'/X} - [\mu^* D]\big), \\
\mathcal{J}(H, D_H) &\;=\; \mu_{H,*} \mathcal{O}_{H'}\big(K_{H'/H} - [\mu_H^* D_H]\big),
\end{aligned}$$

and the issue is to compare the two sheaves appearing on the right.

In the first place, since all the relevant divisors on $X'$ cross normally, we have (Example 9.1.8)

$$\big[\mu_H^* D_H\big] \;=\; \big[\mu^* D\big] \,|\, H'.$$

Next, use the adjunction formula to write

$$\begin{aligned}
K_{H'} &\equiv_{\text{lin}} \big(K_{X'} + H'\big) \,|\, H', \\
K_H &\equiv_{\text{lin}} \big(K_X + H\big) \,|\, H,
\end{aligned} \tag{9.21}$$

so that

$$K_{H'/H} \;\equiv_{\text{lin}}\; \Big(K_{X'/X} - \sum a_j E_j\Big) \,|\, H'. \tag{9.22}$$

We will use the fact that the two sides of (9.22) actually coincide as divisors.[17] This can be checked by using the Poincaré residue to realize the linear equivalences (9.21) as an equality of cycles: see for instance [332].

Consider now the integral divisor

$$B \;=\; K_{X'/X} - [\mu^* D] - \sum a_j E_j$$

on $X'$. Then $B \,|\, H' = K_{H'/H} - [\mu_H^* D_H]$, and hence

---

[17] This amounts to saying that if the restriction of a $\mu$-exceptional component $E_j$ to $H'$ is not $\mu_H$-exceptional, then $\mathrm{ord}_{E_j}(K_{X'/X}) = a_j$.

$$\mathcal{J}(H, D_H) = \mu_{H,*}\mathcal{O}_{H'}(B).$$

On the other hand,

$$\mu_*\mathcal{O}_{X'}(B) \subseteq \mu_*\mathcal{O}_{X'}(K_{X'/X} - [\mu^*D])$$
$$= \mathcal{J}(X, D).$$

So it suffices to prove

$$\mu_{H,*}(\mathcal{O}_{H'}(B)) = \mu_*\mathcal{O}_{X'}(B) \cdot \mathcal{O}_H \qquad (9.23)$$
$$=_{\mathrm{def}} \left(\text{image of } \mu_*\mathcal{O}_{X'}(B) \text{ in } \mathcal{O}_H\right).$$

To this end, observe that

$$B - H' = K_{X'/X} - [\mu^*(D + H)].$$

Therefore the local vanishing theorem 9.4.1, applied to $D + H$, implies that

$$R^1\mu_*\mathcal{O}_{X'}(B - H') = 0.$$

Then (9.23) follows by taking direct images in the exact sequence

$$0 \longrightarrow \mathcal{O}_{X'}(B - H') \xrightarrow{\cdot H'} \mathcal{O}_{X'}(B) \longrightarrow \mathcal{O}_{H'}(B) \longrightarrow 0.$$

This completes the proof. $\qquad\qquad\square$

**Example 9.5.3. (A strengthening).** As Takagi observes, the proof just completed actually yields a stronger statement. Specifically, in the setting of 9.5.1, fix any rational number $0 < t \le 1$. Then

$$\mathcal{J}(H, D_H) \subseteq \mathcal{J}(X, D + (1-t)H)_H. \qquad (9.24)$$

(Keeping the notation of the previous proof, one has

$$B \preccurlyeq K_{X'/X} - [\mu^*(D + (1-t)H)]$$

for every $0 < t \le 1$.) See Proposition 9.5.15 for a related result. $\qquad\square$

**Example 9.5.4. (Restrictions of ideals).** Let $\mathfrak{a} \subseteq \mathcal{O}_X$ be an ideal sheaf on $X$, and let $H \subseteq X$ be a smooth irreducible divisor not contained in the zero-locus of $\mathfrak{a}$. Denote by $\mathfrak{a}_H \subseteq \mathcal{O}_H$ the restriction of $\mathfrak{a}$ to $H$. Then for any rational number $c > 0$,

$$\mathcal{J}(H, c \cdot \mathfrak{a}_H) \subseteq \mathcal{J}(X, c \cdot \mathfrak{a})_H.$$

We leave it to the reader to state and prove the analogue of Takagi's inclusion (9.24). $\qquad\qquad\square$

**Example 9.5.5. (Restrictions of linear series).** Let $V \subseteq H^0(X, L)$ be a linear series on $X$, and let $H \subseteq X$ be a smooth irreducible divisor that is not contained in the base-locus of $|V|$. The restriction $|V|_H$ of $|V|$ to $H$ is a linear series on $H$ associated to the line bundle $L \mid H$. Then for all rational numbers $c > 0$,

$$\mathcal{J}(H, c \cdot |V|_H) \subseteq \mathcal{J}(X, c \cdot |V|)_H. \quad \square$$

Theorem 9.5.1 extends right away to smooth subvarieties of higher codimension:

**Corollary 9.5.6.** *Let $D$ be an effective $\mathbf{Q}$-divisor on the smooth quasi-projective variety $X$, and let $Y \subseteq X$ be a smooth subvariety that is not contained in the support of $D$. Then*

$$\mathcal{J}(Y, D_Y) \subseteq \mathcal{J}(X, D)_Y,$$

*where as usual $D_Y$ and $\mathcal{J}(X, D)_Y$ denote respectively the restrictions of $D$ and $\mathcal{J}(X, D)$ to $Y$.*

*Proof.* In fact $Y \subseteq X$ is a local complete intersection, so the assertion follows from repeated applications of the restriction theorem. $\quad \square$

**Example 9.5.7.** Let $f : X \longrightarrow T$ be a smooth morphism of non-singular varieties, and let $D$ be an effective $\mathbf{Q}$-divisor on $X$. Fix $t \in T$, and assume that the fibre $X_t = f^{-1}(t)$ is not contained in the support of $D$. Then

$$\mathcal{J}(X_t, D|X_t) \subseteq \mathcal{J}(X, D)_{X_t}.$$

Theorem 9.5.35 below shows that in fact equality holds for *general* $t \in T$. $\quad \square$

**Example 9.5.8. (Pullbacks of multiplier ideals).** Let $f : Y \longrightarrow X$ be a morphism of smooth irreducible varieties, and let $D$ be an effective $\mathbf{Q}$-divisor on $X$. Assume that the support of $D$ does not contain $f(Y)$. Then one has an inclusion

$$\mathcal{J}(Y, f^*D) \subseteq f^{-1}\mathcal{J}(X, D)$$

of ideal sheaves on $Y$. (First consider $Y \times X$, with projection $p_2 : Y \times X \longrightarrow X$, and observe that $\mathcal{J}(Y \times X, p_2^*D) = p_2^{-1}\mathcal{J}(X, D)$ (compare Proposition 9.5.22 below). Then apply the previous corollary to the graph $Y = \Gamma_f \subset Y \times X$ of $f$.) $\quad \square$

**Example 9.5.9. (Restriction to general hypersurfaces).** Let $|V|$ be a free linear series, and let $H \in |V|$ be a *general* divisor. Then in the setting of the restriction theorem one has equality of the two ideals in question, i.e.

$$\mathcal{J}(H, D_H) = \mathcal{J}(X, D)_H.$$

(Keeping the notation of the proof of 9.5.1, one can arrange using 9.1.9 that $\mu^*H = H'$, i.e. that $a_i = 0$ for all $i$. Then the assertion follows from (9.23).) This statement is due to Hwang. See Theorem 9.5.35 for a related result. $\quad \square$

**Remark 9.5.10.** Corollary 9.5.6 and Examples 9.5.7, 9.5.8, and 9.5.9 extend in an evident way to multiplier ideals defined by ideal sheaves and linear series. We leave the formulations and verifications to the reader. ☐

The restriction theorem gives rise to various useful criteria for the triviality of multiplier ideals.

**Corollary 9.5.11. (Inversion of adjunction, I).** *In the situation of Theorem 9.5.1, fix a point $x \in H$ and suppose that $\mathcal{J}(H, D_H)$ is trivial at $x$, i.e. assume that $\mathcal{J}(H, D_H)_x = \mathcal{O}_x H$. Then*

$$\mathcal{J}(X, D + (1 - t)H)_x = \mathcal{O}_x X$$

*for any rational number $0 < t \leq 1$, and in particular $\mathcal{J}(X, D)$ is trivial at $x$. Equivalently, if $(H, D_H)$ is KLT at $x$, then so is $(X, D + (1 - t)H)$.*

*Proof.* When $t = 1$ this follows from Theorem 9.5.1, and for $0 < t < 1$ one uses Takagi's strengthening 9.5.3 of the restriction theorem. ☐

**Remark 9.5.12.** Corollary 9.5.11 is a special case of the theorem on inversion of adjunction ([364], [368, Theorem 5.50]). The paper [162] of Ein, Mustață, and Yasuda contains some stronger results involving this circle of ideas. ☐

Corollary 9.5.11 leads in turn to an important numerical criterion for the triviality of a multiplier ideal:

**Proposition 9.5.13. (Divisors of small multiplicity).** *Let $D$ be an effective $\mathbf{Q}$-divisor on a smooth variety $X$. Suppose that $x \in X$ is a point at which*

$$\mathrm{mult}_x D < 1,$$

*the multiplicity of a $\mathbf{Q}$-divisor being understood in the sense of Definition 9.3.1. Then the multiplier ideal $\mathcal{J}(D)$ associated to $D$ is trivial at $x$.*

**Example 9.5.14.** If $A \subseteq X$ is a smooth hypersurface, considered as an integral divisor, then $\mathrm{mult}_x(A) = 1$ for all $x \in A$, while $\mathcal{J}(A) = \mathcal{O}_X(-A)$ is non-trivial at every $x \in A$. So the numerical hypothesis of Proposition 9.5.13 is in general optimal. A more interesting example shows that the proposition is asymptotically the best possible even for divisors $D$ with $[D] = 0$. Specifically, consider in $\mathbf{C}^2$ the plane curve

$$A = \{y^2 - x^n = 0\},$$

and set $D = \left(\frac{n+2}{2n}\right)A$. Then $\mathrm{mult}_0 D = \frac{n+2}{n}$, but it follows from Examples 9.3.31 and 9.3.35 that $\mathcal{J}(D)$ is non-trivial at the origin. ☐

*Proof of Proposition 9.5.13.* We argue by induction on $n = \dim X$, the assertion being evident if $X$ is a smooth curve. Assuming $n > 1$, we may fix a smooth divisor $H \subseteq X$ passing through $x$, not contained in the support of $D$, with the property that if $D_i$ is any component of $D$, then the restriction of $D_i$ to $H$ satisfies $\text{mult}_x(D_i \mid H) = \text{mult}_x(D_i)$. Setting $D_H = D \mid H$, it follows that

$$\text{mult}_x(D_H) = \text{mult}_x(D) < 1.$$

By induction $\mathcal{J}(H, D_H)_x = \mathcal{O}_x H$, and so Corollary 9.5.11 implies that

$$\mathcal{J}(X, D)_x = \mathcal{O}_x X,$$

as required.                                                                                               $\square$

The next result gives roughly speaking a lower bound on how deep the multiplier ideal of a restriction can be. It may be viewed as the generalization to multiplier ideals of some well-known statements for singularities of pairs.

**Proposition 9.5.15.** *Returning to the situation of Theorem 9.5.1, fix a number $0 < s < 1$. Then*

$$\mathcal{J}\big(X\,,\,D + (1-t)H\big)_H \subseteq \mathcal{J}\big(H\,,\,(1-s)D_H\big)$$

*for all sufficiently small $0 < t \ll 1$.*

**Example 9.5.16.** Combining the proposition with Takagi's inclusion (9.24), one finds that if $0 < s < 1$ then

$$\mathcal{J}\big(H\,,\,D_H\big) \subseteq \mathcal{J}\big(X\,,\,D + (1-t)H\big)_H \subseteq \mathcal{J}\big(H\,,\,(1-s)D_H\big)$$

for all sufficiently small $0 < t \ll 1$.                                                                 $\square$

**Corollary 9.5.17. (Inversion of adjunction, II).** *Remaining in the situation of Theorem 9.5.1, suppose that $\mathcal{J}(X, D + (1-t)H)$ is trivial near $x \in H$ for all $0 < t < 1$. Then $\mathcal{J}\big(H, (1-s)D_H\big)$ is trivial near $x$ for all $0 < s < 1$.*     $\square$

**Example 9.5.18.** One cannot take $s = 0$ in 9.5.15. In fact, consider $X = \mathbf{C}^2$ with coordinates $x, y$, let $D = \frac{1}{2}(x^2 - y^2 = 0)$, and take $H$ to be the $x$-axis. Then $\mathcal{J}(X, D + (1-t)H)$ is trivial for all $0 < t < 1$, but $\mathcal{J}(H, D_H)$ vanishes at the origin.                                                                                       $\square$

*Proof of Proposition 9.5.15.* Keeping the notation of the proof of 9.5.1, let $E \subseteq X'$ ($E \neq H'$) be a prime divisor appearing in the support of $\mu^*(D + H) + K_{X'/X}$ that meets $H'$, and put $\overline{E} = E \cap H'$. It is enough to show that

$$\text{ord}_E\left(\big[\,\mu^*\big((1-t)H + D\big)\big] - K_{X'/X}\right) \geq$$
$$\text{ord}_{\overline{E}}\left(\big[\mu_H^*\big((1-s)D_H\big)\big] - K_{H'/H}\right) \qquad (9.25)$$

when $t \ll 1$ and the right side is positive, for then the restriction of a germ $f \in \mathcal{J}(X, D + (1-t)H)$ to $H$ lies in $\mathcal{J}(H, (1-s)D_H)$. To this end, write

$$b = \mathrm{ord}_E\big(K_{X'/X}\big) \ , \quad a = \mathrm{ord}_E\big(\mu^* H\big) \ , \quad r = \mathrm{ord}_E\big(\mu^* D\big).$$

We may assume that $r > 0$, for otherwise the right side of (9.25) is negative. It follows from (9.22) that $\mathrm{ord}_{\overline{E}}(K_{H'/H}) = b - a$, so the issue is to show that

$$\big[(1-t)a + r\big] - b \ \geq \ \big[(1-s)r\big] - (b-a).$$

But it suffices for this that $t \leq \frac{sr}{a}$.                                         $\square$

We conclude this discussion with some remarks on the extension of Theorem 9.5.1 to the case of singular varieties. Suppose then that $(X, \Delta)$ is a pair satisfying the conditions of Definition 9.3.55, with $\Delta$ effective. Let $H \subseteq X$ be a reduced integral Cartier divisor on $X$, with $H \not\subseteq \mathrm{Supp}\Delta$, and assume that $H$ is a normal variety. Then the restriction $\Delta_H = \Delta \,|\, H$ is naturally defined as a Weil divisor on $H$: in fact if $\Delta_i$ is a prime divisor on $X$, then every component of $\Delta_i \cap H$ is generically Cartier on $H$ thanks to the fact that $H$ is non-singular in codimension one, and so determines a unique Weil divisor on $H$. By the same token the adjunction formula $(K_X + H) \,|\, H = K_H$ holds as an equality of Weil divisor classes on $H$ (cf. [368], Remark 5.47). It follows that $K_H + \Delta_H$ is $\mathbf{Q}$-Cartier — so in particular $(H, \Delta_H)$ is a pair in the sense of 9.3.55 — and

$$\big(K_X + \Delta + H\big) \,|\, H \ \equiv_{\mathrm{num}} \ K_H + \Delta_H.$$

The proof of the restriction theorem now goes through with only minor changes to establish:

**Theorem 9.5.19. (Restrictions on singular varieties).** *With $(X, \Delta)$ and $H \subseteq X$ as stated, let $D$ be any effective $\mathbf{Q}$-Cartier $\mathbf{Q}$-divisor on $X$ whose support does not contain $H$. Then*

$$\mathcal{J}\big((H, \Delta_H)\,;\, D_H\big) \ \subseteq \ \mathcal{J}\big((X, \Delta)\,;\, D\big)_H. \quad \square$$

The effectivity hypotheses are imposed in order to avoid fractional ideals. Again we leave to the reader the analogous statements for ideal sheaves and linear series.

### 9.5.B Subadditivity

This section is the first of two addressing the question of how multiplier ideals behave as functions of their argument. The basic result here is the so-called subadditivity theorem, which states that the multiplier ideal of the sum of divisors is contained in the product of the individual multiplier ideals. Several natural geometric constructions lead to ideals — for example the base ideals of linear series — satisfying the reverse inclusions. Once we have developed the machinery of asymptotic multiplier ideals, we will be able to give interesting applications by exploiting this tension. The subadditivity theorem was established by Demailly, Ein, and the author in [130], from which the following discussion is taken.

Here is the statement:

**Theorem 9.5.20. (Subadditivity).** *Let $X$ be a non-singular variety.*

(i). *Suppose that $D_1$ and $D_2$ are any two effective $\mathbf{Q}$-divisors on $X$. Then*

$$\mathcal{J}\big(X, D_1 + D_2\big) \subseteq \mathcal{J}(X, D_1) \cdot \mathcal{J}(X, D_2).$$

(ii). *If $\mathfrak{a}, \mathfrak{b} \subseteq \mathcal{O}_X$ are ideal sheaves, then*

$$\mathcal{J}\big(\mathfrak{a}^c \cdot \mathfrak{b}^d\big) \subseteq \mathcal{J}\big(\mathfrak{a}^c\big) \cdot \mathcal{J}\big(\mathfrak{b}^d\big)$$

*for any $c, d > 0$. In particular, $\mathcal{J}\big(\mathfrak{a} \cdot \mathfrak{b}\big) \subseteq \mathcal{J}(\mathfrak{a}) \cdot \mathcal{J}(\mathfrak{b})$.*

In other words, the multiplier ideal of the sum of divisors, or a product of ideals, is "bounded above" by the product of the corresponding multiplier ideals.

We will focus mostly on the statement involving divisors. The idea of the proof is first of all to compute the multiplier ideal of the exterior sum of the $D_i$ on the product $X \times X$. Then one restricts to the diagonal. This argument was suggested by the proof of [131, Theorem 2.9]. Esnault and Viehweg had previously obtained related results in a similar fashion.

For the first part of the argument, it is clearest to work on a product of possibly different varieties. Suppose then that $X_1$ and $X_2$ are non-singular quasi-projective varieties, and let $D_1$ and $D_2$ be effective $\mathbf{Q}$-divisors on $X_1$ and $X_2$ respectively. Fix a log resolution $\mu_i : X_i' \longrightarrow X_i$ of each of the pairs $(X_i, D_i)$. We consider the product diagram

$$
\begin{array}{ccccc}
X_1' & \xleftarrow{\ q_1\ } & X_1' \times X_2' & \xrightarrow{\ q_2\ } & X_2' \\
\downarrow{\scriptstyle \mu_1} & & \downarrow{\scriptstyle \mu_1 \times \mu_2} & & \downarrow{\scriptstyle \mu_2} \\
X_1 & \xleftarrow{\ p_1\ } & X_1 \times X_2 & \xrightarrow{\ p_2\ } & X_2
\end{array}
$$

where the horizontal maps are projections.

**Lemma 9.5.21.** *The product $\mu_1 \times \mu_2 : X_1' \times X_2' \longrightarrow X_1 \times X_2$ is a log resolution of the pair*

$$( X_1 \times X_2 , \ p_1^* D_1 + p_2^* D_2 ). \quad \square$$

A similar statement holds for log resolutions of two ideals $\mathfrak{a}_1 \subseteq \mathcal{O}_{X_1}$ and $\mathfrak{a}_2 \subseteq \mathcal{O}_{X_2}$. We refer to [130, Lemma 2.1] for the (straight-forward) proof.

**Proposition 9.5.22.** *With notation as above, one has*

$$\mathcal{J}\big(X_1 \times X_2, p_1^* D_1 + p_2^* D_2\big) = p_1^{-1}\mathcal{J}(X_1, D_1) \cdot p_2^{-1}\mathcal{J}(X_2, D_2).$$

*Proof.* The plan is to compute the multiplier ideal on the left using the log resolution $\mu_1 \times \mu_2$. Specifically,

$$\mathcal{J}\big(X_1 \times X_2, p_1^* D_1 + p_2^* D_2\big)$$
$$= (\mu_1 \times \mu_2)_* \mathcal{O}_{X_1' \times X_2'}\big(K_{X_1' \times X_2'/X_1 \times X_2} - [(\mu_1 \times \mu_2)^*(p_1^* D_1 + p_2^* D_2)]\big).$$

Note to begin with that

$$[(\mu_1 \times \mu_2)^*(p_1^* D_1 + p_2^* D_2)] = [q_1^* \mu_1^* D_1] + [q_2^* \mu_2^* D_2]$$

thanks to the fact that $q_1^* \mu_1^* D_1$ and $q_2^* \mu_2^* D_2$ have no common components. Furthermore, as $q_1$ and $q_2$ are smooth:

$$[q_1^* \mu_1^* D_1] = q_1^* [\mu_1^* D_1] \quad \text{and} \quad [q_2^* \mu_2^* D_2] = q_2^* [\mu_2^* D_2].$$

Since $K_{X_1' \times X_2'/X_1 \times X_2} = q_1^*\big(K_{X_1'/X_1}\big) + q_2^*\big(K_{X_2'/X_2}\big)$, it then follows that

$$\mathcal{O}_{X_1' \times X_2'}\big(K_{X_1' \times X_2'/X_1 \times X_2} - [(\mu_1 \times \mu_2)^*(p_1^* D_1 + p_2^* D_2)]\big)$$
$$= q_1^* \mathcal{O}_{X_1'}\big(K_{X_1'/X_1} - [\mu_1^* D_1]\big) \otimes q_2^* \mathcal{O}_{X_2'}\big(K_{X_2'/X_2} - [\mu_2^* D_2]\big).$$

Now compute using the Künneth formula:

$$\mathcal{J}\big(X_1 \times X_2, p_1^* D_1 + p_1^* D_2\big)$$
$$= (\mu_1 \times \mu_2)_*\big(q_1^* \mathcal{O}_{X_1'}\big(K_{X_1'/X_1} - [\mu_1^* D_1]\big) \otimes q_2^* \mathcal{O}_{X_2'}\big(K_{X_2'/X_2} - [\mu_2^* D_2]\big)\big)$$
$$= p_1^* \mu_{1*} \mathcal{O}_{X_1'}\big(K_{X_1'/X_1} - [\mu_1^* D_1]\big) \otimes p_2^* \mu_{2*} \mathcal{O}_{X_2'}\big(K_{X_2'/X_2} - [\mu_2^* D_2]\big)$$
$$= p_1^* \mathcal{J}(X_1, D_1) \otimes p_2^* \mathcal{J}(X_2, D_2).$$

But

$$p_1^* \mathcal{J}(X_1, D_1) = p_1^{-1}\mathcal{J}(X_1, D_1) \quad \text{and} \quad p_2^* \mathcal{J}(X_2, D_2) = p_2^{-1}\mathcal{J}(X_2, D_2)$$

since $p_1$ and $p_2$ are flat. Finally,

$$p_1^{-1}\mathcal{J}(X_1, D_1) \otimes p_2^{-1}\mathcal{J}(X_2, D_2) = p_1^{-1}\mathcal{J}(X_1, D_1) \cdot p_2^{-1}\mathcal{J}(X_2, D_2)$$

by virtue of the fact that $p_1^{-1}\mathcal{J}(X_1, D_1)$ is flat for $p_2$. This completes the proof of the proposition. □

**Remark 9.5.23.** A similar argument shows that if $\mathfrak{a}_1 \subseteq \mathcal{O}_{X_1}$ and $\mathfrak{a}_2 \subseteq \mathcal{O}_{X_2}$ are ideal sheaves, then

$$\mathcal{J}\big(X_1 \times X_2, (p_1^{-1}\mathfrak{a}_1)^c \cdot (p_2^{-1}\mathfrak{a}_2)^d\big) = p_1^{-1}\mathcal{J}(X_1, \mathfrak{a}_1^c) \cdot p_2^{-1}\mathcal{J}(X_2, \mathfrak{a}_2^d)$$

for any $c, d > 0$. □

The subadditivity theorem now follows immediately:

*Proof of Theorem 9.5.20.* The result being local, we can assume that $X$ is quasi-projective. Fix $\mathbf{Q}$-divisors $D_1$, $D_2$ on $X$. Taking $X_1 = X_2 = X$ in the setting of the proposition, we apply the restriction theorem (in the form of Corollary 9.5.6) to the diagonal embedding $\Delta = X \subseteq X \times X$. One finds:

$$\mathcal{J}(X, D_1 + D_2) = \mathcal{J}(\Delta, (p_1^* D_1 + p_2^* D_2)_\Delta)$$
$$\subseteq \mathcal{J}(X \times X, p_1^* D_1 + p_2^* D_2)_\Delta.$$

But 9.5.22 implies that

$$\mathcal{J}(X \times X, p_1^* D_1 + p_2^* D_2)_\Delta = \mathcal{J}(X, D_1) \cdot \mathcal{J}(X, D_2),$$

and (i) follows. The proof of (ii) is the same, except that one replaces Proposition 9.5.22 with Remark 9.5.23. □

**Example 9.5.24. (Subadditivity for linear series).** Given integral divisors $L$ and $M$ on a projective variety $X$, let $|L| + |M|$ denote the linear subseries of $|L + M|$ spanned by $|L|$ and $|M|$. Then the subadditivity theorem implies that

$$\mathcal{J}(|L| + |M|) \subseteq \mathcal{J}(|L|) \cdot \mathcal{J}(|M|).$$

However, it is not in general true that $\mathcal{J}(|L + M|) \subseteq \mathcal{J}(|L|) \cdot \mathcal{J}(|M|)$, or even that $\mathcal{J}(|mL|) \subseteq \mathcal{J}(|L|)^m$, although Theorem 11.2.4 shows that the analogue of the second statement does hold for the asymptotic multiplier ideals introduced below. (For a counter-example, take $X$ to be a curve and $L = M$ divisors such that $|L|$ has a base point but $|2L|$ is free.) □

**Example 9.5.25. (Intersections of ideals).** Let $\mathfrak{a}, \mathfrak{b} \subseteq \mathcal{O}_X$ be ideal sheaves. Then

$$\mathcal{J}(\mathfrak{a} \cap \mathfrak{b}) \subseteq \mathcal{J}(\mathfrak{a}) \cap \mathcal{J}(\mathfrak{b}).$$

(This follows from the elementary fact (Proposition 9.2.32 (iii)) that the formation of multiplier ideals preserves containments.) □

### 9.5.C The Summation Theorem

The subadditivity theorem gives information about the multiplier ideals of a product of two ideals. We present in this subsection a theorem of Mustaţă [456] giving an analogous (but more subtle) statement for the multiplier ideal attached to a sum of ideal sheaves.

**Statement and applications.** The result in question is this:

**Theorem 9.5.26. (Mustaţă's summation theorem).** *Let* $\mathfrak{a}, \mathfrak{b} \subseteq \mathcal{O}_X$ *be two non-zero ideal sheaves on a smooth variety* $X$ *and* $c > 0$ *a rational number. Then*

$$\mathcal{J}\big(X, (\mathfrak{a} + \mathfrak{b})^c\big) \subseteq \sum_{\lambda + \mu = c} \mathcal{J}\big(X, \mathfrak{a}^\lambda\big) \cdot \mathcal{J}\big(X, \mathfrak{b}^\mu\big). \qquad (9.26)$$

Before discussing the proof, we give some applications. To begin with, one can get particularly clean statements by taking into account containments among the terms appearing on the right in (9.26):

**Corollary 9.5.27.** *Let* $\mathfrak{a}, \mathfrak{b} \subseteq \mathcal{O}_X$ *be non-zero ideal sheaves, and fix rational numbers* $c, d > 0$. *Then*

$$\mathcal{J}\big(X, (\mathfrak{a} + \mathfrak{b})^{c+d}\big) \subseteq \mathcal{J}\big(X, \mathfrak{a}^c\big) + \mathcal{J}\big(X, \mathfrak{b}^d\big).$$

*Proof.* The theorem gives the inclusion

$$\mathcal{J}\big((\mathfrak{a} + \mathfrak{b})^{c+d}\big) \subseteq \sum_{\lambda + \mu = c+d} \mathcal{J}\big(\mathfrak{a}^\lambda\big) \cdot \mathcal{J}\big(\mathfrak{b}^\mu\big). \qquad (*)$$

If $\lambda \geq c$ then $\mathcal{J}\big(\mathfrak{a}^\lambda\big) \subseteq \mathcal{J}\big(\mathfrak{a}^c\big)$, while if $\mu \geq d$ then $\mathcal{J}\big(\mathfrak{b}^\mu\big) \subseteq \mathcal{J}\big(\mathfrak{b}^d\big)$. Since one of these inequalities must hold whenever $\lambda + \mu = c + d$, it follows that each of the terms on the right in (*) is contained in $\mathcal{J}\big(\mathfrak{a}^c\big) + \mathcal{J}\big(\mathfrak{b}^d\big)$. $\qquad \square$

The corollary in turn implies a result of Demailly and Kollár [131] concerning the log-canonical threshold of a sum of ideals (Definition 9.3.14).

**Corollary 9.5.28. (Log-canonical threshold of a sum).** *Let* $\mathfrak{a}, \mathfrak{b} \subseteq \mathcal{O}_X$ *be ideal sheaves, and fix a point* $x \in X$. *Then one has the inequality*

$$\mathrm{lct}(\mathfrak{a} + \mathfrak{b}; x) \leq \mathrm{lct}(\mathfrak{a}; x) + \mathrm{lct}(\mathfrak{b}; x)$$

*of log-canonical thresholds at* $x$.

*Proof.* Set $c = \mathrm{lct}(\mathfrak{a}; x)$ and $d = \mathrm{lct}(\mathfrak{b}; x)$. It follows from the corollary that $\mathcal{J}\big((\mathfrak{a} + \mathfrak{b})^{c+d}\big)$ is non-trivial at $x$, and hence $\mathrm{lct}(\mathfrak{a} + \mathfrak{b}; x) \leq c + d$. $\qquad \square$

Still following [456], one can also use Corollary 9.5.27 to study the problem of approximating arbitrary multiplier ideals by ideals with finite co-support.

**Proposition 9.5.29. (Approximating multiplier ideals).** *Let* $\mathfrak{a} \subseteq \mathcal{O}_X$ *be a non-zero ideal, and* $c > 0$ *a rational number. Fix a point* $x \in X$, *and denote by* $\mathfrak{m} = \mathfrak{m}_x$ *the maximal ideal sheaf at* $x$. *Then for any* $\varepsilon > 0$,

$$\mathcal{J}\big(\mathfrak{a}^{c+\varepsilon}\big)_x \subseteq \bigcap_{p > 0} \mathcal{J}\big((\mathfrak{a} + \mathfrak{m}^p)^{c+\varepsilon}\big)_x \subseteq \mathcal{J}\big(\mathfrak{a}^c\big)_x. \qquad (9.27)$$

Recall (Lemma 9.3.21) that if $\varepsilon \ll 1$ is sufficiently small, then $\mathcal{J}\big(\mathfrak{a}^{c+\varepsilon}\big) = \mathcal{J}\big(\mathfrak{a}^c\big)$. Therefore:

**Corollary 9.5.30.** *In the situation of the proposition,*

$$\bigcap_{p>0} \mathcal{J}\big((\mathfrak{a} + \mathfrak{m}^p)^{c+\varepsilon}\big)_x \;=\; \mathcal{J}\big(\mathfrak{a}^c\big)_x$$

*provided that $\varepsilon$ is sufficiently small.*     $\square$

*Proof of Proposition 9.5.29.* Since $\mathfrak{a} \subseteq \mathfrak{a} + \mathfrak{m}^p$ for every $p$, the first inclusion in (9.27) is clear. For the second, we apply Corollary 9.5.27. Adopting the convention that $\mathfrak{m}^q = \mathcal{O}_X$ if $q \le 0$, one has

$$\begin{aligned}
\mathcal{J}\big((\mathfrak{a} + \mathfrak{m}^p)^{c+\varepsilon}\big)_x &\subseteq \mathcal{J}\big(\mathfrak{a}^c\big)_x + \mathcal{J}\big(\mathfrak{m}^{p\varepsilon}\big)_x \\
&= \mathcal{J}\big(\mathfrak{a}^c\big)_x + \mathfrak{m}^{[p\varepsilon]+1-\dim X}.
\end{aligned}$$

Thanks to Krull's theorem this implies

$$\begin{aligned}
\bigcap_{p>0} \mathcal{J}\big((\mathfrak{a} + \mathfrak{m}^p)^{c+\varepsilon}\big)_x &\subseteq \bigcap_{p>0} \left( \mathcal{J}\big(\mathfrak{a}^c\big)_x + \mathfrak{m}^{[p\varepsilon]+1-\dim X} \right) \\
&= \mathcal{J}\big(\mathfrak{a}^c\big)_x,
\end{aligned}$$

as required.     $\square$

**Remark 9.5.31.** Mustaţă [456, §2] uses these ideas to show that the jumping numbers (and in particular the log-canonical threshold) of an arbitrary ideal sheaf $\mathfrak{a} \subseteq \mathcal{O}_X$ actually depends only on the isomorphism class of the scheme defined by $\mathfrak{a}$ and the dimension of the smooth ambient variety $X$.     $\square$

**Proof of Mustaţă's theorem.** We now begin the work of establishing Theorem 9.5.26, following [456]. As in the proof of the subadditivity theorem, the essential point is to show that equality holds on the product of two varieties:

**Proposition 9.5.32.** *Let $X_1$ and $X_2$ be smooth varieties of dimension $n_1$ and $n_2$ respectively, and consider the product $X = X_1 \times X_2$ with projections*

$$p_1 : X \longrightarrow X_1 \quad , \quad p_2 : X \longrightarrow X_2.$$

*Let $\mathfrak{a}_i \subseteq \mathcal{O}_{X_i}$ be non-zero ideal sheaves, and denote by*

$$\mathfrak{a}_1 \stackrel{\circ}{+} \mathfrak{a}_2 \;=\; p_1^{-1}\mathfrak{a}_1 + p_2^{-1}\mathfrak{a}_2 \subseteq \mathcal{O}_{X_1 \times X_2}$$

*their "exterior" sum. Then for any $c > 0$,*

$$\mathcal{J}\big( X_1 \times X_2, (\mathfrak{a}_1 \stackrel{\circ}{+} \mathfrak{a}_2)^c \big) \;=\; \sum_{\lambda+\mu=c} p_1^{-1}\mathcal{J}\big( X_1, \mathfrak{a}_1^\lambda \big) \cdot p_2^{-1}\mathcal{J}\big( X_2, \mathfrak{a}_2^\mu \big). \quad (9.28)$$

Granting 9.5.32, the theorem follows at once by restricting to the diagonal:

*Proof of Theorem 9.5.26.* Applying the proposition with $X_1 = X_2 = X$ yields

$$\mathcal{J}\left(X \times X, (\mathfrak{a}_1 \overset{\circ}{+} \mathfrak{a}_2)^c\right) = \sum_{\lambda+\mu=c} p_1^{-1}\mathcal{J}\left(X, \mathfrak{a}_1^\lambda\right) \cdot p_2^{-1}\mathcal{J}\left(X, \mathfrak{a}_2^\mu\right).$$

Now restrict to the diagonal $X = \Delta \subseteq X \times X$. One has

$$\left(\mathfrak{a}_1 \overset{\circ}{+} \mathfrak{a}_2\right) \cdot \mathcal{O}_\Delta = \mathfrak{a}_1 + \mathfrak{a}_2 \subseteq \mathcal{O}_X$$

and the required inclusion follows from the restriction theorem in the form of Corollary 9.5.6 (and Remark 9.5.10).    □

For the proposition, the plan as before is to pass to a product of log resolutions $X_i' \longrightarrow X_i$ of the given ideals $\mathfrak{a}_i$, but now there are two complications. First, $X' = X_1' \times X_2'$ is not a log resolution of $\mathfrak{a}_1 \overset{\circ}{+} \mathfrak{a}_2$: rather the pullback to $X'$ of this ideal is locally generated by two monomials in independent sets of variables. So we will apply Howald's Theorem 9.3.27 to compute the relevant multiplier ideal on $X'$. The second difficulty is that the right hand side of (9.28) does not behave well with respect to pushforwards. Mustaţă overcomes this by reinterpreting the sum as an intersection.

Turning to details, we start with two lemmas.

**Lemma 9.5.33.** *The statement of Proposition 9.5.32 is true if $\mathfrak{a}_i = \mathcal{O}_X(-F_i)$, where each $F_i$ is an effective divisor on $X_i$ with SNC support.*

*Proof.* The assertion is local on $X_1$ and $X_2$, so we can assume that each $F_i$ is defined by the pullback of a single monomial under an étale morphism

$$\phi_i : X_i \longrightarrow \mathbf{C}^{n_i}.$$

Since the formation of multiplier ideals commutes with étale pullback (Example 9.5.44), we are reduced to the case in which each $\mathfrak{a}_i$ is a monomial ideal on $\mathbf{C}^{n_i}$. But then the statement follows from Howald's Theorem 9.3.27.    □

**Lemma 9.5.34.** *Let $\mathfrak{a}_i \subseteq \mathcal{O}_{X_i}$ (for $i = 1, 2$) be arbitrary non-zero ideal sheaves. Then for every $c > 0$,*

$$\sum_{\lambda+\mu=c} p_1^{-1}\mathcal{J}\left(X_1, \mathfrak{a}_1^\lambda\right) \cdot p_2^{-1}\mathcal{J}\left(X_2, \mathfrak{a}_2^\mu\right)$$

$$= \bigcap_{\lambda+\mu=c} \left(p_1^{-1}\mathcal{J}\left(X_1, \mathfrak{a}_1^\lambda\right) + p_2^{-1}\mathcal{J}\left(X_2, \mathfrak{a}_2^\mu\right)\right).$$

*Sketch of Proof.* This follows formally from the inclusions satisfied by the multiplier ideals in question as the parameters $\lambda$ and $\mu$ vary. In fact, consider any two families of ideal sheaves $\mathcal{J}_i(\nu) \subseteq \mathcal{O}_{X_i}$ depending on a non-negative rational parameter $\nu$, having the properties that $\mathcal{J}_i(\nu) \subseteq \mathcal{J}_i(\nu')$ when $\nu \geq \nu'$ and that there are only finitely many distinct ideals $\mathcal{J}_i(\nu)$ when $0 \le \nu \le c$ for some fixed $c$. Then it is true quite generally that for fixed $c > 0$,

$$\sum_{\lambda+\mu=c} p_1^{-1}\mathcal{J}_1(\lambda) \cdot p_2^{-1}\mathcal{J}_2(\mu) \;=\; \bigcap_{\lambda+\mu=c} \left( p_1^{-1}\mathcal{J}_1(\lambda) + p_2^{-1}\mathcal{J}_2(\mu) \right). \qquad (*)$$

Indeed, the inclusion $\subseteq$ is clear since if $\lambda + \mu = \lambda' + \mu' = c$, then either $\lambda \ge \lambda'$ or $\mu \ge \mu'$. In either case

$$p_1^{-1}\mathcal{J}_1(\lambda) \cdot p_2^{-1}\mathcal{J}_2(\mu) \;\subseteq\; p_1^{-1}\mathcal{J}_1(\lambda') + p_2^{-1}\mathcal{J}_2(\mu'),$$

giving the stated inclusion. For the reverse inclusion in $(*)$, one can assume that $X_i = \mathrm{Spec}(A_i)$ is affine, so that $X_1 \times X_2 = \mathrm{Spec}(A_1 \otimes_{\mathbf{C}} A_2)$. Then one chooses $\mathbf{C}$-bases of $A_i$ compatible with the filtrations determined by the given ideals, and verifies the inclusion $\supseteq$ in $(*)$ on "monomials" (i.e. pure tensors). We refer to [456, Lemma 1.2] or [292, §6] for details. $\qquad \square$

We now turn to the proof of the proposition:

*Proof of Proposition 9.5.32.* Let $\mu_i : X_i' \longrightarrow X_i$ be log resolutions of the given ideals $\mathfrak{a}_i$. We consider the product diagram

$$
\begin{array}{ccccc}
X_1' & \xleftarrow{\;q_1\;} & X_1' \times X_2' & \xrightarrow{\;q_2\;} & X_2' \\[4pt]
{\scriptstyle\mu_1}\big\downarrow & & {\scriptstyle h=\mu_1\times\mu_2}\big\downarrow & & \big\downarrow{\scriptstyle\mu_2} \\[4pt]
X_1 & \xleftarrow{\;p_1\;} & X_1 \times X_2 & \xrightarrow{\;p_2\;} & X_2
\end{array}
$$

Write $X = X_1 \times X_2$, $X' = X_1' \times X_2'$, $K = K_{X'/X}$ and

$$\mathfrak{a}_i' \;=\; \mathfrak{a}_i \cdot \mathcal{O}_{X_i'} \;=\; \mathcal{O}_{X_i}(-F_i)$$

for an SNC-supported divisor $F_i$ on $X_i$. By the birational transformation rule 9.2.33,

$$\mathcal{J}\big( X_1 \times X_2 \,,\, (\mathfrak{a}_1 \overset{\circ}{+} \mathfrak{a}_2)^c \big) \;=\; h_*\Big( \mathcal{J}\big( X_1' \times X_2' \,,\, (\mathfrak{a}_1' \overset{\circ}{+} \mathfrak{a}_2')^c \big) \otimes \mathcal{O}_{X'}(K) \Big). \tag{9.29}$$

Now use Lemmas 9.5.33 and 9.5.34 to compute the multiplier ideal appearing on the right:

$$
\begin{aligned}
\mathcal{J}\big( X_1' \times X_2' \,,\, (\mathfrak{a}_1' \overset{\circ}{+} \mathfrak{a}_2')^c \big) \\
= \sum_{\lambda+\mu=c} q_1^{-1}\mathcal{J}\big( X_1', (\mathfrak{a}_1')^\lambda \big) \cdot q_2^{-1}\mathcal{J}\big( X_2', (\mathfrak{a}_2')^\mu \big) \\
= \bigcap_{\lambda+\mu=c} \Big( q_1^{-1}\mathcal{J}\big( X_1', (\mathfrak{a}_1')^\lambda \big) + q_2^{-1}\mathcal{J}\big( X_2', (\mathfrak{a}_2')^\mu \big) \Big).
\end{aligned}
\tag{9.30}
$$

Taking direct images of these ideals commutes with intersecting them, and so upon combining (9.29) and (9.30) one finds:

$$\mathcal{J}\big(X_1 \times X_2, (\mathfrak{a}_1 \overset{\circ}{+} \mathfrak{a}_2)^c\big)$$
$$= \bigcap_{\lambda+\mu=c} h_* \left( \big( q_1^{-1}\mathcal{J}(X_1', (\mathfrak{a}_1')^{\lambda}) + q_2^{-1}\mathcal{J}(X_2', (\mathfrak{a}_2')^{\mu}) \big) \otimes \mathcal{O}_{X'}(K) \right).$$
(9.31)

We assert that

$$h_* \left( \big( q_1^{-1}\mathcal{J}(X_1', (\mathfrak{a}_1')^{\lambda}) + q_2^{-1}\mathcal{J}(X_2', (\mathfrak{a}_2')^{\mu}) \big) \otimes \mathcal{O}_{X'}(K) \right)$$
$$= h_* \big( q_1^{-1}\mathcal{J}(X_1', (\mathfrak{a}_1')^{\lambda}) \otimes \mathcal{O}_{X'}(K) \big) + h_* \big( q_2^{-1}\mathcal{J}(X_2', (\mathfrak{a}_2')^{\mu}) \otimes \mathcal{O}_{X'}(K) \big).$$
(9.32)

Granting this, using Künneth and Lemma 9.5.34 one finds from (9.31):

$$\mathcal{J}\big(X_1 \times X_2, (\mathfrak{a}_1 \overset{\circ}{+} \mathfrak{a}_2)^c\big) = \bigcap_{\lambda+\mu=c} \big( p_1^{-1}\mathcal{J}(X_1, \mathfrak{a}_1^{\lambda}) + p_2^{-1}\mathcal{J}(X_2, \mathfrak{a}_2^{\mu}) \big)$$
$$= \sum_{\lambda+\mu=c} p_1^{-1}\mathcal{J}(X_1, \mathfrak{a}_1^{\lambda}) \cdot p_2^{-1}\mathcal{J}(X_2, \mathfrak{a}_2^{\mu}),$$

as required.

It remains to verify (9.32). To this end, consider any two ideals $\mathcal{J}_1, \mathcal{J}_2 \subseteq \mathcal{O}_{X'}$ on $X'$. The exact sequence

$$0 \longrightarrow \mathcal{J}_1 \cap \mathcal{J}_2 \longrightarrow \mathcal{J}_1 \oplus \mathcal{J}_2 \longrightarrow \mathcal{J}_1 + \mathcal{J}_2 \longrightarrow 0$$

shows that

$$h_* \big( (\mathcal{J}_1 + \mathcal{J}_2) \otimes \mathcal{O}_{X'}(K) \big) = h_*\big(\mathcal{J}_1 \otimes \mathcal{O}_{X'}(K)\big) + h_*\big(\mathcal{J}_2 \otimes \mathcal{O}_{X'}(K)\big)$$

provided that $R^1 h_* \big( (\mathcal{J}_1 \cap \mathcal{J}_2) \otimes \mathcal{O}_{X'}(K) \big) = 0$. But in the case at hand,

$$\big( q_1^{-1}\mathcal{J}(X_1', (\mathfrak{a}_1')^{\lambda}) \cap q_2^{-1}\mathcal{J}(X_2', (\mathfrak{a}_2')^{\mu}) \big) \otimes \mathcal{O}_{X'}(K)$$
$$= q_1^* \big( \mathcal{J}(X_1', (\mathfrak{a}_1')^{\lambda}) \otimes \mathcal{O}_{X_1'}(K_{X_1'/X_1}) \big) \otimes q_2^* \big( \mathcal{J}(X_2', (\mathfrak{a}_2')^{\mu}) \otimes \mathcal{O}_{X_2'}(K_{X_2'/X_2}) \big),$$

and the higher direct images of this sheaf vanish thanks to Künneth and local vanishing in the form of Example 9.4.6 for ideal sheaves.    □

### 9.5.D Multiplier Ideals in Families

This section is devoted to the question of how multiplier ideals behave in families. We show that "generically" multiplier ideals commute with restriction. Using this we deduce that a family of non-trivial multiplier ideals must have a non-trivial limit.

The central result here asserts that multiplier ideals restrict as nicely as possible to the *general* fibre of a smooth map:

**Theorem 9.5.35. (Generic restrictions).** *Let $X$ and $T$ be non-singular irreducible varieties, and*

$$p : X \longrightarrow T$$

*a smooth surjective mapping. Consider an effective $\mathbf{Q}$-divisor $D$ on $X$ whose support does not contain any of the fibres $X_t = p^{-1}(t)$, so that for each $t \in T$ the restriction $D_t = D \mid X_t$ is defined. Then there is a non-empty Zariski open set $U \subseteq T$ such that*

$$\mathcal{J}(X_t, D_t) \;=\; \mathcal{J}(X, D)_t$$

*for every $t \in U$, where $\mathcal{J}(X,D)_t = \mathcal{J}(X,D) \cdot \mathcal{O}_{X_t}$ denotes the restriction of the indicated multiplier ideal to the fibre $X_t$. More generally, if $t \in U$ then*

$$\mathcal{J}(X_t, c \cdot D_t) \;=\; \mathcal{J}(X, c \cdot D)_t \tag{9.33}$$

*for every $c > 0$.*

**Remark 9.5.36.** Recall from Example 9.5.7 that the inclusion $\mathcal{J}(X_t, D_t) \subseteq \mathcal{J}(X,D)_t$ holds for an arbitrary point $t \in T$. □

*Proof of Theorem 9.5.35.* The question is local on $T$, so we may and do assume that $T$ is affine. Let $\mu : X' \longrightarrow X$ be a log resolution of $(X, D)$, and set $p' = p \circ \mu$:

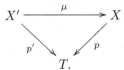

$$T.$$

Denote by $\mu_t : X'_t \longrightarrow X_t$ the evident map on fibres over $t \in T$, so that

$$\mu^*(cD) \mid X'_t \;=\; \mu_t^*(cD_t)$$

for every $t \in T$ and $c > 0$. It follows from the theorem on generic smoothness that in the first place $p'$ is smooth over a non-empty subset of $T$, and consequently that $X'_t$ is non-singular for general $t \in T$. We claim that $\mu_t : X'_t \longrightarrow X_t$ is actually a log resolution of $(X_t, cD_t)$ for sufficiently general $t \in T$ and every $c > 0$. In fact, by definition $\mu^*(cD) + \text{except}(\mu)$ is supported on a normal crossing divisor $\sum E_i$ on $X'$. It is enough to show that

$\sum E_i$ restricts to a normal crossing divisor on $X'_t$ for general $t \in T$.
$$(9.34)$$

But this follows upon applying the theorem on generic smoothness to all of the intersections among the $E_i$ to check that any collection of the components of $\sum (E_i)_t$ on $X'_t$ either meet transversely or not at all (compare Lemma 4.1.11). Let $U \subseteq T$ be the set of points at which (9.34) holds.

We assert that the required equality (9.33) occurs for every $t \in U$. To verify this, fix $c > 0$ and consider the exact sequence of sheaves on $X'$

$$0 \longrightarrow \mathcal{O}_{X'}\big(K_{X'/X} - [\mu^*(cD)]\big) \otimes \mathcal{I}_{X'_t} \longrightarrow \mathcal{O}_{X'}\big(K_{X'/X} - [\mu^*(cD)]\big)$$
$$\longrightarrow \mathcal{O}_{X'_t}\big(K_{X'/X} - [\mu^*(cD)]\big) \longrightarrow 0,$$

where $\mathcal{I}_{X'_t} \subseteq \mathcal{O}_{X'}$ denotes the ideal sheaf of $X'_t$. Since $t \in U$ one has

$$[\mu^* D] \mid X_t = [\mu_t^* D_t] \quad \text{and} \quad (K_{X'/X}) \mid X_t = K_{X'_t/X_t}.$$

Taking direct images under $\mu$, it is therefore enough to prove

$$R^1 \mu_* \Big(\mathcal{O}_{X'}\big(K_{X'/X} - [\mu^*(cD)]\big) \otimes \mathcal{I}_{X'_t}\Big) = 0. \tag{9.35}$$

Now after possibly further shrinking $T$, the maximal ideal $\mathfrak{m}_t \subseteq \mathcal{O}_T$ has a Koszul resolution constructed from a system of parameters at $t$. As $p'$ is flat over $U$, this pulls back to a resolution of $\mathcal{I}_{X'_t}$ having the shape

$$\dots \longrightarrow \mathcal{O}_{X'}^{r_1} \longrightarrow \mathcal{O}_{X'}^{r_0} \longrightarrow \mathcal{I}_{X'_t} \longrightarrow 0.$$

Tensoring through by $\mathcal{O}_{X'}\big(K_{X'/X} - [\mu^*(cD)]\big)$, one then finds that (9.35) is reduced to the vanishing

$$R^i \mu_* \mathcal{O}_{X'}\big(K_{X'/X} - [\mu^*(cD)]\big) = 0 \quad \text{for} \quad i > 0.$$

But this is the assertion of the local vanishing theorem (9.4.1).    $\square$

**Example 9.5.37. (Generic restrictions for ideals and linear series).** Let $p : X \longrightarrow T$ be as in the theorem.

(i).  Let $\mathfrak{a} \subseteq \mathcal{O}_X$ be an ideal, and assume that $\mathfrak{a}_t = \mathfrak{a} \cdot \mathcal{O}_{X_t}$ is non-zero for all $t \in T$. Then $\mathcal{J}(X_t, c \cdot \mathfrak{a}_t) = \mathcal{J}(X, c \cdot \mathfrak{a}) \cdot \mathcal{O}_{X_t}$ for general $t \in T$ and every $c > 0$.

(ii).  Let $|V| \subseteq |L|$ be a linear series on $X$ and assume that the base-locus of $|V|$ does not contain any of the fibres $X_t$. Denote by $|V|_t \subseteq |L_t|$ the restriction of $|V|$ to $X_t$. Then $\mathcal{J}(X_t, c \cdot |V|_t) = \mathcal{J}(X, c \cdot |V|) \cdot \mathcal{O}_{X_t}$ for every $c > 0$ and general $t$.

(When $|V| = |L|$ is a complete linear series, beware that in (ii) we are dealing with the restriction of $|L|$ to $X_t$, which may be smaller than the complete linear series $|L_t|$ associated to the restriction of $L$.)    $\square$

**Remark 9.5.38.** A related result — involving restrictions to the general member of a free linear series — appears in Example 9.5.9.     □

We next discuss a result concerning the semicontinuity of multiplier ideals. Suppose that $\{D_t\}_{t \in T}$ is a family of effective $\mathbf{Q}$-divisors on a smooth projective variety $X$ parametrized by a smooth curve $T$. Assume that we know that for $t \neq 0 \in T$ the corresponding multiplier ideals $\mathcal{J}(D_t)$ vanish at a family of points $y_t \in X$, i.e.

$$y_t \in \text{Zeroes}(\mathcal{J}(D_t)) \quad \text{for all } t \neq 0 \in T.$$

The following corollary, due in effect to Siu, implies that $y_0$ lies in the zeroes of $\mathcal{J}(D_0)$.

**Corollary 9.5.39. (Semicontinuity).** *Let $p : X \longrightarrow T$ be a smooth mapping as in Theorem 9.5.35, and let $D$ be an effective $\mathbf{Q}$-divisor on $X$ satisfying the hypotheses of that statement. Suppose moreover given a section*

$$y : T \longrightarrow X$$

*of $p$, and write $y_t = y(t) \in X_t$. If*

$$y_t \in \text{Zeroes}(\mathcal{J}(X_t, D_t)) \quad \text{for } t \neq 0 \in T,$$

*then $y_0 \in \text{Zeroes}(\mathcal{J}(X_0, D_0))$.*

When $X = Y \times T$ and $p : Y \times T \longrightarrow T$ is the projection, the theorem in particular applies to a family $\{D_t\}$ of $\mathbf{Q}$-divisors on a fixed variety $Y$.

*Proof.* Denote by $\Sigma \subset X$ the image $y(T) \subseteq X$ of $y$, so that $\Sigma_t = y_t$. It follows from Theorem 9.5.35 and the hypotheses of the corollary that

$$\Sigma_t = y_t \in \text{Zeroes}(\mathcal{J}(X_t, D_t)) = \text{Zeroes}(\mathcal{J}(X, D)) \cap X_t$$

for general $t \in T$. Hence $\Sigma \subseteq \text{Zeroes}(\mathcal{J}(X, D))$ and in particular:

$$y_0 = \Sigma_0 \in \text{Zeroes}(\mathcal{J}(X, D)) \cap X_0 = \text{Zeroes}(\mathcal{J}(X, D) \cdot \mathcal{O}_{X_0}).$$

But the restriction theorem (Corollary 9.5.6) implies that

$$\mathcal{J}(X_0, D_0) \subseteq \mathcal{J}(X; D) \cdot \mathcal{O}_{X_0}$$

and hence $\text{Zeroes}(\mathcal{J}(X, D) \cdot \mathcal{O}_{X_0}) \subseteq \text{Zeroes}(\mathcal{J}(X_0, D_0))$. Therefore

$$y_0 \in \text{Zeroes}(\mathcal{J}(X_0, D_0)),$$

as claimed.     □

**Remark 9.5.40.** Once again there are analogous statements for the multiplier ideals associated to linear series and ideal sheaves, whose formulations and verifications we leave to the reader.     □

**Example 9.5.41. (Semicontinuity of log-canonical thresholds).** Let $\{D_t\}_{t\in T}$ be a family of effective **Q**-divisors on a smooth variety $X$, and fix a point $x \in X$. Then

$$\mathrm{lct}(D_0; x) \leq \mathrm{lct}(D_t; x)$$

for general $t \in T$. In other words, the log-canonical threshold of a family of divisors can only "jump down." In a somewhat different context, this is originally due to Varchenko [584]. $\qquad\square$

### 9.5.E Coverings

Finally, we investigate the deportment of multiplier ideals under generically finite coverings.

We start with some notation. Let $f : Y \longrightarrow X$ be a generically finite and surjective projective morphism between irreducible non-singular varieties. Viewing the function fields $\mathbf{C}(X)$ and $\mathbf{C}(Y)$ as constant sheaves on $X$ and $Y$ respectively, there is a natural inclusion $\mathbf{C}(X) \hookrightarrow f_*\mathbf{C}(Y)$ of quasi-coherent $\mathcal{O}_X$-modules. If $\mathcal{J} \subseteq \mathbf{C}(Y)$ is a sheaf of fractional ideals in $Y$, then $f_*\mathcal{J}$ sits as a subsheaf in $f_*\mathbf{C}(Y)$, and one can form the intersection $f_*\mathcal{J} \cap \mathbf{C}(X)$ in $f_*\mathbf{C}(Y)$:

$$
\begin{array}{ccc}
f_*\mathcal{J} \cap \mathbf{C}(X) & \subseteq & f_*\mathcal{J} \\
\cap & & \cap \\
\mathbf{C}(X) & \subseteq & f_*\mathbf{C}(Y).
\end{array}
$$

Viewed as a subsheaf of $\mathbf{C}(X)$, $f_*\mathcal{J} \cap \mathbf{C}(X)$ becomes a sheaf of fractional ideals on $X$. Concretely, given an open subset $U \subseteq X$, $\big(f_*\mathcal{J} \cap \mathbf{C}(X)\big)(U)$ consists of those rational functions $\phi \in \mathbf{C}(X)$ such that $f^*\phi \in \mathcal{J}\big(f^{-1}U\big)$.

The main result of this subsection generalizes the birational transformation rule 9.2.33:

**Theorem 9.5.42. (Multiplier ideals and coverings).** *Consider as above a generically finite projective morphism between smooth varieties of the same dimension, and denote by $K_{Y/X}$ the ramification divisor of $f$. Then for any* **Q**-*divisor $D$ on $X$ one has the equality*

$$\mathcal{J}\big(X, D\big) = f_*\mathcal{J}\big(Y, f^*D - K_{Y/X}\big) \cap \mathbf{C}(X)$$

*of fractional ideals on $X$.*

Even if $D$ is effective, the divisor $f^*D - K_{Y/X}$ appearing on the right might not be. So we are forced here to deal with fractional multiplier ideals associated to possibly ineffective divisors (Remark 9.2.2).

As a corollary, we recover a result due to Reid [515], Ein [145, Prop. 2.8], and Kollár [364, Prop. 3.16].

**Corollary 9.5.43.** *In the situation of the theorem, the multiplier ideal $\mathcal{J}(X, D)$ is non-trivial at $x$ if and only if $\mathcal{J}(Y, f^*D - K_{Y/X})$ is non-trivial at some point $y \in Y$ lying over $x$.* □

Here we say that $\mathcal{J}(X, D)$ is non-trivial at $x \in X$ if $\mathcal{O}_x X \not\subseteq \mathcal{J}(X, D)_x$. Equivalently, if $\mu : X' \longrightarrow X$ is a log resolution of $(X, D)$, there should exist a prime divisor $E \subseteq X'$, whose image passes through $x$, such that

$$\mathrm{ord}_E\big(K_{X'/X} - [\mu^*D]\big) \le -1.$$

The non-triviality of $\mathcal{J}(Y, f^*D - K_{Y/X})$ is of course defined similarly.

*Proof of Theorem 9.5.42.* We can construct log resolutions

$$\mu : X' \longrightarrow X \quad , \quad \nu : Y' \longrightarrow Y,$$

of $(X, D)$ and $(Y, f^*D_{\mathrm{red}} + K_{Y/X})$ sitting in a commutative diagram

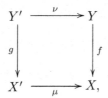

where $g : Y' \longrightarrow X'$ is a generically finite projective and surjective mapping. Suppose for the moment that we are able to prove the theorem under the additional hypothesis that all the relevant divisors have SNC support. Applying this to $g : Y' \longrightarrow X'$ and the divisor $\mu^*D - K_{X'/X}$ on $X'$ gives

$$\mathcal{J}\big(X', \mu^*D - K_{X'/X}\big) = g_* \mathcal{J}\big(Y', g^*\mu^*D - K_{Y'/X}\big) \cap \mathbf{C}(X).$$

Then we find

$$\begin{aligned}
\mathcal{J}(X, D) &= \mu_* \mathcal{J}\big(X', \mu^*D - K_{X'/X}\big) \\
&= f_*\nu_* \mathcal{J}\big(Y', \nu^*(f^*D - K_{Y/X}) - K_{Y'/Y}\big) \cap \mathbf{C}(X') \\
&= f_* \mathcal{J}\big(Y, f^*D - K_{Y/X}\big) \cap \mathbf{C}(X),
\end{aligned}$$

as required. So it is enough to prove the theorem for $f : Y \longrightarrow X$ under the hypothesis that $D$ and $f^*D - K_{Y/X}$ have SNC support.

Making this assumption, fix a prime divisor $E$ on $Y$ appearing in the support of $f^*D$ or $f^*D - K_{Y/X}$. We assert that then

$$\mathrm{ord}_E\big(K_{Y/X} - f^*D\big) > -1 - \mathrm{ord}_E\big(f^*[D]\big). \tag{9.36}$$

For this we argue separately in two cases. Suppose first that $E$ is exceptional for $f$. Noting that (9.36) holds for a **Q**-divisor $D$ if and only if it holds for

$D + A$ for any integral divisor $A$, we are free to suppose in addition that $[D] = 0$. In this case we need to show that $\mathrm{ord}_E(K_{Y/X} - f^*D) > -1$, and here the computations appearing in the proof of Lemma 9.2.19 go through with essentially no change. On the other hand, suppose that $E$ maps onto a prime divisor $F$ on $X$. Then $f$ is locally analytically a cyclic branched covering generically along $E$, and consequently $\mathrm{ord}_E(f^*F) = \mathrm{ord}_E(K_{Y/X})+1$. Writing $a = \mathrm{ord}_F(D)$ and $b = \mathrm{ord}_E(K_{Y/X})$, (9.36) is equivalent to the inequality

$$b - (b+1)a > -1 - (b+1) \cdot [a],$$

which is clear.

The inequality (9.36) implies that $-f^*[D] \preccurlyeq K_{Y/X} - [f^*D]$. Recalling that

$$\mathcal{J}(X, D) = \mathcal{O}_X(-[D]) \quad \text{and} \quad \mathcal{J}(Y, f^*D - K_{Y/X}) = \mathcal{O}_Y(K_{Y/X} - [f^*D])$$

thanks to our normal crossing hypotheses, this gives the inclusion

$$\mathcal{J}(X, D) \subseteq f_*\mathcal{J}(Y, f^*D - K_{Y/X}) \cap \mathbf{C}(X).$$

Conversely, the reverse inclusion fails if and only if there is a prime component $F$ of $D$ together with a rational function $\phi \in \mathbf{C}(X)$ such that

$$\mathrm{ord}_F(\mathrm{div}\,\phi - [D]) < 0 \quad \text{but} \quad \mathrm{ord}_E(K_{Y/X} + f^*(\mathrm{div}\,\phi) - [f^*D]) \geq 0,$$

$E$ being a prime divisor on $Y$ mapping onto $F$. But if $\mathrm{ord}_F(\mathrm{div}\,\phi - [D]) \leq -1$ then as above

$$\mathrm{ord}_E(K_{Y/X} + f^*(\mathrm{div}\,\phi) - f^*[D]) < 0.$$

Since in any event $\mathrm{ord}_E(-[f^*D]) \leq \mathrm{ord}_E(-f^*[D])$, this is impossible.    □

We conclude by observing that multiplier ideals are preserved under pulling back by étale (and more generally smooth) morphisms.

**Example 9.5.44. (Étale coverings).** Let $f : Y \longrightarrow X$ be a finite *étale* covering, and let $D$ be an effective $\mathbf{Q}$-divisor on $X$. Then

$$\mathcal{J}(Y, f^*D) = f^{-1}\mathcal{J}(X, D). \tag{9.37}$$

(If $\mu : X' \longrightarrow X$ is a log resolution of $(X, D)$, then $\nu : Y' = Y \times_X X' \longrightarrow Y$ is a log resolution of $(Y, f^*D)$.) Analogous statements hold for multiplier ideals associated to ideals and linear series.    □

**Example 9.5.45. (Smooth morphisms).** Equation (9.37) remains true more generally if $f : Y \longrightarrow X$ is any smooth morphism of non-singular varieties.    □

## 9.6 Skoda's Theorem

This section revolves around a basic theorem of Skoda concerning the multiplier ideals associated to powers of a given ideal $\mathfrak{a}$. In its simplest form, Skoda's theorem states that if $X$ is a smooth variety of dimension $n$ and $\mathfrak{a} \subseteq \mathcal{O}_X$ is an ideal, then $\mathcal{J}(\mathfrak{a}^n) \subseteq \mathfrak{a}$. Since it is often not hard to check containment in a multiplier ideal one can see this as a non-trivial criterion for membership in $\mathfrak{a}$. This and a related global statement will lead quickly in Section 10.5 to versions of the effective Nullstellensatz.

We start with a review of the circle of ideas surrounding the integral closure of an ideal. Statements of local and global results appear in Section 9.6.B, and are proved in the third subsection. The final subsection is devoted to some variants.

### 9.6.A Integral Closure of Ideals

Skoda's theorem is closely connected to some algebraic concepts involving the integral closure of ideals. For the benefit of the reader not versed in these matters, we provide here a geometrically oriented overview of the basic definitions and results in this direction. Teissier's very nice paper [568] contains a more complete and detailed survey.

**Notation 9.6.1.** Unless otherwise stated, we assume throughout this subsection that $X$ is an irreducible normal variety.

We start with the definition of the integral closure of an ideal:

**Definition 9.6.2. (Integral closure of an ideal).** Given a non-zero ideal $\mathfrak{a} \subseteq \mathcal{O}_X$, let $\nu : X^+ \longrightarrow X$ be the normalization of the blow-up of $X$ along $\mathfrak{a}$. Denote by $E$ the exceptional divisor of $\nu$, so that $\mathfrak{a} \cdot \mathcal{O}_{X^+} = \mathcal{O}_{X^+}(-E)$. The *integral closure* $\overline{\mathfrak{a}}$ of $\mathfrak{a}$ is the ideal

$$\overline{\mathfrak{a}} \;=\; \nu_*\bigl(\mathcal{O}_{X^+}(-E)\bigr).$$

Thus $\mathfrak{a} \subseteq \overline{\mathfrak{a}} \subseteq \sqrt{\mathfrak{a}}$. One says that $\mathfrak{a}$ is *integrally closed* if $\mathfrak{a} = \overline{\mathfrak{a}}$.     □

We will see shortly (Corollary 9.6.12) that $\overline{\mathfrak{a}}$ is itself integrally closed. One can think of $\overline{\mathfrak{a}}$ as being the most natural ideal determined geometrically starting from $\mathfrak{a}$. Many algebraic questions involving integrally closed ideals reduce to elementary computations with divisors.

**Example 9.6.3. (Rees valuations).** Assuming for simplicity that $X$ is affine, the definition may be restated in a particularly concrete fashion using valuations. Specifically, in the situation of 9.6.2 the exceptional divisor $E$ determines a Weil divisor on $X^+$, say $[E] = \sum r_i[E_i]$. Then given a function $f \in \mathbf{C}[X]$,

$$f \in \overline{\mathfrak{a}} \quad \Longleftrightarrow \quad \mathrm{ord}_{E_i}(f) \geq r_i \quad \text{for all } i.$$

The valuations $\mathrm{ord}_{E_i}$ appearing here are known as the *Rees valuations* of $\mathfrak{a}$: they will play a role also later in connection with the effective Nullstellensatz.

□

**Remark 9.6.4. (Alternative computation).** Suppose that

$$f : Y \longrightarrow X$$

is any proper birational mapping from a normal variety $Y$ onto $X$ with the property that $\mathfrak{a} \cdot \mathcal{O}_Y = \mathcal{O}_Y(-F)$ for some effective Cartier divisor $F$ on $Y$. Then $f$ factors through $\nu$ and consequently $f_*\mathcal{O}_Y(-F) = \bar{\mathfrak{a}}$.

□

**Example 9.6.5.** If $\mathfrak{a}, \mathfrak{b} \subseteq \mathcal{O}_X$ are ideals on $X$, then

$$\bar{\mathfrak{a}} \cdot \bar{\mathfrak{b}} \subseteq \overline{\mathfrak{a} \cdot \mathfrak{b}}. \quad □$$

There are several equivalent criteria for membership in the integral closure of an ideal.

**Proposition 9.6.6. (Characterizations of integral closure).** *Assume that $X$ is affine, and view $\mathfrak{a} \subseteq \mathbf{C}[X]$ as an ideal in the coordinate ring of $X$. Then given $f \in \mathbf{C}[X]$, the following are equivalent:*

(i). $f \in \bar{\mathfrak{a}}$.

(ii). *$f$ satisfies an equation*

$$f^k + b_1 f^{k-1} + \ldots + b_{k-1}f + b_k = 0 \tag{9.38}$$

*with $b_i \in \mathfrak{a}^i$.*

(iii). *There exists a non-zero ideal $\mathfrak{b} \subseteq \mathcal{O}_X$ such that $f \cdot \mathfrak{b} \subseteq \mathfrak{a} \cdot \mathfrak{b}$.*

(iv). *There is a non-zero element $c \in \mathbf{C}[X]$ such that $c \cdot f^\ell \in \mathfrak{a}^\ell$ for all (or infinitely many) integers $\ell \gg 0$.*

*Proof.* We first outline the implications (iii) $\Rightarrow$ (ii) $\Rightarrow$ (i) $\Rightarrow$ (iii). That (iii) $\Rightarrow$ (ii) is the familiar determinant trick: choose generators $0 \neq e_1, \ldots, e_p \in \mathfrak{b}$ and write $fe_i = \sum a_{ij}e_j$ with $a_{ij} \in \mathfrak{a}$. Then $\det\left(f\delta_{ij} - a_{ij}\right) = 0$, which gives the required relation (9.38). Supposing next that (ii) holds we deduce (i) as follows. Let $\nu : X^+ \longrightarrow X$ be the normalized blowing-up of $\mathfrak{a}$, and fix any component $E_\alpha$ of the exceptional divisor $E$ of $\nu$. It suffices for (i) to show that

$$\mathrm{ord}_{E_\alpha}(f) \geq m_\alpha =_{\mathrm{def}} \mathrm{ord}_{E_\alpha}(E). \tag{*}$$

But (9.38) implies that

$$k \cdot \mathrm{ord}_{E_\alpha}(f) \geq \min_{1 \leq i \leq k} \left\{\mathrm{ord}_{E_\alpha}\left(b_i f^{k-i}\right)\right\}$$

$$\geq \min_{1 \leq i \leq k} \left\{i \cdot m_\alpha + (k-i) \cdot \mathrm{ord}_{E_\alpha}(f)\right\},$$

from which (*) follows at once. For (i) $\Rightarrow$ (iii) we will show that if $\mathfrak{q} \subseteq \mathcal{O}_X$ is any ideal such that $\mathfrak{q} \cdot \mathcal{O}_{X^+} = \mathcal{O}_{X^+}(-E)$, then

$$\mathfrak{q} \cdot \overline{\mathfrak{a}^{k-1}} = \overline{\mathfrak{a}^k} \quad \text{for } k \gg 0. \tag{**}$$

Bearing in mind Example 9.6.5, this implies that

$$\mathfrak{a} \cdot \overline{\mathfrak{a}^{k-1}} = \overline{\mathfrak{a}} \cdot \overline{\mathfrak{a}^{k-1}} = \overline{\mathfrak{a}^k}$$

for $k \gg 0$. So given $f \in \overline{\mathfrak{a}}$ we can take $\mathfrak{b} = \overline{\mathfrak{a}^{k-1}}$ in (iii). To prove (**), fix generators $e_1, \ldots, e_p \in \mathfrak{q}$ and pull back to the normalized blow-up $V$ to get a surjective mapping $\mathcal{O}_{X^+}^p \longrightarrow \mathcal{O}_{X^+}(-E)$. Twisting the resulting Koszul complex by $\mathcal{O}_{X^+}(-(k-1)E)$ gives the long exact sequence

$$\ldots \longrightarrow \mathcal{O}_{X^+}^{\binom{p}{2}}(-(k-2)E) \longrightarrow \mathcal{O}_{X^+}^p(-(k-1)E) \longrightarrow \mathcal{O}_{X^+}(-kE) \longrightarrow 0. \tag{***}$$

But $\mathcal{O}_{X^+}(-E)$ is ample for $\nu$, so by taking $k \gg 0$ we can arrange that the higher direct images of all terms vanish. Then the direct image under $\nu$ of the map on the right in (***) remains surjective, and (**) follows.[18] To complete the proof, we show that (ii) $\Rightarrow$ (iv) $\Rightarrow$ (i). Assuming (ii), take any $c \in \mathfrak{a}^k$. Then

$$cf^\ell = -c \cdot \left( b_1 f^{\ell-1} + \ldots + b_k f^{\ell-k} \right),$$

and it follows inductively that (iv) holds for $\ell \geq k$. Now suppose that $0 \neq c \in \mathbf{C}[X]$ satisfies (iv). Keeping notation as in the proof that (ii) $\Rightarrow$ (i), the hypothesis $cf^\ell \in \mathfrak{a}^\ell$ implies

$$\mathrm{ord}_{E_\alpha}(c) + \ell \cdot \mathrm{ord}_{E_\alpha}(f) \geq \ell \cdot \mathrm{ord}_{E_\alpha}(E).$$

Letting $\ell \to \infty$ it follows that $\mathrm{ord}_{E_\alpha}(f) \geq \mathrm{ord}_{E_\alpha}(E)$ for every $\alpha$, and hence $f \in \overline{\mathfrak{a}}$. □

**Remark 9.6.7.** Given an ideal $\mathfrak{a} \subseteq A$ in a normal Noetherian domain $A$, the integral closure $\overline{\mathfrak{a}}$ of $\mathfrak{a}$ is usually defined by condition 9.6.6 (ii). The equivalence of statements (i)–(iv) remains valid in this setting. (See [74, Chapter 10.2] or [297, Chapter 5].) □

**Example 9.6.8. (Valuative criterion for integral closure).** Keeping the notation and assumptions of the previous proposition, valuations lead to another useful characterization of integral closure:

Fix an ideal $\mathfrak{a} \subseteq \mathbf{C}[X]$ plus an element $f \in \mathbf{C}[X]$. Then $f \in \overline{\mathfrak{a}}$ if and only if

$$v(f) \geq v(\mathfrak{a})$$

for every rank-one discrete valuation $v$ of the function field $\mathbf{C}(X)$ with center on $X$.[19]

---

[18] Compare with the proof of Theorem 9.6.21.

[19] By $v(\mathfrak{a})$ we understand the minimum of the values of $v$ on all non-zero elements of $\mathfrak{a}$.

Indeed, this follows from characterization (iv) of the previous proposition.    □

**Remark 9.6.9. (Monomial ideals).** Let $\mathfrak{a} \subseteq \mathbf{C}[x_1, \ldots, x_n]$ be a monomial ideal, and denote by $P(\mathfrak{a}) \subseteq \mathbf{R}^n$ the Newton polyhedron of $\mathfrak{a}$ (Section 9.3.C). Then the integral closure $\bar{\mathfrak{a}}$ of $\mathfrak{a}$ is the monomial ideal spanned by all monomials whose exponent vectors lie in $P(\mathfrak{a})$. (See for instance [164, Ex. 4.23].)    □

**Remark 9.6.10. (Metric characterization of integral closure).** Assume that $X$ is affine, fix a point $x \in X$, and say $\mathfrak{a} \subseteq \mathbf{C}[X]$ is the ideal generated by regular functions $g_1, \ldots, g_p \in \mathbf{C}[X]$. Then given a function $h \in \mathbf{C}[X]$, one has $h \in \bar{\mathfrak{a}}$ near $x$ if and only if there is a (classical) neighborhood $U \ni x$ of $x$ in $X$ and a positive constant $C > 0$ such that

$$|h(x')| \le C \cdot \sum_{i=1}^{p} |g_i(x')|$$

for every $x' \in U$. (See [568, 1.3]: the idea is that by passing to the normalized blow-up $X^+$, one reduces to the case in which the $g_i$ generate a principal ideal.)    □

Integrally closed ideals arise naturally from divisors on a blow-up:

**Proposition 9.6.11. (Integral closure of direct images).** *Let* $: Y \longrightarrow X$ *be a proper birational morphism of normal varieties, and let* $D$ *be an effective Cartier divisor on* $Y$. *Then the ideal*

$$\mathfrak{a} =_{\mathrm{def}} f_* \mathcal{O}_Y(-D) \subseteq \mathcal{O}_X$$

*is integrally closed.*

**Corollary 9.6.12.** *If* $\mathfrak{a} \subseteq \mathcal{O}_X$ *is any ideal, then* $\bar{\mathfrak{a}}$ *is integrally closed, i.e.* $\bar{\bar{\mathfrak{a}}} = \bar{\mathfrak{a}}$.    □

**Corollary 9.6.13. (Integral closure of multiplier ideals).** *Assuming that* $X$ *is smooth, let* $\mathfrak{a} \subseteq \mathcal{O}_X$ *be an ideal, and* $c > 0$ *a rational number. Then the multiplier ideal* $\mathcal{J}(X, c \cdot \mathfrak{a}) \subseteq \mathcal{O}_X$ *is integrally closed.*    □

*Proof of Proposition 9.6.11.* We are free without changing the problem to replace $D$ by its pullback to a normal modification of $Y$. Therefore we may suppose that $\mathfrak{a} \cdot \mathcal{O}_Y = \mathcal{O}_Y(-D - E)$ for some effective Cartier divisor $E$ on $Y$. In view of Remark 9.6.4 this implies that $\bar{\mathfrak{a}} \subseteq \mathfrak{a}$, and the reverse inclusion is automatic.    □

**Example 9.6.14. (Primary decomposition of integrally closed ideals).** Let $\mathfrak{a} \subseteq \mathcal{O}_X$ be an integrally closed ideal sheaf. Then $\mathfrak{a}$ has a canonically defined expression

$$\mathfrak{a} = \mathfrak{q}_1 \cap \ldots \cap \mathfrak{q}_t \tag{9.39}$$

as an intersection of primary ideal sheaves. In particular, multiplier ideals admit such a decomposition. (As in Example 9.6.3, write $[E] = \sum_{i=1}^{t} r_i [E_i]$

for the Weil divisor determined by the exceptional divisor $E$ of the normalized blow-up of $\mathfrak{a}$, and set $\mathfrak{q}_i = \nu_* \mathcal{O}_{X^+}(-r_i E_i)$. Then $\mathfrak{q}_i$ is primary, and

$$\mathfrak{a} = \nu_* \mathcal{O}_{X^+}(-E) = \cap_{i=1}^t \mathfrak{q}_i.$$

Note that although there may be inclusions among the $\mathfrak{q}_i$ in (9.39), omitting suitable terms leads to a canonical irredundant primary decomposition of $\mathfrak{a}$.)    □

In dealing with questions that depend only on the integral closure of an ideal $\mathfrak{a}$, it is often convenient to replace $\mathfrak{a}$ by a suitably chosen smaller ideal $\mathfrak{r}$ having the same integral closure. We will take this tack in our proof of Skoda's theorem. This leads to the notion of the reduction of an ideal:

**Definition 9.6.15. (Reduction of an ideal).** Let $\mathfrak{a} \subseteq \mathcal{O}_X$ be a non-zero ideal. A *reduction* of $\mathfrak{a}$ is an ideal $\mathfrak{r} \subseteq \mathfrak{a}$ such that $\bar{\mathfrak{r}} = \bar{\mathfrak{a}}$.    □

**Proposition 9.6.16. (Criterion for reduction).** *Let* $\mathfrak{r} \subseteq \mathfrak{a} \subseteq \mathcal{O}_X$ *be non-zero ideals. Then* $\mathfrak{r}$ *is a reduction of* $\mathfrak{a}$ *if and only if*

$$\mathfrak{r} \cdot \mathcal{O}_{X^+} = \mathfrak{a} \cdot \mathcal{O}_{X^+}, \qquad (*)$$

*where as above* $\nu : X^+ \longrightarrow X$ *is the normalization of the blow-up of* $X$ *along* $\mathfrak{a}$.

**Corollary 9.6.17.** *Assume that* $X$ *is non-singular, and let* $\mathfrak{a} \subseteq \mathcal{O}_X$ *be an ideal. Then for any* $c > 0$,

$$\mathcal{J}(c \cdot \mathfrak{a}) = \mathcal{J}(c \cdot \bar{\mathfrak{a}}).$$

*More generally, if* $\mathfrak{r}$ *is a reduction of* $\mathfrak{a}$, *then* $\mathcal{J}(c \cdot \mathfrak{r}) = \mathcal{J}(c \cdot \mathfrak{a})$ *for every* $c > 0$.    □

*Proof of Proposition 9.6.16.* Write as before

$$\mathfrak{a} \cdot \mathcal{O}_{X^+} = \mathcal{O}_{X^+}(-E).$$

If (*) holds, then Remark 9.6.4 applies to $\mathfrak{r}$ to show that

$$\bar{\mathfrak{r}} = \nu_* \mathcal{O}_{X^+}(-E) = \bar{\mathfrak{a}}.$$

Conversely, suppose that (*) fails, so that $\mathfrak{r} \cdot \mathcal{O}_{X^+} = \mathcal{O}_{X^+}(-E) \cdot \mathfrak{r}^+$ for some non-trivial ideal $\mathfrak{r}^+ \subseteq \mathcal{O}_{X^+}$. Let $X^{++}$ denote the normalization of the blow-up of $\mathfrak{r}^+$. Then

$$\mathfrak{r} \cdot \mathcal{O}_{X^{++}} = \mathcal{O}_{X^{++}}(-F - D) \quad, \quad \mathfrak{a} \cdot \mathcal{O}_{X^{++}} = \mathcal{O}_{X^{++}}(-F)$$

for some effective Cartier divisors $F$ and $D$ on $X^{++}$ with $D \neq 0$. If $v$ is the valuation on $\mathbf{C}(X)$ determined by order of vanishing along an irreducible component of $D$, then it follows that $v(\mathfrak{r}) > v(\mathfrak{a})$. In view of the valuative criterion for integral closure (Example 9.6.8), this implies that $\bar{\mathfrak{r}} \neq \bar{\mathfrak{a}}$.    □

**Example 9.6.18.** Once Proposition 9.6.16 is known, it follows that one can test whether $\mathfrak{r} \subseteq \mathfrak{a}$ is a reduction of $\mathfrak{a}$ by pulling back under any birational proper map $\eta : Y \longrightarrow X$ where $Y$ is normal and $\eta$ factors through the normalized blow-up of $\mathfrak{a}$. (If $\mathfrak{r} \cdot \mathcal{O}_Y = \mathfrak{a} \cdot \mathcal{O}_Y$, then apply 9.6.4 to deduce that $\overline{\mathfrak{r}} = \overline{\mathfrak{a}}$. Conversely, if the pullbacks of $\mathfrak{r}$ and $\mathfrak{a}$ differ on $Y$, then certainly they differ already on $X^+$.)                                                        □

**Example 9.6.19. (Existence of reductions).** Let $X$ be a normal affine variety of dimension $n$, and $\mathfrak{a} \subseteq \mathcal{O}_X$ a non-zero ideal. Then locally $\mathfrak{a}$ has a reduction generated by $n$ elements. More precisely, there exists an open covering $\{U_j\}$ of $X$ such that $\mathfrak{a}|U_j \subseteq \mathcal{O}_{U_j}$ has a reduction $\mathfrak{r}_j \subseteq \mathcal{O}_{U_j}$ that is generated by $n$ elements. (Let $\nu : X^+ \longrightarrow X$ be the normalized blow-up of $\mathfrak{a}$, and $E \subseteq X^+$ the exceptional divisor. Note that all the fibres of $\nu$ have dimension $\leq n - 1$. Fix now a point $x \in X$ and consider $n$ general **C**-linear combinations $s_1, \dots, s_n \in \mathfrak{a}$ of a set of generators of $\mathfrak{a}$. Let

$$s'_1, \dots, s'_n \in \Gamma\big(X^+, \mathcal{O}_{X^+}(-E)\big)$$

be the inverse images of the $s_i$. Then for dimensional reasons, the common zeroes $\text{Zeroes}\big(s'_1, \dots, s'_n\big) \subseteq X^+$ of the $s'_i$ will be disjoint from $\nu^{-1}(x)$ provided that the $s_i$ are chosen sufficiently generally. In short, there is a non-empty open subset $U(x) \subseteq X$ such that the $s'_i$ generate $\mathcal{O}_{X^+}(-E)$ over $U(x)$. But thanks to Proposition 9.6.16 this means that the $s_i$ themselves generate a reduction $\mathfrak{r}_U \subseteq \mathfrak{a}|U \subseteq \mathcal{O}_U$ of $\mathfrak{a}$ on $U = U(x)$. We can in fact cover $X$ by such open sets, and the assertion follows.)                                    □

Finally, there is a basic containment between integral closures and multiplier ideals:

**Lemma 9.6.20.** *Assuming $X$ is non-singular, let $\mathfrak{a} \subseteq \mathcal{O}_X$ be any ideal. Then*

$$\overline{\mathfrak{a}} \subseteq \mathcal{J}(\mathfrak{a}).$$

*Proof.* This follows from Remark 9.6.4. In fact, if $\mu : X' \longrightarrow X$ is a log resolution of $\mathfrak{a}$ with $\mathfrak{a} \cdot \mathcal{O}_{X'} = \mathcal{O}_{X'}(-F)$, then

$$\overline{\mathfrak{a}} = \mu_* \mathcal{O}_{X'}\big(-F\big) \subseteq \mu_* \mathcal{O}_{X'}\big(K_{X'/X} - F\big) = \mathcal{J}(\mathfrak{a}), \tag{9.40}$$

as asserted.                                                                               □

## 9.6.B Skoda's Theorem: Statements

This subsection is devoted to statements of Skoda's theorem. We will present two versions of the result: a local assertion concerning the multiplier ideal of a power of an ideal, and an analogous global division theorem.

**Local Results.** We start with a local statement that computes the multiplier ideal of a high power of an ideal sheaf.

**Theorem 9.6.21. (Skoda's theorem).** *Let $X$ be a non-singular variety of dimension $n$, and let $\mathfrak{a} \subseteq \mathcal{O}_X$ be an ideal sheaf on $X$.*

(i).   *Given any integer $m \geq n$ one has*

$$\mathcal{J}(\mathfrak{a}^m) = \mathfrak{a} \cdot \mathcal{J}(\mathfrak{a}^{m-1}),$$

*and consequently*

$$\mathcal{J}(\mathfrak{a}^m) = \mathfrak{a}^{m-n+1} \cdot \mathcal{J}(\mathfrak{a}^{n-1}) \tag{9.41}$$

*for every $m \geq n$. In particular, $\mathcal{J}(\mathfrak{a}^m) \subseteq \mathfrak{a}^{m-n+1}$.*

(ii).   *More generally, fix a non-zero ideal $\mathfrak{b} \subseteq \mathcal{O}_X$ and a positive rational number $c$. Then*

$$\mathcal{J}(\mathfrak{a}^m \cdot \mathfrak{b}^c) = \mathfrak{a}^{m-n+1} \cdot \mathcal{J}(\mathfrak{a}^{n-1} \cdot \mathfrak{b}^c) \tag{9.42}$$

*for every $m \geq n$.*

**Remark 9.6.22.** More generally still, given finitely many non-zero ideals $\mathfrak{b}_1, \ldots, \mathfrak{b}_t \subseteq \mathcal{O}_X$ together with rational numbers $c_1, \ldots, c_t > 0$, there is an equality

$$\mathcal{J}(\mathfrak{a}^m \cdot \mathfrak{b}_1^{c_1} \cdot \ldots \cdot \mathfrak{b}_t^{c_t}) = \mathfrak{a}^{m-n+1} \cdot \mathcal{J}(\mathfrak{a}^{n-1} \cdot \mathfrak{b}_1^{c_1} \cdot \ldots \cdot \mathfrak{b}_t^{c_t}).$$

However as explained in Example 9.2.9, this is implied by (9.42).    □

**Remark 9.6.23.** It is allowed to take $\mathfrak{b} = \mathfrak{a}$ in statement (ii) of 9.6.21. In particular, $\mathcal{J}(\mathfrak{a}^d) = \mathfrak{a} \cdot \mathcal{J}(\mathfrak{a}^{d-1})$ for any rational (or real) number $d \geq n$.    □

**Remark 9.6.24. (Strengthenings).** Remaining in the setting of Theorem 9.6.21, let $\mathfrak{r} \subseteq \mathfrak{a}$ be a reduction of $\mathfrak{a}$ (Definition 9.6.15). The proof of the theorem will show that in fact

$$\mathcal{J}(\mathfrak{a}^m) = \mathfrak{r}^{m-n+1} \cdot \mathcal{J}(\mathfrak{a}^{n-1}).$$

The equality (9.42) admits an analogous extension. In another direction, suppose that $\mathfrak{a}$ — or more generally a reduction $\mathfrak{r} \subseteq \mathfrak{a}$ — is generated by $p \leq n$ elements. Then the statements in 9.6.21 hold for all $m \geq p$ (Theorem 9.6.36 below).    □

**Example 9.6.25. (Some borderline examples).** The exponents appearing in Skoda's theorem are in general best possible. For instance, take $X = \mathbf{C}^n$, with coordinates $x_1, \ldots, x_n$, and consider the maximal ideal $\mathfrak{m} = (x_1, \ldots, x_n)$. Then $\mathcal{J}(\mathfrak{m}^m) = \mathfrak{m}^{m+1-n}$ provided that $m \geq n$, whereas $\mathcal{J}(\mathfrak{m}^m) = \mathcal{O}_X$ if $0 < m < n$. So (9.41) is sharp in this case. For a somewhat more interesting example, take for instance $\mathfrak{a} = (x_1^2, \cdots, x_n^2)$. Then

$$\mathcal{J}(\mathfrak{a}^m) = \mathcal{J}\big((x_1, \ldots, x_n)^{2m}\big) = (x_1, \ldots, x_n)^{2m+1-n}$$

for $m \geq n$. So here (9.41) asserts that

$$\left(x_1, \ldots, x_n\right)^{2m+1-n} \subseteq \left(x_1^2, \ldots, x_n^2\right)^{m+1-n}$$

(which of course one can also easily check directly). On the other hand,

$$\left(x_1, \ldots, x_n\right)^{2m+1-n} \not\subseteq \left(x_1^2, \ldots, x_n^2\right)^{m+2-n}. \qquad \square$$

Skoda's theorem also implies an important statement of a purely algebraic nature concerning the relation between an ideal and its integral closure:

**Theorem 9.6.26. (Briançon–Skoda theorem).** *As above, let $X$ be a smooth variety of dimension $n$. If $\mathfrak{a} \subseteq \mathcal{O}_X$ is any ideal, and $m \geq n$ is any integer, then*

$$\overline{\mathfrak{a}^m} \subseteq \mathfrak{a}^{m+1-n}.$$

*Proof.* This follows from Skoda's theorem together with Lemma 9.6.20. $\qquad \square$

**Example 9.6.27. (A question of Mather).** Let $f \in \mathbf{C}[x_1, \ldots, x_n]$ be a polynomial having an isolated singularity at the origin, and let

$$j(f) = \left(\frac{\partial f}{\partial x_1}, \ldots, \frac{\partial f}{\partial x_n}\right) \cdot \mathcal{O}$$

denote the ideal generated by the partials of $f$ in the local ring $\mathcal{O} = \mathcal{O}_0 \mathbf{C}^n$ of $X = \mathbf{C}^n$ at the origin. Then $f^N \in j(f)$ for $N \gg 0$, and the genesis of the Briançon–Skoda Theorem was a question of Mather asking whether one could find a uniform value of $N$ that works for all $f$. Noting that $f \in \overline{j(f)}$, it follows from Theorem 9.6.26 that in fact $f^n \in j(f)$ for every $f$. $\qquad \square$

**Remark 9.6.28. (Uniform bounds in commutative algebra).** Given any ideal $\mathfrak{a} \subseteq \mathcal{O}_X$, recall that $\mathfrak{a}$ and $\overline{\mathfrak{a}}$ have the same radical. Therefore it is *a priori* clear that $(\overline{\mathfrak{a}})^s \subseteq \mathfrak{a}$ for some $s \in \mathbf{N}$. The statement coming from the Briançon-Skoda theorem that in fact $(\overline{\mathfrak{a}})^{\dim X} \subseteq \mathfrak{a}$ for every $\mathfrak{a} \subseteq \mathcal{O}_X$ is an early and influential example of a uniform bound in commutative algebra. Huneke [296] found other very interesting and surprising uniform statements involving the Artin–Rees lemma. Some of these were rendered effective in [161] using Skoda's theorem and jumping numbers of multiplier ideals (Example 9.3.26). Asymptotic multiplier ideals led in [159] to some uniform bounds concerning symbolic products: this work is presented in Section 11.3. In Section 10.5 we will show following [155] that Skoda's results lead to versions of the effective Nullstellensatz. $\qquad \square$

**Remark 9.6.29. (Algebraic proofs of Briançon–Skoda).** The problem of giving a purely algebraic proof of the Briançon–Skoda theorem has sparked a great deal of work in commutative algebra. Lipman and Teissier [404] proved that the analogue of Theorem 9.6.26 holds for ideals in any regular local ring. The theory of tight closure developed by Hochster and Huneke (see [297]) —

which centers on the action of the Frobenius — leads to a remarkably simple proof of this fact. Experience shows that algebraic statements established by $L^2$ methods or multiplier ideals can also be understood (in a more general setting) via tight closure. The work of Hara–Yoshida [271] and Takagi [563] mentioned in Remark 9.3.46 establishes a concrete connection between these two viewpoints. □

**Global Results.** In the setting of Theorem 9.6.21, choose local generators $g_1, \ldots, g_p \in \mathfrak{a}$. Skoda's result implies that if $g \in \mathcal{J}(\mathfrak{a}^n)$ then one can write $g$ as a linear combination $g = \sum h_i g_i$ for suitable functions $h_i$. In the global analogue of Skoda's theorem, $g$ and the $g_i$ are replaced by sections of appropriate line bundles. The base ideal determined by the given sections plays the role of $\mathfrak{a}$.

We start with a definition intended to simplify some of the statements:

**Definition 9.6.30. (Sections vanishing along an ideal sheaf).** Let $X$ be an irreducible variety, and let $L$ be an integral divisor on $X$. Consider a global section $s \in \Gamma(X, \mathcal{O}_X(L))$ and an ideal sheaf $\mathfrak{a} \subseteq \mathcal{O}_X$. We say that $s$ *vanishes along* $\mathfrak{a}$ if $s$ lies in the subspace

$$H^0(X, \mathcal{O}_X(L) \otimes \mathfrak{a}) \subseteq H^0(X, \mathcal{O}_X(L)). \quad \Box$$

This involves of course a slight abuse of language in that $s$ actually vanishes along the subscheme defined by $\mathfrak{a}$. However the terminology proposed in the definition is more compact, and we trust that it will not lead to any confusion.

We now fix our set-up. For the remainder of this subsection, $X$ denotes a non-singular complex projective variety of dimension $n$, $\mathfrak{a} \subseteq \mathcal{O}_X$ is an ideal sheaf, and $L$ is an integral divisor on $X$ such that

$$\mathcal{O}_X(L) \otimes \mathfrak{a}$$

is globally generated. We choose finally $p$ sections

$$s_1, \ldots, s_p \in \Gamma(X, \mathcal{O}_X(L) \otimes \mathfrak{a})$$

that generate the sheaf in question. (In fact it is sufficient that the $s_i$ span $\mathcal{O}_X(L) \otimes \mathfrak{r}$ for some reduction $\mathfrak{r} \subseteq \mathfrak{a}$ of $\mathfrak{a}$, but for the time being we do not insist on this.)

The main global result asserts that a section of a suitably positive line bundle vanishing along the multiplier ideal $\mathcal{J}(\mathfrak{a}^{n+1})$ can be written as a linear combination of the $s_i$.

**Theorem 9.6.31. (Global division theorem).** *With notation as above, fix a nef and big divisor $P$ on $X$, and a positive integer $m \geq n + 1$. Then any section*

$$s \in \Gamma\Big(X, \mathcal{O}_X(K_X + mL + P) \otimes \mathcal{J}(\mathfrak{a}^m)\Big)$$

*can be expressed as a linear combination*

$$s = \sum h_i \, s_i$$

*with $h_i \in \Gamma\Big(X, \mathcal{O}_X(K_X + (m-1)L + P) \otimes \mathcal{J}(\mathfrak{a}^{m-1})\Big)$.*

In other words, if $s \in \Gamma(X, \mathcal{O}_X(K_X + mL + P))$ vanishes along the multiplier ideal $\mathcal{J}(\mathfrak{a}^m)$, then $s$ can be expressed as a linear combination of fixed generators of $\mathcal{O}_X(L) \otimes \mathfrak{a}$ provided that $m \geq n + 1$. We will see in Example 9.6.34 that this can fail when $m = n$. As in the local setting there is a more general result involving multiplier ideals of the form $\mathcal{J}(\mathfrak{a}^m \cdot \mathfrak{b}^c)$. However the statement becomes a bit involved, so we postpone it until Section 9.6.D.

Note that as in 9.6.21, the theorem inductively yields:

**Corollary 9.6.32.** *In the situation of the theorem, one can write*

$$s = \sum_{|I|=m-n} h_I \, s_I$$

*where the sum ranges over all multi-indices $I = (i_1, \ldots, i_p)$ of weight $m - n$, with $s_I$ denoting the corresponding monomial in the sections $s_i$, and*

$$h_I \in \Gamma\Big(X, \mathcal{O}_X(K_X + nL + P) \otimes \mathcal{J}(\mathfrak{a}^n)\Big). \quad \Box$$

**Remark 9.6.33. (Few generators).** It is enough for Theorem 9.6.31 and Corollary 9.6.32 to assume that $m \geq p$. Typically it requires $p \geq n+1$ sections $s_i$ to generate the sheaf $\mathcal{O}_X(L) \otimes \mathfrak{a}$ or $\mathcal{O}_X(L) \otimes \mathfrak{r}$ for a reduction $\mathfrak{r}$ of $\mathfrak{a}$, so this is not an improvement. However, one does get a stronger statement when the sheaves in question happen to be spanned by an unusually small number of sections. $\quad \Box$

**Example 9.6.34. (Connection with Castelnuovo–Mumford regularity).** Suppose that $|L|$ is ample and free and that the $s_i$ span $\Gamma(X, \mathcal{O}_X(L))$. Then Theorem 9.6.31 reduces to the assertion that the map

$$H^0(X, \mathcal{O}_X(L)) \otimes H^0(X, \mathcal{O}_X(K_X + nL + P))$$
$$\longrightarrow H^0(X, \mathcal{O}_X(K_X + (n+1)L + P))$$

is surjective. This is a special case of Theorem 1.8.5 (using Kodaira vanishing), and the proof will show that 9.6.31 is quite closely related to the circle of ideas surrounding regularity. By the same token, the example of $X = \mathbf{P}^n$ and $L = P$ a hyperplane divisor shows that the conclusion of 9.6.31 can fail when $m = n$. $\quad \Box$

**Example 9.6.35. (Surjectivity of multiplication maps).** Continuing the train of thought of the previous example, let $\mathfrak{b} = \mathfrak{b}(|L|)$ be the base ideal of a complete linear series $|L|$. Then the natural mapping

$$S^k H^0\left(X, \mathcal{O}_X(L) \otimes \mathfrak{b}\right) \otimes H^0\left(X, \mathcal{O}_X(K_X + nL + P) \otimes \mathcal{J}(\mathfrak{b}^n)\right)$$
$$\longrightarrow H^0\left(X, \mathcal{O}_X(K_X + (n+k)L + P) \otimes \mathfrak{b}^k \cdot \mathcal{J}(\mathfrak{b}^n)\right)$$

determined by multiplication is surjective for every $k \geq 0$. (Observe that $\mathfrak{b}^k \cdot \mathcal{J}(\mathfrak{b}^n) = \mathcal{J}(\mathfrak{b}^{k+n})$ thanks to the local Skoda theorem, and then apply 9.6.32.)  □

### 9.6.C Skoda's Theorem: Proofs

This subsection is devoted to the proof of Theorems 9.6.21 and 9.6.31. We will deduce these results using certain "Skoda complexes" that one can define in the present setting. The main technical point is the exactness of these complexes, which will be verified at the end of the subsection. In a general way this line of attack goes back to Lipman and Teissier [404]. It was taken up again in the paper [155] of Ein and the author.

We start by fixing notation and hypotheses. Denote by $X$ a non-singular variety of dimension $n$, and let $\mathfrak{a}, \mathfrak{b} \subseteq \mathcal{O}_X$ be non-zero ideal sheaves on $X$. Fix also an integral divisor $L$ on $X$ together with $p$ global sections

$$s_1, \ldots, s_p \in \Gamma(X, \mathcal{O}_X(L) \otimes \mathfrak{a})$$

generating $\mathcal{O}_X(L) \otimes \mathfrak{r}$ for some reduction $\mathfrak{r} \subseteq \mathfrak{a}$ of $\mathfrak{a}$. To rephrase the last assumption more concretely, consider any proper birational map $\eta : Y \longrightarrow X$ from a normal variety $Y$ onto $X$ that dominates the blow-up of $X$ along $\mathfrak{a}$. Then $\mathfrak{a} \cdot \mathcal{O}_Y = \mathcal{O}_Y(-F)$ for some effective Cartier divisor $F$ on $Y$, and the condition on the sections $s_i$ is just that their pullbacks $\eta^* s_i$ generate the line bundle $\mathcal{O}_Y(\eta^* L - F)$ (Example 9.6.18). In particular, this automatically holds if the $s_i$ generate $\mathcal{O}_X(L) \otimes \mathfrak{a}$. Finally, denote by

$$V = \langle s_1, \ldots, s_p \rangle \subseteq \Gamma(X, \mathcal{O}_X(L) \otimes \mathfrak{a})$$

the subspace spanned by the $s_i$. Note that if in the local setting $X$ is affine, then one can take $L = \mathcal{O}_X$.

Fix $m \in \mathbf{N}$ and $c > 0$. Recalling that $\mathfrak{a} \cdot \mathcal{J}(\mathfrak{a}^{m-1} \cdot \mathfrak{b}^c) \subseteq \mathcal{J}(\mathfrak{a}^m \cdot \mathfrak{b}^c)$, there is a natural map

$$\tau : V \otimes_{\mathbf{C}} \mathcal{J}(\mathfrak{a}^{m-1} \cdot \mathfrak{b}^c) \otimes \mathcal{O}_X(-L) \longrightarrow \mathcal{J}(\mathfrak{a}^m \cdot \mathfrak{b}^c), \qquad (9.43)$$

and by construction one has

$$\operatorname{im}\tau \; = \; \mathfrak{r} \cdot \mathcal{J}\big(\mathfrak{a}^{m-1} \cdot \mathfrak{b}^c\big).$$

Moreover, these maps give rise to a Koszul-type complex

$$0 \to \Lambda^p V \otimes \mathcal{J}\big(\mathfrak{a}^{m-p} \cdot \mathfrak{b}^c\big) \otimes \mathcal{O}_X(-pL) \to \dots$$
$$\dots \to \Lambda^2 V \otimes \mathcal{J}\big(\mathfrak{a}^{m-2} \cdot \mathfrak{b}^c\big) \otimes \mathcal{O}_X(-2L) \to V \otimes \mathcal{J}\big(\mathfrak{a}^{m-1} \cdot \mathfrak{b}^c\big) \otimes \mathcal{O}_X(-L)$$
$$\xrightarrow{\;\tau\;} \mathcal{J}\big(\mathfrak{a}^m \cdot \mathfrak{b}^c\big) \to 0$$

of sheaves on $X$. We call this the $m^{\text{th}}$ Skoda complex $(\text{Skod}_m)$: it will be constructed formally during the proof of 9.6.36.

The main technical result asserts that $(\text{Skod}_m)$ is an exact sequence of sheaves provided that $m \ge p$.

**Theorem 9.6.36. (Exactness of Skoda complex).** *If $m \ge p$, then the complex $(\text{Skod}_m)$ is exact. In particular,*

$$\mathcal{J}\big(\mathfrak{a}^m \cdot \mathfrak{b}^c\big) = \mathfrak{r} \cdot \mathcal{J}\big(\mathfrak{a}^{m-1} \cdot \mathfrak{b}^c\big).$$

The proof appears at the end of the subsection. In the meantime we deduce the results stated in the previous subsection.

*Proof of Theorem 9.6.21.* The question being local, we are free to replace $X$ by an open subset. But locally $\mathfrak{a}$ has a reduction $\mathfrak{r}$ that is generated by $\le n$ elements (Example 9.6.19). So we may suppose that $p \le n$, and then Theorem 9.6.36 applies. The strengthening mentioned in Remark 9.6.24 is treated similarly: given a reduction $\mathfrak{r}$ of $\mathfrak{a}$, we localize and pass to a further reduction $\mathfrak{r}' \subseteq \mathfrak{r}$ that is generated by $\le n$ elements, and then apply the Theorem to $\mathfrak{r}'$. $\qquad\square$

*Proof of Theorem 9.6.31.* We assert to begin with that $\mathfrak{a}$ admits a reduction $\mathfrak{r} \subseteq \mathfrak{a}$ such that $\mathcal{O}_X(L) \otimes \mathfrak{r}$ is globally generated by $(n + 1)$ **C**-linear combinations $t_1, \dots, t_{n+1}$ of the given sections $s_i$. In fact, consider as before the pullback $\mathfrak{a} \cdot \mathcal{O}_{X^+} = \mathcal{O}_{X^+}(-E)$ of $\mathfrak{a}$ to the normalized blow-up of $\nu : X^+ \longrightarrow X$ of $\mathfrak{a}$. By assumption $\mathcal{O}_{X^+}(\nu^* L - E)$ is globally generated by the pullbacks of the $s_i$. So for dimensional reasons the bundle in question is already generated by $n + 1$ general **C**-linear combinations of those pullbacks. In view of Proposition 9.6.16, the ideal $\mathfrak{r} \subseteq \mathcal{O}_X$ spanned by the $t_i$ is a reduction of $\mathfrak{a}$.

We now apply Theorem 9.6.36 with $\mathfrak{b} = \mathcal{O}_X$ and $p = n + 1$. Specifically, tensor $(\text{Skod}_m)$ through by $\mathcal{O}_X(K_X + mL + P)$. The $i^{\text{th}}$ term of the resulting exact complex has the form

$$\Lambda^i V \otimes \mathcal{O}_X\big(K_X + (m - i)L + P\big) \otimes \mathcal{J}\big(\mathfrak{a}^{m-1}\big),$$

which has vanishing higher cohomology thanks to Nadel vanishing. This implies that the complex obtained by taking global sections in the indicated twist of $(\text{Skod}_m)$ remains exact. In particular, the right-most map

$$V \otimes H^0\big(\mathcal{O}_X(K_X + (m-1)L + P) \otimes \mathcal{J}(\mathfrak{a}^{m-1})\big) \longrightarrow$$
$$H^0\big(\mathcal{O}_X(K_X + mL + P) \otimes \mathcal{J}(\mathfrak{a}^m)\big)$$

is surjective, and the theorem follows.                                         □

We now turn to the

*Proof of Theorem 9.6.36.* Take a common log resolution $\mu : X' \longrightarrow X$ of $\mathfrak{a}, \mathfrak{b} \subseteq \mathcal{O}_X$, and write

$$\mathfrak{a} \cdot \mathcal{O}_{X'} = \mathcal{O}_{X'}(-A) \quad \text{and} \quad \mathfrak{b} \cdot \mathcal{O}_{X'} = \mathcal{O}_{X'}(-B).$$

Let $s_i' = \mu^* s_i$ be the pullback of the given section $s_i$. By hypothesis the $s_i$ generate $\mathcal{O}_X(L) \otimes \mathfrak{r}$, so it follows from Example 9.6.18 that the $s_i'$ span $\mathcal{O}_{X'}(\mu^* L - A)$. Set

$$\Lambda_m = K_{X'/X} - mA - [cB],$$

and consider the Koszul complex determined by the $s_i$. Twisting through by $\Lambda_m$, one obtains the long exact sequence

$$\cdots \longrightarrow \Lambda^2 V \otimes \mathcal{O}_{X'}\big(\Lambda_{m-2} - 2\mu^* L\big) \longrightarrow V \otimes \mathcal{O}_{X'}\big(\Lambda_{m-1} - \mu^* L\big)$$
$$\longrightarrow \mathcal{O}_{X'}\big(\Lambda_m\big) \longrightarrow 0 \qquad (\text{Kos}_m)$$

of sheaves on $X'$. The Skoda complex $(\text{Skod}_m)$ is obtained by taking direct images in $(\text{Kos}_m)$. Therefore the exactness of $(\text{Skod}_m)$ is implied by the vanishing of the higher direct images of each of the terms in $(\text{Kos}_m)$. But this follows from local vanishing (Variant 9.4.4). In fact, if $m \ge p \ge i$ then

$$R^j \mu_* \Big( \mathcal{O}_{X'}\big(\Lambda_{m-i} - i \cdot \mu^* L\big) \Big)$$
$$= R^j \mu_* \Big( \mathcal{O}_{X'}\big(K_{X'/X} - [(m-i)A + cB]\big) \Big) \otimes \mathcal{O}_X(-iL)$$
$$= 0$$

for $j > 0$, as required.                                                       □

## 9.6.D  Variants

This subsection contains some variants of the previous results.

The first statement involves the primary decomposition of a given ideal. Consider a non-singular variety $X$ and a non-zero ideal sheaf $\mathfrak{a} \subseteq \mathcal{O}_X$. Recall that (at least locally) one can write

$$\mathfrak{a} = \mathfrak{q}_1 \cap \ldots \cap \mathfrak{q}_h \qquad (9.44)$$

where each $\mathfrak{q}_i \subseteq \mathcal{O}_X$ is a sheaf of primary ideals. This decomposition is not unique, but the prime ideals $\mathfrak{p}_i =_{\mathrm{def}} \sqrt{\mathfrak{q}_i}$ — called the *associated primes* of $\mathfrak{a}$ — are canonically associated to $\mathfrak{a}$ and hence globally defined. We will refer to the zero loci of the $\mathfrak{p}_i$ as the *associated subvarieties* of $\mathfrak{a}$.

**Variant 9.6.37.** *Suppose that all the associated subvarieties of $\mathfrak{a}$ have codimension $\le e$ in $X$. Then*
$$\mathcal{J}\left(X, \mathfrak{a}^m\right) \subseteq \mathfrak{a}$$
*for all $m \ge e$ (and hence also $\overline{\mathfrak{a}^m} \subseteq \mathfrak{a}$ for $m \ge e$).*

*Proof.* The statement is local so we may suppose that $X$ is affine. Taking a primary decomposition (9.44) of $\mathfrak{a}$, it is enough to show that
$$\mathcal{J}\left(\mathfrak{a}^e\right) \subseteq \mathfrak{q}_i \quad \text{for each } 1 \le i \le h. \tag{*}$$

To this end, fix an index $i$. Since $\mathfrak{q}_i$ is primary, membership in $\mathfrak{q}_i$ is tested locally at the general point of $Y_i =_{\mathrm{def}} \mathrm{Zeroes}(\mathfrak{p}_i)$. So to verify (*) for the given index $i$ we are free to replace $X$ by any Zariski-open subset meeting $Y_i$. Hence to begin with we can suppose that $Y_i$ is a minimal associated subvariety[20] of $\mathfrak{a}$ having dimension $\ge n - e$ (where as usual $\dim X = n$).

Now consider the normalized blowing-up $\nu : X^+ \longrightarrow X$ of $X$ along $\mathfrak{a}$. Note that if $e > 0$ then
$$\dim \left\{x \in X \mid \mu^{-1}(x) \text{ has dimension } \ge e\right\} \le n - e - 1.$$

Therefore the set in question cannot contain $Y_i$. So after further localizing, we can suppose that every fibre of $\nu : X^+ \longrightarrow X$ has dimension $\le e - 1$. Arguing as in Example 9.6.19, after perhaps still further localizing we can construct a reduction $\mathfrak{r}$ of $\mathfrak{a}$ that is generated by $p \le e$ elements. Then Theorem 9.6.36 implies that
$$\mathcal{J}\left(\mathfrak{a}^m\right) \subseteq \mathfrak{a} \subseteq \mathfrak{q}_i$$
as soon as $m \ge e$. The required inclusion (*) follows.   $\square$

We next record a generalization of Theorem 9.6.31 involving "mixed" multiplier ideals. We leave the proof to the interested reader.

**Variant 9.6.38.** *In the situation of Theorem 9.6.31, fix a non-zero ideal $\mathfrak{b}$ and a positive rational number $c > 0$. Suppose that $D$ is a divisor on $X$ having the property that the sheaf*
$$\mathcal{O}_X(d\,D) \otimes \mathfrak{b}^{dc}$$
*is generated by its global sections for some large natural number $d$ that clears the denominator of $c$ (i.e. that is sufficiently divisible so that $dc \in \mathbf{N}$). Then any section*

---

[20] In other words, $Y_i$ is defined by a maximal element of the family of associated prime ideals of $\mathfrak{a}$.

$$s \in \Gamma\big(X, \mathcal{O}_X(K_X + mL + P + D)\big)$$

that vanishes along $\mathcal{J}\big(\mathfrak{a}^m \cdot \mathfrak{b}^c\big)$ can be expressed as a linear combination

$$s = \sum h_i\, s_i$$

where $h_i \in \Gamma\big(X, \mathcal{O}_X(K_X + (m-1)L + P + D)\big)$ vanishes along the multiplier ideal $\mathcal{J}\big(\mathfrak{a}^{m-1} \cdot \mathfrak{b}^c\big)$, and where as above the $s_i$ are sections generating $\mathcal{O}_X(L) \otimes \mathfrak{a}$. □

Finally we discuss very briefly the extension of Skoda's theorem to singular varieties. We will restrict ourselves to the simplest statements.

Suppose then that $X$ is a normal variety of dimension $n$ and $\Delta$ is a Weil **Q**-divisor on $X$ such that the pair $(X, \Delta)$ satisfies the requirements of Definition 9.3.55. Then for any ideal $\mathfrak{a} \subseteq \mathcal{O}_X$ we have defined a multiplier ideal $\mathcal{J}\big((X, \Delta); \mathfrak{a}\big)$, and using these, Skoda's theorem generalizes without difficulty. For example:

**Variant 9.6.39. (Local Skoda on singular varieties).** *If $m \ge n$ then*

$$\mathcal{J}\big((X, \Delta); \mathfrak{a}^m\big) = \mathfrak{a}^{m+1-n} \cdot \mathcal{J}\big((X, \Delta); \mathfrak{a}^{n-1}\big).\quad \square$$

One can either prove this directly by applying local vanishing (Theorem 9.4.17) to a Koszul complex on a log resolution, or else by using this line of argument to first generalize Theorem 9.6.36 to the present setting. One can also extend the global statements to the singular case by drawing on Theorem 9.4.17 (ii). We leave the statements and proofs to the reader.

**Remark 9.6.40. (Failure of Briançon–Skoda on singular varieties).** Although Skoda-type results extend with very little change to singular varieties, it is no longer true in general that $\bar{\mathfrak{a}} \subseteq \mathcal{J}\big((X, \Delta); \mathfrak{a}\big)$. Because of this, statements à la Skoda do not imply inclusions for integral closures along the lines of Theorem 9.6.26. In fact, it is known that the Briançon–Skoda theorem can fail on singular varieties. □

**Remark 9.6.41. (Skoda's theorem for canonical modules).** A different extension of Skoda's theorem to singular varieties appears in work of Hyry–Smith [305, §4]. The idea is to deal directly with submodules of the canonical module of a ring. □

## Notes

As we indicated in the Introduction to Part III, many of the basic properties of multiplier ideals were worked out in passing by Esnault and Viehweg in [172], [173], [174]: see [174, §7, especially (7.4)–(7.10)]. Proposition 9.4.26 is

[174, Corollary 7.10]; it was rediscovered from an analytic viewpoint by Siu in [539]. Some of the results for multiplier ideals generalize — and are established by the same arguments as — related properties involving singularities of pairs. For example, the crucial restriction theorem occurs as Theorem 7.5(d) in [174], but very closely related ideas had appeared in connection with the inversion of adjunction (see [364, §7] and the references therein). Proposition 9.5.13 appears in [364, Exercise 3.14].

Besides [174], we have drawn on [145] and [323], which contain some of the basic facts of the theory. Demailly's notes [126] and their successor [129] are the standard references for the analytic viewpoint.

Our presentation of Mustaţă's summation theorem draws on the exposition in [292]. The semicontinuity of multiplier ideals is used by Siu and Angehrn in [10]. The algebro-geometric approach appears in [364, (6.8.3)].

Theorem 9.5.42 was worked out with Karen E. Smith. Example 9.5.8 was observed in discussions with Ein.

# 10

# Some Applications of Multiplier Ideals

The machinery developed in the previous chapter already has many substantial applications, and we present a number of these here. We start in the first section with several results involving singularities of divisors. In Section 10.2 we prove Matsusaka's theorem following the approach of Siu and Demailly. The next section is devoted to a theorem of Nakamaye describing the base loci of big and nef divisors. Results and conjectures concerning global generation of adjoint linear series occupy Section 10.4, where in particular we prove the theorem of Angehrn and Siu. Finally, in Section 10.5 we discuss applications of Skoda's theorem to the effective Nullstellensatz.

Further applications, drawing on asymptotic constructions, appear in Chapter 11.

## 10.1 Singularities

Multiplier ideals are well suited to studying singularities of divisors. The idea is that singular divisors produce non-trivial multiplier ideals, to which one can apply vanishing theorems. We begin in the first subsection with some statements concerning projective hypersurfaces. Singularities of theta divisors in abelian varieties occupy Section 10.1.B, and we conclude in 10.1.C with a criterion for adjoint linear series to separate a given number of jets.

### 10.1.A Singularities of Projective Hypersurfaces

The first application involves a very concrete question about homogeneous polynomials. Specifically, let $S \subset \mathbf{P}^n$ be a finite set of points. In the course of certain arithmetic arguments, one constructs a hypersurface of degree $d$ that has large multiplicity at every point of $S$. One wants to deduce from this that $S$ lies on a hypersurface of small degree. In this direction one has a statement going back to Bombieri, Skoda, Demailly, Waldschmidt, Esnault–Viehweg and others:[1]

**Proposition 10.1.1.** *Suppose that $A \subset \mathbf{P}^n$ is a hypersurface of degree $d$ such that*

$$\text{mult}_x A \;\geq\; k \quad \text{for all} \;\; x \in S.$$

*Then $S$ lies on a hypersurface of degree $\leq \left[\frac{dn}{k}\right]$.*

**Remark 10.1.2.** It is conjectured by Chudnovsky that in the situation of the Proposition $S$ actually lies on a hypersurface of degree $\leq \left[\frac{dn}{k}\right] - (n-1)$. This is known for $n = 2$, but is open for $n \geq 3$. We refer to the paper [168] of Esnault and Viehweg for references and an overview.     □

*Proof of Proposition 10.1.1.* Consider the effective **Q**-divisor $D = \frac{n}{k} A$ on $\mathbf{P}^n$. Then

$$\text{mult}_x D \;\geq\; n \quad \text{for all} \;\; x \in S,$$

and hence $\mathcal{J}(D) \subseteq \mathcal{I}_S$ thanks to Proposition 9.3.2, where $\mathcal{I}_S$ here denotes the ideal sheaf of $S$, considered as a reduced subscheme of $\mathbf{P}^n$. Set $\delta = \left[\frac{dn}{k}\right]$ and let $H$ be a hyperplane divisor. Then

$$\ell H - D \;\equiv_{\text{num}}\; \left(\ell - \frac{dn}{k}\right) H$$

is ample if $\ell \geq \delta + 1$. Recalling that $K_{\mathbf{P}^n} \equiv_{\text{num}} -(n+1)H$, Corollary 9.4.26 implies that $\mathcal{O}_{\mathbf{P}^n}(\delta) \otimes \mathcal{J}(D)$ is globally generated. But $\mathcal{J}(D) \subseteq \mathcal{I}_S$, so we deduce in particular that $S$ lies on a hypersurface of degree $\delta$.     □

**Example 10.1.3.** Consider as in the previous proposition a finite set $S \subset \mathbf{P}^n$. Given $t \geq 1$, denote by $\omega_t(S)$ the least degree of a hypersurface that has multiplicity $\geq t$ at every $x \in S$. If $t' < t$ then

$$\frac{\omega_{t'}(S)}{t' + n - 1} \;\leq\; \frac{\omega_t(S)}{t}.$$

This is a result of Skoda generalizing the previous proposition, which is the case $t' = 1$. (If $A$ is a hypersurface of degree $d = \omega_t(S)$ having multiplicity $\geq t$ along $S$, then $D = \left(\frac{t'+n-1}{t}\right) A$ has multiplicity $\geq t' + n - 1$ at every point of $S$, and hence $\mathcal{J}(D) \subseteq \mathcal{I}_S^{t'}$.)     □

**Example 10.1.4.** Let $S \subset \mathbf{P}^n$ be an algebraic subset all of whose irreducible components have codimension $\leq e$. Suppose that there is a hypersurface $A \subset \mathbf{P}^n$ of degree $d$ passing through $S$ such that

$$\text{mult}_x A \;\geq\; k \quad \text{for all} \;\; x \in S.$$

Then $S$ lies on a hypersurface of degree $\leq \left[\frac{de}{k}\right]$.     □

---

[1] It is interesting to observe that the early work by Bombieri and Skoda used some of the analytic techniques out of which the theory of multiplier ideals later emerged.

**Example 10.1.5. (Castelnuovo–Mumford regularity of varieties with isolated singularities).** This example reconsiders one of the regularity bounds discussed in Section 1.8.C. Specifically, let $X \subseteq \mathbf{P}^r$ be an irreducible variety of dimension $n$ and codimension $e = r - n$ that is scheme-theoretically cut out by hypersurfaces of degree $d$.[2] Assuming that $X$ has at worst isolated singularities, the statement of Theorem 1.8.40 remains valid, i.e. $X$ is $(de - e + 1)$-regular. (Consider the multiplier ideal $\mathcal{J} = \mathcal{J}(\mathbf{P}^r, \mathcal{I}_X^e)$. Since $X$ is reduced we have $\mathcal{J} \subseteq \mathcal{I}_X$, and the quotient $\mathcal{I}_X/\mathcal{J}$ is supported in the finitely many singular points of $X$. On the other hand, Nadel vanishing 9.4.15 implies that
$$H^i(\mathbf{P}^r, \mathcal{J}(k)) = 0 \quad \text{for } i \geq 1 \text{ and } k \geq ed - r.$$
The assertion follows.) □

### 10.1.B Singularities of Theta Divisors

It was observed by Kollár [362, Chapter 17] that one could use Kawamata–Viehweg vanishing to study the singularities of theta divisors on principally polarized abelian varieties. We present Kollár's theorem here, and outline some later developments due to Ein and the author [154].

Let $(A, \Theta)$ be a principally polarized abelian variety (PPAV) of dimension $g$. In other words, $A$ is a $g$-dimensional complex torus, and $\Theta \subset A$ is an ample divisor such that $h^0(A, \mathcal{O}_A(\Theta)) = 1$. Starting with the classical work of Andreotti and Mayer [9] on the Schottky problem, there has been interest in understanding what sort of singularities $\Theta$ can have. Kollár's result is the following:

**Theorem 10.1.6.** *The pair* $(A, \Theta)$ *is log-canonical, i.e.* $\mathcal{J}(A, (1 - \varepsilon)\Theta) = \mathcal{O}_A$ *for* $0 < \varepsilon < 1$. *In particular,* $\mathrm{mult}_x \Theta \leq g$ *for every* $x \in \Theta$. *More generally, if*
$$\Sigma_k(\Theta) = \{x \in A \mid \mathrm{mult}_x \Theta \geq k\},$$
*then every irreducible component of* $\Sigma_k$ *has codimension* $\geq k$ *in* $A$.

*Proof.* In view of Proposition 9.3.2 and Example 9.3.5, the statements on multiplicity follow from the triviality of $\mathcal{J}(A, (1 - \varepsilon)\Theta)$. So we focus on the first assertion of the theorem. Suppose to the contrary that $\mathcal{J}((1 - \varepsilon)\Theta) \neq \mathcal{O}_A$ for some $\varepsilon > 0$. Set
$$Z = \mathrm{Zeroes}(\mathcal{J}((1 - \varepsilon)\Theta)),$$
and consider the exact sequence
$$0 \longrightarrow \mathcal{O}_A(\Theta) \otimes \mathcal{J}((1 - \varepsilon)\Theta) \longrightarrow \mathcal{O}_A(\Theta) \longrightarrow \mathcal{O}_Z(\Theta) \longrightarrow 0.$$

---

[2] Recall that by definition this means that $\mathcal{I}_X(d)$ is globally generated, $\mathcal{I}_X$ being the ideal sheaf of $X$ in $\mathbf{P}^r$.

Recalling that $\mathcal{O}_A(K_A) = \mathcal{O}_A$, Nadel vanishing implies that the sheaf on the left has vanishing $H^1$. Therefore the restriction mapping

$$\rho : H^0\big(A, \mathcal{O}_A(\Theta)\big) \longrightarrow H^0\big(Z, \mathcal{O}_Z(\Theta)\big)$$

is surjective. But evidently $Z \subseteq \Theta$, and consequently the unique section of $\mathcal{O}_A(\Theta)$ vanishes on $Z$, i.e. $\rho$ is the zero map. We conclude that

$$H^0\big(Z, \mathcal{O}_Z(\Theta)\big) \; = \; 0,$$

and then an elementary lemma provides the required contradiction. $\qquad\square$

**Lemma 10.1.7.** *Let $(A, \Theta)$ be a principally polarized abelian variety, and let $Z \subset A$ be a non-empty subset (or subscheme). Then $H^0\big(Z, \mathcal{O}_Z(\Theta)\big) \neq 0$.*

*Proof.* Given $a \in A$, denote by $\Theta_a = \Theta + a$ the translate of $\Theta$ by $a$. If $a$ is sufficiently general then $\Theta_a$ meets $Z$ properly, and consequently $H^0\big(Z, \mathcal{O}_Z(\Theta_a)\big) \neq 0$. Letting $a \to 0$, the Lemma follows by semicontinuity. $\qquad\square$

Note that if $(A, \Theta)$ splits as a non-trivial product of two smaller PPAVs, then $\Sigma_2 \subset A$ has codimension 2 in $A$. However, a result of Ein and the author [154] shows that this is the only way equality can hold in Kollár's bounds:

**Theorem 10.1.8. (Irreducible theta divisors).** *Let $(A, \Theta)$ be a principally polarized abelian variety, and assume that $\Theta$ is irreducible. Then $\Theta$ is normal and has only rational singularities. Consequently, $\Sigma_k(\Theta)$ has a component of codimension exactly $k$ in $A$ if and only if $(A, \Theta)$ splits as a $k$-fold product of PPAVs.*

**Remark 10.1.9.** Smith and Varley [545] proved prior to Theorem 10.1.8 that if $\Sigma_g \neq \varnothing$, then $(A, \Theta)$ is a product of $g$ elliptic curves. The fact that an irreducible theta divisor is normal had been conjectured by Arbarello and DeConcini [16]. $\qquad\square$

*Indication of Proof of Theorem 10.1.8.* The statements concerning multiplicity are established by induction using the first assertion of the theorem, so the issue is to prove that an irreducible theta divisor $\Theta$ has no worse than rational singularities. The idea is to study the adjoint ideal of $\Theta$ (Section 9.3.E). Specifically, it follows from Proposition 9.3.48 (ii) that $\Theta$ is normal with at worst rational singularities if and only if its adjoint ideal $\mathrm{adj}\big(A, \Theta\big)$ is trivial. So we need to show that $\mathrm{adj}(\Theta) = \mathcal{O}_A$.

To this end, take a desingularization $\nu : X \longrightarrow \Theta$ and consider the exact sequence (9.16) appearing in Proposition 9.3.48:

$$0 \longrightarrow \mathcal{O}_A \overset{\Theta}{\longrightarrow} \mathcal{O}_A(\Theta) \otimes \mathrm{adj}(\Theta) \longrightarrow \nu_* \mathcal{O}_X(K_X) \longrightarrow 0.$$

Twisting this by a topologically trivial line bundle $P \in \mathrm{Pic}^0(A)$, we arrive at the sequence

$$0 \longrightarrow P \xrightarrow{\Theta} \mathcal{O}_A(\Theta) \otimes P \otimes \mathrm{adj}(\Theta) \longrightarrow \nu_* \mathcal{O}_X(K_X) \otimes P \longrightarrow 0. \qquad (*)$$

We study this using the generic vanishing theorem from [242] discussed in Section 4.4. Since evidently $X$ has maximal Albanese dimension, it follows from Theorem 4.4.3 and Remark 4.4.4 that

$$H^i\big(X, \mathcal{O}_X(K_X + \nu^* P)\big) = 0$$

for $i > 0$ and generic $P$. Therefore

$$\begin{aligned}
H^0\big(A, \nu_* \mathcal{O}_X(K_X) \otimes P\big) &= H^0\big(X, \mathcal{O}_X(K_X) \otimes \nu^* P\big) \\
&= \chi\big(X, \mathcal{O}_X(K_X) \otimes \nu^* P\big) \\
&= \chi\big(X, \mathcal{O}_X(K_X)\big)
\end{aligned}$$

for general $P$.

The fact that $\Theta$ is irreducible implies by a classical argument that $X$ is of general type (see [439]). Moreover, one sees from $(*)$ that $h^0\big(X, \mathcal{O}_X(K_X)\big) \le h^1(A, \mathcal{O}_A) = g$. But a theorem of Kawamata and Viehweg [327] states that if $W$ is birational to divisor in an abelian variety, and if $W$ is of general type and satisfies $p_g(W) \le g$, then $\chi\big(W, \mathcal{O}_X(K_W)\big) \ne 0$. This applies in particular to our variety $X$. Putting all this together, we conclude that $H^0\big(A, \mathcal{O}_A(\Theta) \otimes P \otimes \mathrm{adj}(\Theta)\big) \ne 0$ for general $P$, which in turn implies that

$$H^0\big(A, \mathcal{O}_A(\Theta + a) \otimes \mathrm{adj}(\Theta)\big) \ne 0 \qquad (**)$$

for general $a \in A$.

Now suppose for a contradiction that $\mathrm{adj}(\Theta) \ne \mathcal{O}_A$. Then the corresponding zero locus $Z = \mathrm{Zeroes}\big(\mathrm{adj}(\Theta)\big)$ is non-empty. Since $h^0(A, \mathcal{O}_A(\Theta + a)) = 1$ for all $a \in A$, $(**)$ implies that $Z \subseteq (\Theta + a)$ for all $a$. But the translates of $\Theta$ don't all have any points in common, so this is impossible. $\qquad \square$

**Example 10.1.10. (Pluri-theta divisors).** Kollár's Theorem 10.1.6 is generalized in [154, 3.5] to the case of pluri-theta divisors. Specifically, let $(A, \Theta)$ be a PPAV, and for $m \ge 1$ fix a divisor $E \in |m\Theta|$. Then the pair

$$\big(A, \tfrac{1}{m} E\big)$$

is log-canonical. In particular, every component of $\Sigma_{mk}(E)$ has codimension $\ge k$ in $A$. (Consider for $0 \le \varepsilon \ll 1$ the $\mathbf{Q}$-divisor

$$D = \tfrac{1-\varepsilon}{m} E \equiv_{\mathrm{num}} (1 - \varepsilon)\Theta.$$

Then Nadel vanishing implies that $H^i\big(A, \mathcal{O}_A(\Theta) \otimes P \otimes \mathcal{J}(D)\big) = 0$ for $i > 0$ and $P \in \mathrm{Pic}^0(A)$. On the other hand, if $\mathcal{J}(D) \ne \mathcal{O}_A$ then arguing as above one finds that $H^0\big(A, \mathcal{O}_A(\Theta) \otimes P \otimes \mathcal{J}(D)\big) = 0$ first for generic $P$ and hence also for every $P$. But a lemma of Mukai [444] asserts that if $\mathcal{F}$ is a coherent sheaf on $A$ such that $H^i\big(A, \mathcal{F} \otimes P\big) = 0$ for all $i \ge 0$ and every $P \in \mathrm{Pic}^0(A)$, then $\mathcal{F} = 0$, and this gives the required contradiction.) Hacon [261] has generalized Theorem 10.1.8 to the setting of pluri-theta divisors. $\qquad \square$

### 10.1.C  A Criterion for Separation of Jets of Adjoint Series

As a final example of the interaction between multiplier ideals and singularities, we give a criterion for the separation of jets of adjoint linear series, due to Demailly and Siu.

Let $X$ be a smooth projective variety of dimension $n$, let $x \in X$ be a fixed point, and let $L$ be an integral ample divisor on $X$. There has been a great deal of interest in recent years in trying to understand the geometry of the *adjoint linear series* $|K_X + L|$, which one can view as the higher-dimensional analogues of line bundles of large degree on curves. For example, one would like to know when these series are base-point free or very ample. We will discuss some results and conjectures in this direction in Section 10.4.

Given a fixed integer $s \geq 0$, we explain here a criterion to guarantee that the linear series $|K_X + L|$ separates $s$-jets at $x$, i.e. that the natural mapping

$$H^0\big(X, \mathcal{O}_X(K_X + L)\big) \longrightarrow H^0\big(X, \mathcal{O}_X(K_X + L) \otimes \mathcal{O}_X/\mathfrak{m}_x^{s+1}\big)$$

is surjective, $\mathfrak{m}_x$ denoting as usual the maximal ideal of $x$. For example, when $s = 0$ this amounts to asking that $\mathcal{O}_X(K_X + L)$ have a section that is non-zero at $x$, whereas generation of 1-jets means that the corresponding rational map $\Phi_{|K_X+L|}$ separates tangent directions at $x$.

**Proposition 10.1.11.** *Suppose that for some $k > 0$ there exists a divisor $A \in |kL|$ such that*
$$\mathrm{mult}_x(A) > k(n + s),$$
*while there is a neighborhood $U(x)$ of $x$ such that*
$$\mathrm{mult}_y(A) \leq k \quad \text{for} \quad y \in U(x) - \{x\}.$$
*Then $\mathcal{O}_X(K_X + L)$ separates $s$-jets at $x$.*

**Remark 10.1.12.** Suppose that $\int_X c_1(L)^n > (n + s)^n$. Then a dimension count using the asymptotic Riemann–Roch formula shows that one can always find a divisor $A \in |kL|$ $(k \gg 0)$ having multiplicity $> k(n+s)$ at $x$ (Proposition 1.1.31). However, it is very hard to control the singularities of $A$ away from $x$, and for this reason the proposition is not easy to apply in practice. As we shall see, the results discussed in Section 10.4 require more subtle arguments. Nonetheless the statement does give an instructive first indication of how one can use multiplier ideals and singularities of divisors to study adjoint linear series. □

*Proof of Proposition 10.1.11.* Set $q = \mathrm{mult}_x(A) > k(n + s)$, and consider the $\mathbf{Q}$-divisor $D = \frac{n+s}{q}A$. Then

$$L - D \equiv_{\mathrm{num}} \left(1 - \frac{n+s}{q}k\right) \cdot L$$

is ample. So by Nadel vanishing,

$$H^1\big(X, \mathcal{O}_X(K_X + L) \otimes \mathcal{J}(D)\big) \; = \; 0,$$

and hence the natural mapping

$$H^0\big(X, \mathcal{O}_X(K_X + L)\big) \longrightarrow H^0\big(X, \mathcal{O}_X(K_X + L) \otimes \mathcal{O}_X/\mathcal{J}(D)\big) \qquad (*)$$

is surjective. But $\mathrm{mult}_x(D) = n + s$, while $\mathrm{mult}_y(D) < 1$ for $y \in U(x) - \{x\}$. So on the one hand $\mathcal{J}(D) \subseteq \mathfrak{m}_x^{s+1}$ thanks to Proposition 9.3.2, whereas on the other hand $\mathcal{J}(D)_y = \mathcal{O}_y X$ for $y \in U(X) - \{x\}$ by virtue of Proposition 9.5.13. It follows that the subscheme of $X$ defined by $\mathcal{J}(D)$ contains $x$ as a connected component. Therefore there is a surjective mapping

$$H^0\big(X, \mathcal{O}_X(K_X + L) \otimes \mathcal{O}_X/\mathcal{J}(D)\big) \longrightarrow H^0\big(X, \mathcal{O}_X(K_X + L) \otimes \mathcal{O}_X/\mathfrak{m}_x^{s+1}\big),$$

which coupled with the surjectivity of $(*)$ shows that $|K_X + L|$ does indeed generate $s$-jets at $x$.    □

## 10.2 Matsusaka's Theorem

In this section, following Siu and Demailly, we apply the uniform non-vanishing statement Corollary 9.4.24 to prove Matsusaka's theorem.

Let $X$ be a non-singular complex projective variety of dimension $n$, and $L$ an ample divisor on $X$. A fundamental theorem of Matsusaka [420], [401] asserts that there is an integer $M$, depending only on the Hilbert polynomial $P(m) = \chi\big(X, \mathcal{O}_X(mL)\big)$ of $L$, such that $mL$ is very ample for $m \geq M$. A refinement by Kollár and Matsusaka [366] states that in fact one can arrange that $M$ depends only on the intersection numbers $(L^n)$ and $(K_X \cdot L^{n-1})$. The original proofs of these theorems were ineffective and fundamentally non-cohomological.

Siu [537] realized that one could use the non-vanishing statement Corollary 9.4.24 to give an effective version of Matsusaka's theorem. The argument was subsequently streamlined and generalized by Demailly [126], [125] and Siu himself [541]. We present here an exposition of their approach. In dimension $n \geq 3$, one does not expect the particular statements coming out of the argument — which become rather complicated — to have the optimal shape (see Remark 10.2.9). Therefore, in the interests of transparency, we have felt it best not to keep explicit track of the bounds (although it will be clear that one could do so).[3] In the case of surfaces, Fernandez del Busto [185], [394,

---

[3] Keeping the presentation ineffective also helps us avoid the temptation of trying to estimate efficiently.

Exercise 7.7] has used related methods to obtain an essentially optimal result (which also comes out of the present approach: see [126, §13]).

As above, let $X$ be a smooth complex projective variety of dimension $n$, and let $L$ be any ample divisor on $X$. We will have occasion to draw on the following:

**Lemma 10.2.1.** *There are positive integers* $\lambda_n, \beta_n$ *depending only on* $n$ *having the property that*

$$\lambda_n K_X + \beta_n L + P$$

*is very ample for every nef* $P$ *and ample* $L$.

*Proof.* To begin with, a result of Angehrn–Siu (appearing below as Theorem 10.4.2) asserts that

$$B =_{\text{def}} K_X + \left( \binom{n+1}{2} + 2 \right) L$$

is ample and globally generated. Then it follows from Example 1.8.23 that $K_X + (n+2)B + P$ is very ample for every nef $P$. So it suffices to take $\lambda_n = (n+3)$ and $\beta_n = (n+2)\left( \binom{n+1}{2} + 2 \right)$.    □

**Remark 10.2.2.** A conjecture of Fujita predicts that in fact one should be able to take $\lambda_n = 1$ and $\beta_n = (n+2)$: see Section 10.4.A.    □

**Notation 10.2.3.** Denoting as above by $L$ a fixed ample divisor on $X$, set

$$\rho_L = (L^n) \quad , \quad \rho_K = (K_X \cdot L^{n-1}).$$

We also write

$$A = A_n = \lambda_n K_X + \beta_n L$$

for the divisor in 10.2.1. Thus $A + P$ is very ample for any nef bundle $P$ on $X$. Note moreover that the intersection number $(A \cdot L^{n-1})$ depends only on $\rho_L$ and $\rho_K$ (and the dimension $n$).    □

Now suppose given any nef (integral) divisor $B$ on $X$. The approach of Siu-Demailly is to look for an explicit bound on $m$ in order that $mL - B$ be first nef and subsequently very ample. The shape of these statements allows one to subtract off multiples of $K_X$ that arise through applications of Corollary 9.4.24.

The essential result is the following:

**Theorem 10.2.4.** *Assume as above that* $L$ *is ample and that* $B$ *is nef, and set*

$$\rho_B = (B \cdot L^{n-1}).$$

*Then there is an integer*

$$M_0 = M_0(\rho_L, \rho_K, \rho_B)$$

*depending only on the indicated intersection numbers (and also on the dimension n of X) such that $mL - B$ is nef whenever $m \geq M_0$.*

**Corollary 10.2.5.** *In the setting of the previous theorem, there exists an integer $M_1 = M_1(\rho_L, \rho_K, \rho_B)$ such that $mL - B$ is very ample for all $m \geq M_1$.*

One recovers the statement of Kollár–Matsusaka by taking $B$ to be trivial.

*Proof of Corollary 10.2.5.* Apply the theorem with $B$ replaced by $A + B$, $A = A_n$ being the divisor fixed in 10.2.3. It yields a constant

$$M_1 = M_1(\rho_L, \rho_K, \rho_{(A+B)})$$

such that $mL - B - A$ is nef whenever $m \geq M_1$. But then

$$A + \left(mL - (A + B)\right) = mL - B$$

is very ample whenever $m \geq M_1$ thanks to 10.2.1 and 10.2.3. It remains only to observe that $\rho_{(A+B)} = \left((A + B) \cdot L^{n-1}\right)$ depends only on $\rho_L, \rho_K$, and $\rho_B$. □

*Proof of Theorem 10.2.4.* Fix an irreducible curve $C \subset X$. We will construct a diagram

$$
\begin{array}{ccccc}
C' & \subseteq & Y' & & \\
\downarrow & & \downarrow{\scriptstyle\mu} & & \\
C & \subseteq & Y & \subseteq & X
\end{array}
\qquad (10.1)
$$

consisting of:

- an irreducible subvariety $Y = Y_C$ of $X$ with $C \subseteq Y$;
- a resolution of singularities $\mu : Y' \longrightarrow Y$ of $Y$;
- an irreducible curve $C' \subseteq Y'$ mapping finitely onto $C$.

We will also produce an integer

$$\nu_p = \nu_p(\rho_L, \rho_K, \rho_B),$$

depending only on the dimension $p$ of $Y$ and the indicated parameters, in such a way that these data satisfy:

**Property 10.2.6.** There is a positive integer

$$k = k_Y \leq \nu_p$$

(depending on $Y$) such that $\mathcal{O}_{Y'}\left(\mu^*(kL - B)\right)$ has a section that does not vanish identically along $C'$.

Here and in the sequel we (somewhat abusively) write $\mu^*(kL - B)$ to denote the pullback to $Y'$ of the indicated divisor class on $X$.

Granting the existence of these data, it follows that

$$\left( (\nu_p L - B) \cdot C \right) \geq \left( (kL - B) \cdot C \right)$$

$$= \frac{1}{\deg(C' \longrightarrow C)} \cdot \left( \mu^*(kL - B) \cdot C' \right)$$

$$\geq 0.$$

Now set

$$M_0 = \max \{ \, \nu_p \mid 1 \leq p \leq n \, \}.$$

Then $\left( (mL - B) \cdot C \right) \geq 0$ whenever $m \geq M_0$, and the theorem will be proved.

For the construction of $Y$ and the diagram (10.1), the plan is to proceed inductively to build a chain

$$X = Y_n \supset Y_{n-1} \supset \cdots \supset Y_p = Y \supseteq C \qquad (10.2)$$

of subvarieties of $X$. Given $Y_i$, the idea roughly speaking is to apply the non-vanishing criterion (Corollary 9.4.24) to produce a non-zero section $s_i \in H^0(Y_i, \mathcal{O}_{Y_i}(kL - B))$ for suitable $k$. If $s_i$ doesn't vanish on $C$, then we set $Y = Y_i$. Otherwise we take $Y_{i-1}$ to be an irreducible component of $\mathrm{Zeroes}(s_i)$ that contains $C$, and continue.[4]

We package the application of Corollary 9.4.24 into a lemma.

**Lemma 10.2.7.** *Let $Y \subseteq X$ be an irreducible subvariety of $X$ having dimension $p$, and let*

$$\delta_Y = \left( A^p \cdot Y \right)$$

*be the degree of $Y$ with respect to the divisor $A$ fixed in 10.2.3. Set*

$$\nu(Y) = p \cdot \left( L^{p-1} \cdot (B + \delta_Y A) \cdot Y \right) + (p + 2), \qquad (10.3)$$

*and consider any resolution of singularities $\mu : Y' \longrightarrow Y$. Then there exists a positive integer $k_Y \leq \nu(Y)$ such that the pullback $\mathcal{O}_{Y'}\left( \mu^*(k_Y L - B) \right)$ has a non-zero section.*

---

[4] To circumvent some technical issues that arise if $Y_i$ is non-normal, we will actually pass to a resolution of singularities at each stage. We remark also that we fixed $C$ at the outset of the argument only in order to guide the choice of an irreducible component $Y_{i-1} \subseteq \mathrm{Zeroes}(s_i)$. As in [537] and [125] one could equivalently phrase the construction as producing a chain of algebraic subsets

$$X = W_n \supset W_{n-1} \supset \cdots \supset W_1,$$

each $W_p$ being a finite union $W_p = \cup Y_{p,\alpha}$ of $p$-dimensional irreducible varieties $Y_{p,\alpha}$ having the property that for some $m_\alpha \leq \nu_p$ the pullback of $m_\alpha L - B$ to a desingularization $Y'_{p,\alpha} \longrightarrow Y_{p,\alpha}$ has a non-zero section.

*Proof of Lemma 10.2.7.* Put

$$\nu' = \nu'(Y) = p \cdot \left(L^{p-1} \cdot (B + \delta_Y A) \cdot Y\right) + 1.$$

We observe first that if $k \geq \nu'$, then $\mathcal{O}_Y(kL - B - \delta_Y A)$ is big. In fact, by applying Siu's numerical criterion (Theorem 2.2.15) for the bigness of a difference to the restrictions $(kL)|Y$ and $(B + \delta_Y A)|Y$, we find that $\mu^*(kL - B - \delta_Y A)$ is big as soon as

$$k \cdot \left(L^p \cdot Y\right) \geq p \cdot \left(L^{p-1} \cdot (B + \delta_Y A) \cdot Y\right) + 1.$$

It follows from Corollary 9.4.24 that

$$H^0\left(Y', \mathcal{O}_{Y'}\left(K_{Y'} + \mu^*(kL - B - \delta_Y A)\right)\right) \neq 0$$

for some $k = k_Y \leq \nu' + p = \nu(Y) - 1$. We will prove that there exists an effective $\mu$-exceptional divisor $E$ on $Y'$ such that

$$H^0\left(Y', \mathcal{O}_{Y'}(\mu^*(\delta_Y A) - K_{Y'} + E)\right) \neq 0. \tag{*}$$

It follows that
$$H^0\left(Y', \mathcal{O}_{Y'}\left(\mu^*(kL - B) + E\right)\right) \neq 0,$$

and then 2.1.16 yields

$$H^0\left(Y', \mathcal{O}_{Y'}(\mu^*(kL - B))\right) \neq 0,$$

as required.

For (*), consider the embedding $Y \hookrightarrow \mathbf{P}$ defined by the complete linear series $|A|$, and take a general projection to $\mathbf{P}^{p+1}$ mapping $Y$ finitely and birationally onto a hypersurface of degree $\delta_Y$ in that projective space. Composing with $\mu$ we have a morphism

$$f : Y' \longrightarrow \mathbf{P}^{p+1}$$

that is birational onto its hypersurface image. Lemma 10.2.8 at the end of this section uses the double-point formula to produce effective divisors $D$ and $E$ on $Y'$, where $E$ is $f$-exceptional — and hence also $\mu$-exceptional — such that

$$(\delta_Y - p - 2)\mu^*(A) - K_{Y'} \equiv_{\mathrm{lin}} D - E.$$

In particular,

$$H^0\left(Y', \mathcal{O}_{Y'}\left((\delta_Y - p - 2)\mu^*(A) - K_{Y'} + E\right)\right) \neq 0,$$

and (*) follows since $\mu^*(A)$ itself is effective. $\qquad\square$

Returning to the proof of the theorem, we now begin the construction of the chain of subvarieties

$$X = Y_n \supset Y_{n-1} \supset \cdots \supset Y_p = Y \supseteq C$$

appearing in (10.2), together with integers $\nu_i = \nu_i(\rho_L, \rho_K, \rho_B)$ such that the pullback of $kL - B$ to a desingularization $\mu_i : Y_i' \longrightarrow Y_i$ has a section for some $k = k_{Y_i} \le \nu_i$. To this end, the first step is to apply the Lemma to $X$ itself. We deduce the existence of an integer

$$k \le \nu(X) = n \cdot \left( L^{n-1} \cdot (B + \delta_X A) \right) + (n+2)$$

such that $H^0(X, \mathcal{O}_X(kL - B)) \ne 0$. Now as it stands, $\nu(X)$ doesn't depend only on the intersection numbers $\rho_L, \rho_K$ and $\rho_B$ due to the presence of the term $\delta_X = (A^n)$. However, the Hodge-type inequalities of Section 1.6 yield

$$(A^n)(L^n)^{n-1} \le (A \cdot L^{n-1})^n$$

(Corollary 1.6.3 (ii)). So $(A^n)$ is (somewhat wastefully) bounded above by $(A \cdot L^{n-1})^n$, and hence we arrive at an integer $\nu_n = \nu_n(\rho_L, \rho_K, \rho_B)$ with the required dependence such that

$$H^0(X, \mathcal{O}_X(kL - B)) \ne 0 \quad \text{for some } k \le \nu_n.$$

Let $s_n \in H^0(X, \mathcal{O}_X(kL - B))$ be a non-zero section.

If $s_n$ doesn't vanish identically on our fixed curve $C \subset X$, then we get (10.1) by taking $Y = Y' = X$. So we may suppose that there is an irreducible component $Y_{n-1}$ of $\text{Zeroes}(s_n) \subset X$ that contains $C$. Fix a desingularization

$$\mu_{n-1} : Y_{n-1}' \longrightarrow Y_{n-1}$$

and an irreducible curve $C_{n-1}' \subseteq Y_{n-1}'$ lifting $C$. We then apply the lemma to $Y_{n-1}$ to conclude that there exists an integer

$$\begin{aligned} k = k_{Y_{n-1}} &\le \nu(Y_{n-1}) \\ &= (n-1) \cdot \left( L^{n-2} \cdot (B + \delta_{Y_{n-1}} A) \cdot Y_{n-1} \right) + (n+1) \end{aligned}$$

$$(10.4)$$

such that the pullback of $kL - B$ to $Y_{n-1}'$ has a non-zero section.

The next step is to produce an integer $\nu_{n-1} = \nu_{n-1}(\rho_L, \rho_K, \rho_B)$ bounding $\nu(Y_{n-1})$. To this end note that since $Y$ is a component of a divisor in $|kL - B|$, and since $A, L,$ and $B$ are nef, one has the inequalities

$$\begin{aligned} \delta_{Y_{n-1}} &= (A^{n-1} \cdot Y_{n-1}) \\ &\le (A^{n-1} \cdot (kL - B)) \\ &\le (A^{n-1} \cdot (\nu_n L - B)) \\ &\le \nu_n \cdot (A^{n-1} \cdot L) \\ &\le \nu_n \cdot (A \cdot L^{n-1})^{n-1}, \end{aligned}$$

where in the last line we have used the Hodge-type inequality

$$\left(A^{n-1} \cdot L\right)\left(L^n\right)^{n-2} \le \left(A \cdot L^{n-1}\right)^{n-1}$$

(Corollary 1.6.3 (i) with $q = 1$ and $p = n - 1$). Similarly,

$$\left(L^{n-1} \cdot Y_{n-1}\right) \le \nu_n \cdot \left(L^n\right),$$
$$\left(L^{n-2} \cdot B \cdot Y_{n-1}\right) \le \nu_n \cdot \left(L^{n-1} \cdot B\right),$$
$$\left(L^{n-2} \cdot A \cdot Y_{n-1}\right) \le \nu_n \cdot \left(L^{n-1} \cdot A\right).$$

Substituting each of these bounds for the corresponding term on the right in (10.4), and recalling that $A = \lambda_n K_X + \beta_n L$, we get the required bound $\nu_{n-1}(\rho_L, \rho_K, \rho_B)$.

Let

$$s_{n-1} \in H^0\left(Y'_{n-1}, \mathcal{O}_{Y'_{n-1}}\left(\mu^*_{n-1}(kL - B)\right)\right)$$

($k \le \nu_{n-1}$) denote the non-zero section produced by the lemma. If $s_{n-1}$ does not vanish identically on $C'_{n-1} \subseteq Y'_{n-1}$, then we take

$$Y' = Y_{n-1} \quad , \quad C' = C'_{n-1},$$

and we are done. Otherwise, choose an irreducible component $Y''_{n-2}$ of the zero-locus $\mathrm{Zeroes}(s_{n-1})$ that contains $C'_{n-1}$. Set $Y_{n-2} = \mu_{n-1}(Y''_{n-2})$, and fix a resolution $\mu_{n-1} : Y'_{n-2} \longrightarrow Y_{n-2}$ together with an irreducible curve $C'_{n-2} \subseteq Y'_{n-2}$ lifting $C$. The degree and intersection numbers of $Y_{n-2}$ with the relevant divisors are bounded just as above, allowing us to compute $\nu_{n-2}(\rho_L, \rho_K, \rho_B)$. The construction now continues step by step, and this completes the proof $\square$

Finally, we state and prove the lemma used in the course of 10.2.7.

**Lemma 10.2.8. (Birational double-point formula).** *Consider a morphism*

$$f : V \longrightarrow M$$

*of smooth irreducible projective varieties of dimensions $m$ and $m + 1$ respectively, and assume that $f$ maps $V$ birationally onto a (singular) hypersurface $W \subseteq M$. Then there exist effective divisors $D$ and $E$ on $V$, where $E$ is $f$-exceptional, such that*

$$f^*(K_M + W) - K_V \equiv_{\mathrm{lin}} D - E.$$

By definition, to say that $E$ is $f$-exceptional means that every irreducible component of $E$ is mapped by $f$ to a variety of dimension $\le m - 2$.

*Proof of Lemma 10.2.8.* This is a consequence of the general double-point formula of [208, Theorem 9.3], which produces a divisor class

$$\mathbf{D}(f) \equiv_{\mathrm{lin}} f^*(K_M + W) - K_V$$

computing the double-point cycle of $f$. It follows from the discussion at the top of page 166 of [208] that $\mathbf{D}(f)$ is represented by an effective cycle away from the exceptional locus of $f$, and the lemma follows.

Alternatively, one can argue directly using adjoint ideals (Section 9.3.E). In fact, the exact sequence (9.16) shows that there is an ideal sheaf $\mathfrak{a} \subseteq \mathcal{O}_W$ with the property that

$$\mathcal{O}_W(K_M + W) \otimes \mathfrak{a} = f_* \mathcal{O}_V(K_V).$$

Viewing $f$ as a map from $V$ to $W$, we pull this back to $V$ and take double duals. Recalling that a reflexive sheaf of rank one on a smooth variety is invertible ([488, II.1.1.15]), one finds that

$$f^* \mathcal{O}_W(K_M + W) \otimes \mathcal{O}_V(-D) = \left(f^* f_* \mathcal{O}_V(K_V)\right)^{**}$$

for some effective divisor $D$ on $V$. On the other hand, the natural map $f^* f_* \mathcal{O}_V(K_V) \longrightarrow \mathcal{O}_V(K_V)$ is surjective off the exceptional locus of $f$, and this map factors through $\left(f^* f_* \mathcal{O}_V(K_V)\right)^{**}$. Therefore

$$\left(f^* f_* \mathcal{O}_V(K_V)\right)^{**} = \mathcal{O}_V(K_V - E)$$

for some $f$-exceptional divisor $E$, and the lemma follows.    $\square$

**Remark 10.2.9.** At several points in his presentation [125], Demailly takes pains to estimate more carefully than we have here. The statement that comes out is that in the setting of Corollary 10.2.5 one can take

$$M_1 = (2n)^{(3^{n-1}-1)/2} \frac{\left(L^{n-1} \cdot (B+H)\right)^{(3^{n-1}+1)/2} \left(L^{n-1} \cdot H\right)^{3^{n-2}(n/2-3/4)-1/4}}{\left(L^n\right)^{3^{n-2}(n/2-1/4)-1/4}},$$

where $H = (n^3 - n^2 - n - 1)\left(K_X + (n+2)L\right)$. Siu obtains a somewhat stronger bound in [541].    $\square$

**Example 10.2.10. (Effective Serre vanishing).** In the situation of the theorem, there is an integer $M_2 = M_2(\rho_L, \rho_K)$ such that

$$H^i\left(X, \mathcal{O}_X(mL)\right) = 0 \quad \text{for} \quad i > 0$$

as soon as $m \geq M_2$. (Take $B$ as in the proof of Lemma 10.2.1, and apply 10.2.4 together with Kodaira vanishing.)    $\square$

## 10.3 Nakamaye's Theorem on Base Loci

We discuss a theorem of Nakamaye [468] describing numerically the stable base locus of (a small perturbation of) a nef and big divisor.

Let $X$ be a smooth projective variety of dimension $n$, and let $L$ be a big divisor on $X$. Recall (Definition 2.1.20) that the stable base locus of $L$ consists of those points appearing in the base locus $\operatorname{Bs}(|mL|)$ for all $m \gg 0$:

$$\mathbf{B}(L) \;=\; \bigcap_{m \geq 1} \operatorname{Bs}(|mL|).$$

Equivalently, $\mathbf{B}(L) = \operatorname{Bs}(|mL|)$ for sufficiently large and divisible $m$ (Proposition 2.1.21). In the evident way, one gives a natural meaning to $\mathbf{B}(D)$ for an arbitrary $\mathbf{Q}$-divisor $D$ (Example 2.1.24).

In general these loci can behave rather unpredictably: for example, $\mathbf{B}(L)$ does not depend only on the numerical class of $L$ (Example 10.3.3). However, Nakamaye realized that the situation becomes much cleaner if one perturbs $L$ slightly by subtracting off a small ample divisor.

**Lemma 10.3.1.** *Let $L$ be a big divisor, and $A$ an ample $\mathbf{Q}$-divisor, on $X$. Then the stable base locus $\mathbf{B}(L - \varepsilon A)$ is independent of $\varepsilon$ provided that $0 < \varepsilon \ll 1$. If $A'$ is a second ample divisor, then*

$$\mathbf{B}(L - \varepsilon A) \;=\; \mathbf{B}(L - \varepsilon' A') \quad \text{for } 0 < \varepsilon, \varepsilon' \ll 1.$$

*Moreover these loci depend only on the numerical class of $L$.*

**Definition 10.3.2. (Augmented base locus).** The *augmented base locus* of $L$ is the Zariski-closed set

$$\mathbf{B}_+(L) \;=\; \mathbf{B}(L - \varepsilon A)$$

for any ample $A$ and sufficiently small $\varepsilon > 0$. Thus

$$\mathbf{B}(L) \;\subseteq\; \mathbf{B}_+(L).$$

The augmented base locus of a $\mathbf{Q}$-divisor is defined analogously. If $\delta \in N^1(X)_{\mathbf{Q}}$ is a big divisor class, we write $\mathbf{B}_+(\delta)$ for the augmented base locus of any divisor in the given numerical equivalence class. $\qquad\square$

*Proof of Lemma 10.3.1.* Note first that if $0 < \varepsilon_1 < \varepsilon_2$ then evidently

$$\mathbf{B}(L - \varepsilon_1 A) \;\subseteq\; \mathbf{B}(L - \varepsilon_2 A).$$

Since closed subsets of $X$ satisfy the descending chain condition, the loci $\mathbf{B}(L - \varepsilon A)$ must therefore stabilize for sufficiently small $\varepsilon > 0$. For the second statement, fix $\delta > 0$ such that $A'' = A - \delta A'$ is ample. Then

$$\mathbf{B}(L - \varepsilon A) \;=\; \mathbf{B}(L - \varepsilon\delta A' - \varepsilon A'') \;\supseteq\; \mathbf{B}(L - \varepsilon\delta A').$$

Therefore $\mathbf{B}(L - \varepsilon' A') \subseteq \mathbf{B}(L - \varepsilon A)$ for $0 < \varepsilon, \varepsilon' \ll 1$, and the reverse inclusion follows by symmetry. Given a numerically trivial bundle $P$, take $A' = A - \frac{1}{\varepsilon}P$. Then $L + P - \varepsilon A = L - \varepsilon A'$, and so by what we just showed

$$\mathbf{B}(L + P - \varepsilon A) \;\supseteq\; \mathbf{B}(L - \varepsilon' A') \;=\; \mathbf{B}(L - \varepsilon A)$$

for $0 < \varepsilon, \varepsilon' \ll 1$. We get the reverse inclusion by replacing $L$ by $L - P$. $\qquad\square$

**Example 10.3.3. (Stable base loci on a ruled surface).** We give some examples using Cutkosky's construction (Section 2.3.B). Let $V$ be a smooth curve of genus $\geq 1$, and fix divisors $A$ and $P$ on $V$ of degrees 1 and 0 respectively. Consider the rank-two vector bundle

$$\mathcal{E} = \mathcal{E}_P = \mathcal{O}_V(P) \oplus \mathcal{O}_V(A+P)$$

on $V$. Put $X_P = \mathbf{P}(\mathcal{E})$ and let $L_P$ be a divisor on $X_P$ representing $\mathcal{O}_{\mathbf{P}(\mathcal{E})}(1)$. The bundles $\mathcal{E}_P$ differ only by a twist from bundle $\mathcal{E}_0$ corresponding to $P = 0$: $\mathcal{E}_p = \mathcal{E}_0 \otimes \mathcal{O}_V(P)$. Therefore the varieties $X_P$ are all naturally isomorphic to $X = X_0$. Making this identification, the $L_P$ form a family of numerically equivalent nef and big divisors on $X$.

There are now two cases to consider. If $P$ determines a torsion class in $\mathrm{Pic}^0(V)$, then $L_P$ is semiample thanks to Lemma 2.3.2 (iv). Therefore $\mathbf{B}(L_P) = \varnothing$ in this case. On the other hand, if $P$ is non-torsion then $\mathbf{B}(L_P)$ consists of the curve $C \subseteq X$ of negative self-intersection given by the inclusion $\mathbf{P}(\mathcal{O}_V(P)) \subseteq \mathbf{P}(\mathcal{E}_P)$. In either case, $\mathbf{B}_+(L_P) = C$.    □

Suppose henceforth that $L$ is nef as well as big, so that in particular $(L^n) > 0$ (Theorem 2.2.16). Then the restriction of $\mathcal{O}_X(L)$ to any subvariety $V \subseteq X$ is a nef bundle on $V$, but it may not be big, i.e. it may happen that $(L^{\dim V} \cdot V) = 0$.

**Definition 10.3.4. (Null locus of a nef and big divisor).** Given a nef and big divisor $L$ on $X$, the *null locus* $\mathrm{Null}(L) \subseteq X$ of $L$ is the union of all positive dimensional subvarieties $V \subseteq X$ with

$$(L^{\dim V} \cdot V) = 0. \tag{10.5}$$

This is a proper algebraic subset of $X$ (Lemma 10.3.6).    □

Nakamaye's theorem states that such null subvarieties account for every irreducible component of $\mathbf{B}_+(L)$:

**Theorem 10.3.5. (Nakamaye's theorem).** *If $L$ is any nef and big divisor on $X$, then*

$$\mathbf{B}_+(L) = \mathrm{Null}(L).$$

Note that the inclusion $\mathrm{Null}(L) \subseteq \mathbf{B}_+(L)$ is clear. Indeed, if $(L^{\dim V} \cdot V) = 0$ then the restriction $L \mid V$ of $L$ to $V$ lies on the boundary of the pseudoeffective cone $\overline{\mathrm{Eff}}(V)$ of $V$ (e.g. since $\mathrm{vol}_V(L|V) = 0$). Therefore given any ample divisor $A$, $(L - \varepsilon A) \mid V$ lies in the exterior of the pseudoeffective cone of $V$, and hence $V \subseteq \mathbf{B}(L - \varepsilon A)$ for every $\varepsilon > 0$. The essential (and surprising) content of Nakamaye's theorem is that conversely every irreducible component $V$ of $\mathbf{B}_+(L)$ must be as in (10.5).

Following [468], we start by verifying that $\mathrm{Null}(L)$ is Zariski-closed.

**Lemma 10.3.6. (Zariski-closedness of null locus).** *Let $L$ be a big and nef divisor on $X$. Then $\mathrm{Null}(L)$ is a Zariski-closed subset of $X$, and every irreducible component of $\mathrm{Null}(L)$ satisfies* (10.5). *Furthermore,* $\mathrm{Null}(L) \ne X$.

*Proof.* Let $V \subseteq X$ be any subvariety arising as the Zariski closure of the union of a collection $\{V_i\}_{i \in I}$ of subvarieties of $X$ having the property that

$$\left(L^{\dim V_i} \cdot V_i\right) \;=\; 0 \;\text{ for each }\; i \in I. \tag{*}$$

We will show that then $\left(L^{\dim V} \cdot V\right) = 0$, which implies the first statement of the lemma. $\mathrm{Null}(L)$ is then a proper subvariety of $X$ since $\left(L^{\dim X}\right) > 0$ ($L$ being big).

Suppose to the contrary that $\left(L^{\dim V} \cdot V\right) > 0$, so that $\mathcal{O}_V(L)$ is big. Then given an ample divisor $A$ there is an integer $p \gg 0$ such that $\mathcal{O}_V(pL - A)$ has a non-vanishing section. Let $W \subseteq V$ be the zero-locus of this section. Then for any subvariety $T \not\subseteq W$, the restriction $\mathcal{O}_T(pL - A)$ of this bundle to $T$ likewise has a non-zero section, and hence $\mathcal{O}_T(L)$ is big. Therefore all the $V_i$ appearing in (*) must actually lie in $W$. But in this case $V$ is not the Zariski closure of their union.  $\square$

*Proof of Theorem 10.3.5.* As explained following the statement of the theorem, the issue is to prove that $\mathbf{B}_+(L) \subseteq \mathrm{Null}(L)$. We begin by realizing $\mathbf{B}_+(L)$ as the zero-locus of a suitable multiplier ideal. To this end, fix first a very ample divisor $A$ such that $A - K_X$ is ample. Choose next integers $a, p \gg 0$ such that

$$\mathbf{B}_+(L) \;=\; \mathbf{B}(aL - 2A) \;=\; \mathrm{Bs}(|paL - 2pA|).$$

Now choose $n + 1$ general divisors $E_1, \dots, E_{n+1} \in |paL - 2pA|$, and set

$$D \;=\; \tfrac{n}{n+1} \cdot \left(E_1 + \dots + E_{n+1}\right).$$

Then $\mathrm{mult}_x(D) \ge n$ if $x \in \mathbf{B}_+(L)$ while $\mathrm{mult}_x(D) < 1$ for $x \in X - \mathbf{B}_+(L)$. Therefore Propositions 9.3.2 and 9.5.13 yield the set-theoretic equality

$$\mathrm{Zeroes}(\mathcal{J}(D)) \;=\; \mathbf{B}_+(L).$$

Setting $q = np$, we have $D \equiv_{\mathrm{num}} qaL - 2qA$. Recalling that $L$ is nef and $A - K_X$ is ample, Nadel vanishing (Theorem 9.4.8) implies

$$H^1\Big(X, \mathcal{O}_X(mL - qA) \otimes \mathcal{J}(D)\Big) \;=\; 0 \quad \text{for} \quad m \ge qa. \tag{10.6}$$

Write $Z \subseteq X$ for the subscheme defined by $\mathcal{J}(D)$. It follows from (10.6) that the restriction map

$$H^0\big(X, \mathcal{O}_X(mL - qA)\big) \longrightarrow H^0\big(Z, \mathcal{O}_Z(mL - qA)\big) \qquad (10.7)$$

is surjective for all $m \geq qa$.

We assert next that

$$\mathbf{B}_+(L) \;=\; \mathrm{Bs}\big(|mL - qA|\big) \quad \text{for} \quad m \geq qa. \qquad (10.8)$$

In fact, $\big((m-qa)L+qA\big)$ is 0-regular with respect to $A$ thanks to Kodaira vanishing and the nefness of $L$. Therefore $\big((m-qa)L+qA\big)$ is globally generated (Theorem 1.8.5), and hence

$$\mathrm{Bs}\big(|mL - qA|\big) \;\subseteq\; \mathrm{Bs}\big(|qaL - 2qA|\big) \;=\; \mathbf{B}\big(aL - 2A\big) \;=\; \mathbf{B}_+(L)$$

provided that $m \geq qa$. On the other hand, $\mathbf{B}_+(L) \subseteq \mathrm{Bs}\big(|mL - qA|\big)$ for $\frac{q}{m} \leq \frac{2}{a}$ by construction of the augmented base locus.

Now suppose for a contradiction that $\mathbf{B}_+(L)$ has an irreducible component $V$ that is not contained in $\mathrm{Null}(L)$. We view $V$ as a reduced subscheme of $X$. Then on the one hand, $\mathcal{O}_V(mL)$ is big and nef (Lemma 10.3.6). On the other hand, it follows from (10.8) that $V \subseteq \mathrm{Bs}\big(|mL - qA|\big)$ for all $m \gg 0$. In light of the surjectivity (10.7) it is enough for the required contradiction to establish that for sufficiently large $m$, the restriction map

$$H^0\big(Z, \mathcal{O}_Z(mL - qA)\big) \longrightarrow H^0\big(V, \mathcal{O}_V(mL - qA)\big)$$

is non-zero. To complete the proof we will establish more generally that for any fixed divisor $M$ on $X$, the restriction map

$$H^0\big(Z, \mathcal{O}_Z(mL + M)\big) \longrightarrow H^0\big(V, \mathcal{O}_V(mL + M)\big) \qquad (10.9)$$

is non-zero provided that $m \gg 0$.

To this end, start with a primary decomposition of $\mathcal{I}_Z = \mathcal{J}(D)$ (Example 9.6.14) to construct subschemes $Y, W \subseteq X$, with $Y_{\mathrm{red}} = V$, having the properties that

$$\mathcal{I}_Z \;=\; \mathcal{I}_Y \cap \mathcal{I}_W \quad \text{and} \quad V \cap W_{\mathrm{red}} \subsetneq V.$$

(Take $Y$ to be the subscheme defined by the $\mathcal{I}_V$-primary component of $\mathcal{I}_Z$, and $\mathcal{I}_W$ to be the intersection of the remaining ideals in the primary decomposition.) Then there is an exact sequence

$$0 \longrightarrow \mathcal{O}_Z \longrightarrow \mathcal{O}_Y \oplus \mathcal{O}_W \longrightarrow \mathcal{O}_{Y \cap W}.$$

Twisting by $\mathcal{O}_X(mL + M)$, we find that it suffices to produce a section

$$s \;\in\; \ker\Big( H^0\big(Y, \mathcal{O}_Y(mL + M)\big) \longrightarrow H^0\big(Y \cap W, \mathcal{O}_{Y \cap W}(mL + M)\big)\Big) \qquad (10.10)$$

such that $s_{\mathrm{red}} \in H^0\big(V, \mathcal{O}_V(mL + M)\big)$ is non-vanishing: for then $(s, 0)$ determines a section of $\mathcal{O}_Z(mL + M)$ restricting to $s_{\mathrm{red}}$, and we are done. To produce the required section $s$, the plan is to estimate the dimensions of the relevant cohomology groups, exploiting the assumption that $\mathcal{O}_V(L)$ is big.

Write $K_m \subseteq H^0\big(Y, \mathcal{O}_Y(mL + M)\big)$ for the kernel in (10.10) and set

$$K'_m = \ker\Big(H^0\big(Y, \mathcal{O}_Y(mL + M)\big) \longrightarrow H^0\big(V, \mathcal{O}_V(mL + M)\big)\Big).$$

We are required to show that $K_m \not\subseteq K'_m$ for $m \gg 0$. To this end note first that since the group on the right in (10.10) grows at most like $m^{\dim(Y \cap W)}$, the codimension of $K_m$ in $H^0\big(Y, \mathcal{O}_Y(mL + M)\big)$ is $O(m^{(\dim V - 1)})$. On the other hand, consider the exact sequence of ideal sheaves

$$0 \longrightarrow \mathcal{I}_{V/Y}(mL + M) \longrightarrow \mathcal{O}_Y(mL + M) \longrightarrow \mathcal{O}_V(mL + M) \longrightarrow 0.$$

Since $\mathcal{O}_V(L)$ is nef and big, it follows from asymptotic Riemann–Roch (Corollary 1.4.41) that $h^0\big(V, \mathcal{O}_V(mL + M)\big) \geq C \cdot m^{\dim V}$ for some positive constant $C$ and all $m \gg 0$. On the other hand, $h^1\big(Y, H^0\big(Y, \mathcal{O}_Y(mL + M)\big)\big) = O(m^{\dim V - 1})$ thanks to Theorem 1.4.40. Therefore codim $K'_m = C' \cdot m^{\dim V}$ for some $C' > 0$, and in particular it follows that $K'_m$ cannot contain $K_m$ for sufficiently large $m$, as required.    □

**Remark 10.3.7. (Results for arbitrary big divisors).** In his paper [469], Nakamaye studies the stable base loci of big linear series that may not be nef. Given a big divisor $L$, he introduces the "moving Seshadri constant" $\varepsilon(\|L\|; x)$, which measures asymptotically how many jets the linear series $|kL|$ separate at a point $x$ (compare Theorem 5.1.17). Nakamaye proves roughly speaking that one can characterize the irreducible components $V$ of $\mathbf{B}_+(L)$ in terms of the behavior of $\varepsilon(\|L\|; x)$ for variable $L$ and general $x \in V$.    □

**Remark 10.3.8. (Asymptotic invariants of base loci).** Motivated by Nakamaye's papers [468] and [469], as well as an unpublished preprint [472] of Nakayama, the paper [157] of Ein, Nakamaye, Mustaţă, Popa, and the author undertakes a systematic exploration of how the stable base locus $\mathbf{B}(D)$ varies with $D$. Some natural asymptotically defined invariants are introduced, and — by analogy with Theorem 2.2.44 — are shown to vary continuously with $D$. Some related questions are studied by Boucksom [67] from an analytic perspective.    □

# 10.4 Global Generation of Adjoint Linear Series

This section revolves around some conjectures and results concerning the geometry of adjoint linear series. We start in the first subsection with Fujita's conjectures and their background, and then state the theorem of Angehrn and

Siu. The second subsection is devoted to some technical results concerning loci of log-canonical singularities. These are applied in Section 10.4.C to prove the Angehrn–Siu theorem.

### 10.4.A Fujita's Conjecture and the Theorem of Angehrn–Siu

Consider a smooth projective curve $C$ of genus $g$, and suppose that $D$ is a divisor of degree $d$ on $C$. A classical (and elementary) theorem states that if $d \geq 2g$ then $|D|$ is free, while if $d \geq 2g + 1$ then $D$ is very ample. It is natural to ask whether this extends to smooth varieties of arbitrary dimension.

Interestingly enough, the shape that such a statement might take came into focus only in the late 1980s. The starting point is to rephrase the result for curves without explicitly mentioning the genus $g$ of $C$. Specifically, note that

$$\deg D \geq 2g \iff D \equiv_{\mathrm{num}} K_C + L$$

where $L$ is an ample divisor of degree $\geq 2$, and similarly $\deg D \geq 2g + 1$ if and only if $D$ admits the same expression with $\deg L \geq 3$. This suggests that bundles of large degree on a curve generalize to *adjoint divisors* on a smooth projective variety $X$ of arbitrary dimension, i.e. divisors of the form $K_X + L$ for a suitably positive divisor $L$ on $X$.

These considerations are made precise in a conjecture due to Fujita:

**Conjecture 10.4.1. (Fujita's conjecture).** Let $X$ be a smooth projective variety of dimension $n$, and $A$ an ample divisor on $X$. Then

(i). If $m \geq n + 1$ then $K_X + mA$ is free.

(ii). If $m \geq n + 2$ then $K_X + mA$ is very ample.

The case of the hyperplane divisor on $X = \mathbf{P}^n$ shows that the statement would be sharp.[5] Recall that it follows from Mori's cone theorem that in any event the divisors appearing in (i) and (ii) are nef and ample respectively (Example 1.5.35).

As we have seen the statement is classical in dimension one, and for surfaces it follows from a more precise result due to Reider [518] (cf. [394]). The first serious attack in higher dimension was mounted by Demailly [124], who used sophisticated analytic methods involving the Monge–Ampère equation to show for instance that if $X$ and $A$ are as in 10.4.1, then $2K_X + 12n^n A$ is very ample.[6] While the statement is rather far from what one expects, Demailly's paper was very influential, and it initiated an active body of work aimed at establishing effective results for higher-dimensional varieties. It also

---

[5] See for instance [364, Example 5.3.3] for other examples on the boundary of Fujita's conjectures.

[6] An algebro-geometric analogue of Demailly's method appears in [158].

introduced multiplier ideals for the first time as a tool in the study of linear series.

The next step was taken by Ein and the author in [151]. This paper brought the techniques of Kawamata, Shokurov, and Reid into the picture, and used them to prove 10.4.1 (i) in dimension three. Arguably the biggest progress to date occurred in [10], where Angehrn and Siu combined the Kawamata–Shokurov–Reid methods in [151] with some new ideas involving multiplier ideals to prove a quadratic result for global generation in all dimensions:

**Theorem 10.4.2. (Theorem of Angehrn and Siu).** *Let $X$ be a smooth projective variety of dimension $n$, and let $L$ be an ample divisor on $X$. Fix a point $x \in X$, and assume that*

$$\left( L^{\dim Z} \cdot Z \right) > \binom{n+1}{2}^{\dim Z}$$

*for every irreducible subvariety $Z \subseteq X$ passing through $x$ (including of course $X$ itself). Then $K_X + L$ is free at $x$, i.e. $\mathcal{O}_X(K_X + L)$ has a section that doesn't vanish at $x$.*

In particular, if $L \equiv_{\mathrm{num}} mA$ for some $m > \binom{n+1}{2}$ and some ample $A$, then $K_X + L$ is free. However it is characteristic of essentially all the work on these questions starting with Demailly's paper [124] that $L$ is not actually required to be a multiple of some other divisor: rather one imposes quantitative hypotheses of Nakai-type.

The argument of Angehrn and Siu was algebrized by Kollár [364], who showed moreover that it suffices to assume in the statement that $L$ is nef and big. Kollár also gives a statement allowing $X$ to have mild singularities. Further improvements were made by Helmke in [285], [286], and Kawamata [320] subsequently proved 10.4.1 (i) in dimension four.

The next two subsections contain a proof of Theorem 10.4.2 following [364, §6]. Although 10.4.2 is not the strongest known statement, the argument gives a good idea of how one can use multiplier ideals to attack problems of this sort.

We conclude this discussion by stating a few variants and generalizations.

**Remark 10.4.3. (Very ample adjoint series).** The papers [10], [364] give statements analogous to 10.4.2 allowing one to conclude that $|K_X + L|$ separates a finite number of distinct points. However, existing methods do not seem very adept at separating tangent vectors, and for this reason much less is known about when $K_X + L$ is very ample. Siu gives a statement with exponential bounds in [538]. $\quad\quad\quad\square$

**Remark 10.4.4. (Generalized Fujita conjecture).** One can ask for an analogue of Fujita's Conjecture 10.4.1 involving only numerical hypotheses. The generally accepted statement here (cf. [364, Conjecture 5.4]) is that if $L$ is a nef and big line bundle on a smooth projective $n$-fold $X$ that satisfies

$$\left(L^{\dim Z} \cdot Z\right) \; > \; n^{\dim Z}$$

for every irreducible subvariety $Z$ passing through a fixed point $x \in X$, then $\mathcal{O}_X(K_X + L)$ should have a section that doesn't vanish at $x$.[7] Kawamata [325, Conjecture 1.4] has proposed an analogous conjecture in which $K_X$ is replaced by any of the direct images $R^q f_* \mathcal{O}_Y(K_Y)$ of the canonical bundle of $Y$ under a morphism $f : Y \longrightarrow X$ of non-singular projective varieties that is smooth on the complement of an SNC divisor $B \subseteq X$.                              □

**Remark 10.4.5. (Adjoint bundles in positive characteristics).** If $B$ is ample and *free* on a smooth complex $n$-fold $X$, then the global generation of $K_X + (n+1)B$ follows easily from Kodaira vanishing (Example 1.8.23). When $X$ is instead a variety defined over an algebraically closed field of characteristic $p > 0$ the question becomes much more subtle. In this case the statement was established by K. E. Smith in [543] using tight closure methods.

However when $B$ is *very* ample, Viehweg points out that one can give a quick proof, valid in all characteristics, as follows. Consider the embedding

$$X \; \subseteq \; \mathbf{P} \; = \; \mathbf{P}H^0\big(X, \mathcal{O}_X(B)\big)$$

defined by $B$, and let $N = N_{X/\mathbf{P}}$ denote the normal bundle to $X$ in $\mathbf{P}$. Then $N(-1) = N \otimes B^*$ is globally generated (since $T_{\mathbf{P}}(-1)$ is). But this implies that

$$\det N(-1) \; = \; \mathcal{O}_X(K_X + (n+1)B)$$

is likewise free.                              □

**Remark 10.4.6. (Fujita's conjecture on toric varieties).** The toric case of Fujita's conjecture has been studied by several authors, notably Fujino [190], Mustaţă [459], and Payne [498]. On smooth toric varieties, a torus-invariant divisor is nef (or ample) if and only it is globally generated (or very ample), so the toric analogue of 10.4.1 follows as in Example 1.8.23 from Mori's cone theorem in the form of Example 1.5.35. The cited papers give quite precise results in the case of singular toric varieties: here the picture is more subtle.                              □

## 10.4.B Loci of Log-Canonical Singularities

In this subsection we establish some terminology and auxiliary results that will be helpful in the proof of the theorem of Angehrn and Siu. We follow the approach of [364]; slightly different arguments appear in [145].

Let $X$ be a smooth variety and $D$ an effective **Q**-divisor on $X$. Given a point $x \in X$ contained in the support of $D$, recall (Definition 9.3.12) that the log-canonical threshold of $D$ at $x$ is defined to be

---

[7] If $Z \subsetneq X$ is a proper subvariety, then it is expected to be enough to require that
$\left(L^{\dim Z} \cdot Z\right) \geq n^{\dim Z}$.

$$\operatorname{lct}(D;x) \;=\; \inf\big\{\, c>0 \mid \mathcal{J}(c \cdot D) \text{ is non-trivial at } x \,\big\}. \qquad (10.11)$$

Thus if $c = \operatorname{lct}(D;x)$, then $\mathcal{J}(X, c \cdot D)$ is a radical ideal in a neighborhood of $x$ (Example 9.3.18).

**Definition 10.4.7. (Locus of log-canonical singularities).** The *locus of log-canonical singularities* of $D$ at $x$ — written

$$\operatorname{LC}(D;x) \;\subseteq\; X$$

and sometimes called the *LC-locus* of $D$ at $x$ for short — is the union of all irreducible components of the zero-locus

$$\operatorname{Zeroes}\big(\mathcal{J}(c \cdot D)\big) \;\subseteq\; X$$

that pass through $x$, where $c = \operatorname{lct}(D;x)$ We view $\operatorname{LC}(D;x)$ as a reduced subscheme of $X$, and by construction $x \in \operatorname{LC}(D;x)$. $\qquad\square$

It is often useful to be able to assume that the LC-locus of a divisor $D$ arising in the course of an argument is actually irreducible at $x$. The next lemma asserts that one can achieve this after an arbitrarily small perturbation of $D$.

**Lemma 10.4.8. (Making the LC-locus irreducible).** *Let $X$ be a non-singular variety, $x \in X$ a fixed point and $D$ an effective $\mathbf{Q}$-divisor whose support contains $x$. Then there exists an effective $\mathbf{Q}$-divisor $E$ such that*

$$\operatorname{LC}\big((D+tE);x\big) \;\subseteq\; X$$

*is irreducible at $x$ for all $0 < t \ll 1$. Moreover if $X$ is projective and $L$ is any ample divisor, then one can take $E \equiv_{\mathrm{num}} pL$ for some $p > 0$.*

*Proof.* Assuming first that $X$ is affine, we argue by decreasing induction on the dimension of $Z = \operatorname{LC}(D;x)$. If $\dim Z = 0$ there is nothing further to prove. Otherwise fix an irreducible component $Z_1$ of $Z$ of minimal dimension that passes through $x$. Pick a general element $f \in \mathcal{I}_{Z_1}$ in the ideal of $Z_1$, and set

$$E \;=\; \operatorname{div}(f).$$

We suppose in particular that $E$ does not contain any other component of $Z$. Then for fixed $t > 0$, considerations of discrepancies as in Example 9.3.16 show that the multiplier ideal

$$\mathcal{J}\big(X, c \cdot (D+tE)\big)$$

still vanishes on $Z_1$ whenever $c < \operatorname{lct}(D;x)$ is sufficiently close to $\operatorname{lct}(D;x)$. But by construction this ideal does not vanish on any other components of $Z$. Therefore either

$$\mathrm{LC}\big((D+tE)\,;\,x\big) \;=\; Z_1 \quad \text{for all } 0 < t \ll 1,$$

or else these LC-loci are all contained in some fixed proper algebraic subset $Z' \subseteq Z_1$ passing through $x$ (in which case we conclude by induction). The global case is similar: one first fixes $p \gg 0$ such that $\mathcal{I}_{Z_1} \otimes \mathcal{O}_X(pL)$ is globally generated, and then takes

$$f \;\in\; \Gamma\big(X, \mathcal{I}_{Z_1} \otimes \mathcal{O}_X(pL)\big)$$

to be a general section.    □

**Remark 10.4.9. (Kawamata's subadjunction theorem).** Suppose that $D$ is an effective **Q**-divisor on a smooth projective variety $X$, and let $Z = \mathrm{LC}(D; x)$. One says that $Z$ is a *minimal center of log canonical singularities* if $Z$ is irreducible and if in addition it satisfies a minimality condition guaranteeing roughly speaking that it cannot be further cut down by the process of Lemma 10.4.8.[8] In his fundamental paper [322], Kawamata proves a deep and remarkable theorem about the structure of such $Z$. Specifically, assuming (without loss of generality) that $\mathrm{lct}(D; x) = 1$, Kawamata shows that given any ample divisor $H$ on $X$ and any rational number $\varepsilon > 0$, there exists an effective **Q**-divisor $D_Z$ on $Z$ such that

$$\big(K_X + D + \varepsilon H\big) \,|\, Z \;\equiv_{\mathrm{num}}\; K_Z + D_Z,$$

and $(Z, D_Z)$ is KLT. In particular, $Z$ has only rational singularities.    □

For the study of adjoint series, one begins by constructing a divisor $D$ whose LC-locus $Z = \mathrm{LC}(D; x)$ passes through a given point $x$, and the essential difficulty is to deal with the possibility that $Z$ has dimension $\geq 1$. Angehrn and Siu developed a method by which one can "cut down" the dimension of the locus in question. The idea is to start with an effective **Q**-divisor $D$ passing through $x$, and a general point $y \in Z = \mathrm{LC}(D; x)$. The following result shows that if $B$ is an effective **Q**-divisor on $X$ such that the restriction $B_Z = B \,|\, Z$ has high multiplicity at $y$, then $\mathcal{J}\big(X, (1 - \varepsilon)D + B\big)$ is non-trivial at $y$ for $\varepsilon \ll 1$. Under mild hypotheses, the LC-locus of this divisor is then a proper subvariety of $Z$:

**Proposition 10.4.10. (Cutting down the LC locus).** *Let $X$ be a smooth variety and $x \in X$ a fixed point. Consider an effective **Q**-divisor $D$ on $X$ with $\mathrm{lct}(D; x) = 1$, and suppose that*

$$Z \;=_{def}\; \mathrm{LC}(D; x)$$

*is irreducible of dimension $d$ at $x$. Fix a general smooth point $y \in Z$, and let $B$ be any effective **Q**-divisor on $X$, with $Z \not\subseteq \mathrm{Supp} B$. Assume that*

$$\mathrm{mult}_y\big(Z; B_Z\big) \;>\; d \;=\; \dim Z,$$

---

[8] We refer to [321] or [322] for the precise definitions and statements.

*where $B_Z$ denotes the restriction of $B$ to $Z$, and as indicated the multiplicity is computed on $Z$. Then for $0 < \varepsilon \ll 1$,*

$$\mathcal{J}\big( X , (1 - \varepsilon)D + B \big)$$

*is non-trivial at $y$. If moreover $\mathcal{J}(X, D + B) = \mathcal{J}(X, D)$ away from $Z$, then*

$$Z' =_{\mathrm{def}} \mathrm{Zeroes}\big(\mathcal{J}\big( (1 - \varepsilon)D + B \big)\big)$$

*is a proper algebraic subset of $Z$ in a neighborhood of $y$.*

**Remark 10.4.11.** For the application to global generation of adjoint series, one needs to arrange that in fact $\mathcal{J}\big( X , (1 - \varepsilon)D + B \big)$ vanishes at the given point $x$. This will be achieved in the next section by first choosing $B = B(y)$ as in the proposition, and then taking the limit of these divisors as $y$ approaches $x$.    □

*Proof of Proposition 10.4.10.* The function $z \mapsto \mathrm{lct}(D; z)$ is Zariski lower-semicontinuous (Example 9.3.17), so the hypothesis $\mathrm{lct}(D; x) = 1$ implies that $\mathrm{lct}(D; y) = 1$ for general $y \in Z$. On the other hand, since $Z \not\subseteq \mathrm{Supp}B$, the multiplier ideal $\mathcal{J}\big( (1 - \varepsilon)D + B \big)$ cannot vanish everywhere along $Z$. So if $\mathcal{J}(D + B) = \mathcal{J}(D)$ off $Z$, then the zero-locus of $\mathcal{J}\big( (1 - \varepsilon)D + B \big)$ is a proper subset of $Z$.

The essential point is therefore to prove that this ideal is non-trivial at $y$. To this end, we start by fixing some notation. Let $\mu : X' \longrightarrow X$ be a log-resolution of $D$. Thus there exists a prime divisor $E$ on $X'$, with $\mu(E) = Z$, such that

$$\mathrm{ord}_E\big(K_{X'/X} - \mu^* D\big) = -1.$$

Write

$$K_{X'/X} = qE + K' \quad , \quad \mu^* D = (q + 1)E + D',$$

where $K'$ and $D'$ are effective divisors on $X'$ that do not contain $E$. Denote by $\mu_Z : E \longrightarrow Z$ the restriction of $\mu$ to $E$, and choose $y \in Z$ such that $\mu_Z$ is smooth over a neighborhood of $y$. Put $F = F_y = \mu_Z^{-1}(y)$, so that $F \subseteq E$ is a smooth subvariety of codimension $d = \dim Z$ in $E$. The situation is summarized in the following diagram:

$$
\begin{array}{ccccc}
F & \subseteq & E & \subseteq & X' \\
\downarrow & & \downarrow{\scriptstyle \mu_Z} & & \downarrow{\scriptstyle \mu} \\
\{y\} & \in & Z & \subseteq & X.
\end{array}
$$

By taking $y \in Z$ sufficiently general, we can suppose in addition that $F$ is not contained in the support of $D'$ or of $K'$.

Now fix any positive $\varepsilon \ll 1$. The birational transformation rule (Proposition 9.2.33) gives:

$$\mathcal{J}\big(X, (1-\varepsilon)D + B\big) \;=\; \mu_*\mathcal{J}\big(X', \mu^*((1-\varepsilon)D + B) - K_{X'/X}\big).$$

So it is enough to show that the multiplier ideal appearing on the right is non-trivial at a general point $z \in F$. Now

$$\mu^*\big((1-\varepsilon)D + B\big) - K_{X'/X} \;=\; (1-\varepsilon')E + \mu^*B + (1-\varepsilon)D' - K',$$

with $\varepsilon' = \varepsilon(q+1)$. Since $F$ is not contained in the supports of $D'$ or $K'$, it is sufficient to show that

$$\mathcal{J}\big(X', (1-\varepsilon')E + \mu^*B\big)_z \;\not\subseteq\; \mathcal{O}_z X' \tag{*}$$

at a general point $z \in F$ for some — and consequently for all — small $\varepsilon' > 0$.

To this end the essential point is to apply Corollary 9.5.17. Taking $H = E$ in that statement, one finds that if the multiplier ideal appearing on the left in (*) is trivial for all $\varepsilon' \ll 1$, then

$$\mathcal{J}\big(E, (1-t)(\mu^*B)|E\big)_z \;=\; \mathcal{O}_z E \tag{**}$$

is likewise trivial for all $t \ll 1$. But $(\mu^*B)\,|\,E = \mu_Z^*(B_Z)$ has multiplicity $> d$ at every point of $F$. Hence $(1-t)(\mu^*B)$ still has multiplicity $\ge d = \mathrm{codim}_E F$ along $F$ for $t \ll 1$. Therefore it follows from 9.3.5 that the ideal $\mathcal{J}\big(E, (1-t)(\mu^*B)|E\big)_z$ appearing in (**) is indeed non-trivial, and we are done.  □

Finally, we recall for convenient reference a special case of Proposition 1.1.31, which will allow us to construct divisors having high multiplicity at a point.

**Lemma 10.4.12. (Constructing singular divisors).** *Let $V$ be an irreducible projective variety of dimension $d$, $L$ an ample divisor on $V$, and $x \in V$ a smooth point. Assume that*

$$\big(L^d\big) \;>\; \alpha^d$$

*for some positive rational number $\alpha$. Then for $k \gg 0$ there exists a divisor*

$$A \;=\; A_x \;\in\; |kL| \quad with \quad \mathrm{mult}_x(A) \;>\; k\alpha.$$

*Moreover we can take $k$ to be independent of the smooth point $x$.*  □

## 10.4.C Proof of the Theorem of Angehrn and Siu

We turn now to the proof of Theorem 10.4.2. In an effort to highlight the essential geometric ideas, we will keep the presentation somewhat informal. Fully detailed proofs appear in the original paper [10] of Angehrn and Siu, and in Kollár's notes [364].

*Set-up and plan.*

We fix once and for all a point $x \in X$ and an ample divisor $L$ satisfying the numerical hypothesis of 10.4.2. The plan is to construct step by step an effective $\mathbf{Q}$-divisor $D$ on $X$, with

$$D \equiv_{\text{num}} \lambda L \quad \text{for some } \lambda < 1,$$

such that $\mathrm{lct}(D; x) = 1$ and $x$ is an isolated point of the LC-locus $\mathrm{LC}(D; x)$. Granting the existence of $D$, consider the scheme $Z \subseteq X$ defined by the multiplier ideal $\mathcal{J}(D)$. Then

$$H^1\big(X, \mathcal{O}_X(K_X + L) \otimes \mathcal{J}(D)\big) \; = \; 0$$

thanks to Nadel vanishing, and consequently the restriction map

$$H^0\big(X, \mathcal{O}_X(K_X + L)\big) \longrightarrow H^0\big(Z, \mathcal{O}_Z(K_X + L)\big)$$

is surjective. On the other hand, since $x$ is an isolated point of $Z$, there is a natural surjection

$$H^0\big(X, \mathcal{O}_X(K_X + L)\big) \longrightarrow H^0\big(Z, \mathcal{O}_Z(K_X + L) \otimes \mathbf{C}(x)\big),$$

$\mathbf{C}(x)$ denoting the one-dimensional sky-scraper sheaf at $x$. All told, the mapping

$$\mathrm{eval}_x \; : \; H^0\big(X, \mathcal{O}_X(K_X + L)\big) \longrightarrow H^0\big(X, \mathcal{O}_X(K_X + L) \otimes \mathbf{C}(x)\big)$$

determined by evaluation at $x$ is surjective, and the theorem follows.

*Step 1. Creating the initial divisor.*

For the remainder of the proof, write $M = \binom{n+1}{2}$. We start by producing an effective $\mathbf{Q}$-divisor $D_1$ on $X$, with

$$D_1 \equiv_{\text{num}} \lambda_1 L \quad \text{for some} \quad \lambda_1 < \frac{n}{M},$$

such that $\mathrm{lct}(D_1; x) = 1$ and $\mathrm{LC}(D_1; x)$ is irreducible at $x$. In fact,

$$\left( \left( \tfrac{n}{M} L \right)^n \right) > n^n$$

by hypothesis, so for sufficiently divisible $k \gg 0$, Lemma 10.4.12 guarantees the existence of a divisor

$$A \in \big| \tfrac{kn}{M} L \big| \quad \text{with} \quad \mathrm{mult}_x A > kn.$$

Then $\mathrm{lct}(A; x) < \tfrac{1}{k}$, and by perturbing $A$ slightly one can assume in addition that $\mathrm{LC}(A; x)$ is irreducible (Lemma 10.4.8). We then take

$$D_1 \; = \; \mathrm{lct}(A; x) \cdot A,$$

and put $Z_1 = \mathrm{LC}(D_1; x)$, $d_1 = \dim Z_1$.

*Step 2. Constructing a family of divisors on $Z_1$.*

To lighten notation, write

$$Z = Z_1 \quad , \quad d = d_1 = \dim Z_1 \quad , \quad \text{and} \quad \overline{L} = L \mid Z.$$

Consider to begin with a smooth point $y \in Z$. The numerical hypothesis of the theorem implies that $\left( \left( \frac{d}{M} \overline{L} \right)^d \right) > d^d$. So we can invoke 10.4.12 on $Z$ to produce a suitably divisible integer $k \gg 0$ (independent of $y$) plus a divisor

$$\overline{A}_y \in \left| \frac{kd}{M} \overline{L} \right| \quad \text{with} \quad \mathrm{mult}_y \overline{A}_y > kd, \tag{10.12}$$

where here we are dealing with linear series and multiplicities on $Z$. We also suppose that we've chosen $k$ sufficiently large so that the restriction map

$$H^0 \left( X, \mathcal{O}_X \left( \tfrac{kd}{M} L \right) \right) \longrightarrow H^0 \left( Z, \mathcal{O}_Z \left( \tfrac{kd}{M} \overline{L} \right) \right) \tag{10.13}$$

is surjective, and so that moreover,

the twisted ideal sheaf $\mathcal{I}_{Z/X} \left( \tfrac{kd}{M} L \right)$ is globally generated. (10.14)

One would like to produce an analogous divisor based at $x$ itself. However, since $Z$ may be singular at $x$, the construction in (10.12) does not work directly. The idea of Angehrn and Siu is to use instead a limiting process.

Specifically, let $T$ be a smooth curve and $u : T \longrightarrow Z$ a morphism with $u(0) = x$ for some point $0 \in T$. Writing

$$y_t = u(t) \in Z,$$

we suppose that $y_t$ is a smooth point of $Z$ for $t \neq 0$, and that it is sufficiently general so that Proposition 10.4.10 applies. We can also assume that for $t \neq 0$ the divisors $\overline{A}_t =_{\mathrm{def}} \overline{A}_{y_t}$ constructed in (10.12) also vary in a flat family parametrized by $T - \{0\}$. Then set

$$\overline{A}_0 = \lim_{t \to 0} \overline{A}_t.$$

Thus $\overline{A}_0$ is an effective Cartier divisor on $Z$ that lies in the linear series $\left| \frac{kd}{M} \overline{L} \right|$ and passes through $x$.

*Step 3. Lifting the $\overline{A}_t$ and cutting down $Z_1$.*

In this step we will produce an effective **Q**-divisor $D_2$ on $X$, with

$$D_2 \equiv_{\mathrm{num}} \lambda_2 L \quad \text{for some} \quad \lambda_2 < \frac{n}{M} + \frac{n-1}{M}, \tag{10.15}$$

such that $\mathrm{lct}(D_2; x) = 1$ and $Z_2 = \mathrm{LC}(D_2; x) \subsetneq Z$ is a proper irreducible subvariety of $Z$.

We start by producing a family of divisors $A_t \in |\frac{kd}{M}L|$ on $X$ lifting the $\overline{A}_t$ constructed in Step 2. First, by the surjectivity of (10.13) we may lift $\overline{A}_0$ to a divisor $A_0$ on $X$ which does not contain $Z$. Thanks to (10.14), the possible choices of $A_0$ form a free linear series off $Z$. So by Example 9.2.29, we may assume moreover that $\mathcal{J}(D_1 + cA_0) = \mathcal{J}(D_1)$ away from $Z$ for every $0 < c < 1$. Now for $t \in T$ near 0, we extend $A_0$ to a flat family of divisors

$$A_t \in |\tfrac{kd}{M}L|$$

lifting $\overline{A}_t$.

Set $B_t = \frac{1}{k}A_t$ and $\overline{B}_t = B_t \,|\, Z$. It follows from (10.12) that

$$\mathrm{mult}_{y_t}\overline{B}_t > d = \dim Z$$

for $t \neq 0$. So Proposition 10.4.10 shows that

$$y_t \in \mathrm{Zeroes}\big(\mathcal{J}\big((1-\varepsilon)D_1 + B_t\big)\big)$$

for $t \neq 0$ and $\varepsilon \ll 1$. By the semi-continuity of multiplier ideals (Corollary 9.5.39) the analogous statement holds at $x = y_0$:

$$x \in \mathrm{Zeroes}\big(\mathcal{J}\big((1-\varepsilon)D_1 + B_0\big)\big),$$

i.e. the log-canonical threshold of this divisor satisfies

$$c_2 =_{\mathrm{def}} \mathrm{lct}\big((1-\varepsilon)D_1 + B_0; x\big) \leq 1.$$

Put
$$D_2 = c_2 \cdot \big((1-\varepsilon)D_1 + B_0\big).$$

By construction $\mathcal{J}(D_1 + B_0) = \mathcal{J}(D_1)$ away from $Z$, and therefore $\mathrm{LC}(D_2; x)$ is a proper subset of $Z$ containing $x$. Replacing $D_2$ by a slight perturbation if necessary we can assume also that this LC-locus is irreducible at $x$.

It remains only to show that $D \equiv_{\mathrm{num}} \lambda_2 L$ where $\lambda_2$ satisfies the inequality on the right in (10.15). But

$$\lambda_2 = c_2 \cdot \left((1-\varepsilon)\lambda_1 + \frac{d}{M}\right)$$
$$< \frac{n}{M} + \frac{d}{M}$$
$$\leq \frac{n}{M} + \frac{n-1}{M},$$

and the required divisor is at hand.

*Step 4. Completion of the Proof.*

We now repeat this construction: after at most $n$ steps one arrives at an effective **Q**-divisor $D_n \equiv_{\mathrm{num}} \lambda_n L$ with

$$\lambda_n < \frac{n}{M} + \frac{n-1}{M} + \ldots + \frac{1}{M} = 1,$$

so that $\mathrm{lct}(D_n; x) = 1$ and $x$ is isolated in $\mathrm{LC}(D_n; x)$. As explained at the outset of the argument, we are then done.

## 10.5 The Effective Nullstellensatz

It was shown by Ein and the author in [155] — and independently observed by Hickel [287] — that one can use Skoda's theorem and its variants (Section 9.6) to establish geometric statements in the direction of the effective Nullstellensatz. The present section is devoted to an exposition of this circle of ideas, following quite closely the discussion in [155].

Let $X$ be a non-singular complex quasi-projective variety of dimension $n$. Fix a non-zero ideal sheaf $\mathfrak{a} \subseteq \mathcal{O}_X$, and denote by

$$Z = \mathrm{Zeroes}(\mathfrak{a})_{\mathrm{red}} \subseteq X$$

the zero-locus of $\mathfrak{a}$, considered as a *reduced* subscheme of $X$. Then Hilbert's Nullstellensatz states that the ideal $\mathcal{I}_Z$ of $Z$ is the radical of $\mathfrak{a}$, so that $\mathcal{I}_Z^s \subseteq \mathfrak{a}$ for some large integer $s$. The first results we discuss address the problem of finding an effective bound on what power $s$ is required here. We begin by deriving an "abstract" local statement (Theorem 10.5.3) involving a further invariant of $\mathfrak{a}$. This leads to more concrete bounds (Corollary 10.5.6, Theorem 10.5.8) when $X$ is projective.

**Remark 10.5.1. (The "classical" effective Nullstellensatz).** We reproduce from [155] a brief account of work on the effective Nullstellensatz in the case of polynomial rings. Consider polynomials

$$g_1, \ldots, g_m \in \mathbf{C}[t_1, \ldots, t_n],$$

having no common zeroes in $\mathbf{C}^n$. Then one version of the classical Nullstellensatz states that the $g_j$ generate the unit ideal, i.e. that there exist $h_j \in \mathbf{C}[t_1, \ldots, t_n]$ such that

$$\sum h_j g_j = 1. \tag{*}$$

A first formulation of the effective problem is to bound the degrees of the $h_j$ in terms of those of the $g_j$. Modern work in this area started with a theorem of Brownawell [73], who showed that if $\deg g_j \le d$ for all $j$, then one can find $h_j$ as in (*) such that $\deg h_j \le n^2 d^n + nd$. Brownawell's argument was arithmetic and analytic in nature, drawing on height inequalities from transcendence theory and the classical theorem of Skoda. Shortly thereafter, Kollár [359] gave a more elementary and entirely algebraic proof of the optimal statement that in the situation above, one can in fact take

$$\deg(h_j g_j) \leq d^n \qquad (10.16)$$

provided that $d \neq 2$.

Kollár deduces (10.16) as an immediate consequence of a very appealing statement in the projective setting. Specifically, consider a homogeneous ideal $J \subseteq \mathbf{C}[T_0, \ldots, T_n]$. Then of course $J$ contains some power of its radical. The main theorem of [359] is the effective statement that if $J$ is generated by forms of degree $\leq d$ ($d \neq 2$), then already

$$\left(\sqrt{J}\right)^{d^n} \subseteq J. \qquad (10.17)$$

To see that this implies (10.16), let $G_j \in \mathbf{C}[T_0, \ldots, T_n]$ be the homogenization of $g_j$. Then the common zeroes of the $G_j$ lie in the hyperplane at infinity $\{T_0 = 0\}$, and consequently $T_0 \in \sqrt{(G_1, \ldots, G_m)}$. Therefore $(T_0)^{d^n} = \sum H_j G_j$ thanks to (10.17), and (10.16) follows upon dehomogenizing. We refer to [569] for an excellent survey of this body of work, and to [51] for a discussion of some analytic approaches to these questions. Among the more recent work in this area, it is worth mentioning especially [365] and [547]. $\qquad \square$

Returning to the smooth variety $X$, and the ideal sheaf $\mathfrak{a} \subseteq \mathcal{O}_X$, we recall from Section 5.4.B the construction of the *distinguished subvarieties* of $X$ associated to $\mathfrak{a}$. Let

$$\nu : X^+ \longrightarrow X$$

be the normalization of the blow-up of $\mathfrak{a}$ with exceptional divisor $F$, so that $\mathfrak{a} \cdot \mathcal{O}_{X^+} = \mathcal{O}_{X^+}(-F)$. Then $F$ determines a Weil divisor on $X^+$, say

$$[F] = \sum_{i=1}^{t} r_i \cdot [F_i],$$

where the $F_i$ are the irreducible components of the support of $F$, and $r_i > 0$. The Fulton-MacPherson distinguished subvarieties associated to $\mathfrak{a}$ are the images

$$Z_i =_{\text{def}} \nu(F_i) \subseteq X,$$

considered as reduced subschemes of $X$.[9] Thus one has the decomposition $Z = \cup Z_i$ of $Z = \text{Zeroes}(\sqrt{\mathfrak{a}})$ as the union of its distinguished subvarieties. In particular, each irreducible component of $Z$ is distinguished, but there may be "embedded" distinguished components as well.[10]

The connection with Skoda's theorem flows from the elementary

---

[9] Note that several of the $F_i$ may have the same image in $X$, i.e. there may be repetitions among the $Z_i$. But this doesn't cause any problems.

[10] In general there is no clear connection between the distinguished subvarieties $Z_i$ and the subvarieties defined by the associated primes of $\mathfrak{a}$ in the sense of primary decomposition. However if $\mathfrak{a}$ is integrally closed, then it is elementary that every associated subvariety is distinguished: see Example 9.6.14 or [99, §2].

**Lemma 10.5.2.** *For any $m \geq 1$, there is an inclusion*

$$\mathcal{I}_{Z_1}^{<r_1 m>} \cap \ldots \cap \mathcal{I}_{Z_t}^{<r_t m>} \subseteq \mathcal{J}(X, \mathfrak{a}^m).$$

Given an irreducible subvariety $V \subseteq X$, recall (Definition 9.3.4) that $\mathcal{I}_V^{<m>} \subseteq \mathcal{O}_V$ denotes the $m^{\text{th}}$ symbolic power of the ideal of $V$, consisting of germs of functions that vanish to order $\geq m$ at a general point of $V$.

*Proof of Lemma 10.5.2.* It suffices in view of Lemma 9.6.20 to prove that

$$\mathcal{I}_{Z_1}^{<r_1 m>} \cap \ldots \cap \mathcal{I}_{Z_t}^{<r_t m>} \subseteq \overline{\mathfrak{a}^m}.$$

The assertion is local on $X$, so we assume that $X$ is affine and that the ideals in question lie in the coordinate ring $\mathbf{C}[X]$ of $X$. Fix

$$\phi \in \mathcal{I}_{Z_1}^{<r_1 m>} \cap \ldots \cap \mathcal{I}_{Z_t}^{<r_t m>},$$

and consider the normalized blow-up $\nu : X^+ \longrightarrow X$ of $\mathfrak{a}$. Then $\phi$ has multiplicity $\geq r_i m$ at each point of $Z_i$, and consequently $\nu^* \phi$ has multiplicity $\geq r_i m$ at a general point of $F_i$. This implies that $\mathrm{ord}_{F_i}(\nu^* \phi) \geq r_i m$ for every $i$, and hence that $\mathrm{div}(\nu^* \phi) \succcurlyeq mF$. But this means exactly that

$$\phi \in \nu_* \mathcal{O}_{X^+}(-mF) = \overline{\mathfrak{a}^m},$$

as required.    □

As an immediate consequence of Skoda's Theorem 9.6.21 one then obtains:

**Theorem 10.5.3. (Local effective Nullstellensatz).** *Keeping notation as above, set*

$$r = r(\mathfrak{a}) = \max_{1 \leq i \leq t} \{ r_i \}.$$

*Then*

$$\mathcal{I}_Z^{<rn>} = \left( \sqrt{\mathfrak{a}} \right)^{<rn>} \subseteq \mathfrak{a}, \tag{10.18}$$

*where the symbolic power denotes the set of all (germs of) functions having multiplicity $\geq rn$ at a general point of each irreducible component of $Z$. In particular, $\left( \sqrt{\mathfrak{a}} \right)^{rn} \subseteq \mathfrak{a}$.*

*Proof.* In fact, by definition of $r = r(\mathfrak{a})$ together with the preceding lemma,

$$\mathcal{I}_Z^{<rn>} \subseteq \mathcal{I}_{Z_1}^{<r_1 n>} \cap \ldots \cap \mathcal{I}_{Z_t}^{<r_t n>} \subseteq \mathcal{J}(X, \mathfrak{a}^n). \tag{10.19}$$

But $\mathcal{J}(\mathfrak{a}^n) \subseteq \mathfrak{a}$ by Skoda's theorem.    □

**Remark 10.5.4.** It follows similarly from (9.41) or 9.6.26 that

$$\mathcal{I}_Z^{<rm>} \subseteq \mathfrak{a}^{m+1-n}$$

for every $m \geq n$. The variant 9.6.37 leads to an analogous improvement of Theorem 10.5.3.    □

In order to have a truly convincing statement, one needs of course to control the integer $r = r(\mathfrak{a})$ appearing in 10.5.3. It would be interesting to know whether there are any useful local results in this direction involving algebraic invariants of $\mathfrak{a}$. However in the global setting, to which we now turn, such bounds follow from elementary computations of intersection numbers.

We assume for the remainder of this subsection that $X$ is a non-singular projective variety of dimension $n$. As in the second part of Section 9.6.B, choose an integral divisor $L$ on $X$ such that $\mathcal{O}_X(L) \otimes \mathfrak{a}$ is globally generated. The bounds on $r(\mathfrak{a})$ come from a Bézout-type inequality involving the distinguished subvarieties of $\mathfrak{a}$.

**Proposition 10.5.5. (Degree bound for distinguished subvarieties).**
*If $L$ is ample — or merely nef — then*

$$\sum_{i=1}^{t} r_i \cdot \deg_L (Z_i) \; \leq \; \deg_L (X), \tag{10.20}$$

*where for a subvariety $V \subseteq X$, $\deg_L(V)$ denotes as usual the degree $(L^{\dim V} \cdot V)$ of $V$ with respect to $L$.*

**Corollary 10.5.6.** *Assume in the situation of Proposition 10.5.5 that $L$ is ample. Then $r(\mathfrak{a}) \leq \deg_L X$. In particular,*

$$\left( \sqrt{\mathfrak{a}} \right)^{n \cdot \deg_L X} \subseteq \left( \sqrt{\mathfrak{a}} \right)^{<n \cdot \deg_L X>} \subseteq \mathfrak{a}.$$

*Proof.* If $L$ is ample then all the degrees appearing in (10.20) are positive integers. □

**Remark 10.5.7.** We do not know whether one can drop the factor of $n = \dim X$ in the exponents appearing in the Corollary. However simple examples ([155], Example 2.3) show that it cannot be omitted in equation (10.19). If in fact it were true that $\left( \sqrt{\mathfrak{a}} \right)^{\deg_L X} \subseteq \mathfrak{a}$ then the inclusion involving the symbolic power would be a consequence of the results discussed in Section 11.3. □

*Indication of Proof of Proposition 10.5.5.* The stated inequality was established in the course of the proof of Theorem 5.4.18: since $L \otimes \mathfrak{a}$ is globally generated, one can take $s = 1$ in equation (5.20). Alternatively, see [155, §3]. □

Finally, combining the proposition and Lemma 10.5.2 with the global variant (Theorem 9.6.31) of Skoda's theorem, one arrives at a global effective Nullstellensatz.

**Theorem 10.5.8. (Geometric effective Nullstellensatz).** *Keeping the notation and assumptions of 10.5.6, fix generating sections*

$$g_1, \ldots, g_p \in \Gamma(X, \mathcal{O}_X(L) \otimes \mathfrak{a}),$$

*together with a big and nef integral divisor $P$ on $X$. Suppose that*

$$g \in \Gamma(X, \mathcal{O}_X(K_X + mL + P))$$

*is a section vanishing to order $\geq (n+1) \cdot \deg_L X$ at a general point of each irreducible component of $Z$. If $m \geq n+1$, then $g$ can be expressed as a linear combination*

$$g = \sum h_j g_j$$

*with $h_j \in \Gamma(X, \mathcal{O}_X(K_X + (m-1)L + P))$.* □

**Example 10.5.9. (The case $X = \mathbf{P}^n$).** Consider the "classical" case $X = \mathbf{P}^n$ and $L = \mathcal{O}_{\mathbf{P}^n}(d)$, so that we are dealing with $m$ homogeneous polynomials

$$g_1, \ldots, g_m \in \mathbf{C}[T_0, \ldots, T_n]$$

of degree $d$. Then the previous theorem states that if $g$ is a homogeneous polynomial of degree $\geq dn + d - n$ vanishing to order $\geq (n+1)d^n$ on each irreducible component of $Z$, then $g$ lies in the homogeneous ideal $J$ spanned by the $g_j$. Because of the presence of the factor $n+1$, one does not quite recover Kollár's result (10.17). However, one can make more refined statements involving order of vanishing along each of the distinguished subvarieties $Z_j$, and then Proposition 10.5.5 leads to stronger results provided that not all of the components of $Z$ have very small degree. We refer to [155] for a fuller discussion of how Theorem 10.5.8 compares with earlier bounds. □

**Example 10.5.10.** M. Rojas has observed that following the model of Sombra's paper [547] one can apply the theorem to suitable toric compactifications $X$ of $\mathbf{C}^n$ to obtain extensions of the results of Brownawell and Kollár to certain sparse systems of polynomials. In some settings, the numerical bounds that come out strengthen Sombra's. We illustrate this in a special case. Consider polynomials $g_j \in \mathbf{C}[t_1, \ldots, t_n]$ and suppose that one is given separate degree bounds in each of the variables $t_k$:

$$\deg_{t_k}(g_j) \leq d_k \quad \forall\, j.$$

Assuming that the $g_j$ have no common zeroes in $\mathbf{C}^n$, then one can find $h_j$ with $\sum h_j g_j = 1$ where now

$$\deg(h_j g_j) \leq (n+1)!\, d_1 \cdots\cdots d_n. \tag{*}$$

If for instance one thinks of $d_1, \ldots, d_{n-1}$ as being fixed, then (*) gives a linear bound in the remaining input degree $d_n$. (To prove (*), one applies the Theorem to $X = \mathbf{P}^1 \times \cdots \times \mathbf{P}^1$ and $L = \mathcal{O}(d_1, \ldots, d_n)$, and argues as in the proof that (10.17) implies (10.16).) □

# Notes

The exposition in Section 10.1.A closely follows the presentation in [174, §7].
Section 10.2 is a somewhat modified version of the argument in [126, §13].
Our account of Nakamaye's theorem draws on some notes by M. Popa. As
indicated in the text, Section 10.5 follows parts of [155] quite closely.

# 11

## Asymptotic Constructions

As we have suggested on a number of occasions, an important use of multiplier ideals is to make it possible to apply vanishing theorems for $\mathbf{Q}$-divisors without first passing to a normal crossing situation. While this can be extremely valuable, in many cases it constitutes mainly a conceptual and technical simplification of the direct approach: several of the theorems presented in Chapter 10, for example, were originally proven without the language of multiplier ideals.

However multiplier ideals also open the door to asymptotic constructions that could not easily be made directly, and this has led to some very interesting applications. For example, suppose that one wishes to study the behavior of the linear series $|mL|$ associated to high multiples of a big divisor $L$ on a smooth projective variety $X$. The "classical" approach is to pass to a log resolution of $|mL|$, but this runs into the fundamental difficulty that there needn't exist a single modification $X' \longrightarrow X$ that simultaneously resolves all the $|mL|$. This makes it hard to compare the situation for different $m$. On the other hand, it turns out that for given $c$ the multiplier ideals $\mathcal{J}\left(X, \frac{c}{m} \cdot |mL|\right)$ stabilize for $m \gg 0$ to a fixed ideal, which we denote by $\mathcal{J}\left(X, c \cdot \|L\|\right)$. In other words, there is some finiteness built into multiplier ideals that need not be present for the underlying linear series: any feature of the asymptotic geometry of $|mL|$ that can be detected by $\mathcal{J}\left(\frac{c}{m} \cdot |mL|\right)$ is captured already for some one large $m$. These asymptotic ideals $\mathcal{J}\left(c \cdot \|L\|\right)$ behave better in many respects than their parents $\mathcal{J}\left(c \cdot |L|\right)$ (cf. Theorems 11.1.8, 11.2.12(ii), 11.2.16 and Corollary 11.2.4). Similar constructions can be made in an algebraic setting: for example, when $\mathfrak{q} \subseteq \mathcal{O}_X$ is the ideal sheaf of an irreducible variety, asymptotic multiplier ideals associated to the symbolic powers $\mathfrak{q}^{<m>}$ lead to interesting results. While some of the constructions were implicitly known previously, the importance of asymptotic multiplier ideals really crystallized first in Siu's work [539] on deformation-invariance of plurigenera.

The present chapter is devoted to these asymptotic ideals and their applications. The first section gives the construction, while the basic properties are worked out in Section 11.2. The remaining three sections present some applications. We discuss in Section 11.3 a theorem of Ein, Smith, and the author [159] comparing the symbolic and ordinary powers of a radical ideal. Section 11.4 revolves around a result of Fujita to the effect that one can approximate the volume of a big line bundle by an ample divisor on a modification. Fujita's theorem has turned out to have several interesting consequences, and we present in particular very recent work of Boucksom, Demailly, Paun and Peternell [68] in which the result in question is used to determine the dual of the cone of effective divisors on a projective variety. We conclude in 11.5 with an account of Siu's theorem [539] on the deformation-invariance of plurigenera for varieties of general type.

# 11.1 Construction of Asymptotic Multiplier Ideals

This section gives the definition and basic properties of the asymptotic multiplier ideals. We will work in three contexts: complete linear series on a projective variety, graded linear series on a possibly incomplete variety, and graded families of ideal sheaves (Section 2.4). The latter settings are actually equivalent, and contain the first as a special case. Nonetheless, in the interests of familiarity we start with complete linear series.

## 11.1.A Complete Linear Series

We begin by attaching to a complete linear series $|L|$ a multiplier ideal that measures the asymptotic behavior of the complete series $|pL|$ as $p$ goes to infinity. In one guise or another, these have been considered by several authors ([144], [539], [128], [574]), and a closely related construction was used by Kollár in [361], [364]. However, it was especially Siu's work [539] that rendered clear their utility and importance. Ein [144] and independently Kawamata [324], [323] worked out aspects of the theory in the algebro-geometric setting, and at a few points we draw here on Ein's unpublished notes [144].

Throughout this subsection, $X$ denotes unless otherwise stated an irreducible non-singular projective variety.

For the convenience of the reader, we start by recalling some basic notation and definitions from Section 2.1 concerning linear series. Let $L$ be an integral divisor on $X$. One says that $L$ has *non-negative Iitaka dimension*, written $\kappa(X, L) \geq 0$, if $H^0\big(X, \mathcal{O}_X(mL)\big) \neq 0$ for some $m > 0$. In this case the *semigroup* of $L$ consists of those non-negative multiples of $L$ that are effective:

$$\mathbf{N}(L) \ = \ \mathbf{N}(X, L) \ = \ \big\{m \geq 0 \ | \ H^0\big(X, \mathcal{O}_X(mL)\big) \neq 0\big\}.$$

All sufficiently large elements of $\mathbf{N}(L)$ are multiples of a largest single natural number $e = e(L) \geq 1$, called the *exponent* of $L$, and all sufficiently large multiples of $e(L)$ appear in $\mathbf{N}(L)$.[1] We define the *Iitaka threshold* of $L$ to be the least integer $m_0 = m_0(L)$ such that $H^0\big(X, \mathcal{O}_X(mL)\big) \neq 0$ for all $m \geq m_0(L)$ with $e(L) \mid m$. If $L$ is big (Definition 2.2.1) — i.e. if $h^0\big(X, \mathcal{O}_X(mL)\big)$ grows like $m^{\dim X}$ — then $e(L) = 1$, i.e. every large multiple of $L$ is effective (Corollary 2.2.10).

We now turn to the construction of the asymptotic multiplier ideals. Let $L$ be a divisor on $X$ with $\kappa(X, L) \geq 0$, and fix a positive rational or real number $c > 0$.[2] For $p > 0$ consider the complete linear series $|pL|$, and form the multiplier ideal

$$\mathcal{J}\big(\tfrac{c}{p} \cdot |pL|\big) \subseteq \mathcal{O}_X$$

constructed in Definition 9.2.10.[3]

**Lemma 11.1.1.** *For every integer $k \geq 1$ one has the inclusion*

$$\mathcal{J}\big(\tfrac{c}{p} \cdot |\, pL\,|\big) \subseteq \mathcal{J}\big(\tfrac{c}{pk} \cdot |\, pkL\,|\big).$$

Grant this for the moment. We assert that then the family of ideals

$$\Big\{ \mathcal{J}\big(\tfrac{c}{p} \cdot |pL|\big) \Big\}_{(p\,\geq 0)}$$

has a *unique* maximal element. In fact, the existence of at least one maximal member follows from the ascending chain condition on ideals. On the other hand, if $\mathcal{J}\big(\tfrac{c}{p} \cdot |pL|\big)$ and $\mathcal{J}\big(\tfrac{c}{q} \cdot |qL|\big)$ are each maximal, then by the Lemma they must both coincide with $\mathcal{J}\big(\tfrac{c}{pq} \cdot |pqL|\big)$.

**Definition 11.1.2. (Asymptotic multiplier ideal associated to a complete linear series).** The *asymptotic multiplier ideal sheaf* associated to $c$ and $|L|$,

$$\mathcal{J}\big(c \cdot \|L\|\big) \;=\; \mathcal{J}\big(X, c \cdot \|L\|\big) \subseteq \mathcal{O}_X,$$

is defined to be the unique maximal member among the family of ideals $\big\{ \mathcal{J}\big(\tfrac{c}{p} \cdot |pL|\big) \big\}$.  □

To complete the construction, it remains only to give the

*Proof of Lemma 11.1.1.* We may suppose that $|pL| \neq \varnothing$ (and hence also $|pkL| \neq \varnothing$). Choose a simultaneous log resolution

$$\mu : X' \longrightarrow X$$

for both $|pL|$ and $|pkL|$. Write in the usual way

---

[1] The exponent $e$ is the g.c.d. of all the elements of $\mathbf{N}(L)$.

[2] It is occasionally important to be able to deal with real coefficients: see Example 11.1.22.

[3] Recall that by convention $\mathcal{J}\big(\tfrac{c}{p} \cdot |pL|\big) = (0)$ if $|pL| = \varnothing$.

$$\mu^* \left( |pL| \right) \;=\; | \, W_p \, | + F_p,$$
$$\mu^* \left( |pkL| \right) \;=\; | \, W_{pk} \, | + F_{pk},$$

where $F_p$ and $F_{pk}$ are the fixed divisors of $\mu^*\left(|pL|\right)$ and $\mu^*\left(|pkL|\right)$ respectively, while $W_p$ and $W_{pk}$ are free linear series. The image of the natural map

$$S^k W_p \longrightarrow W_{pk}$$

is a free linear subseries of $|\mu^*(pkL)|$ whose fixed divisor is $kF_p$. Therefore

$$k \cdot F_p \;\succcurlyeq\; F_{pk},$$

and consequently

$$\begin{aligned}
\mathcal{J}\!\left(\tfrac{c}{p} \cdot |pL|\right) \;&=\; \mu_* \mathcal{O}_{X'}\!\left(K_{X'/X} - \left[\tfrac{c}{p} F_p\right]\right) \\
&\subseteq\; \mu_* \mathcal{O}_{X'}\!\left(K_{X'/X} - \left[\tfrac{c}{pk} F_{pk}\right]\right) \\
&=\; \mathcal{J}\!\left(\tfrac{c}{pk} \cdot |pkL|\right)
\end{aligned}$$

as required.                                                                    □

**Example 11.1.3. (Finitely generated linear series).** Suppose that the section ring $R(X, L) = \oplus H^0\left(X, \mathcal{O}_X(mL)\right)$ is finitely generated. Then there exists an integer $r > 0$ such that

$$\mathcal{J}\big(X, c \cdot |rkL|\big) \;=\; \mathcal{J}\big(X, c \cdot \|rkL\|\big)$$

for every $c > 0$ and $k \geq 1$, so in this case the theory essentially reverts to the constructions made in Chapter 9. Thus asymptotic multiplier ideals become interesting precisely in the situation where $R(X, L)$ is not finitely generated, or at least not known to be so. (If $R = R(X, L)$ is finitely generated, then there exists a positive integer $r > 0$ such that the Veronese subring $R^{(rk)}$ is generated by $R_{rk}$ for every $k \geq 1$ [69, Chapter III, §1, Proposition 3]. This implies that the homomorphisms

$$\mathrm{Sym}^p\Big(H^0\big(X, \mathcal{O}_X(rkL)\big)\Big) \longrightarrow H^0\big(X, \mathcal{O}_X(prkL)\big)$$

are surjective for $p > 0$, and the assertion follows as in the proof of the previous lemma.)                                                                    □

It follows from the construction that $\mathcal{J}\big(c \cdot |L|\big) = \mathcal{J}\big(\tfrac{c}{p} \cdot |pL|\big)$ for some large multiple $p$ of $e(L)$, and then $\mathcal{J}\big(c \cdot |L|\big) = \mathcal{J}\big(\tfrac{c}{pk} \cdot |pkL|\big)$ whenever $k \geq 1$. We will say in this case that $p$ *computes the multiplier ideal* $\mathcal{J}\big(c \cdot \|L\|\big)$. In fact the divisibility condition is unnecessary for divisors of exponent 1:

**Proposition 11.1.4.** *In the situation of Definition 11.1.2 assume that* $e(L) = 1$, *so that every sufficiently large multiple of* $L$ *is effective. Then there exists a positive integer* $p_0 = p_0(c, L)$ *such that*

$$\mathcal{J}\big(c \cdot \|L\|\big) \;=\; \mathcal{J}\big(\tfrac{c}{p} \cdot |pL|\big) \quad \text{for all} \ \ p \geq p_0.$$

This is a special case of the corresponding statement (Proposition 11.1.18) for graded families of ideals. The proof is notationally simpler in that setting, so we defer it until then.

**Example 11.1.5. (Divisors of exponent greater than one).** Suppose that $L$ has non-negative Iitaka dimension and exponent $e(L) = e$. Then $eL$ has exponent one, and

$$\mathcal{J}\big(c \cdot \|L\|\big) \; = \; \mathcal{J}\big(\tfrac{c}{e} \cdot \|eL\|\big).$$

It follows from 11.1.4 that there exists an integer $p_0$ such that $\mathcal{J}\big(c \cdot \|L\|\big) = \mathcal{J}\big(\tfrac{c}{pe} \cdot |peL|\big)$ for all $p \geq p_0$. $\qquad\square$

**Example 11.1.6. (Semiample divisors).** If $|kL|$ is free for some $k > 0$, then

$$\mathcal{J}\big(\|mL\|\big) \; = \; \mathcal{O}_X$$

for all $m > 0$. (See Proposition 11.2.18 for a generalization when $L$ is big.) $\quad\square$

**Example 11.1.7.** Given rational or real numbers $d \geq c > 0$, one has

$$\mathcal{J}\big(d \cdot \|L\|\big) \; \subseteq \; \mathcal{J}\big(c \cdot \|L\|\big). \quad\square$$

Note that it is not true in general that $\mathcal{J}\big(X, m \cdot |L|\big) = \mathcal{J}\big(X, |mL|\big)$ for $m \geq 2$, since for instance $|mL|$ may be free while $|L|$ has base points of high multiplicity. The first statement of the next theorem shows that this sort of difficulty disappears when one passes to asymptotic ideals. As we will see on a number of occasions, this and related simplifications render the asymptotic theory quite supple.

**Theorem 11.1.8. (Elementary formal properties).** *Let $L$ be an integral divisor of non-negative Iitaka dimension, and $c > 0$ a fixed rational or real number.*

(i). *For any integer $m \geq 1$,*

$$\mathcal{J}\big(c \cdot \|mL\|\big) \; = \; \mathcal{J}\big(mc \cdot \|L\|\big).$$

(ii). *The ideals $\mathcal{J}\big(c \cdot \|mL\|\big)$ form a decreasing sequence in $m$, i.e.*

$$\mathcal{J}\big(c \cdot \|mL\|\big) \; \supseteq \; \mathcal{J}\big(c \cdot \|(m+1)L\|\big)$$

*for every $m$.*

(iii). *Let $\mathfrak{b}_m = \mathfrak{b}\big(|mL|\big) \subseteq \mathcal{O}_X$ be the base ideal of $|mL|$, where by convention we set $\mathfrak{b}_m = (0)$ if $|mL| = \varnothing$. Then*

$$\mathfrak{b}_m \cdot \mathcal{J}\big(\|\ell L\|\big) \; \subseteq \; \mathcal{J}\big(\|(m+\ell)L\|\big).$$

(iv). *One has*
$$\mathfrak{b}_m \subseteq \mathcal{J}\big(\|mL\|\big) \quad \text{for all } m \ge 1.$$

*Proof.* Fix $p \gg 0$ that computes both of the multiplier ideals $\mathcal{J}\big(cm \cdot \|L\|\big)$ and $\mathcal{J}\big(c \cdot \|mL\|\big)$. Then

$$
\begin{aligned}
\mathcal{J}\big(c \cdot \|mL\|\big) &= \mathcal{J}\big(\tfrac{c}{p} \cdot |pmL|\big) \\
&= \mathcal{J}\big(\tfrac{cm}{pm} \cdot |pmL|\big) \\
&= \mathcal{J}\big(cm \cdot \|L\|\big),
\end{aligned}
$$

which gives (i). Statement (ii) follows from (i) and the preceding example. In fact:

$$
\begin{aligned}
\mathcal{J}\big(c \cdot \|(m+1)L\|\big) &= \mathcal{J}\big(c(m+1) \cdot \|L\|\big) \\
&\subseteq \mathcal{J}\big(cm \cdot \|L\|\big) \\
&= \mathcal{J}\big(c \cdot \|mL\|\big).
\end{aligned}
$$

For (iii) we can assume that $|mL| \ne \varnothing$. Fix $p \gg 0$ and a common log resolution $\mu : X' \longrightarrow X$ of $|p\ell L|, |mL|, |pmL|, |p(m+\ell)L|$, and as usual denote by

$$F_{p\ell} \, , \ F_m \, , \ F_{pm} \, , \ F_{p(m+\ell)} \ \subseteq \ X'$$

the fixed divisors of $\mu^*|p\ell L|$, $\mu^*|mL|$, $\mu^*|pmL|$, and $\mu^*|p(m+\ell)L|$ respectively. Then

$$pF_m + F_{p\ell} \ \succcurlyeq \ F_{pm} + F_{p\ell} \ \succcurlyeq \ F_{p(m+\ell)}$$

and consequently

$$-F_m - \big[\tfrac{1}{p}F_{\ell p}\big] \ \preccurlyeq \ -\big[\tfrac{1}{p}F_{p(m+\ell)}\big].$$

Thus

$$
\begin{aligned}
\mathfrak{b}_m \cdot \mathcal{J}\big(\|\ell L\|\big) &\subseteq \mu_* \mathcal{O}_{X'}\big( -F_m + K_{X'/X} - \big[\tfrac{1}{p}F_{\ell p}\big]\big) \\
&\subseteq \mu_* \mathcal{O}_{X'}\big(K_{X'/X} - \big[\tfrac{1}{p}F_{p(m+\ell)}\big]\big) \\
&= \mathcal{J}\big(\|(m+\ell)L\|\big),
\end{aligned}
$$

as required. Finally, (iv) follows from (iii) by taking $\ell = 0$. $\qquad\square$

**Remark 11.1.9.** It is established in [157] that if $L$ is big, then the asymptotic multiplier ideals $\mathcal{J}\big(c \cdot \|L\|\big)$ depend only on the numerical equivalence class of $L$: see Example 11.3.12. $\qquad\square$

**Remark 11.1.10. (Divisors of sub-maximal Iitaka dimension).** When $L$ isn't big, the asymptotic multiplier ideal $\mathcal{J}\big(\|L\|\big)$ can fail to have some of the properties for which one might hope (Remarks 11.1.11 and 11.2.20). This suggests that in this case the construction might not yet be in its final form. $\qquad\square$

**Remark 11.1.11. (Singular metrics with minimal singularities).** Demailly [128] has shown that if $L$ is any pseudo-effective divisor, then up to "equivalence of singularities" $\mathcal{O}_X(L)$ carries a unique singular metric $h_{\min}$ with minimal singularities having non-negative curvature. If $L$ has non-negative Iitaka dimension, then

$$\mathcal{J}\big(\|mL\|\big) \subseteq \mathcal{J}\big(h_{\min}^m\big)$$

for every $m > 0$: see [130, §1]. Examples given there show that in general the inclusion can be strict, but one conjectures that equality holds if $L$ is big.   □

**Variant 11.1.12. (Relative linear series).**   Let $X$ be a smooth variety, let $f : X \longrightarrow T$ be a surjective projective mapping, and consider an integral divisor $L$ on $X$ whose restriction to a general fibre of $f$ has non-zero Iitaka dimension. Then the exponent $e(L, f)$ and Iitaka threshold $m_0(L, f)$ of $L$ relative to $f$ are defined in the evident manner (e.g. by restriction to a general fibre of $f$). We assume for simplicity that $e(L, f) = 1$. Given a rational number $c > 0$ we constructed in Variant 9.2.21 multiplier ideals

$$\mathcal{J}\big(f, c \cdot |kL|\big) \subseteq \mathcal{O}_X$$

starting from the canonical mapping $\rho_k : f^* f_* \mathcal{O}_X(kL) \longrightarrow \mathcal{O}_X(kL)$. Just as above, these give rise to asymptotic ideals $\mathcal{J}\big(f, c \cdot \|kL\|\big) \subseteq \mathcal{O}_X$, defined e.g. by the formula

$$\mathcal{J}\big(f, c \cdot \|kL\|\big) \;=\; \mathcal{J}\big(f, \tfrac{c}{p} \cdot |pkL|\big)$$

for sufficiently divisible $p \gg 0$. The evident analogue of Theorem 11.1.8 — whose formulation we leave to the reader — remains true in this relative setting. This can be verified by adapting the arguments in the proof of that statement. More compactly, one can observe that the base ideals $\mathfrak{b}_k = \mathfrak{b}\big(f, |kL|\big)$ associated to the maps $\rho_k$ form a graded family of ideals in the sense of Section 11.1.B.[4] Then $\mathcal{J}\big(f, c \cdot \|kL\|\big) = \mathcal{J}\big(\mathfrak{b}_\bullet^{ck}\big)$ is just the corresponding asymptotic multiplier ideal constructed in the next subsection.   □

**Remark 11.1.13. (Singular varieties).** One can avoid the non-singularity assumption on $X$ by working instead with a pair $(X, \Delta)$ satisfying the conditions in Definition 9.3.55. Then with the exception of statement (iv) in Theorem 11.1.8, the constructions and results go through with only minor modifications.[5] We leave this extension to the interested reader.   □

---

[4] Recall from (2.4.B) that $\mathfrak{b}\big(f, |kL|\big)$ is by definition the image of the homomorphism

$$f^* f_* \mathcal{O}_X(kL) \otimes \mathcal{O}_X(-kL) \longrightarrow \mathcal{O}_X$$

determined by $\rho_k$.

[5] The proof of Theorem 11.1.8 (iv) uses that $\mathcal{J}\big(X, \mathcal{O}_X\big) = \mathcal{O}_X$, and the analogous statement can fail for $(X, \Delta)$ when $X$ is singular.

### 11.1.B Graded Systems of Ideals and Linear Series

The construction of the previous section did not draw in a fundamental way on global properties of complete linear series. The essential point was rather that the base ideals $\mathfrak{b}_k = \mathfrak{b}(|kL|)$ satisfy the defining property

$$\mathfrak{b}_m \cdot \mathfrak{b}_\ell \subseteq \mathfrak{b}_{m+\ell} \quad \text{for all} \quad m, \ell \geq 1$$

of a graded family of ideals (Section 2.4). In the present subsection we attach multiplier ideals to an arbitrary graded system $\mathfrak{a}_\bullet = \{\mathfrak{a}_m\}$, as well as to graded linear series. This construction contains the multiplier ideals introduced in the previous subsection as a special case, and it is often cleaner and simpler to work directly with systems of ideals. By the same token, it is generally easier to give non-trivial examples of the theory in this algebraic setting.

For the remainder of this subsection, $X$ is (unless otherwise stated) a non-singular complex algebraic variety, and $\mathfrak{a}_\bullet = \{\mathfrak{a}_k\}$ is a graded system of ideals on $X$. In other words, $\mathfrak{a}_\bullet$ consists of a collection of ideal sheaves $\mathfrak{a}_k \subseteq \mathcal{O}_X$, satisfying $\mathfrak{a}_0 = \mathcal{O}_X$ and $\mathfrak{a}_m \cdot \mathfrak{a}_\ell \subseteq \mathfrak{a}_{m+\ell}$ for all $m, \ell \geq 1$. We refer to Section 2.4.B for further definitions and examples.

The first point to observe is that the analogue of Lemma 11.1.1 remains true in the present setting:

**Lemma 11.1.14.** *Let $\mathfrak{a}_\bullet = \{\mathfrak{a}_k\}$ be a graded system of ideals. Then given any fixed real or rational number $c > 0$ one has the inclusion*

$$\mathcal{J}\left(\tfrac{c}{p} \cdot \mathfrak{a}_p\right) \subseteq \mathcal{J}\left(\tfrac{c}{pk} \cdot \mathfrak{a}_{pk}\right)$$

*for all integers $p, k \geq 1$.*

Granting this, we proceed just as in the previous subsection. Specifically, the lemma implies that the family of ideals

$$\left\{\mathcal{J}\left(\tfrac{c}{p} \cdot \mathfrak{a}_p\right)\right\}_{(p \geq 1)}$$

has a unique maximal element: the existence of at least one maximal member follows as before from the ACC, whereas if $\mathcal{J}\left(\tfrac{c}{p} \cdot \mathfrak{a}_p\right)$ and $\mathcal{J}\left(\tfrac{c}{q} \cdot \mathfrak{a}_q\right)$ are each maximal, then they coincide with $\mathcal{J}\left(\tfrac{c}{pq} \cdot \mathfrak{a}_{pq}\right)$ thanks to 11.1.14.

**Definition 11.1.15. (Multiplier ideal associated to a graded system of ideals).** The *asymptotic multiplier ideal sheaf* of $\mathfrak{a}_\bullet$ with *coefficient* or *exponent* $c$, written either

$$\mathcal{J}\left(c \cdot \mathfrak{a}_\bullet\right) \quad \text{or} \quad \mathcal{J}\left(\mathfrak{a}_\bullet^c\right),$$

is defined to be the unique maximal member among the family of ideals $\left\{\mathcal{J}\left(\tfrac{c}{p} \cdot \mathfrak{a}_p\right)\right\}$ for $p \geq 1$. Thus $\mathcal{J}\left(c \cdot \mathfrak{a}_\bullet\right) = \mathcal{J}\left(\tfrac{c}{p} \cdot \mathfrak{a}_p\right)$ for all sufficiently large and divisible integers $p \gg 0$. $\qquad\square$

When there is some possibility of confusion about the underlying variety $X$ we write $\mathcal{J}(X, c \cdot \mathfrak{a}_\bullet)$ or $\mathcal{J}(X, \mathfrak{a}_\bullet^c)$, and when $c = 1$ we speak simply of $\mathcal{J}(\mathfrak{a}_\bullet)$.

**Remark 11.1.16. (Obsolete notation).** In the paper [159], as well as preliminary drafts of the present book, the notation $\mathcal{J}(c \cdot \|\mathfrak{a}_m\|)$ appeared for what we would call here $\mathcal{J}(cm \cdot \mathfrak{a}_\bullet)$ or $\mathcal{J}(\mathfrak{a}_\bullet^{cm})$: see 11.1.19. Intended to emphasize the analogy with the multiplier ideals constructed in the previous subsection, this notation has come to seem cumbersome and we propose to retire it.     □

**Example 11.1.17. (Examples of multiplier ideals attached to graded systems).** We illustrate the construction with several examples.

(i).  Suppose that $\mathfrak{a}_k = \mathfrak{b}^k$ is the trivial family consisting of powers of a fixed ideal $\mathfrak{b} \subseteq \mathcal{O}_X$. Then $\mathcal{J}(\mathfrak{a}_\bullet^m) = \mathcal{J}(\mathfrak{b}^m)$ for every $m \in \mathbf{N}$. So here we don't get anything new.

(ii).  If $\mathfrak{b}_\bullet$ is the system of base ideals $\mathfrak{b}_k = \mathfrak{b}(|kL|)$ determined by a divisor $L$ on a projective variety, then it follows immediately from the definitions that
$$\mathcal{J}(c \cdot \mathfrak{b}_\bullet) = \mathcal{J}(c \cdot \|L\|).$$

(iii).  Consider the ideals $\mathfrak{a}_k \subseteq \mathbf{C}[x, y]$ associated to the valuation $v(f) = \mathrm{ord}_t \, f(t, e^t - 1)$ discussed in Example 2.4.18 (i). Then
$$\mathcal{J}(c \cdot \mathfrak{a}_\bullet) = \mathbf{C}[x, y]$$
for every $c > 0$. This follows for instance from the observation that $\mathfrak{a}_k$ contains a polynomial (viz. $y - p_{k-1}(x)$ in the notation of 2.4.18) whose divisor is a non-singular curve. Proposition 9.2.26 then shows that $\mathcal{J}(c \cdot \mathfrak{a}_k) = \mathcal{O}_X$ whenever $0 < c < 1$. (From a more sophisticated point of view, the triviality of $\mathcal{J}(\mathfrak{a}_\bullet^m)$ is implied by the fact that the colength of $\mathfrak{a}_k$ in $\mathbf{C}[x, y]$ grows linearly rather than quadratically in $k$: see Theorem 11.3.1 and Example 11.3.2.)

(iv).  Let $\mathfrak{a}_\bullet$ be the "diagonal" family of monomial ideals figuring in Example 2.4.19, so that $\mathfrak{a}_k \subseteq \mathbf{C}[x_1, \ldots, x_n]$ is generated by all monomials $x_1^{i_1} \cdots \cdots x_n^{i_n}$ with $\frac{i_1}{\delta_1} + \ldots + \frac{i_n}{\delta_n} \geq k$ for fixed real numbers $\delta_1, \ldots, \delta_n > 0$. Then it follows via Howald's Theorem 9.3.27 that $\mathcal{J}(c \cdot \mathfrak{a}_\bullet)$ is the monomial ideal spanned by all monomials $x_1^{i_1} \cdot \ldots \cdot x_n^{i_n}$ whose exponent vectors satisfy the inequality
$$\frac{i_1 + 1}{\delta_1} + \ldots + \frac{i_n + 1}{\delta_n} \geq c.$$

(Compare Example 9.3.29.)     □

The proof of Lemma 11.1.14 is essentially identical to the argument used to establish the corresponding fact (Lemma 11.1.1) for linear series:

*Proof of Lemma 11.1.14.* We can assume that $\mathfrak{a}_p \neq (0)$. Let $\mu : X' \longrightarrow X$ be a common log resolution of the two ideals $\mathfrak{a}_p$, $\mathfrak{a}_{pk} \subseteq \mathcal{O}_X$, with

$$\mathfrak{a}_p \cdot \mathcal{O}_{X'} = \mathcal{O}_{X'}(-F_p) \quad , \quad \mathfrak{a}_{pk} \cdot \mathcal{O}_{X'} = \mathcal{O}_{X'}(-F_{pk}).$$

Now $\mathfrak{a}_p^k \subseteq \mathfrak{a}_{pk}$ by definition of a system of ideals, and hence $-k \cdot F_p \preccurlyeq -F_{pk}$. Therefore

$$\mu_* \mathcal{O}_{X'}\big(K_{X'/X} - \lceil \tfrac{c}{p} F_p \rceil\big) \subseteq \mu_* \mathcal{O}_{X'}\big(K_{X'/X} - \lceil \tfrac{c}{pk} F_{pk} \rceil\big),$$

as required.                                                                 ☐

As promised above, we show next that the asymptotic multiplier ideals are computed by any sufficiently large integer $p$ provided that the graded system in question has exponent $= 1$:

**Proposition 11.1.18.** *Let $\mathfrak{a}_\bullet = \{\mathfrak{a}_k\}$ be a graded family of ideals, and assume that $e(\mathfrak{a}_\bullet) = 1$, i.e. that $\mathfrak{a}_k \neq (0)$ for all $k \gg 0$. Then there is an integer $p_0 > 0$ such that*

$$\mathcal{J}(c \cdot \mathfrak{a}_\bullet) = \mathcal{J}\big(\tfrac{c}{p} \cdot \mathfrak{a}_p\big)$$

*for every $p \geq p_0$.*

Thanks to Example 11.1.17 (ii), one obtains Proposition 11.1.4 by applying 11.1.18 to the family of base ideals associated to a divisor of exponent one.

*Proof of Proposition 11.1.18.* First fix $p \gg 0$ such that

$$\mathcal{J}(c \cdot \mathfrak{a}_\bullet) = \mathcal{J}\big(\tfrac{c}{rp} \cdot \mathfrak{a}_{rp}\big) \quad \text{for all } r \geq 1.$$

It is enough to show that if $k \gg 0$, then $\mathcal{J}\big(\tfrac{c}{p} \cdot \mathfrak{a}_p\big) \subseteq \mathcal{J}\big(\tfrac{c}{k} \cdot \mathfrak{a}_k\big)$. To this end, fix a large integer $q$ relatively prime to $p$ having the property that $\mathfrak{a}_q \neq (0)$. Then we can write any $k \gg 0$ in the form

$$k = rp + sq \quad \text{with} \quad r > 0 \ , \ 0 \leq s \leq p - 1.$$

Then $\mathfrak{a}_p^r \cdot \mathfrak{a}_q^s \subseteq \mathfrak{a}_k$ and therefore

$$\mathcal{J}\big(\tfrac{c}{k} \cdot (\mathfrak{a}_p^r \mathfrak{a}_q^s)\big) \subseteq \mathcal{J}\big(\tfrac{c}{k} \cdot \mathfrak{a}_k\big).$$

So we are reduced to proving that if $k \gg 0$ (and hence $r \gg 0$), then $\mathcal{J}\big(\tfrac{c}{p} \cdot \mathfrak{a}_p\big) \subseteq \mathcal{J}\big(\tfrac{c}{k} \cdot (\mathfrak{a}_p^r \mathfrak{a}_q^s)\big)$.

Now fix $\mu : X' \longrightarrow X$ which is a log resolution of both $\mathfrak{a}_p$ and $\mathfrak{a}_q$, and in the usual way write

$$\mu^{-1} \mathfrak{a}_p = \mathcal{O}_{X'}(-F_p) \quad , \quad \mu^{-1} \mathfrak{a}_q = \mathcal{O}_{X'}(-F_q).$$

Then

$$\mu^{-1}\left(\mathfrak{a}_p^r \cdot \mathfrak{a}_q^s\right) \;=\; \mathcal{O}_{X'}(-rF_p - sF_q).$$

We assert that if $k \gg 0$, then

$$\left[\, \tfrac{c}{k} \cdot (rF_p + sF_q)\,\right] \;\preccurlyeq\; \left[\, \tfrac{c}{p} \cdot F_p\,\right] \qquad (*)$$

Granting this, it follows that

$$
\begin{aligned}
\mathcal{J}\!\left(\tfrac{c}{p} \cdot \mathfrak{a}_p\right) 
&= \mu_* \mathcal{O}_{X'}\!\left(\, K_{X'/X} - [\tfrac{c}{p} \cdot F_p]\,\right) \\
&\subseteq \mu_* \mathcal{O}_{X'}\!\left(\, K_{X'/X} - [\tfrac{c}{k} \cdot (rF_p + sF_q)]\,\right) \\
&= \mathcal{J}\!\left(\tfrac{c}{k} \cdot (\mathfrak{a}_p^r \mathfrak{a}_q^s)\right),
\end{aligned}
$$

and we will be done. For $(*)$, fix a prime divisor $E \subset X'$ appearing in $F_p$ or $F_q$, and write

$$x = \mathrm{ord}_E(cF_p) \;,\quad y = \mathrm{ord}_E(cF_q).$$

The question is equivalent to the inequality

$$\left[\, \frac{rx}{rp + sq} + \frac{sy}{rp + sq}\,\right] \;\le\; \left[\, \frac{x}{p}\,\right]. \qquad (**)$$

But by construction $0 \le s \le p - 1$ is bounded, and then $(**)$ holds for all $r \gg 0$, as required.[6] □

The remaining formal properties of the multiplier ideals associated to a graded system mirror the analogous statements from the previous subsection:

**Theorem 11.1.19. (Formal properties).** *Let* $\mathfrak{a}_\bullet = \{\mathfrak{a}_k\}$ *be a graded family of ideals on a smooth variety* $X$, *and fix a rational or real number* $c > 0$.

(i). *For any* $m \in \mathbf{N}$,

$$\mathcal{J}\!\left(cm \cdot \mathfrak{a}_\bullet\right) \;=\; \mathcal{J}\!\left(\tfrac{c}{p} \cdot \mathfrak{a}_{pm}\right)$$

*for all sufficiently large and divisible* $p$.

(ii). *The multiplier ideals* $\mathcal{J}\left(c \cdot \mathfrak{a}_\bullet\right)$ *are decreasing in* $c$ *in the sense that*

$$\mathcal{J}\!\left(c \cdot \mathfrak{a}_\bullet\right) \;\supseteq\; \mathcal{J}\!\left(d \cdot \mathfrak{a}_\bullet\right) \quad \text{for} \;\; d \ge c.$$

(iii). *For every* $\ell, m > 0$,

$$\mathfrak{a}_m \cdot \mathcal{J}\!\left(\mathfrak{a}_\bullet^\ell\right) \;\subseteq\; \mathcal{J}\!\left(\mathfrak{a}_\bullet^{m+\ell}\right).$$

---

[6] We are using here the observation that if $f(r) \to \lambda$ and $g(r) \to 0$ as $r \to \infty$ for non-negative functions $f(r)$ and $g(r)$, then the inequality

$$[\, f(r) + g(r)\,] \;\le\; [\,\lambda\,]$$

on integer parts holds for all $r \gg 0$.

(iv). *For every $m \geq 0$,*

$$\mathfrak{a}_m \subseteq \mathcal{J}(\mathfrak{a}_\bullet^m).$$

**Remark 11.1.20.** Statement (iii) of the theorem shows that the direct sum $\oplus_{\ell \geq 0} \mathcal{J}(\mathfrak{a}_\bullet^\ell)$ is a graded module over the Rees ring $R(\mathfrak{a}_\bullet) = \oplus_{m \geq 0} \mathfrak{a}_m$.    □

*Proof of Theorem 11.1.19.* If $p$ is sufficiently large and divisible then

$$
\begin{aligned}
\mathcal{J}(cm \cdot \mathfrak{a}_\bullet) &= \mathcal{J}\left(\tfrac{cm}{p} \cdot \mathfrak{a}_p\right) \\
&= \mathcal{J}\left(\tfrac{cm}{pm} \cdot \mathfrak{a}_{pm}\right) \\
&= \mathcal{J}\left(\tfrac{c}{p} \cdot \mathfrak{a}_{pm}\right),
\end{aligned}
$$

where the second equality is obtained by taking $pm$ in place of $p$ for the large index in Definition 11.1.15. This gives (i). Statement (ii) is immediate from the definitions, while (iii) is proved in the same way as Theorem 11.1.8 (iii). Finally, (iv) follows as before from (iii) by taking $\ell = 0$.    □

**Remark 11.1.21. (Singular varieties).** One can again replace the non-singular base variety $X$ by a pair $(X, \Delta)$ satisfying the conditions in Definition 9.3.55: as above, all the constructions and results go through with the exception of statement (iv) in Theorem 11.1.19.    □

**Example 11.1.22. (Log-canonical threshold and jumping numbers of a graded system).** These asymptotic multiplier ideals permit one to discuss the log-canonical threshold and jumping coefficients (Section 9.3.B) of a graded system. Let $\mathfrak{a}_\bullet = \{\mathfrak{a}_k\}$ be a graded family of ideals. A real number $\xi > 0$ is called a *jumping number* of $\mathfrak{a}_\bullet$ at a point $x \in X$ if

$$\mathcal{J}\left(\mathfrak{a}_\bullet^{\xi + \varepsilon}\right)_x \subsetneq \mathcal{J}\left(\mathfrak{a}_\bullet^{\xi - \varepsilon}\right)_x$$

for every $\varepsilon > 0$. We denote by $\mathrm{Jump}(\mathfrak{a}_\bullet; x) \subseteq \mathbf{R}$ the collection of all such numbers. The *log-canonical threshold* $\mathrm{lct}(\mathfrak{a}_\bullet; x)$ is the infimum of all elements in $\mathrm{Jump}(\mathfrak{a}_\bullet; x)$ (or $\infty$ if $\mathrm{Jump}(\mathfrak{a}_\bullet; x) = \varnothing$).

(i). Let $\mathfrak{a}_\bullet$ be the "diagonal" system of monomial ideals appearing in Examples 2.4.19 and 11.1.17 (iv). Then $\mathrm{lct}(\mathfrak{a}_\bullet) = \frac{1}{\delta_1} + \ldots + \frac{1}{\delta_n}$ and $\mathrm{Jump}(\mathfrak{a}_\bullet; x)$ consists of all real numbers of the form

$$\frac{e_1 + 1}{\delta_1} + \ldots + \frac{e_n + 1}{\delta_n}$$

as $(e_1, \ldots, e_n)$ ranges over all vectors in $\mathbf{N}^n$ (compare Example 9.3.32). Note that these invariants can be irrational, and that the periodicity of Example 9.3.24 fails.

(ii). Let $\mathfrak{a}_\bullet$ be the graded family from Examples 2.4.18 (i) and 11.1.17 (iii) determined by valuation $v(f) = \mathrm{ord}_t f(t, e^t - 1)$ on $\mathbf{C}[x, y]$. Then $\mathrm{lct}(\mathfrak{a}_\bullet) = \infty$.

(iii). It can happen that the collection of jumping coefficients of a graded system contains cluster points. However, $\text{Jump}(\mathfrak{a}_\bullet; x)$ satisfies the descending chain condition: every decreasing sequence of jumping coefficients stabilizes. In particular, $\text{lct}(\mathfrak{a}_\bullet)$ is actually the minimum of the jumping numbers of $\mathfrak{a}_\bullet$. (See [161, §5].) $\qquad\square$

**Example 11.1.23. (Mixed multiplier ideals).** Let $\mathfrak{a}_\bullet = \{\mathfrak{a}_k\}$ and $\mathfrak{b}_\bullet = \{\mathfrak{b}_k\}$ be two graded families of ideals on $X$, and fix real numbers $c, d > 0$. Then there is a unique maximal element among the family of ideals

$$\left\{ \mathcal{J}\left(\mathfrak{a}_p^{c/p} \cdot \mathfrak{b}_p^{d/p}\right) \right\}_{(p>0)}.$$

This defines the "mixed" multiplier ideal $\mathcal{J}\left(\mathfrak{a}_\bullet^c \cdot \mathfrak{b}_\bullet^d\right)$. $\qquad\square$

We observe finally that one can attach multiplier ideals to graded linear series. Specifically, let $V_\bullet = \{V_k\}$ be a graded linear series belonging to a divisor $L$ on $X$ (Section 2.4.A). Thus $V_k \subseteq H^0(X, \mathcal{O}_X(kL))$ is a finite-dimensional subspace, and

$$V_m \cdot V_\ell \subseteq V_{\ell+m} \quad \text{for all } \ell, m > 0,$$

where $V_m \cdot V_\ell$ denotes the image of $V_m \otimes V_\ell$ under the homomorphism

$$H^0(X, \mathcal{O}_X(mL)) \otimes H^0(X, \mathcal{O}_X(\ell L)) \longrightarrow H^0(X, \mathcal{O}_X((m+\ell)L))$$

determined by multiplication.

Given such a graded linear series, fix a rational or real coefficient $c > 0$ and an index $m$. Then one checks as before that

$$\mathcal{J}\left(X, \tfrac{c}{p} \cdot |V_{pm}|\right) \subseteq \mathcal{J}\left(X, \tfrac{c}{pk} \cdot |V_{pkm}|\right)$$

for every $p, k > 0$. This once again gives rise to asymptotic multiplier ideals:

**Definition 11.1.24. (Multiplier ideal of a graded linear series).** The *asymptotic multiplier ideal*

$$\mathcal{J}\left(X, c \cdot \|V_m\|\right)$$

is the unique maximal element among the ideals $\mathcal{J}\left(X, \tfrac{c}{p} \cdot |V_{pm}|\right)$. When $m = 1$ we use also the alternative notation

$$\mathcal{J}(c \cdot V_\bullet) = \mathcal{J}(c \cdot \|V_1\|). \quad \square$$

Equivalently, let $\mathfrak{b}_\bullet = \{\mathfrak{b}_k\}$ be the graded system of base ideals $\mathfrak{b}_k = \mathfrak{b}(|V_k|)$ of $V_\bullet$. Then it follows from the definitions that $\mathcal{J}(c \cdot V_\bullet) = \mathcal{J}(\mathfrak{b}_\bullet^c)$ and consequently

$$\mathcal{J}(c \cdot \|V_m\|) = \mathcal{J}(\mathfrak{b}_\bullet^{cm})$$

for every index $m$. We leave it to the reader to state and verify the evident analogue of Theorems 11.1.8 and 11.1.19.

## 11.2 Properties of Asymptotic Multiplier Ideals

This section spells out some basic geometric properties of asymptotic multiplier ideals. We start with local results: here it is most natural to work in the context of graded families of ideals. The second subsection focuses on global results, particularly vanishing theorems. As an application, we give in 11.2.C a quick proof of Kollár's theorem [361] on multiplicativity of plurigenera.

### 11.2.A Local Statements

We extend to the asymptotic setting some of the properties established in Section 9.5. Throughout this subsection, $X$ denotes unless otherwise stated a non-singular complex variety, and $\mathfrak{a}_\bullet = \{\mathfrak{a}_k\}$ is a graded family of ideals on $X$.

We start with restrictions. Fix a subvariety $Y \subseteq X$. Given an ideal $\mathfrak{b} \subseteq \mathcal{O}_X$, we denote by $\mathfrak{b}_Y =_{\mathrm{def}} \mathfrak{b} \cdot \mathcal{O}_Y$ the ideal on $Y$ obtained by restricting $\mathfrak{b}$. In particular, the restrictions of the ideals $\mathfrak{a}_k$ determine a graded system $\mathfrak{a}_{\bullet,Y}$ of ideals on $Y$.

**Theorem 11.2.1. (Restrictions of asymptotic ideals).** *Let $Y \subseteq X$ be a smooth subvariety. Then*

$$\mathcal{J}\left(Y, c \cdot \mathfrak{a}_{\bullet,Y}\right) \subseteq \mathcal{J}\left(X, c \cdot \mathfrak{a}_\bullet\right)_Y$$

*for every rational or real number $c > 0$.*

*Proof.* Fix $p \gg 0$ computing each of the asymptotic multiplier ideals in question. Then applying the restriction theorem 9.5.1 to $\mathfrak{a}_p$ one obtains:

$$\begin{aligned}
\mathcal{J}\left(Y, c \cdot \mathfrak{a}_{\bullet,Y}\right) &= \mathcal{J}\left(Y, \tfrac{c}{p} \cdot \mathfrak{a}_{p,Y}\right) \\
&\subseteq \mathcal{J}\left(X, \tfrac{c}{p} \cdot \mathfrak{a}_p\right)_Y \\
&= \mathcal{J}\left(X, c \cdot \mathfrak{a}_\bullet\right)_Y,
\end{aligned}$$

as required.    □

**Example 11.2.2. (Restriction of linear series).** Assuming that $X$ is projective let $L$ be an integral divisor on $X$ of non-negative Iitaka dimension, and let $Y \subseteq X$ be a smooth subvariety not contained in the stable base-locus of $L$.

(i). Consider the graded linear series $\mathrm{Tr}_Y(L)_\bullet$ on $Y$ given by

$$\mathrm{Tr}_Y(L)_k = \mathrm{im}\left(H^0\left(X, \mathcal{O}_X(kL)\right) \longrightarrow H^0\left(Y, \mathcal{O}_Y(kL)\right)\right)$$

(Example 2.4.3 (iv)), and write $\mathcal{J}\left(Y, \|L\|_Y\right)$ for the corresponding multiplier ideals. Then

$$\mathcal{J}\big(Y,\|mL\|_Y\big) \;\subseteq\; \mathcal{J}\big(X,\|mL\|\big)_Y$$

for every $m \geq 0$. This sort of "trace" ideal will play an important role in Siu's theorem on plurigenera in Section 11.5.

(ii). Write $L_Y$ for the restriction of $L$ to $Y$. Then it can happen that

$$\mathcal{J}\big(Y,\|mL_Y\|\big) \;=\; \mathcal{O}_Y \quad \text{for all} \;\; m \geq 1,$$

whereas each of the restricted ideals $\mathcal{J}\big(X,\|mL\|\big)_Y$ is non-trivial. (For instance, $\mathcal{J}\big(X,\|mL\|\big)$ might be non-trivial while $L_Y$ is ample.) $\qquad \square$

We turn next to the subadditivity theorem. As before, $\mathfrak{a}_\bullet = \{\mathfrak{a}_k\}$ is a graded family of ideals on the smooth variety $X$.

**Theorem 11.2.3. (Subadditivity for graded systems of ideals).** *Fix positive integers $\ell$ and $m$, and a rational or real number $c > 0$. Then*

$$\mathcal{J}\big(\mathfrak{a}_\bullet^{c(\ell+m)}\big) \;\subseteq\; \mathcal{J}\big(\mathfrak{a}_\bullet^{c\ell}\big) \cdot \mathcal{J}\big(\mathfrak{a}_\bullet^{cm}\big).$$

*In particular,*

$$\mathcal{J}\big(\mathfrak{a}_\bullet^{cm\ell}\big) \;\subseteq\; \mathcal{J}\big(\mathfrak{a}_\bullet^{cm}\big)^\ell$$

*for every $\ell > 0$.*

In view of Example 11.1.17 (ii) and Theorem 11.1.8 (i), this immediately yields:

**Corollary 11.2.4. (Subadditivity for complete linear series).** *Assuming that $X$ is projective, let $L$ be a divisor on $X$ of non-negative Iitaka dimension, and fix a rational or real number $c > 0$. Then*

$$\mathcal{J}\big(c \cdot \|(m+k)L\|\big) \;\subseteq\; \mathcal{J}\big(c \cdot \|mL\|\big) \cdot \mathcal{J}\big(c \cdot \|kL\|\big)$$

*for any positive integers $m, k > 0$. In particular,*

$$\mathcal{J}\big(\|mL\|\big) \;\subseteq\; \mathcal{J}\big(\|L\|\big)^m. \quad \square$$

**Remark 11.2.5. (Sums of divisors).** In the setting of the corollary, it is not in general the case that $\mathcal{J}\big(\|L+M\|\big) \subseteq \mathcal{J}\big(\|L\|\big) \cdot \mathcal{J}\big(\|M\|\big)$ even when $L$ and $M$ are big divisors on $X$. For instance if $\mathcal{J}\big(\|L\|\big)$ is non-trivial then one builds a counter-example by simply taking $M$ to be sufficiently positive so that both $M$ and $L + M$ are ample. $\qquad \square$

*Proof of Theorem 11.2.3.* This again follows directly from the corresponding statement (Theorem 9.5.20 (ii)) for individual ideals. In fact, fix $p \gg 0$ that computes the relevant multiplier ideals. Then

$$\mathcal{J}\left(\mathfrak{a}_{\bullet}^{c(\ell+m)}\right) = \mathcal{J}\left(\mathfrak{a}_{p}^{c(\ell+m)/p}\right)$$
$$\subseteq \mathcal{J}\left(\mathfrak{a}_{p}^{c\ell/p}\right) \cdot \mathcal{J}\left(\mathfrak{a}_{p}^{cm/p}\right)$$
$$= \mathcal{J}\left(\mathfrak{a}_{\bullet}^{c\ell}\right) \cdot \mathcal{J}\left(\mathfrak{a}_{\bullet}^{cm}\right),$$

as asserted.     □

**Example 11.2.6. (Subadditivity for mixed multiplier ideals).** Let $\mathfrak{a}_{\bullet}$ and $\mathfrak{b}_{\bullet}$ be two graded systems of ideals on $X$, and fix real numbers $c, d > 0$. Then

$$\mathcal{J}\left(\mathfrak{a}_{\bullet}^{c} \cdot \mathfrak{b}_{\bullet}^{d}\right) \subseteq \mathcal{J}\left(\mathfrak{a}_{\bullet}^{c}\right) \cdot \mathcal{J}\left(\mathfrak{b}_{\bullet}^{d}\right),$$

the ideal on the left being the "mixed" multiplier ideal from Example 11.1.23.

□

Given graded families $\mathfrak{a}_{\bullet} = \{\mathfrak{a}_{k}\}$ and $\mathfrak{b}_{\bullet} = \{\mathfrak{b}_{k}\}$ on $X$, recall that their sum $\mathfrak{a}_{\bullet} + \mathfrak{b}_{\bullet}$ is the graded series given by

$$\left(\mathfrak{a}_{\bullet} + \mathfrak{b}_{\bullet}\right)_{k} = \sum_{\ell+m=k} \mathfrak{a}_{\ell} \cdot \mathfrak{b}_{m}$$

(Definition 2.4.24). Mustaţă [456] has shown that his summation theorem 9.5.26 extends to the present setting:

**Theorem 11.2.7. (Summation theorem for asymptotic ideals).** *Given graded families* $\mathfrak{a}_{\bullet}$, $\mathfrak{b}_{\bullet}$ *and any rational* $c > 0$ *one has*

$$\mathcal{J}\left((\mathfrak{a}_{\bullet} + \mathfrak{b}_{\bullet})^{c}\right) \subseteq \sum_{\lambda+\mu=c} \mathcal{J}\left(\mathfrak{a}_{\bullet}^{\lambda}\right) \cdot \mathcal{J}\left(\mathfrak{b}_{\bullet}^{\mu}\right). \quad \square$$

We refer to [456] for the proof: we will not draw on this result.

Finally, we reproduce from [161, §5] an analogue of Theorem 9.5.35 concerning the continuity of multiplier ideals in families.

**Theorem 11.2.8. (Generic restrictions of asymptotic ideals).** *Let*

$$f : X \longrightarrow T$$

*be a surjective smooth mapping between non-singular varieties, and fix a graded system* $\mathfrak{a}_{\bullet} = \{\mathfrak{a}_{k}\}$ *of ideals on* $X$. *Given* $t \in T$ *write* $X_{t}$ *for the fibre of* $X$ *over* $t \in T$, *and denote by* $\mathfrak{a}_{\bullet,t}$ *the restricted graded family on* $X_{t}$. *There is a countable union of proper closed subvarieties* $\mathcal{B} \subsetneq T$ *such that if* $t \in T - \mathcal{B}$ *then*

$$\mathcal{J}\left(X_{t}, c \cdot \mathfrak{a}_{\bullet,t}\right) = \mathcal{J}\left(X, c \cdot \mathfrak{a}_{\bullet}\right)_{t} \tag{11.1}$$

*for every* $c > 0$.

The ideal on the right in (11.1) is the restriction to $X_{t}$ of the multiplier ideal $\mathcal{J}\left(X, c \cdot \mathfrak{a}_{\bullet}\right)$.

*Proof of Theorem 11.2.8.* Fix any $p > 0$. By Theorem 9.5.35, there exists a proper closed subset $B_p \subsetneq T$ such that

$$\mathcal{J}\big( X , d \cdot \mathfrak{a}_p \big)_t \;=\; \mathcal{J}\big( X_t , d \cdot \mathfrak{a}_{p,t} \big)$$

for all $t \in T - B_p$ and all $d > 0$. Take $\mathcal{B} = \cup_{p \geq 1} B_p$, and fix $c > 0$. There is a natural number $p \gg 0$ (depending on $c$) such that $\mathcal{J}\big(X, c \cdot \mathfrak{a}_\bullet\big) = \mathcal{J}\big(X, \frac{c}{p} \cdot \mathfrak{a}_p\big)$. If $t \in T - \mathcal{B}$ then by construction $\mathcal{J}\big(X, \frac{c}{p} \cdot \mathfrak{a}_p\big)_t = \mathcal{J}\big(X_t, \frac{c}{p} \cdot \mathfrak{a}_{p,t}\big)$, and so

$$\mathcal{J}\big( X , c \cdot \mathfrak{a}_\bullet \big)_t \;=\; \mathcal{J}\big( X_t , \tfrac{c}{p} \cdot \mathfrak{a}_{p,t} \big) \;\subseteq\; \mathcal{J}\big( X_t , c \cdot \mathfrak{a}_{\bullet,t} \big).$$

On the other hand, we have the reverse inclusion from Theorem 11.2.1 and consequently $\mathcal{J}\big(X, c \cdot \mathfrak{a}_\bullet\big)_t = \mathcal{J}\big(X_t, c \cdot \mathfrak{a}_{\bullet,t}\big)$, as required.    □

**Remark 11.2.9. (Graded linear series).** We leave it to the reader to formulate the analogous assertions for graded linear series on a possibly non-complete variety.    □

### 11.2.B Global Results

We study next some global properties of the asymptotic multiplier ideals associated to a complete linear series. Throughout this subsection $X$ denotes unless otherwise stated a non-singular projective variety.

The first point is that all the sections of a complete linear series vanish along the corresponding asymptotic multiplier ideal.

**Proposition 11.2.10. (Multiplier vs. base ideals, I).** *Let $L$ be any integral divisor of non-negative Iitaka dimension on $X$. Then the natural inclusion*

$$H^0\big(X, \mathcal{O}_X(mL) \otimes \mathcal{J}\big( \|mL\| \big)\big) \longrightarrow H^0\big(X, \mathcal{O}_X(mL)\big)$$

*is an isomorphism for every $m \geq 1$. More generally, if $V_\bullet$ is a graded linear series belonging to $L$, then any section in $V_m$ vanishes along $\mathcal{J}\big( \|V_m\| \big)$.*

**Remark 11.2.11. (Multiplier ideals and Zariski decompositions).** The fact that the asymptotic multiplier ideals capture all the sections of $|mL|$ suggests that they determine at least a weak analogue of a Zariski decomposition (Section 2.3.E). Indeed, Tsuji [574], [575], refers to the corresponding singular metric (Remark 11.1.11) as giving an analytic Zariski decomposition on $\mathcal{O}_X(L)$. However as Kawamata remarks, the existence of an actual Zariski decomposition often has important structural consequences (cf. [326, §7.3]) that do not follow from Proposition 11.2.10 or its counterpart in the analytic setting. So this analogy, while suggestive, should be taken with a large grain of salt.    □

*Proof of Proposition 11.2.10.* The first assertion of the proposition is equivalent to the claim that $\mathfrak{b}(|mL|) \subseteq \mathcal{J}(\|mL\|)$ for all $m$, where $\mathfrak{b}(|mL|)$ denotes the base ideal of $|mL|$. But this was established in Theorem 11.1.8 (iv). The second statement follows similarly from the inclusion $\mathfrak{b}(|V_m|) \subseteq \mathcal{J}(\|V_m\|)$. □

Now we come to the vanishing properties of these ideals.

**Theorem 11.2.12. (Vanishing for asymptotic multiplier ideals).** *Let $X$ be a smooth non-singular complex projective variety, and $L$ an integral divisor on $X$ of non-negative Iitaka dimension.*

(i). *If $A$ is any big and nef integral divisor on $X$, then*
$$H^i\big(X, \mathcal{O}_X(K_X + mL + A) \otimes \mathcal{J}(\|mL\|)\big) = 0 \quad \text{for } i > 0.$$

(ii). *If $L$ is big, then the statement holds assuming only that $A$ is nef. In particular, if $L$ is big then*
$$H^i\big(X, \mathcal{O}_X(K_X + mL) \otimes \mathcal{J}(\|mL\|)\big) = 0 \quad \text{for } i > 0.$$

(iii). *More generally, let $V_\bullet$ be a graded linear series belonging to $L$. Then*
$$H^i\big(X, \mathcal{O}_X(K_X + mL + A) \otimes \mathcal{J}(\|V_m\|)\big) = 0 \quad \text{for } i > 0$$

*whenever $A$ is big and nef. If $V_\bullet$ is big in the sense that $\kappa(V_\bullet) = \dim X$ (Definition 2.4.8), then the same vanishing holds assuming only that $A$ is nef.*

*Proof.* Statement (i), and the first assertion in (iii), are immediate consequences of Nadel vanishing (Theorem 9.4.8 or Corollary 9.4.15). For (ii), we go back to the proof of that result. Fix $p \gg 0$ such that $\mathcal{J}(\|mL\|) = \mathcal{J}(\frac{1}{p} \cdot |pmL|)$ and such that the rational mapping
$$\phi_{|pmL|} : X \dashrightarrow \mathbf{P}$$
defined by the complete linear series $|pmL|$ is generically finite over its image. Let $\mu : X' \longrightarrow X$ be a log resolution of $|pmL|$, with
$$\mu^* |pmL| = |W_{pm}| + F_{pm}.$$

Then the morphism $\phi_{|W_{pm}|} : X' \longrightarrow \mathbf{P}$ defined by the free linear series $|W_{pm}|$ resolves the indeterminacies of $\phi_{|pmL|}$, and in particular it is generically finite over its image. Therefore the linear series $|W_{pm}|$ is big (and of course nef). Let $B \in |W_{pm}|$ be a general divisor. Then
$$\mu^*(mL + A) - \big[ \tfrac{1}{p} F_{pm} \big] \equiv_{\text{num}} \mu^* A + \tfrac{1}{p} B + \Delta,$$

where $\Delta$ is a fractional divisor with normal crossing support, and $D = (\mu^* A + \frac{1}{p} B)$ is nef and big. Therefore by Kawamata–Viehweg vanishing (Theorem 9.1.18) we find that

$$H^i\left( X', \mathcal{O}_{X'}\left(K_{X'/X} + \mu^*(K_X + mL + A) - [\tfrac{1}{p} F_{pm}]\right)\right) \;=\; 0.$$

The stated vanishing now follows by taking direct images as in the proof of Theorem 9.4.8. The second assertion in (iii) is similar, with $|V_{pm}|$ replacing $|pmL|$.  $\square$

The argument leading to Proposition 9.4.26 then gives:

**Corollary 11.2.13. (Global generation of asymptotic multiplier ideals).** *In the situation of Theorem 11.2.12 (i) or (ii), assume that $B$ is a globally generated ample line bundle on $X$. Then for any $m \geq 1$,*

$$\mathcal{O}_X\left(K_X + nB + A + mL\right) \otimes \mathcal{J}\left(m \cdot \|L\|\right)$$

*is globally generated, where $n = \dim X$.*  $\square$

As we shall see, the important point here is that the twisting factor $\mathcal{O}_X(K_X + nB)$ is independent of $m$.

**Example 11.2.14. (Non-vanishing for big divisors).** One can use Theorem 11.2.12 (ii) to give a slight improvement of Corollary 9.4.24. In fact, if $L$ is big then in the situation of that corollary one can find $k$ with the stated property lying in the range $k \in [0, n]$.  $\square$

**Generalization 11.2.15. (Relative linear series).** The previous results extend in the natural way to a relative setting. Keeping the notation of Variants 9.4.16 and 11.1.12, one has the following

**Theorem.** *Let $f : X \longrightarrow T$ be a surjective projective morphism, and $L$ a divisor on $X$ whose restriction to a general fibre of $f$ has non-negative Iitaka dimension, and assume for simplicity that $e(L, f) = 1$.*

(i). *For every $m \geq m_0(L, f)$, the canonical map $\rho_m : f^* f_* \mathcal{O}_X(mL) \longrightarrow \mathcal{O}_X(mL)$ factors through the inclusion $\mathcal{O}_X(mL) \otimes \mathcal{J}\left(f, \|mL\|\right)$, i.e.*

$$\mathfrak{b}\left(f, |mL|\right) \;\subseteq\; \mathcal{J}\left(f, \|mL\|\right),$$

*where $\mathfrak{b}\left(f, |mL|\right)$ denotes the base ideal of $|mL|$ relative to $f$. Equivalently, the natural map*

$$f_*\left(\mathcal{O}_X(mL) \otimes \mathcal{J}\left(f, \|mL\|\right)\right) \longrightarrow f_*\left(\mathcal{O}_X(mL)\right)$$

*is an isomorphism.*

(ii). *If $A$ is a divisor on $X$ that is nef and big for $f$, then*

$$R^i f_* \Big( \mathcal{O}_X(K_X + mL + A) \otimes \mathcal{J}(f, \|mL\|) \Big) = 0 \quad \text{for} \ i > 0.$$

(iii). *If $L$ is big for $f$, then the same vanishing holds assuming only that $A$ is nef for $f$. In particular,*

$$R^i f_* \Big( \mathcal{O}_X(K_X + mL) \otimes \mathcal{J}(f, \|mL\|) \Big) = 0 \quad \text{for} \ i > 0$$

*in this case.*

The second and third statements follow from the relative analogue of Nadel vanishing (Generalization 9.4.16) by the same argument used to establish Theorem 11.2.12. We leave details to the reader. □

The next result analyzes the behavior of asymptotic multiplier ideals under étale covers. This will be applied in the next subsection to prove a theorem of Kollár concerning the multiplicativity of plurigenera of varieties of general type.

**Theorem 11.2.16. (Étale pullbacks of asymptotic multiplier ideals).** *Let*

$$f : Y \longrightarrow X$$

*be a finite étale cover of smooth projective varieties, and let $L$ be a divisor on $X$ having non-negative Iitaka dimension. Then*

$$\mathcal{J}(Y, \|f^*L\|) \ = \ f^* \mathcal{J}(X, \|L\|).$$

**Remark 11.2.17.** Note that once again the corresponding statement for the (ordinary) multiplier ideal associated to a linear series can fail. For example, suppose that $f : E' \longrightarrow E$ is an isogeny of degree $> 1$ between elliptic curves, and take $L = \mathcal{O}_X(P)$ to be the line bundle of degree one corresponding to a point $P \in E$. Then $\mathcal{J}(E, |L|) = \mathcal{O}_E(-P)$, but $\mathcal{J}(E', |f^*L|) = \mathcal{O}_{E'}$ since $f^*L$ moves in a free linear series. □

*Proof.* By considering the Galois closure of the given cover, one reduces to treating the case when $f$ is Galois. Denote by $G = \mathrm{Gal}(Y/X)$ the corresponding Galois group and by $g = \#G$ its order. Fix $p_0 \gg 0$ such that

$$\mathcal{J}(X, \|L\|) \ = \ \mathcal{J}(X, \tfrac{1}{p}|pL|),$$
$$\mathcal{J}(Y, \|f^*L\|) \ = \ \mathcal{J}(Y, \tfrac{1}{p}|f^*pL|),$$

for all positive multiples $p$ of $p_0$.

Since $f^*|p_0 L|$ is a subseries of $|f^*(p_0 L)|$, it follows from 9.2.32 (ii) that

$$f^* \mathcal{J}(X, \|L\|) \ \subseteq \ \mathcal{J}(Y, \|f^*L\|).$$

For the reverse inclusion, let $\nu : Y' \longrightarrow Y$ be a log resolution of $|f^*(p_0 L)|$, with $\nu^*|f^*(p_0 L)| = |M| + F$ for a free linear series $|M|$ on $Y'$. Assume that we have chosen $\nu$ to be $G$-equivariant, so that $G$ acts on $M$. We assert that

$$|gM|^G \quad \text{is free.} \qquad (*)$$

Granting this for the moment, it follows that if $A' \in |gM|^G$ is a general divisor, then $A' + gF$ has SNC support (Lemma 9.1.9). Hence

$$
\begin{aligned}
\mathcal{J}\big(Y\,,\,\|f^*L\|\big) &= \nu_* \mathcal{O}_{Y'}\big(K_{Y'/Y} - [\tfrac{1}{p_0}F]\big) \\
&= \nu_* \mathcal{O}_{Y'}\big(K_{Y'/Y} - [\tfrac{1}{gp_0}(A' + gF)]\big) \\
&= \mathcal{J}\big(Y\,,\,\tfrac{1}{gp_0}A\big),
\end{aligned}
$$

where $A \in |f^*(gp_0 L)|$ is the unique effective divisor on $Y$ such that $\nu^* A = A' + gF$. On the other hand, $A'$ and $F$ and hence $A$ are $G$-invariant, and so $A = f^*B$ for some $B \equiv_{\text{num}} gp_0 L$. Therefore (Example 9.5.44)

$$\mathcal{J}\big(Y\,,\,\tfrac{1}{gp_0}A\big) = f^* \mathcal{J}\big(X\,,\,\tfrac{1}{gp_0}B\big) \subseteq f^* \mathcal{J}\big(X,\tfrac{1}{gp_0}|gp_0 L|\big),$$

and so all told, $\mathcal{J}\big(Y, \|f^*L\|\big) \subseteq f^* \mathcal{J}\big(X, \|L\|\big)$, as required.

Turning next to $(*)$, pick any point $y \in Y'$. We need to produce a $G$-invariant divisor $D \in |gM|$ not passing through $y$. To this end, fix any $D_0 \in |M|$ that does not contain any of the finitely many points in the $G$-orbit $G \cdot y$ of $y$. Then $D = \sum_{\sigma \in G} \sigma^* D \in |gM|$ is $G$-invariant and avoids $y$.

Finally, one can avoid using a $G$-equivariant log resolution $\nu : Y' \longrightarrow Y$ as follows. Let $\mathfrak{b} \subset \mathcal{O}_Y$ be the base ideal of the linear series $|f^*(p_0 L)|$. Then $\mathfrak{b}$ is $G$-invariant, and hence $\mathfrak{b} = f^{-1}\mathfrak{a}$ for some ideal $\mathfrak{a} \subset \mathcal{O}_X$ containing the base ideal of $|p_0 L|$. Let $\nu_0 : Y_0 \longrightarrow Y$ be the normalization of the blow-up of $\mathfrak{b}$, so that one can write

$$\nu_0^*|f^*(p_0 L)| = |M_0| + F_0$$

for a free linear series $M_0$ on $Y_0$. Then $G$ acts on $M_0$, and arguing as above one sees that $|gM_0|^G$ is free. Now take $\nu_1 : Y' \longrightarrow Y_0$ giving a log resolution of $|f^*(p_0 L)|$. The argument above works for $A' \in \nu_1^*\big(|(gM_0)|^G\big)$. □

We conclude with a few additional results concerning the multiplier ideals attached to a big divisor. To begin with, we prove that among big divisors, the triviality of all the multiplier ideals $\mathcal{J}\big(\|mL\|\big)$ characterizes those that are nef.

**Proposition 11.2.18. (Characterization of nefness for big divisors).**
*Let $L$ be a big line bundle on a smooth projective variety $X$ of dimension $n$. Then $L$ is nef if and only if*

$$\mathcal{J}\big(X, \|mL\|\big) = \mathcal{O}_X \quad \text{for all} \ \ m \geq 1. \qquad (11.2)$$

*Proof.* Suppose first that (11.2) holds. Setting $A = K_X + (n+1)B$ for some very ample divisor $B$ on $X$, it follows from Corollary 11.2.13 and the triviality of $\mathcal{J}(\|mL\|)$ that

$$\mathcal{O}_X(A + mL)$$

is globally generated for all $m \gg 0$. This implies that if $C \subseteq X$ is any irreducible curve, then

$$(L \cdot C) \geq -\tfrac{1}{m} \cdot (A \cdot C).$$

Letting $m \to \infty$, we find that $L$ is nef.

Conversely, assume that $L$ is nef. Aiming for a contradiction, suppose that there exists a point

$$x \in \mathrm{Zeroes}\big(\mathcal{J}(\|mL\|)\big)$$

for some $m \geq 1$. Choose $k_0$ such that

$$\mathcal{J}(\|mL\|) = \mathcal{J}\big(\tfrac{1}{k} \cdot |kmL|\big) \quad \text{for all } k \geq k_0$$

(Lemma 11.1.4), and let $D_k \in |kmL|$ be a general divisor. Then it follows from Proposition 9.5.13 that

$$\mathrm{mult}_x D_k \geq k.$$

But this contradicts the boundedness of the base loci of nef and big divisors (Corollary 2.3.12). □

**Example 11.2.19. (Goodman's theorem).** Given a divisor $L$ of nonnegative Kodaira dimension on a smooth projective variety $X$, suppose there is a constant $M > 0$ such that

$$\mathrm{mult}_x |kL| \leq M$$

for all $x \in X$ and all $k \gg 0$. Then $L$ is nef. (The hypothesis implies that $\mathcal{J}(\|mL\|) = \mathcal{O}_X$ for $m \geq 0$, and the proof of Proposition 11.2.18 applies.) This is a result of Goodman [226], who in fact proves the analogous statement for an arbitrary projective variety. □

**Remark 11.2.20. (Nef and good divisors).** Just as in Example 2.3.16, the proposition can fail if $L$ is not big. However, Russo [520] shows that this sort of pathology does not occur if $L$ is abundant (Example 2.3.17), i.e. if its numerical dimension coincides with its Iitaka dimension. In fact, he generalizes Proposition 11.2.18 by proving that if $L$ is any divisor having non-negative Iitaka dimension, then

$$\mathcal{J}(X, \|mL\|) = \mathcal{O}_X$$

for all $m$ if and only if $L$ is nef and abundant. □

Finally, we show in effect that the multiplier ideals and base ideals of a big divisor differ by a "uniformly bounded" amount:

**Theorem 11.2.21. (Multiplier vs. base ideal, II).** *Let $L$ be a big divisor on the smooth projective variety $X$. Then there exists an effective divisor $E$ on $X$, together with a positive integer $t_0 = t_0(L)$, such that*

$$\mathcal{J}(\|mL\|) \otimes \mathcal{O}_X(-E) \subseteq \mathfrak{b}(|mL|) \tag{11.3}$$

*for every $m \geq t_0$. Moreover, if $|L| \neq \varnothing$, then one can take $t_0 = 1$.*

The crucial point here is that $E$ is independent of $m$. This statement is inspired by one of the steps in Siu's proof of deformation-invariance of plurigenera: see Section 11.5.

The theorem yields a comparison among different base ideals:

**Corollary 11.2.22.** *In the situation of the theorem, fix an integer $m \geq t_0(L)$. Then*

$$\mathfrak{b}(|m\ell L|) \otimes \mathcal{O}_X(-\ell E) \subseteq \mathfrak{b}(|mL|)^\ell$$

*for every $\ell \geq 1$.*

*Proof.* This follows from Proposition 11.2.10, Corollary 11.2.4, and Theorem 11.2.21. In fact,

$$\begin{aligned}
\mathfrak{b}(|m\ell L|) \otimes \mathcal{O}_X(-\ell E) &\subseteq \mathcal{J}(\|m\ell L\|) \otimes \mathcal{O}_X(-\ell E) \\
&\subseteq \left( \mathcal{J}(\|mL\|) \otimes \mathcal{O}_X(-E) \right)^\ell \\
&\subseteq \mathfrak{b}(|mL|)^\ell,
\end{aligned}$$

as stated. $\qquad\square$

*Proof of Theorem 11.2.21.* Write $n = \dim X$ and fix a very ample divisor $B$. Since $L$ is big, we can pick a positive integer $a$ sufficiently large so that

$$E \equiv_{\mathrm{num}} aL - \big( K_X + (n+1)B \big)$$

is effective. Then

$$mL - E \equiv_{\mathrm{num}} (m-a)L + \big( K_X + (n+1)B \big),$$

and so it follows from Corollary 11.2.13 that

$$\mathcal{O}_X(mL) \otimes \Big( \mathcal{J}(\|(m-a)L\|) \otimes \mathcal{O}_X(-E) \Big)$$

is globally generated provided that $m \geq a$. Therefore

$$\mathcal{J}(\|(m-a)L\|) \otimes \mathcal{O}_X(-E) \subseteq \mathfrak{b}(|mL|)$$

for $m \geq a$, and since $\mathcal{J}(\|mL\|) \subseteq \mathcal{J}(\|(m-a)L\|)$ we get the required inclusion for $m \geq a$. If $|L| \neq \varnothing$, then by adding divisors in this linear series to $E$ we can suppose that (11.3) holds also for $m < a$. $\qquad\square$

## 11.2.C Multiplicativity of Plurigenera

As an application of the machinery just developed, we discuss a theorem of Kollár concerning the multiplicative behavior of plurigenera under étale coverings. Other applications appear in the remaining three sections of this chapter.

Recall that if $X$ is a smooth projective variety, its $m^{\text{th}}$ plurigenus $P_m(X)$ is the dimension of the space of $m$-canonical forms on $X$:

$$P_m(X) \; = \; h^0\big(X, \mathcal{O}_X(mK_X)\big).$$

**Theorem 11.2.23. (Kollár, [361]).** *Let $f : Y \longrightarrow X$ be an étale covering of degree $d$ between smooth complex projective varieties of general type. Then for any $m \geq 2$,*

$$P_m(Y) \; = \; d \cdot P_m(X).$$

**Remark 11.2.24.** The case of curves shows already that the geometric genus $P_1(X)$ is not multiplicative in covers.     □

**Remark 11.2.25.** The argument that follows is essentially the one appearing in [361] and [364]. However, the language of multiplier ideals considerably simplifies the presentation.     □

*Proof of Theorem 11.2.23.* It follows from Proposition 11.2.10 that

$$
\begin{aligned}
H^0\Big(X, \mathcal{O}_X(mK_X)\Big) &= H^0\Big(X, \mathcal{O}_X(mK_X) \otimes \mathcal{J}\big(\|(m-1)K_X\|\big)\Big) \\
&= H^0\Big(X, \mathcal{O}_X(mK_X) \otimes \mathcal{J}\big(\|mK_X\|\big)\Big),
\end{aligned}
$$

where we are using the evident fact that $\mathcal{J}\big(\|mK_X\|\big) \subseteq \mathcal{J}\big(\|(m-1)K_X\|\big)$. On the other hand, $K_X$ is big since $X$ is of general type, and consequently

$$H^i\Big(X, \mathcal{O}_X(mK_X) \otimes \mathcal{J}\big(\|(m-1)K_X\|\big)\Big) \; = \; 0 \quad \text{for} \;\; i > 0$$

thanks to Theorem 11.2.12 (ii). In short, so long as $m \geq 2$, the $m^{\text{th}}$ plurigenus of $X$ is computed by an Euler characteristic:

$$h^0\Big(X, \mathcal{O}_X(mK_X)\Big) \; = \; \chi\Big(X, \mathcal{O}_X(mK_X) \otimes \mathcal{J}\big(\|(m-1)K_X\|\big)\Big).$$

Similarly,

$$h^0\Big(Y, \mathcal{O}_Y(mK_Y)\Big) \; = \; \chi\Big(Y, \mathcal{O}_Y(mK_Y) \otimes \mathcal{J}\big(\|(m-1)K_Y\|\big)\Big).$$

But since $f$ is étale,

$$\mathcal{O}_Y(mK_Y) \otimes \mathcal{J}\big(\|(m-1)K_Y\|\big) \; = \; f^*\Big(\mathcal{O}_X(mK_X) \otimes \mathcal{J}\big(\|(m-1)K_X\|\big)\Big)$$

thanks to Theorem 11.2.16. The theorem then follows from the fact (Proposition 1.1.28) that Euler characteristics are multiplicative in étale covers.     □

**Example 11.2.26. (Extension for adjoint-type bundles).** Let $L$ be a big divisor on a smooth projective variety $X$. Then for every $m \geq 1$ the quantity $h^0\big(X, \mathcal{O}_X(K_X + mL)\big)$ is multiplicative under étale coverings. (The proof of Kollár's theorem goes through with little change.) This statement is also due to Kollár [362, Proposition 9.4].                                        □

## 11.3 Growth of Graded Families and Symbolic Powers

Following the paper [159] of Ein, Smith, and the author, we use the subadditivity theorem to obtain some effective uniform bounds on the multiplicative behavior of graded families of ideals. The basic technical result (Theorem 11.3.1) follows immediately from the theory developed in Section 11.2, but it has some quite surprising corollaries concerning symbolic powers.

Throughout this section $X$ denotes a smooth complex variety of dimension $n$. Since the results are local, we will find it convenient to assume occasionally that $X$ is actually affine.

**Multiplier ideals and growth of graded systems.** The following statement shows that asymptotic ideals capture uniform multiplicative behavior of graded families:

**Theorem 11.3.1. (Multiplicative growth of graded systems).** *Let* $\mathfrak{a}_\bullet = \{\mathfrak{a}_k\}$ *be a graded family of ideals on* $X$, *and fix any index* $\ell \in \mathbf{N}$. *Then for every* $m \geq 1$ *one has*

$$\mathfrak{a}_\ell^m \subseteq \mathfrak{a}_{\ell m} \subseteq \mathcal{J}\big(\mathfrak{a}_\bullet^{\ell m}\big) \subseteq \mathcal{J}\big(\mathfrak{a}_\bullet^\ell\big)^m. \tag{11.4}$$

*In particular, if* $\mathcal{J}\big(\mathfrak{a}_\bullet^\ell\big) \subseteq \mathfrak{b}$ *for some index* $\ell$ *and some ideal* $\mathfrak{b}$, *then*

$$\mathfrak{a}_{m\ell} \subseteq \mathfrak{b}^m$$

*for every* $m \geq 1$, *and more generally* $\overline{\mathfrak{a}_{m\ell}} \subseteq \mathfrak{b}^m$ *for all* $m$.

One can view (11.4) as showing that passing to multiplier ideals "reverses" the natural inclusion $\mathfrak{a}_\ell^m \subseteq \mathfrak{a}_{\ell m}$ for graded families.

*Proof of Theorem 11.3.1.* This follows immediately from Theorem 11.1.19 (iv) — which shows that $\mathfrak{a}_{\ell m} \subseteq \mathcal{J}\big(\mathfrak{a}_\bullet^{\ell m}\big)$ — and the subadditivity theorem 11.2.3, which gives the last inclusion in (11.4). The final statement is a consequence of the fact that multiplier ideals are integrally closed (Corollary 9.6.13).                                                                        □

**Example 11.3.2.** Consider the ideals $\mathfrak{a}_k \subseteq \mathbf{C}[x, y]$ discussed in Examples 2.4.18 (i) and 11.1.17 (iii). These ideals are getting deeper in the sense that their colength in $\mathbf{C}[x, y]$ grows (linearly) in $k$. But there is no index $\ell > 0$ or non-trivial ideal $\mathfrak{b}$ such that $\mathfrak{a}_{m\ell} \subseteq \mathfrak{b}^m$ for all $m$. In view of the theorem, this reflects the fact that $\mathcal{J}\big(\mathfrak{a}_\bullet^\ell\big) = (1)$ for every $\ell$.                        □

**Application to symbolic powers.** Perhaps the most interesting application of Theorem 11.3.1 involves symbolic powers. Suppose that $\mathfrak{q} \subseteq \mathcal{O}_X$ is a non-zero ideal sheaf on $X$, and suppose for the moment that $\mathfrak{q}$ is radical, with zeroes $Z = \text{Zeroes}(\mathfrak{q})$. Recall (Definition 9.3.4) that the symbolic powers $\mathfrak{q}^{<m>}$ of $\mathfrak{q}$ can be described as the sheaf of all function germs vanishing to order $\geq m$ at a general point (or equivalently at every point) of each irreducible component of $Z$. It is evident that

$$\mathfrak{q}^m \subseteq \mathfrak{q}^{<m>}, \tag{11.5}$$

and it is elementary that equality holds if $Z$ is non-singular. In general, however, the inclusion can be strict:

**Example 11.3.3. (Symbolic square of coordinate axes).** Working in $X = \mathbf{C}^3$ with coordinates $x, y, z$, let $Z \subseteq X$ be the union of the three coordinate axes, defined by the ideal

$$\mathfrak{q} = (xy, yz, xz).$$

Then $xyz \in \mathfrak{q}^{<2>}$, since evidently the union of the coordinate planes has multiplicity 2 at a general point of each irreducible component of $Z$. On the other hand, since $\mathfrak{q}$ is homogeneous all the generators of $\mathfrak{q}^2$ have degree 4. Therefore $xyz \notin \mathfrak{q}^2$. Speaking somewhat vaguely, this computation suggests that the difference between $\mathfrak{q}^{<m>}$ and $\mathfrak{q}^m$ is accounted for by the singular points of $Z$. See [164, §3.9] for related examples and further information.   □

Swanson [561] established (in a much less restrictive setting) that there exists an integer $k = k(Z)$ depending on $Z$ such that

$$\mathfrak{q}^{<km>} \subseteq \mathfrak{q}^m \quad \text{for all } m \in \mathbf{N}.$$

This is already striking on geometric grounds since membership in the symbolic power on the left is tested at general smooth points of $Z$, while the actual power on the right involves also its singular points. So one's first guess might be that the worse the singularities of $Z$, the larger one will have to take the coefficient $k(Z)$ to be. Surprisingly enough this is not the case, and in fact the main result of [159] shows that there is a uniform statement depending only on the codimension of $Z$:

**Theorem 11.3.4. (Comparison theorem for symbolic powers).** *Assume that every irreducible component of $Z$ has codimension $\leq e$ in $X$. Then*

$$\mathfrak{q}^{<me>} \subseteq \mathfrak{q}^m \quad \text{for all } m \in \mathbf{N}.$$

*In particular, $\mathfrak{q}^{<m \cdot \dim X>} \subseteq \mathfrak{q}^m$ for every radical ideal $\mathfrak{q} \subseteq \mathcal{O}_X$.*

**Example 11.3.5. (Points in the plane).** Let $T \subseteq \mathbf{P}^2$ be a finite set of points, considered as a reduced scheme, and let $I$ be the homogeneous ideal of $T$. If $F$ is a homogeneous polynomial that has multiplicity $\geq 2m$ at every point of $T$, then $F \in I^m$. (Apply the theorem to the affine cone over $T$.) In spite of the very classical nature of this statement, we do not know a direct elementary proof.   □

*Proof of Theorem 11.3.4.* We consider the graded family $\mathfrak{q}^\bullet = \{\mathfrak{q}^{<k>}\}$ of symbolic powers of $\mathfrak{q}$ (Example 2.4.16 (iv)).[7] Applying Theorem 11.3.1, it is enough to show that
$$\mathcal{J}\left(e \cdot \mathfrak{q}^{<\bullet>}\right) \subseteq \mathfrak{q}. \qquad (*)$$
But since $\mathfrak{q}$ is radical, membership in $\mathfrak{q}$ is tested at a general point of each irreducible component of $Z = \mathrm{Zeroes}(\mathfrak{q})$. So after replacing $X$ by a suitable open subset we can assume for (*) that $Z$ is smooth and irreducible, of codimension $e$ in $X$. But in this case (*) is clear. For then $\mathfrak{q}^{<k>} = \mathfrak{q}^k$, and $\mathfrak{q} = \mathcal{I}_Z$ is resolved by taking $\mu : X' \longrightarrow X$ to be the blow-up of $X$ along $Z$. Writing $E$ for the corresponding exceptional divisor one has
$$K_{X'/X} = (e-1)E \quad \text{and} \quad \mathfrak{q}^k \cdot \mathcal{O}_{X'} = \mathcal{O}_{X'}(-kE).$$
Consequently
$$\mathcal{J}\left(e \cdot \mathfrak{q}^{<\bullet>}\right) = \mu_* \mathcal{O}_{X'}(K_{X'/X} - eE) = \mu_* \mathcal{O}_{X'}(-E) = \mathfrak{q},$$
as asserted. □

**Example 11.3.6. (Higher symbolic powers).** In the situation of Theorem 11.3.4, one has more generally :
$$\mathfrak{q}^{<m\ell>} \subseteq \left(\mathfrak{q}^{<\ell+1-e>}\right)^m \qquad \text{for all } \ell \geq e \text{ and } m \geq 1.$$

(Arguing as above, one finds that $\mathcal{J}\left(\ell \cdot \mathfrak{q}^{<\bullet>}\right) \subseteq \mathfrak{q}^{<\ell+1-e>}$.) □

It was observed by Hochster and Huneke [290] that one does not need to assume that $\mathfrak{q}$ is radical in Theorem 11.3.4. Still following [159], we explain the extension to the case of unmixed ideals, referring to [290] for more general statements.

Assume for simplicity of exposition that $X$ is affine. Given an ideal $\mathfrak{q} \subseteq \mathbf{C}[X]$, fix a primary decomposition
$$\mathfrak{q} = \mathfrak{q}_1 \cap \ldots \cap \mathfrak{q}_h \qquad (*)$$
of $\mathfrak{q}$, and let $Z_i = \mathrm{Zeroes}\left(\sqrt{\mathfrak{q}_i}\right)$ be the subvarieties of $X$ corresponding to the associated primes $\mathfrak{p}_i = \sqrt{\mathfrak{q}_i}$ of $\mathfrak{q}$. We assume that $\mathfrak{q}$ is *unmixed*, i.e. that none of the associated primes $\mathfrak{p}_i$ are embedded (or equivalently that there are no inclusions among the $Z_i$). Then the symbolic powers $\mathfrak{q}^{<k>} \subseteq \mathbf{C}[X]$ of $\mathfrak{q}$ are defined as follows. For each associated subvariety $Z_i$ of $\mathfrak{q}$, there is a natural map $\phi_i : \mathbf{C}[X] \longrightarrow \mathcal{O}_{Z_i} X$ from the coordinate ring of $X$ to the local ring of $X$ along $Z_i$. One then sets

---

[7] In an effort to avoid confusion, we will write $\mathcal{J}\left(c \cdot \mathfrak{q}^{<\bullet>}\right)$ for the corresponding multiplier ideals, reserving the exponent for the power in question. However, in light of Theorem 11.1.19 (i), confusing the indexing variable and the weighting coefficient is actually harmless.

$$\mathfrak{q}^{<k>} = \bigcap_{i=1}^{h} \phi_i^{-1}\Big(\mathfrak{q}^k \cdot \mathcal{O}_{Z_i}X\Big).$$

In other words, $f \in \mathfrak{q}^{<k>}$ if and only if there is an element $s \in \mathbf{C}[X]$, not lying in any of the associated primes $\mathfrak{p}_i$ of $\mathfrak{q}$, such that $fs \in \mathfrak{q}^k$. (It is a theorem of Nagata and Zariski that this agrees with the alternative definition of symbolic powers given above when $\mathfrak{q}$ is radical. See [164, Chapter 3.9].)

Theorem 11.3.4 generalizes in the natural manner:

**Variant 11.3.7. (Symbolic powers of unmixed ideals).** *Let $\mathfrak{q} \subseteq \mathbf{C}[X]$ be an unmixed ideal, and assume that every associated subvariety $Z_i$ of $\mathfrak{q}$ has codimension $\leq e$ in $X$. Then $\mathfrak{q}^{<me>} \subseteq \mathfrak{q}^m$.*

*Proof.* The symbolic powers $\mathfrak{q}^{<\bullet>} = \{\mathfrak{q}^{<k>}\}$ again form a graded family of ideals, so Theorem 11.3.1 will apply as soon as we establish that $\mathcal{J}\big(e \cdot \mathfrak{q}^{<\bullet>}\big) \subseteq \mathfrak{q}$. For this we argue much as in the proof of Variant 9.6.37 of Skoda's Theorem. Namely, referring to the primary decomposition (*), it is enough to show that

$$\mathcal{J}\big(e \cdot \mathfrak{q}^{<\bullet>}\big) \subseteq \mathfrak{q}_i \quad \text{for each } 1 \leq i \leq h. \tag{**}$$

For a given index $i$, inclusion in $\mathfrak{q}_i$ is tested at a generic point of $Z_i$. So having fixed $i$ we are free to replace $X$ by any open subset meeting $Z_i$. Therefore, by definition of the symbolic powers, we may assume after localizing that $\mathfrak{q}^{<k>} = \mathfrak{q}^k$. But in this case $\mathcal{J}\big(e \cdot \mathfrak{q}^{<\bullet>}\big) = \mathcal{J}\big(\mathfrak{q}^e\big)$, and then $\mathcal{J}\big(\mathfrak{q}^e\big) \subseteq \mathfrak{q} \subseteq \mathfrak{q}_i$ thanks to Variant 9.6.37 of Skoda's theorem. □

**Remark 11.3.8.** Theorem 11.3.4 was reproved and generalized by Hochster and Huneke [290] using the theory of tight closure. In fact, these authors show that 11.3.4 and 11.3.7 remain valid in any Noetherian regular local ring containing a field. While there are certain similarities of spirit between the two arguments — e.g. both reduce to the situation in which $\mathfrak{q}^{<k>} = \mathfrak{q}^k$ — the precise connections between the approaches were for a long time quite mysterious. However, a concrete link has now been provided by the recent work of Hara–Yoshida [271] and Takagi [563] giving a tight closure interpretation of multiplier ideals. □

**Example 11.3.9. (Izumi's theorem).** One can use Theorem 11.3.1 to render effective and extend in certain directions a formulation due to Hübl and Swanson [295, (1.4)] of a theorem of Izumi [307]. Specifically, let $\nu : Y \longrightarrow X$ be a birational mapping of non-singular projective varieties. Let $E \subseteq Y$ be a prime divisor, set

$$\ell = 1 + \operatorname{ord}_E(K_{Y/X}),$$

and for $k \geq 1$ put $\mathfrak{o}_k = \nu_* \mathcal{O}_Y(-kE)$. Fix an irreducible subvariety $Z \subseteq X$ such that $Z \subseteq \nu(E)$ and denote by $\mathfrak{p} = \mathcal{I}_Z$ the ideal of $Z$. Then

$$\mathfrak{o}_{\ell m} \subseteq \mathfrak{p}^m \quad \text{for all } m \geq 1.$$

(A straight-forward calculation shows that $\mathcal{J}\left(\|\mathfrak{o}_\ell\|\right) \subseteq \mathfrak{p}$. This result appears in [159, §2], to which we refer for details.)    $\square$

**Remark 11.3.10. (Multiplicative approximation of Abhyankar valuations on smooth varieties).** The paper [160] of Ein, Smith, and the author contains another algebraic application of Theorem 11.3.1. Let $X$ be a smooth affine variety of dimension $n$, and $x \in X$ a fixed point. Referring to [160] for the definitions, let $\nu$ be an Abhyankar valuation centered at $x$: this is an **R**-valued valuation of $\mathbf{C}(X)$ centered at $x$ that satisfies the inequality

$$\text{trans. deg}\, \nu + \text{rat. rk.}\, \nu = n.$$

For example, a divisorial valuation centered at $x$ is Abhyankar, as are the valuations on $\mathbf{C}^n$ given by weighted degree with **Q**-independent weights (cf. Examples 2.4.18 (ii) and 2.4.19). Then $\nu$ determines ideals

$$\mathfrak{a}_k = \left\{ f \in \mathbf{C}[X] \,\middle|\, \nu(f) \geq k \right\}.$$

The main result of [160] is that there exists a fixed integer $k > 0$ such that

$$\mathfrak{a}_\ell^m \subseteq \mathfrak{a}_{m\ell} \subseteq \mathfrak{a}_{\ell-k}^m$$

for every $\ell \geq k$ and every $m \geq 1$. For the proof, the essential point is to produce an integer $k$ with the property that $\mathcal{J}\left(\mathfrak{a}_\bullet^\ell\right) \subseteq \mathfrak{a}_{\ell-k}$ for all $\ell \geq k$; then 11.3.1 applies.    $\square$

**Comparisons among graded systems.** Finally, we record a result comparing two graded systems and the corresponding multiplier ideals. The first statement is from [157]. A global analogue of (ii) will come into the proof of Siu's theorem on deformation-invariance of plurigenera in Section 11.5.

**Theorem 11.3.11. (Comparing graded systems).** *Let* $\mathfrak{a}_\bullet = \{\mathfrak{a}_k\}$ *and* $\mathfrak{b}_\bullet = \{\mathfrak{b}_k\}$ *be two graded families of ideals on a smooth affine variety* $X$.

(i). *Suppose that there exists a fixed non-zero element* $\phi \in \mathbf{C}[X]$ *such that*

$$\phi \cdot \mathfrak{a}_k \subseteq \mathfrak{b}_k \quad \text{for all } k \gg 0.$$

*Then* $\mathcal{J}\left(\mathfrak{a}_\bullet^c\right) \subseteq \mathcal{J}\left(\mathfrak{b}_\bullet^c\right)$ *for every* $c > 0$.

(ii). *Suppose that there exists a fixed non-zero element* $\phi \in \mathbf{C}[X]$ *such that*

$$\phi \cdot \mathfrak{a}_k \subseteq \mathcal{J}\left(\mathfrak{b}_\bullet^k\right) \quad \text{for all } k \gg 0.$$

*Then* $\mathfrak{a}_m \subseteq \mathcal{J}\left(\mathfrak{b}_\bullet^m\right)$ *for every* $m \geq 1$.

*Proof.* For (i), fix $c$ and then choose $p > 0$ computing the asymptotic multiplier ideals in question. It follows from Example 9.2.30 that for $\ell \gg 0$,

$$\mathcal{J}(\mathfrak{a}_\bullet^c) \; = \; \mathcal{J}(\mathfrak{a}_p^{c/p})$$
$$= \; \mathcal{J}(\phi^{c/p\ell} \cdot \mathfrak{a}_p^{c/p})$$
$$= \; \mathcal{J}((\phi^p \cdot \mathfrak{a}_p^{p\ell})^{c/p^2\ell}).$$

On the other hand, $\mathfrak{a}_p^{p\ell} \subseteq \mathfrak{a}_{p\ell}^p$. So applying the hypothesis with $k = p\ell$, we find:

$$\mathcal{J}((\phi^p \cdot \mathfrak{a}_p^{p\ell})^{c/p^2\ell}) \; \subseteq \; \mathcal{J}((\phi \cdot \mathfrak{a}_{p\ell})^{pc/p^2\ell})$$
$$\subseteq \; \mathcal{J}(\mathfrak{b}_{p\ell}^{c/p\ell})$$
$$= \; \mathcal{J}(\mathfrak{b}_\bullet^c),$$

as required. Turning to (ii), fix $m \geq 1$. Apply the hypothesis with $k = \ell m$ to deduce that for $\ell \gg 0$:

$$\phi \cdot \mathfrak{a}_m^\ell \; \subseteq \; \phi \cdot \mathfrak{a}_{\ell m}$$
$$\subseteq \; \mathcal{J}(\mathfrak{b}_\bullet^{\ell m})$$
$$\subseteq \; \mathcal{J}(\mathfrak{b}_\bullet^m)^\ell,$$

the last inclusion coming from subadditivity (Theorem 11.2.3). Thus $\mathfrak{a}_m \subseteq \overline{\mathcal{J}(\mathfrak{b}_\bullet^m)}$ thanks to Proposition 9.6.6. But the multiplier ideal $\mathcal{J}(\mathfrak{b}_\bullet^m)$ is integrally closed (Example 9.6.13), and hence $\mathfrak{a}_m \subseteq \mathcal{J}(\mathfrak{b}_\bullet^m)$, as asserted.    □

**Example 11.3.12. (Numerical nature of asymptotic multiplier ideals attached to big divisors, [157]).** Let $L_1$ and $L_2$ be numerically equivalent big divisors on a smooth projective variety $X$. Then

$$\mathcal{J}(X, c \cdot \|L_1\|) \; = \; \mathcal{J}(X, c \cdot \|L_2\|)$$

for every $c > 0$. (Arguing as in the proof of Theorem 11.2.21, one finds a fixed effective divisor $E$ and a fixed integer $a \geq 0$ such that

$$\mathcal{J}(\|(k - a)L_1\|) \otimes \mathcal{O}_X(-E) \; \subseteq \; \mathfrak{b}(|kL_2|)$$

for all $k \geq a$. On the other hand,

$$\mathfrak{b}(|kL_1|) \; \subseteq \; \mathcal{J}(\|kL_1\|) \; \subseteq \; \mathcal{J}(\|(k - a)L_1\|)$$

for all $k \geq a$. Putting these together, we conclude that

$$\mathfrak{b}(|kL_1|) \otimes \mathcal{O}_X(-E) \; \subseteq \; \mathfrak{b}(|kL_2|)$$

for $k \gg 0$, and then 11.3.11 (i) implies that

$$\mathcal{J}(X, c \cdot \|L_1\|) \; \subseteq \; \mathcal{J}(X, c \cdot \|L_2\|)$$

for every $c > 0$. The reverse inclusion follows symmetrically.)    □

## 11.4 Fujita's Approximation Theorem

We discuss in this section a very interesting theorem of Fujita [199] stating that the volume of a big divisor can be approximated arbitrarily closely by the self-intersection of an ample line bundle on a modification. These approximations lead to several results showing that arbitrary big linear series display more structure than one might have expected. For example, Fujita's theorem implies the log-concavity of the volume function (Theorem 11.4.9), and Boucksom, Demailly, Paun, and Peternell have recently used it to describe the dual of the cone of pseudoeffective divisors (Section 11.4.C).

   We start in Section 11.4.A with the statement and some first consequences of Fujita's theorem. A proof of Fujita's result — closely following [130] — appears in Section 11.4.B. The work of Boucksom, Demailly, Paun, and Peternell is presented in 11.4.C.

**Notation 11.4.1.** Throughout this section, $X$ denotes unless otherwise stated an irreducible projective variety of dimension $n$.                            □

### 11.4.A Statement and First Consequences

Let $L$ be an integral divisor on $X$. Recall from Section 2.2.C that the volume of $L$ measures the rate of growth of $H^0(X, \mathcal{O}_X(mL))$:

**Definition 11.4.2. (Volume of a divisor).** The *volume* of $L$ is the real number
$$\operatorname{vol}(L) = \operatorname{vol}_X(L) = \limsup_{m \to \infty} \frac{h^0(X, \mathcal{O}(mL))}{m^n/n!}. \quad \square$$

Thus $L$ is big iff $\operatorname{vol}(L) > 0$. The volume of a **Q**-divisor $D$ is defined analogously, the limit being taken over $m$ such that $mD$ is an integral divisor. Numerically equivalent divisors have the same volume, and the volume extends uniquely to a continuous function
$$\operatorname{vol}_X : N^1(X)_{\mathbf{R}} \longrightarrow \mathbf{R}.$$

If $\mu : X' \longrightarrow X$ is a birational morphism of irreducible projective varieties, then $\operatorname{vol}_{X'}(\mu^*\xi) = \operatorname{vol}_X(\xi)$ for every $\xi \in N^1(X)_{\mathbf{R}}$. We refer to Section 2.2.C for proofs and examples.

   Fujita's idea is to approximate an arbitrary big divisor by an ample one. To streamline the discussion, it is convenient to introduce some terminology:

**Definition 11.4.3. (Fujita approximation of a big class).** Let $\xi \in N^1(X)_{\mathbf{R}}$ be a big class. A *Fujita approximation* for $\xi$ consists of a projective birational morphism $\mu : X' \longrightarrow X$, with $X'$ irreducible, together with a decomposition
$$\mu^*(\xi) = a + e \tag{11.6}$$

in $N^1(X')_{\mathbf{R}}$, such that $a$ is ample and $e$ is effective.[8]                    □

Note that given such a decomposition one has

$$\mathrm{vol}_X(\xi) \; = \; \mathrm{vol}_{X'}(\mu^*\xi) \; \geq \; \mathrm{vol}_{X'}(a) \; = \; \left(a^n\right)_{X'}$$

thanks to the effectivity of $e$ (Example 2.2.48). Here and subsequently, when confusion seems possible we use a subscript to indicate the space on which an intersection number is computed.

The main theorem asserts that "most of" the volume of a big divisor is accounted for by the volume of the ample class in a Fujita approximation:

**Theorem 11.4.4. (Fujita's approximation theorem).** *Let $\xi$ be a big class on $X$, and fix any positive $\varepsilon > 0$. Then there exists a Fujita approximation*

$$\mu : X' \longrightarrow X \quad , \quad \mu^*(\xi) = a + e \tag{11.7}$$

*having the property that*

$$\mathrm{vol}_{X'}(a) \; > \; \mathrm{vol}_X(\xi) - \varepsilon.$$

*Moreover, if $\xi$ is a rational class, then one can take $a$ and $e$ to be rational as well.*

We will refer to the output of the theorem as a *Fujita $\varepsilon$-approximation* of $\xi$. We stress that the data in (11.7) depend on $\varepsilon$. The proof of 11.4.4 appears in the next subsection.

The next examples sketch some variations and formal consequences of Fujita's theorem. More substantial applications appear later in the subsection.

**Example 11.4.5. (Modifications).** In the situation of the Theorem, let

$$\mu : X' \longrightarrow X \quad , \quad \mu^*(\xi) = a + e$$

be a Fujita $\varepsilon$-approximation of a big class $\xi \in N^1(X)_{\mathbf{R}}$, and let $\nu : X'' \longrightarrow X'$ be a projective birational morphism. Then $\mu_1 = \mu \circ \nu$ is again a Fujita $\varepsilon$-approximation to $\xi$. (There exist arbitrarily small effective classes $e_0$ on $X''$ such that $a_1 = \nu^*(a) - e_0$ is ample (Example 2.2.19). Setting $e_1 = \nu^*(e) + e_0$ one has $\mu_1^*(\xi) = a_1 + e_1$, and by taking $e_0$ sufficiently small we can arrange that $\mathrm{vol}_{X''}(a_1) > \mathrm{vol}_X(\xi) - \varepsilon$.) It follows that we can assume in 11.4.4 that $X'$ is normal or even smooth, and that moreover one can find a simultaneous $\varepsilon$-approximation to any finite collection of big classes.                    □

**Example 11.4.6. (Fujita approximation of integral divisors).** Suppose that $L$ is a big integral divisor on the projective variety $X$, and fix $\varepsilon > 0$.

---

[8] Recall (Definition 2.2.30) that a class $e \in N^1(X')_{\mathbf{R}}$ is effective if $e$ is represented by an effective $\mathbf{R}$-divisor on $X'$.

In this setting, Fujita's theorem gives the existence of a birational morphism $\mu : X' \longrightarrow X$ and an integer $p > 0$ such that

$$\mu^*(pL) \equiv_{\text{lin}} A + E,$$

where $A$ and $E$ are respectively an ample and an effective divisor, both integral, and

$$\left(A^n\right) > p^n \cdot \left(\text{vol}_X(L) - \varepsilon\right).$$

(This follows from the last statement of 11.4.4.) By the previous example we are free to assume in addition that $X'$ is normal or non-singular.    □

**Example 11.4.7. (Volume as a limit).** Let $L$ be an integral divisor on $X$. Then

$$\liminf_{m \to \infty} \frac{h^0\left(X, \mathcal{O}(mL)\right)}{m^n/n!} = \limsup_{m \to \infty} \frac{h^0\left(X, \mathcal{O}(mL)\right)}{m^n/n!},$$

i.e. the $\limsup$ computing $\text{vol}_X(L)$ is actually a limit. (One can suppose that $L$ is big, the statement being trivial otherwise. Arguing as in Proposition 2.2.43, one can also assume without loss of generality that $X$ is normal. Fix $\varepsilon > 0$, and let

$$\mu : X' \longrightarrow X \quad , \quad \mu^*(pL) \equiv_{\text{lin}} A + E$$

be the Fujita approximation described in Example 11.4.6, with $X'$ normal. Since $L$ is big, there exists an integer $q_0 > 0$ such that $rL$ is effective for each $r$ with $q_0 p \le r \le (q_0 + 1)p$. Then

$$h^0\left(X, \mathcal{O}_X\left((kp+r)L\right)\right) \ge h^0\left(X, \mathcal{O}_X\left(kpL\right)\right) \ge h^0\left(X', \mathcal{O}_{X'}\left(kA\right)\right)$$

for all $k \ge 0$ and $r \in [q_0 p, (q_0 + 1)p]$. On the other hand, since $A$ is ample, there exists some $k_0 \gg 0$ such that

$$\begin{aligned}
h^0\left(X', kA\right) &\ge \left(\left(A^n\right)_{X'} - p^n \varepsilon\right) \cdot \frac{k^n}{n!} \\
&> \left(p^n \cdot \left(\text{vol}_X(L) - 2\varepsilon\right)\right) \cdot \frac{k^n}{n!}
\end{aligned}$$

for $k \ge k_0$ (Example 2.3.26). The assertion follows.)    □

**Remark 11.4.8. (Connection with Zariski decompositions).** Let $L$ be an integral divisor on a smooth projective variety $X$. Suppose that $L$ admits a Zariski decomposition in the sense of Cutkosky, Kawamata, and Moriwaki (Section 2.3.E), i.e. assume that there exists a birational modification $\mu : X' \longrightarrow X$, together with a decomposition $\mu^* L = P + N$, where $P$ and $N$ are $\mathbf{Q}$-divisors, with $P$ nef and $N$ effective, having the property that

$$H^0\left(X, \mathcal{O}_X(mL)\right) = H^0\left(X', \mathcal{O}_{X'}([mP])\right)$$

for all $m \ge 0$. Then

$$\mathrm{vol}_X(L) \;=\; \mathrm{vol}_{X'}(P) \;=\; \left(P^n\right)_{X'},$$

i.e. the volume of $L$ is computed by the volume of a nef divisor on a modification. While it is known that such decompositions do not exist in general ([97], Section 2.3.E), one can view 11.4.4 as asserting that an approximate asymptotic statement does hold. For this reason, the output of the theorem is sometimes called an approximate Zariski decomposition.     □

We next use Fujita's theorem to show that volumes satisfy a basic inequality of Brunn–Minkowski type:

**Theorem 11.4.9. (Log-concavity of volume).** *One has*

$$\mathrm{vol}\!\left(\xi + \xi'\right)^{1/n} \;\ge\; \mathrm{vol}\!\left(\xi\right)^{1/n} + \mathrm{vol}\!\left(\xi'\right)^{1/n}$$

*for any two big classes $\xi, \xi' \in N^1(X)_{\mathbf{R}}$. In other words, the volume function*

$$\mathrm{vol}_X : N^1(X)_{\mathbf{R}} \longrightarrow \mathbf{R}$$

*is log-concave on the big cone of $X$.*

*Proof.* For ample classes the stated inequality follows from 1.6.3 (iii), and the plan is to use 11.4.4 to reduce to this case. To this end, fix $\varepsilon > 0$ and construct a simultaneous Fujita-approximation

$$\mu : X' \longrightarrow X \quad, \quad \mu^*(\xi) = a + e \; , \quad \mu^*(\xi') = a' + e'$$

to the two classes in question, with

$$\mathrm{vol}_{X'}(a)^{1/n} \ge \mathrm{vol}_X(\xi)^{1/n} - \tfrac{\varepsilon}{2} \quad , \quad \mathrm{vol}_{X'}(a')^{1/n} \ge \mathrm{vol}_X(\xi')^{1/n} - \tfrac{\varepsilon}{2}.$$

Observing that $\mathrm{vol}_X(\xi + \xi') \ge \mathrm{vol}_{X'}(a + a')$ since the difference $\mu^*(\xi + \xi') - (a + a')$ is effective, one finds:

$$\begin{aligned}
\mathrm{vol}_X(\xi + \xi')^{1/n} &\ge \mathrm{vol}_{X'}(a + a')^{1/n} \\
&\ge \mathrm{vol}_{X'}(a)^{1/n} + \mathrm{vol}_{X'}(a')^{1/n} \\
&\ge \mathrm{vol}_X(\xi)^{1/n} + \mathrm{vol}_X(\xi')^{1/n} - \varepsilon.
\end{aligned}$$

As this holds for all $\varepsilon > 0$, the theorem follows.     □

Recall that if $L$ is ample, then

$$\mathrm{vol}_X(L) \;=\; \left(L^n\right)$$

is the top self-intersection number of $L$. So in this case one has a concrete geometric interpretation of the volume of $L$: fixing $m$ so that $mL$ is very ample, one picks $n$ general divisors $D_1, \ldots, D_n \in |mL|$, and then

$$\operatorname{vol}_X(L) = \frac{1}{m^n} \cdot \#\Big(D_1 \cap \ldots \cap D_n\Big).$$

Our next goal is to use Theorem 11.4.4 to establish an extension of this statement to arbitrary big divisors.

First, a definition:

**Definition 11.4.10. (Moving self-intersection number).** Let $L$ be a big divisor on $X$. Fix a positive integer $m > 0$ sufficiently large so that that linear series $|mL|$ defines a birational mapping of $X$, and denote by $B_m = \operatorname{Bs}\big(|mL|\big)$ the base locus of $|mL|$. The *moving self-intersection number* $(mL)^{[n]}$ of $|mL|$ is then defined by choosing $n$ general divisors $D_1, \ldots, D_n \in |mL|$ and putting

$$\big(mL\big)^{[n]} = \#\Big(D_1 \cap \ldots \cap D_n \cap (X - B_m)\Big). \qquad \square$$

In other words, we simply count the number of intersection points away from the base locus of $n$ general divisors in the linear series $|mL|$. This notion arises for example in Matsusaka's proof of his "big theorem" (cf. [401]).

At least when $X$ is normal, the volume $\operatorname{vol}_X(L)$ of an arbitrary big divisor measures the rate of growth with respect to $m$ of these moving self-intersection numbers:

**Theorem 11.4.11. (Growth of moving self-intersection numbers).** *Let $L$ be a big divisor on a normal projective variety $X$. Then*

$$\operatorname{vol}_X(L) = \limsup_{m \to \infty} \frac{\big(mL\big)^{[n]}}{m^n}.$$

**Remark 11.4.12.** The intuition underlying Theorem 11.4.11 goes back to Demailly's paper [124] (see especially the remark following Corollary 10.8 of that paper). In fact, Fujita seems to have been led to 11.4.4 by the project of algebrizing [124]: see also [158] and [394, §7].                                         $\square$

*Proof of Theorem 11.4.11.* We start by interpreting $(mL)^{[n]}$ geometrically. Let

$$\nu_m : X_m \longrightarrow X$$

be a resolution of the blowing-up of $X$ along the base ideal $\mathfrak{b}\big(|mL|\big)$ of $|mL|$. Thus $\nu_m^*|mL| = |P_m| + F_m$, where

$$P_m =_{\mathrm{def}} \nu_m^*(mL) - F_m$$

is free,

$$H^0\big(X, \mathscr{O}_X(mL)\big) = H^0\big(X_m, \mathscr{O}_{X_m}(\nu_m^*(mL))\big) = H^0\big(X_m, \mathscr{O}_{X_m}(P_m)\big)$$

(the first equality by normality), and $B_m = \nu_m(F_m)$. Then evidently $(mL)^{[n]}$ counts the number of intersection points of $n$ general divisors in $|P_m|$, and consequently

$$\left(mL\right)^{[n]} \;=\; \left(\left(P_m\right)^n\right)_{X_m}.$$

We have $\left(\left(P_m\right)^n\right) = \mathrm{vol}_{X_m}(P_m)$ since $P_m$ is nef, and $\mathrm{vol}_X(mL) \ge \mathrm{vol}_{X_m}(P_m)$ since $kP_m$ embeds in $\mu_m^*(mkL)$. Hence

$$\mathrm{vol}_X(mL) \;\ge\; \left(mL\right)^{[n]} \quad \text{for } m \gg 0.$$

On the other hand, $\mathrm{vol}_X(mL) = m^n \cdot \mathrm{vol}_X(L)$ (Proposition 2.2.35), and so we conclude that

$$\mathrm{vol}_X(L) \;\ge\; \frac{\left(mL\right)^{[n]}}{m^n}. \tag{11.8}$$

for every $m \gg 0$.

For the reverse inequality we use Fujita's theorem. Fix $\varepsilon > 0$, and construct an $\varepsilon$-approximation

$$\mu : X' \longrightarrow X \quad , \quad \mu^*L \;=\; A + E$$

to $L$, where $A$ and $E$ are $\mathbf{Q}$-divisors on $X'$ with $A$ ample and $E$ effective. Let $k$ be any positive integer such that $kA$ is integral and globally generated, and consider the corresponding modification $\nu_k : X_k \longrightarrow X$ constructed in the previous paragraph, so that $\nu_k^*(kL) \equiv_{\mathrm{lin}} P_k + F_k$. By passing to a common resolution we can assume that $X_k$ and $X'$ coincide. So we can write

$$\nu_k^*(kL) \;\equiv_{\mathrm{lin}}\; A_k + E_k$$

with $A_k = kA$ globally generated, $E_k = kE$ effective, and

$$\left(\left(A_k\right)^n\right)_{X_k} \;\ge\; k^n \cdot \left(\mathrm{vol}_X(L) - \varepsilon\right).$$

But since $|A_k|$ is free, we have $E_k \succcurlyeq F_k$. Consequently

$$\left(\left(A_k\right)^n\right)_{X_k} \;\le\; \left(\left(P_k\right)^n\right)_{X_k} \;=\; \left(kL\right)^{[n]}$$

(compare Lemma 11.4.22). Therefore

$$\frac{\left(kL\right)^{[n]}}{k^n} \;\ge\; \mathrm{vol}_X(L) - \varepsilon. \tag{*}$$

But $(*)$ holds for any sufficiently large and divisible $k$, and in view of (11.8) the theorem follows. $\qquad\square$

**Remark 11.4.13. (Equivariant analogues).** Paoletti [493] has extended some of these results to the setting of a smooth projective variety $X$ on which a finite group $G$ acts faithfully. In particular, he establishes a $G$-equivariant analogue of Fujita's theorem. Given a big line bundle $\mathcal{O}_X(L)$ linearized by $G$ — so that $G$ acts on all the spaces $H^0(X, \mathcal{O}_X(mL))$ — Paoletti deduces from this an appealing formula for the asymptotic rate of growth of the isotypical component of $H^0(X, \mathcal{O}_X(mL))$ corresponding to an irreducible representation of $G$. $\qquad\square$

**Remark 11.4.14. (Fujita approximations on Kähler manifolds).** In another direction, Boucksom [66] has extended many of the ideas and results of this subsection to an arbitrary compact Kähler manifold $X$. One can use an analogue of 11.4.11 to define the volume of a big class $\alpha \in H^{1,1}(X, \mathbf{R})$. Boucksom establishes a version of Fujita's theorem in this setting, and he applies it to investigate the nature of the resulting volume function.          □

## 11.4.B Proof of Fujita's Theorem

We now turn to the proof of Fujita's Theorem 11.4.4, following the approach of Demailly, Ein and the author in [130]. This is where multiplier ideals enter into the picture. The presentation in this subsection closely follows [130, §3].

*Proof of Theorem 11.4.4.* We begin with some reductions. By passing first to a desingularization, we can suppose that $X$ is smooth. It is also enough to prove the statement for rational classes, for the general case then follows by perturbation. So we can assume that $\xi$ is the class of a big integral divisor $L$, and we need to construct a modification $\mu : X' \longrightarrow X$, together with a decomposition

$$\mu^*(L) \equiv_{\text{num}} A + E, \tag{11.9}$$

where $A$ and $E$ are $\mathbf{Q}$-divisors on $X'$ with $A$ ample and $E$ effective, and $\text{vol}_{X'}(A) \geq \text{vol}_X(L) - \varepsilon$. Finally, observe that it suffices to produce a big and nef divisor $A$ in (11.9). For then there exists an effective divisor $F$ on $X'$ such that $A - \delta F$ is ample for arbitrarily small $\delta > 0$ (Example 2.2.19), and writing

$$\mu^*(L) \equiv_{\text{num}} (A - \delta F) + (E + \delta F)$$

gives a decomposition in which the first term is actually ample. On the other hand, by taking $\delta$ sufficiently small, we can arrange that $\text{vol}_{X'}(A - \delta F)$ comes arbitrarily close to $\text{vol}_{X'}(A)$.

Fix now a very ample bundle $B$ on $X$, and set $G = K_X + (n+1)B$. We suppose that $G$ is very ample, and for $p \geq 0$ put

$$M_p = pL - G.$$

Given $\varepsilon > 0$, use Proposition 2.2.35 (ii) to fix $p \gg 0$ such that

$$\text{vol}(M_p) \geq p^n (\text{vol}(L) - \varepsilon). \tag{11.10}$$

We further assume that $p$ is sufficiently large so that $M_p$ is big.

Having chosen $p \gg 0$ satisfying (11.10), we will produce an ideal sheaf $\mathcal{J} = \mathcal{J}_p \subset \mathcal{O}_X$ (depending on $p$) such that

$$\mathcal{O}_X(pL) \otimes \mathcal{J} \text{ is globally generated;} \tag{11.11a}$$

$$H^0\big(X, \mathcal{O}_X(\ell M_p)\big) \subseteq H^0\big(X, \mathcal{O}_X(\ell pL) \otimes \mathcal{J}^\ell\big) \quad \text{for all } \ell \geq 1. \tag{11.11b}$$

Granting for the time being the existence of $\mathcal{J}$, we complete the proof. Let $\mu : X' \longrightarrow X$ be a resolution of the blowing up of $\mathcal{J}$, so that $\mathcal{J} \cdot \mathcal{O}_{X'} = \mathcal{O}_{X'}(-E_p)$ for some effective divisor $E_p$ on $X'$. It follows from (11.11a) that

$$A_p =_{\text{def}} \mu^*(pL) - E_p$$

is globally generated, and hence nef. Using (11.11b) we find:

$$\begin{aligned} H^0\big(X, \mathcal{O}_X(\ell M_p)\big) &\subseteq H^0\big(X, \mathcal{O}_X(\ell pL) \otimes \mathcal{J}^\ell\big) \\ &\subseteq H^0\big(X', \mathcal{O}_{X'}(\mu^*(\ell pL) - \ell E_p)\big) \\ &= H^0\big(X', \mathcal{O}_{X'}(\ell A_p)\big) \end{aligned}$$

(which shows in particular that $A_p$ is big). This implies that

$$\begin{aligned} \big((A_p)^n\big)_{X'} &= \text{vol}_{X'}(A_p) \\ &\geq \text{vol}_X(M_p) \\ &\geq p^n\big(\text{vol}(L) - \varepsilon\big), \end{aligned}$$

so we get the decomposition (11.9) upon setting $A = \frac{1}{p}A_p$ and $E = \frac{1}{p}E_p$.

Turning to the construction of $\mathcal{J}$, set

$$\mathcal{J} = \mathcal{J}\big(X, \|M_p\|\big).$$

Since $pL = M_p + \big(K_X + (n+1)B\big)$, (11.11a) follows from Corollary 11.2.13 applied to $M_p$. As for (11.11b), it follows in the first place from Proposition 11.2.10 and subadditivity (Theorem 11.2.4) that

$$\begin{aligned} H^0\Big(X, \mathcal{O}_X(\ell M_p)\Big) &= H^0\Big(X, \mathcal{O}_X(\ell M_p) \otimes \mathcal{J}\big(\|\ell M_p\|\big)\Big) \\ &\subseteq H^0\Big(X, \mathcal{O}_X(\ell M_p) \otimes \mathcal{J}\big(\|M_p\|\big)^\ell\Big). \end{aligned} \qquad (11.12)$$

Now fix a divisor $D \in |G|$. Then multiplication by (a local equation for) $\ell D$ gives rise to an injective sheaf homomorphism

$$\mathcal{O}_X(\ell M_p) \otimes \mathcal{J}\big(\|M_p\|\big)^\ell \xrightarrow{\cdot \ell D} \mathcal{O}_X(\ell pL) \otimes \mathcal{J}\big(\|M_p\|\big)^\ell,$$

and consequently an inclusion

$$H^0\Big(X, \mathcal{O}_X(\ell M_p) \otimes \mathcal{J}\big(\|M_p\|\big)^\ell\Big) \subseteq H^0\Big(X, \mathcal{O}_X(\ell pL) \otimes \mathcal{J}\big(\|M_p\|\big)^\ell\Big).$$

The required embedding (11.11b) is deduced by combining (11.12) with the previous equation. This completes the proof of Fujita's theorem.    □

**Remark 11.4.15. (Nakamaye's proof of Fujita's theorem).** Nakamaye [469, §2] has given another proof of Theorem 11.4.4. His interesting argument is based on the analysis of a Seshadri-like constant that measures how many jets an arbitrary big line bundle generates at a general point of a variety (Remark 10.3.7).    □

### 11.4.C The Dual of the Pseudoeffective Cone

Boucksom, Demailly, Paun, and Peternell [68] have used the Fujita approximation theorem to identify the cone of curves dual to the closed cone of pseudoeffective divisors (Definition 2.2.25). The present subsection is devoted to an account of their work, closely following their paper.

We start with a definition:

**Definition 11.4.16. (Mobile curves, movable cone).** Let $X$ be an irreducible projective variety of dimension $n$. A class $\gamma \in N_1(X)_{\mathbf{R}}$ is *movable* or *mobile* if there exists a projective birational mapping $\mu : X' \longrightarrow X$, together with ample classes $a_1, \ldots, a_{n-1} \in N^1(X')_{\mathbf{R}}$, such that

$$\gamma = \mu_*(a_1 \cdot \ldots \cdot a_{n-1}).$$

The *movable cone*

$$\overline{\mathrm{Mov}}(X) \subseteq N_1(X)_{\mathbf{R}}$$

of $X$ is the closed convex cone spanned by all movable classes.    □

Here $\mu_* : N_1(X')_{\mathbf{R}} \longrightarrow N_1(X)_{\mathbf{R}}$ is the homomorphism determined by proper pushforward of numerical equivalence classes (cf. [208, Chapter 1.4]).

**Example 11.4.17.** If $X$ is a smooth projective surface, so that $N_1(X)_{\mathbf{R}} = N^1(X)_{\mathbf{R}}$, then it follows from the definition that $\overline{\mathrm{Mov}}(X) = \mathrm{Nef}(X)$.    □

The terminology stems from the fact that a mobile class $\gamma$ can be represented by an effective cycle whose support meets any given divisor properly:

**Lemma 11.4.18. (Supports of movable classes).** *Let $\gamma \in N_1(X)_{\mathbf{R}}$ be a movable class. Then given any codimension-one subvariety $E \subseteq X$, $\gamma$ is represented by an effective 1-cycle $z_E \in Z_1(X)_{\mathbf{R}}$ whose support meets $E$ properly. In particular,*

$$(\gamma \cdot e) \geq 0$$

*for every effective class $e \in N^1(X)_{\mathbf{R}}$.*

*Proof.* Say $\gamma = \mu_*(a_1 \cdot \ldots \cdot a_{n-1})$ with $\mu : X' \longrightarrow X$ and $a_i \in N^1(X')_{\mathbf{R}}$ as in the definition. Each $a_i$ is a positive $\mathbf{R}$-linear combination of ample integral divisor classes, so one may suppose without loss of generality that the $a_i$ are themselves integral. In this case the statement follows from the fact that if $F$ is any effective divisor on $X'$, and if $A_1, \ldots, A_{n-1}$ are very ample divisors that are general in their linear series, then $A_1 \cap \ldots \cap A_{n-1}$ meets $F$ properly.    □

Recall (Section 2.2.B) that the pseudoeffective cone

$$\overline{\mathrm{Eff}}(X) \subseteq N^1(X)_{\mathbf{R}}$$

is defined to be the closure of the cone of effective divisor classes on $X$. It coincides with the closure of the cone of big classes. This being said, it follows from the Lemma that

$$(\xi \cdot \gamma) \geq 0 \quad \text{for all } \xi \in \overline{\mathrm{Eff}}(X) \, , \, \gamma \in \overline{\mathrm{Mov}}(X),$$

or equivalently that

$$\overline{\mathrm{Eff}}(X) \subseteq \overline{\mathrm{Mov}}(X)^*. \tag{11.13}$$

The very appealing discovery of Boucksom–Demailly–Paun–Peternell is that in fact these two cones coincide:

**Theorem 11.4.19. (Theorem of BDPP, [68]).** *Let $X$ be an irreducible projective variety of dimension $n$. Then the cones*

$$\overline{\mathrm{Mov}}(X) \quad and \quad \overline{\mathrm{Eff}}(X)$$

*are dual.*

The proof appears at the end of this subsection.

As an important application, Boucksom–Demailly–Paun–Peternell deduce:

**Corollary 11.4.20. (Characterization of uniruled varieties).** *Let $X$ be a smooth projective variety. Then $X$ is uniruled — i.e. covered by rational curves — if and only if $K_X$ is not pseudoeffective.*

It was known previously that the corollary would follow from the minimal model program.

*Sketch of Proof of Corollary 11.4.20.* Assume that $K_X$ lies in the exterior of $\overline{\mathrm{Eff}}(X)$. Then by the theorem, there exists a movable class $\gamma \in N_1(X)$ such that $(K_X \cdot \gamma) < 0$. As in the proof of Lemma 11.4.18, this implies the existence of a family $\{C_t\}_{t \in T}$ of irreducible curves, covering a dense open subset of $X$, with $(K_X \cdot C_t) < 0$. But by a result of Miyaoka and Mori [431, Theorem 1], this is equivalent to $X$ being uniruled. $\square$

The crucial ingredient underlying Theorem 11.4.19 is an orthogonality property of Fujita approximations:

**Theorem 11.4.21. (Asymptotic orthogonality of Fujita approxima- tions, [68]).** *Let $X$ be a projective variety of dimension $n$, $\xi \in N^1(X)_{\mathbf{R}}$ a big class on $X$, and consider a Fujita approximation*

$$\mu : X' \longrightarrow X \, , \quad \mu^*(\xi) = a + e$$

*of $\xi$. Fix an ample class $h$ on $X$ which is sufficiently positive so that $h \pm \xi$ is ample. Then there is a universal constant $C$ so that*

$$\left(a^{n-1} \cdot e\right)^2_{X'} \leq C \cdot \left(h^n\right)_X \cdot \left(\mathrm{vol}_X(\xi) - \mathrm{vol}_{X'}(a)\right),$$

*where as usual we indicate with a subscript the space on which an intersection number is computed.*

In other words, as one forms a better and better Fujita approximation of $\xi$ — so that $\left(a^n\right)_{X'}$ comes closer and closer to the volume of $\xi$ — the intersection number $\left(a^{n-1} \cdot e\right)_{X'}$ approaches $0$. A statement along these lines had been conjectured by Nakamaye [469, p. 566]. The proof of [68], which we reproduce here, will show that in fact one can take $C = 20$.

Our immediate goal is to prove Theorem 11.4.21. To this end, it will be useful to note an elementary inequality on intersection numbers:

**Lemma 11.4.22.** *Let $V$ be a projective variety of dimension $n$, and let $b_1, b_2 \in N^1(V)_{\mathbf{R}}$ be nef classes with $b_1 - b_2$ effective. Then*

$$\left(b_1^k \cdot \delta_1 \cdot \ldots \cdot \delta_{n-k}\right) \geq \left(b_2^k \cdot \delta_1 \cdot \ldots \cdot \delta_{n-k}\right) \tag{11.14}$$

*for any nef classes $\delta_1, \ldots, \delta_{n-k}$.*

*Proof.* Set $e = b_1 - b_2$. Then

$$b_1^k - b_2^k = e \cdot \left(b_1^{k-1} + b_1^{k-2} \cdot b_2 + \ldots + b_1 \cdot b_2^{k-2} + b_2^{k-1}\right).$$

It follows that the difference of the two sides in (11.14) is represented by the intersection of an effective divisor with a positive linear combination of products of nef classes. Hence this difference is non-negative. $\qquad\square$

*Proof of Theorem 11.4.21.* For simplicity of notation put $\omega = \mu^*(h)$, so that in particular $\int_X h^n = \int_{X'} \omega^n$. The plan is to study the function $\mathrm{vol}_{X'}(a + te)$ for small $t \in [0, 1]$. Specifically, we will prove that if $t \leq \frac{1}{10n}$, then

$$\mathrm{vol}_{X'}(a + te) \geq \left(a^n\right) + n\left(a^{n-1}e\right) \cdot t - 5n^2\left(\omega^n\right) \cdot t^2. \tag{11.15}$$

Grant this for the time being. The right-hand side of (11.15) is maximized when

$$t = t_0 =_{\mathrm{def}} \frac{1}{10n} \cdot \frac{\left(a^{n-1} \cdot e\right)}{\left(\omega^n\right)}.$$

Moreover, since $\omega - a$ and $\omega - e$ are effective, one has $\left(\omega^n\right) \geq \left(\omega \cdot a^{n-1}\right) \geq \left(e \cdot a^{n-1}\right)$, i.e. $t_0 \leq \frac{1}{10n}$. So we can take $t = t_0$ in (11.15), and we find:

$$\mathrm{vol}_{X'}(a + te) \geq \left(a^n\right) + \frac{\left(a^{n-1} \cdot e\right)^2}{20 \cdot \left(\omega^n\right)}.$$

But $\mathrm{vol}_X(\xi) = \mathrm{vol}_{X'}(a + e) \geq \mathrm{vol}(a + t_0 e)$ since $t_0 \leq 1$ and $e$ is effective, and the theorem follows with $C = 20$.

It remains to prove (11.15). To this end, write $a + te = p_t - q_t$ as the difference of the nef classes

$$p_t = a + t \cdot \mu^*(\xi + h) \quad , \quad q_t = t(a + \omega).$$

Then thanks to 2.2.47,

$$\mathrm{vol}_{X'}(a + te) \geq \left(p_t^n\right) - n \cdot \left(p_t^{n-1} \cdot q_t\right). \tag{11.16}$$

The next step is to estimate the two terms on the right. To begin with, since $a$ and $\xi + h$ are ample, and $\mu^*(\xi + h) = a + e + w$, one has

$$\left(p_t^n\right) \geq \left(a^n\right) + nt \cdot \left(a^{n-1} \cdot (a + e + w)\right). \tag{11.17}$$

On the other hand,

$$n \cdot \left(p_t^{n-1} \cdot q_t\right) = n\left(a^{n-1} \cdot (a + w)\right) \cdot t + nS \cdot t^2,$$

where

$$S = \sum_{k=1}^{n-1} t^{k-1} \binom{n-1}{k} \cdot \left(a^{n-1-k} \cdot (a + e + w)^k \cdot (a + w)\right).$$

Now each of $w - a$, $2w - (a + e + w)$ and $2w - (a + w)$ is an effective difference of nef classes, to which Lemma 11.4.22 applies. Using the inequality $\binom{n-1}{k+1} \leq (n-1)\binom{n-2}{k}$ one then obtains:

$$S \leq \sum_{k=1}^{n-1} 2^{k+1} t^{k-1} \binom{n-1}{k} (w^n)$$

$$\leq 4(n-1) \sum_{k=0}^{n-2} (2t)^k \binom{n-2}{k} (w^n)$$

$$= 4(n-1)(1 + 2t)^{n-2} (w^n).$$

But if $0 \leq t \leq \frac{1}{10n}$, then $4(n-1)(1 + 2t)^{n-2} \leq 5n$, and so

$$n \cdot \left(p_t^{n-1} \cdot q_t\right) \leq n\left(a^{n-1} \cdot (a + w)\right) \cdot t + 5n^2 (w^n) \cdot t^2$$

in this case. The required estimate (11.15) then follows from (11.16) and (11.17). □

We can now establish the theorem of Boucksom–Demailly–Paun–Peternell:

*Proof of Theorem 11.4.19.* Aiming for a contradiction, we assume that the inclusion $\overline{\mathrm{Eff}}(X) \subseteq \overline{\mathrm{Mov}}(X)^*$ noted in (11.13) is strict. Then there exists a class

$$\xi \in \mathrm{boundary}\left(\overline{\mathrm{Eff}}(X)\right) , \quad \xi \in \mathrm{interior}\left(\overline{\mathrm{Mov}}(X)^*\right).$$

Fix an ample class $h$ such that $h \pm 2\xi$ is ample. On the one hand, since $\xi$ lies in the interior of $\overline{\mathrm{Mov}}(X)^*$ there exists $\varepsilon > 0$ such that $\xi - \varepsilon h \in \overline{\mathrm{Mov}}(X)^*$. In particular,

$$\frac{(\xi \cdot \gamma)}{(h \cdot \gamma)} \geq \varepsilon \tag{11.18}$$

for every mobile class $\gamma$ on $X$. On the other hand, consider for small $\delta > 0$ the class

$$\xi_\delta \quad =_{\text{def}} \quad \xi + \delta h.$$

Since $\xi$ is pseudo-effective, $\xi_\delta$ is big for all $\delta > 0$. The plan is to apply the asymptotic orthogonality theorem (11.4.21) to a Fujita approximation of $\xi_\delta$.

To this end, for small $\delta > 0$ let

$$\mu_\delta \colon X'_\delta \longrightarrow X \quad , \quad \mu^*(\xi_\delta) \ = \ a_\delta + e_\delta$$

be a Fujita approximation with

$$\text{vol}_{X'_\delta}(a_\delta) \ \geq \ \text{vol}_X(\xi_\delta) - \delta^{2n}. \tag{11.19}$$

We may suppose also that

$$\text{vol}_{X'_\delta}(a_\delta) \ \geq \ \frac{1}{2} \cdot \text{vol}_X(\xi_\delta) \ \geq \ \frac{\delta^n}{2} \cdot (h^n). \tag{11.20}$$

Consider now the movable class

$$\gamma_\delta \quad =_{\text{def}} \quad \mu_{\delta,*}(a_\delta^{n-1}).$$

Then by the projection formula and the Hodge-type inequality 1.6.3 (ii),

$$\begin{aligned}
(h \cdot \gamma_\delta)_X \ &= \ (\mu_\delta^*(h) \cdot a_\delta^{n-1})_{X'_\delta} \\
&\geq \ (h^n)_X^{1/n} \cdot (a_\delta^n)_{X'_\delta}^{(n-1)/n}.
\end{aligned} \tag{11.21}$$

On the other hand,

$$\begin{aligned}
(\xi \cdot \gamma_\delta)_X \ &\leq \ (\xi_\delta \cdot \gamma_\delta)_X \\
&= \ (\mu_\delta^*(\xi_\delta) \cdot a_\delta^{n-1})_{X\delta'} \\
&= \ (a_\delta^n)_{X'_\delta} + (e_\delta \cdot a_\delta^{n-1})_{X'_\delta}.
\end{aligned} \tag{11.22}$$

Now $h \pm \xi_\delta$ is ample provided that $\delta < \frac{1}{2}$. So Theorem 11.4.21 applies to show that

$$\begin{aligned}
(e_\delta \cdot a_\delta^{n-1})_{X'_\delta} \ &\leq \ \Big( C_1 \cdot (h^n)_X \cdot \big(\text{vol}_X(\xi_\delta) - \text{vol}_{X'_\delta}(a_\delta)\big) \Big)^{1/2} \\
&\leq \ C_2 \cdot \delta^n
\end{aligned} \tag{11.23}$$

for suitable constants $C_1$ and $C_2$ independent of $\delta$. Bearing in mind (11.20), one finds from (11.21), (11.22), and (11.23) that

$$\frac{(\xi \cdot \gamma_\delta)}{(h \cdot \gamma_\delta)} \;\leq\; \frac{\left(a_\delta^n\right)_{X_\delta'} + C_2 \cdot \delta^n}{\left(h^n\right)_X^{1/n} \cdot \left(a_\delta^n\right)_{X_\delta'}^{(n-1)/n}} \tag{11.24}$$

$$\leq\; C_3 \cdot \left(a_\delta^n\right)_{X_\delta'}^{1/n} + C_4 \cdot \delta,$$

where again the constants $C_3, C_4$ are independent of $\delta$.

But now recall that $\xi$ was chosen to lie on the boundary of the big cone. Therefore, by the continuity of volume (Theorem 2.2.44),

$$\lim_{\delta \to 0} \mathrm{vol}_X(\xi_\delta) \;=\; \mathrm{vol}_X(\xi) \;=\; 0,$$

and hence also

$$\lim_{\delta \to 0} \mathrm{vol}_{X_\delta'}(a_\delta) \;=\; \lim_{\delta \to 0} \left(a_\delta^n\right)_{X_\delta'} \;=\; 0.$$

Thus both sides of (11.24) go to zero with $\delta$. But this contradicts (11.18), and completes the proof. □


## 11.5 Siu's Theorem on Plurigenera

In one of the most striking applications of the theory to date, Siu [539] used multiplier ideals to give a remarkably simple proof of the deformation invariance of plurigenera of varieties of general type. This section is devoted to a presentation of Siu's theorem. In a general way we follow the approach of [539], although the language and details are different.

Let $V$ be a smooth projective variety. Recall again that for $m \geq 1$ the $m^{\text{th}}$ plurigenus $P_m(V)$ of $V$ is the dimension of the space of $m$-canonical forms on $V$:

$$P_m(V) \;=\; \dim H^0\big(V, \mathcal{O}_V(mK_V)\big).$$

These plurigenera are basic birational invariants, and it was a long-standing open problem to determine whether they are constant under deformations of $V$. Siu proved in [539] that this is indeed the case for varieties of general type:

**Theorem 11.5.1. (Deformation invariance of plurigenera).** *Consider a smooth projective morphism*

$$\pi : X \longrightarrow T$$

*of non-singular irreducible quasi-projective varieties, and for $t \in T$ write $X_t$ for the corresponding fibre of $\pi$. Assume that each $X_t$ is irreducible and of general type. Then for every $m \geq 1$ the plurigenera*

$$P_m(X_t) \;=\; h^0\big(X_t, \mathcal{O}_{X_t}(mK_{X_t})\big)$$

*are independent of $t$.*

Siu [540] subsequently established that the same statement holds even when $X_t$ is not of general type, but we will not discuss this here.

It is useful to make right away some elementary comments and reductions. In the first place, since any two points on an irreducible variety can be joined by a smooth connected curve mapping to the variety (Example 3.3.5), we may — and do — assume that $\dim T = 1$. Note next that each of the plurigenera $P_m(X_t)$ is in any event upper semi-continuous in the Zariski topology. So fixing a point $0 \in T$ and an integer $m \geq 1$, the issue is to show that $P_m(X_0) \leq P_m(X_t)$ for any $m \geq 0$ for $t$ in some neighborhood $U_m$ of $0 \in T$.

Keeping the notation of the theorem, we next give a few examples to help bring into relief some of the essential issues involved in 11.5.1.

**Example 11.5.2. (Deformation invariance of geometric genus).** The Hodge decomposition and the semicontinuity theorem combine to show that all of the Hodge numbers $h^{p,q}(X_t)$ are constant. In particular, $P_1(X_t) = h^{\dim X_t, 0}(X_t)$ is independent of $t$. However, $P_m(X_t)$ does not have a naive Hodge-theoretic interpretation when $m \geq 2$. □

**Example 11.5.3. (Nef and big canonical bundles).** Fixing as above a point $0 \in T$, suppose that $K_{X_0}$ is big and nef.[9] Then it follows from vanishing (Theorem 4.3.1) that

$$H^i\big(X_0, \mathcal{O}_{X_0}(mK_{X_0})\big) \;=\; 0 \quad \text{for } i > 0 \, , \; m \geq 2. \tag{11.25}$$

By semicontinuity the analogous statement holds on $X_t$ for $t$ near 0, and in particular $P_m(X_t) = \chi\big(X_t, \mathcal{O}_{X_t}(mK_{X_t})\big)$ for $t$ near 0. So in this case, 11.5.1 is a consequence of the deformation invariance of Euler characteristics. This gives a first hint that the question at hand is related to vanishing theorems. □

**Example 11.5.4. (Levine's theorem).** Suppose that for some $p > 0$ there is a divisor $D \in |pK_{X_0}|$ whose singularities are sufficiently mild so that $\mathcal{J}\big(X_0, c \cdot D\big) = 0$ for $c < 1$. Then Nadel vanishing shows that (11.25) holds for $m \leq p$, which as above implies that $P_m(X_t)$ is locally constant near 0 when $m \leq p$. This is (a slight generalization of) a theorem of Levine [398], and Siu's observation that one could quickly recover Levine's result via multiplier ideals was one of the starting points for his work on these questions. □

The remainder of this section is devoted to the proof of Siu's theorem. As a matter of terminology, recall from Definition 9.6.30 that given an integer divisor $L$ and an ideal sheaf $\mathfrak{a} \subseteq \mathcal{O}_X$, we say that a section $s \in \Gamma\big(X, \mathcal{O}_X(L)\big)$ *vanishes along* $\mathfrak{a}$ if $s$ lies in the subspace $\Gamma\big(X, \mathcal{O}_X(L) \otimes \mathfrak{a}\big) \subseteq \Gamma\big(X, \mathcal{O}_X(L)\big)$.

Returning to the setting of 11.5.1, recall that we assume that $T$ is a smooth curve. Write $K$ for the relative canonical bundle $K_{X/T}$ of $X$ over $T$, so that $K \mid X_t = K_{X_t}$ for all $t \in T$. By a slight abuse of terminology, we refer to sections of $\mathcal{O}_X(mK)$ as $m$-canonical forms on $X$. Consider the sheaves

---

[9] Recall that this holds in particular if $X_0$ is a smooth curve of genus $\geq 2$ or a minimal surface of general type (Example 1.4.18).

$$\mathcal{P}_m =_{\text{def}} \pi_*\mathcal{O}_X(mK)$$

on $T$. Being a torsion-free sheaf on a smooth curve, each $\mathcal{P}_m$ is a vector bundle. At a general point $t \in T$, the base-change theorem gives an isomorphism

$$\phi_m(t) : \mathcal{P}_m(t) = \mathcal{P}_m \otimes \mathbf{C}(t) \longrightarrow H^0\big(X_t, \mathcal{O}_{X_t}(mK_{X_t})\big)$$

from the fibre of $\mathcal{P}_m$ at $t$ to the space of $m$-canonical forms on $X_t$.[10] In particular, the rank of $\mathcal{P}_m$ is the generic value of the plurigenus $P_m(X_t)$. To prove the theorem, it therefore suffices (and in fact is equivalent) to show that the corresponding map

$$\phi_m = \phi_m(0) \; : \; \mathcal{P}_m(0) \longrightarrow H^0\big(X_0, \mathcal{O}_{X_0}(mK_{X_0})\big) \qquad (11.26)$$

at $t = 0$ is surjective.[11]

To say the same thing more geometrically, note that $\mathcal{P}_m$ is globally generated since we are supposing that $T$ is affine, and of course $\Gamma(T, \mathcal{P}_m) = \Gamma(X, \mathcal{O}_X(mK))$. Therefore we may identify $\operatorname{im} \phi_m \subseteq H^0\big(X_0, \mathcal{O}_{X_0}(mK_{X_0})\big)$ as the subspace of $m$-canonical forms on $X_0$ that extend to $m$-canonical forms on all of $X$. So the required surjectivity in (11.26) is equivalent to establishing:

any section $s \in \Gamma\big(X_0, \mathcal{O}_{X_0}(mK_{X_0})\big)$ extends to a section

$$\tilde{s} \in \Gamma\big(X, \mathcal{O}_X(mK)\big). \qquad (11.27)$$

To prove (11.27), the plan is to compare two multiplier ideal sheaves that can be defined in the present setting.

It will clarify matters to introduce the first of these multiplier ideals in a more general context. Consider a smooth irreducible variety $Y$, and a surjective projective mapping

$$\pi \; : \; Y \longrightarrow T$$

from $Y$ onto a smooth affine curve. We fix a point $0 \in T$ and assume that the fibre $Y_0 = \pi^{-1}(0) \subseteq Y$ is non-singular and irreducible. Suppose now that $L$ is

---

[10] Quite generally, if $f : V \longrightarrow W$ is any proper morphism of varieties, with $W$ reduced, and if $\mathcal{F}$ is any coherent sheaf on $V$, then there is a non-empty Zariski-open subset $U \subset W$ such that the natural maps

$$R^j f_*\mathcal{F} \otimes \mathbf{C}(t) \longrightarrow H^j\big(V_t, \mathcal{F}_t\big)$$

are isomorphisms for all $j \geq 0$ and $t \in U$. In fact, thanks to [280, Corollary III.12.9] it suffices to choose $U$ such that $\mathcal{F}$ is flat over $U$ and all the direct images $R^j f_*\mathcal{F}$ are locally free on $U$.

[11] Observe that — at least after shrinking $T$ to ensure that there is a regular function vanishing simply at $0$ and nowhere else — the map $\phi_m$ is non-zero provided that $\mathcal{P}_m \neq 0$, i.e. $X_0$ does not appear in the base ideal of $\rho_m : \pi^*\pi_*\mathcal{O}_X(mK) \longrightarrow \mathcal{O}_X(mK)$.

an integral divisor on $Y$ that is big for $\pi$ (meaning that the restriction of $L$ to a general fibre of $\pi$ is big), and assume that the restriction $L_0 = L \,|\, Y_0$ is also big. Then for every $k \geq 1$ there is a natural map

$$\phi_k : \Gamma(Y, \mathcal{O}_Y(kL)) \longrightarrow \Gamma(Y_0, \mathcal{O}_{Y_0}(kL_0)),$$

and since $T$ is affine, $\phi_k \neq 0$ for $k \gg 0$. The images of these maps form a graded linear series on $Y_0$ in the sense of Definition 2.4.1, and so give rise to an asymptotic multiplier ideal on $Y_0$, which we shall denote by

$$\mathcal{J}(\,\|L\|_0\,) \;=\; \mathcal{J}(\,Y_0\,,\,\|L\|_0\,).$$

Very concretely, if $\mathfrak{c}_k \subseteq \mathcal{O}_{Y_0}$ is the base ideal of the linear series

$$\mathrm{im}(\phi_k) \;\subseteq\; H^0(Y_0, \mathcal{O}_{Y_0}(kL))$$

(so that $\mathfrak{c}_k$ is the restriction to $Y_0$ of the base ideal of $\Gamma(Y, \mathcal{O}_Y(kL))$ on $Y$) then $\mathcal{J}(\,\|L\|_0\,) = \mathcal{J}(\mathfrak{c}_\bullet)$ is the asymptotic multiplier ideal determined by this graded family.

The fact we shall need is the following:

**Lemma 11.5.5.** *In the situation just described there is an inclusion*

$$H^0\!\left(\,Y_0\,,\,\mathcal{O}_{Y_0}(K_{Y_0} + L_0) \otimes \mathcal{J}(\,\|L\|_0\,)\,\right)$$
$$\subseteq \,\mathrm{im}\!\left(\,H^0(Y, \mathcal{O}_Y(K_{Y/T} + L)) \longrightarrow H^0(Y_0, \mathcal{O}_{Y_0}(K_{Y_0} + L_0))\,\right),$$

*where the map appearing here is the natural restriction.*

In other words, any section of $\mathcal{O}_{Y_0}(K_{Y_0} + L_0)$ vanishing along $\mathcal{J}(\,\|L\|_0\,)$ extends to a section of $\mathcal{O}_Y(K_{Y/T} + L)$.

*Proof.* The argument mirrors the proof of the restriction theorem. Fix $p \gg 0$ computing the multiplier ideal in question, and let $\mathfrak{b} = \mathfrak{b}(\pi, \mathcal{O}_Y(pL))$ be the base ideal of $\mathcal{O}_Y(pL)$ with respect to $\pi$; since $T$ is affine, $\mathfrak{b}$ is equivalently the base ideal of the full linear series $\Gamma(Y, \mathcal{O}_Y(pL))$. Let $f : V \longrightarrow Y$ be a log resolution of $\mathfrak{b}$, and write

$$\mathfrak{b} \cdot \mathcal{O}_V \;=\; \mathcal{O}_V(-F),$$
$$f^* Y_0 \;=\; V_0 + \sum a_i E_i,$$

$V_0 \subseteq V$ being the proper transform of $Y_0$. The situation is summarized in the following diagram:

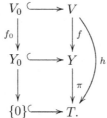

Put

$$B = (K_V + V_0) - f^*(K_Y + Y_0) - \left[\tfrac{1}{p}F\right] = K_{V/Y} - \left[\tfrac{1}{p}F\right] - \sum a_i E_i,$$

so that $\mathcal{J}(Y_0, \|L\|_0) = f_{0,*}\mathcal{O}_{V_0}(B \mid V_0)$. Since

$$h_*\mathcal{O}_V(B + f^*(K_{Y/T} + L)) \subseteq \pi_*\mathcal{O}_Y(K_{Y/T} + L),$$

it is enough to show that the map

$$h_*\Big(\mathcal{O}_V(B + f^*(K_{Y/T} + L))\Big) \longrightarrow h_*\Big(\mathcal{O}_{V_0}(B + f^*(K_{Y/T} + L))\Big)$$

is surjective. Now

$$f^*(K_{Y/T} + L) + B - V_0 \equiv_{\mathrm{lin}} K_{V/T} + f^*L - \left[\tfrac{1}{p}F\right] - h^*(0),$$

and $f^*L - \tfrac{1}{p}F$ is nef and big for $h$. So the required surjectivity follows by vanishing for $h$ (Generalization 9.1.22). ☐

Returning to the situation of the theorem, we apply this construction (for fixed $m \geq 2$) to the map $\pi : X \longrightarrow T$ and the divisor $L = (m-1)K$. Thus we have multiplier ideals

$$\mathcal{J}(\|mK\|_0) = \mathcal{J}(X_0, \|mK\|_0) \subseteq \mathcal{O}_{X_0}.$$

Like any multiplier ideal, $\mathcal{J}(\|mK\|_0)$ is integrally closed (Example 9.6.13), and every $m$-canonical form on $X_0$ lying in the subspace

$$\mathrm{im}\ \phi_m \subseteq H^0(X_0, \mathcal{O}_{X_0}(mK_{X_0}))$$

vanishes along $\mathcal{J}(\|mK\|_0)$. Note for later reference that subadditivity (Corollary 11.2.4) implies that

$$\mathcal{J}(\|\ell mK\|_0) \subseteq \mathcal{J}(\|mK\|_0)^\ell \tag{11.28}$$

for every $\ell, m \geq 1$. The preceding lemma shows that any section

$$s \in H^0\Big(X_0, \mathcal{O}_{X_0}(mK_{X_0}) \otimes \mathcal{J}(\|(m-1)K\|_0)\Big) \subseteq H^0(X_0, \mathcal{O}_{X_0}(mK_{X_0}))$$

extends to an $m$-canonical form $\tilde{s} \in H^0(X, \mathcal{O}_X(mK))$. Consequently:

> To prove the theorem it suffices to show that every $m$-canonical form $s \in \Gamma(X_0, \mathcal{O}_{X_0}(mK_{X_0}))$ on $X_0$ vanishes along the multiplier ideal $\mathcal{J}(\|(m-1)K\|_0)$. $\qquad(11.29)$

The second multiplier ideal we consider is the asymptotic multiplier ideal

$$\mathcal{J}(X_0, \|mK_{X_0}\|) \; = \; \mathcal{J}(\|mK_{X_0}\|) \subseteq \mathcal{O}_{X_0}$$

associated to the pluricanonical linear series $|mK_{X_0}|$ on $X_0$. Recall from Proposition 11.2.10 the isomorphism

$$H^0\Big(X_0, \mathcal{O}_{X_0}(mK_{X_0}) \otimes \mathcal{J}(\|mK_{X_0}\|)\Big) \xrightarrow{\;\cong\;} H^0\Big(X_0, \mathcal{O}_{X_0}(mK_{X_0})\Big)$$
$$(11.30)$$

for each $m \geq 1$, i.e. every $m$-canonical form on $X_0$ vanishes along the multiplier ideal $\mathcal{J}(\|mK_{X_0}\|)$. It is clear that

$$\mathcal{J}(X_0, \|mK\|_0) \subseteq \mathcal{J}(X_0, \|mK_{X_0}\|) \qquad (11.31)$$

since the ideal on the left is defined by a graded subseries of the full pluri-canonical series that figures on the right.

We now come to the heart of the proof. By way of motivation, imagine for the moment that one had an inclusion

$$\mathcal{J}(X_0, \|mK_{X_0}\|) \subseteq \mathcal{J}(X_0, \|(m-1)K\|_0). \qquad (*)$$

Then the criterion specified in (11.29) would follow immediately from (11.30). While in reality (*) may be too much to hope for, Siu's beautiful observation is that one can establish a slightly weaker statement that — as it turns out — is just as good.

Specifically, we will prove the following

**Main Claim.** There exists a positive integer $a \geq 1$, together with an effective divisor $D_0 \subseteq X_0$, such that

$$\mathcal{J}(X_0, \|kK_{X_0}\|)(-D_0) \subseteq \mathcal{J}(X_0, \|(k+a-1)K\|_0) \qquad (11.32)$$

for every $k \geq 1$. Moreover, there is a non-zero section $s_D \in \Gamma(X_0, \mathcal{O}_{X_0}(aK_{X_0}))$ vanishing on $D_0$.

The crucial point here is that $a$, $D_0$, and $s_D$ are independent of $k$. One can think of (11.32) and (11.31) as asserting roughly speaking that the ideals $\mathcal{J}(\|kK_{X_0}\|)$ and $\mathcal{J}(\|kK\|_0)$ have "bounded difference".

Granting the Main Claim, we now complete the proof of the Theorem (via a global analogue of Proposition 11.3.11). Specifically, fix $m \geq 1$ and a pluricanonical form $s \in \Gamma(X_0, \mathcal{O}_{X_0}(mK_{X_0}))$. According to (11.29) the issue is to prove that $s$ vanishes along $\mathcal{J}(X_0, \|(m-1)K\|_0)$. To this end, fix $\ell > 0$ and consider the section

$$s^\ell \cdot s_D \in \Gamma\Big(X_0, \mathcal{O}_{X_0}((m\ell + a)K_{X_0}))\Big).$$

Now $s^\ell$ vanishes along $\mathcal{J}(\|\ell m K_{X_0}\|)$ thanks to (11.30) and $s_D$ vanishes on $D_0$. Therefore $s^\ell s_D$ vanishes along the ideal

$$\mathcal{J}\big(\|\, m\ell K_{X_0}\,\|\big)\big(-D_0\big) \subseteq \mathcal{O}_{X_0}.$$

But using the Main Claim and the subadditivity relation (11.28) we find:

$$\begin{aligned}
\mathcal{J}\big(\|\, m\ell K_{X_0}\,\|\big)\big(-D_0\big) &\subseteq \mathcal{J}\big(\|\,(m\ell + a - 1)K\,\|_0\big)\\
&\subseteq \mathcal{J}\big(\|\, m\ell\, K\,\|_0\big)\\
&\subseteq \mathcal{J}\big(\|\, mK\,\|_0\big)^{\ell},
\end{aligned}$$

where we have used the fact that $\mathcal{J}\big(\|\, pK\,\|_0\big) \subseteq \mathcal{J}\big(\|\, qK\,\|_0\big)$ whenever $p \geq q$. In other words,

$$s_D \cdot s^{\ell} \;\in\; \Gamma\Big(X_0\,,\, \mathcal{O}_{X_0}\big((m\ell + a)K_{X_0}\big) \otimes \mathcal{J}\big(\|\, mK\,\|_0\big)^{\ell}\,\Big).$$

Since this holds for all $\ell \gg 0$, it follows from Example 11.5.6 below — which globalizes the characterization 9.6.6 (iv) of integral closure — that $s$ vanishes along the integral closure $\overline{\mathcal{J}\big(\|\, mK\,\|_0\big)}$ of $\mathcal{J}\big(\|\, mK\,\|_0\big)$. But $\mathcal{J}\big(\|\, mK\,\|_0\big)$, being a multiplier ideal, is integrally closed. Since in any event

$$\mathcal{J}\big(\|\, mK\,\|_0\big) \subseteq \mathcal{J}\big(\|\,(m-1)K\,\|_0\big),$$

it follows that $s$ vanishes along $\mathcal{J}\big(\|\,(m-1)K\,\|_0\big)$, as required. This completes the proof modulo Example 11.5.6 and the Main Claim.

In the course of the argument just given, we drew on the following fact:

**Example 11.5.6. (Global analogue of Proposition 9.6.6 (iv)).** Let $M$ and $D$ be integral divisors on $X$, and let $\mathfrak{a} \subseteq \mathcal{O}_X$ be a fixed ideal sheaf. Suppose that

$$s \;\in\; H^0\big(X, \mathcal{O}_X(M)\big) \quad,\quad t \;\in\; H^0\big(X, \mathcal{O}_X(D)\big)$$

are non-zero sections having the property that

$$s^{\ell} \cdot t \;\in\; H^0\big(X, \mathcal{O}_X(\ell M + D)\big)$$

vanishes along the ideal $\mathfrak{a}^{\ell}$ for all (or infinitely many) $\ell \gg 0$. Then $s$ vanishes along the integral closure $\overline{\mathfrak{a}}$ of $\mathfrak{a}$, i.e.

$$s \;\in\; H^0\big(X, \mathcal{O}_X(M) \otimes \overline{\mathfrak{a}}\big).$$

(It suffices to prove the assertion locally on $X$, so we can assume that $X$ is affine and that $M$ and $L$ are principal divisors on $X$. Choose nowhere zero sections

$$\xi \;\in\; \Gamma\big(X, \mathcal{O}_X(M)\big) \quad,\quad \zeta \;\in\; \Gamma\big(X, \mathcal{O}_X(D)\big).$$

Then we can write $s = f \cdot \xi$, $t = c \cdot \zeta$ for functions $f, c \in \mathbf{C}[X]$. The hypothesis implies that $cf^{\ell} \in \mathfrak{a}^{\ell}$ for all (or infinitely many) $\ell \gg 0$, and therefore $f \in \overline{\mathfrak{a}}$. The assertion follows.) □

Finally, we give the

*Proof of Main Claim.* Note at the outset that we are free whenever convenient to replace $T$ by any Zariski-open neighborhood of $0 \in T$, and to shrink $X$ accordingly. This being said, fix a divisor $B$ on $X$ that is very ample for $\pi$, so that in particular $B \mid X_t$ is a very ample divisor on $X_t$ for every $t \in T$. We may assume in addition that $2K + (n+1)B$ is represented by an effective divisor $G$ on $X$, where $n$ is the relative dimension of $\pi$ (i.e. $n = \dim X - 1$), and that $B$ is sufficiently positive so that $\mathcal{O}_X(G)$ is free. Our hypothesis that each of the fibres of $\pi$ has general type implies that $K$ is $\pi$-big. Thus by Kodaira's Lemma we can fix an integer $a \geq 1$ such that

$$aK - G \equiv_{\mathrm{lin}} aK - \big(2K + (n+1)B\big) \equiv_{\mathrm{lin}} D$$

for some effective divisor $D$ on $X$. After possibly replacing $T$ by a neighborhood of $0$, we may assume moreover that $D$ and $G$ do not contain any fibres of $\pi$ as components.[12] The linear equivalence $D + G \equiv_{\mathrm{lin}} aK$ then determines a section $\tilde{s}_D \in \Gamma\big(X, \mathcal{O}_X(aK)\big)$ vanishing on $D$ but not on any fibre of $\pi$.

Set $D_0 = D \mid X_0$ and $B_0 = B \mid X_0$. We now prove by induction on $k$ the stated inclusion

$$\mathcal{J}\big(\|\,kK_{X_0}\,\|\big)\,(-D_0) \;\subseteq\; \mathcal{J}\big(\|\,(k+a-1)\,K\,\|_0\big) \qquad (*)$$

of ideal sheaves on $X_0$. Consider first the case $k = 1$. It is certainly enough to show that $\mathcal{O}_{X_0}(-D_0) \subseteq \mathcal{J}\big(\|\,aK\,\|_0\big)$. But $aK - D \equiv_{\mathrm{lin}} G$ is free, so if $N \in |2G|$ is a general divisor then (Example 9.2.29)

$$\mathcal{J}\big(X_0, D_0\big) \;=\; \mathcal{J}\big(X_0, \tfrac{1}{2}(2D + N)\mid X_0\big) \;\subseteq\; \mathcal{J}\big(X_0,\, \|\,aK\,\|_0\big),$$

as required.

Assuming that $(*)$ holds for given $k$, we will prove that

$$\mathcal{O}_{X_0}\big((k+a)K_{X_0} - D_0\big) \otimes \mathcal{J}\big(\|\,(k+1)K_{X_0}\,\|\big)$$
$$\subseteq\; \mathcal{O}_{X_0}\big((k+a)K_{X_0}\big) \otimes \mathcal{J}\big(\|\,(k+a)K\,\|_0\big), \qquad (**)$$

which implies $(*)$ for $k+1$. To this end, observe first that the sheaf appearing on the left in $(**)$ is globally generated. In fact,

$$(k+a)K_{X_0} - D_0 \equiv_{\mathrm{lin}} (k+1)K_{X_0} + \big(K_{X_0} + (n+1)B_0\big),$$

and

---

[12] Start by collecting together any fibral components to write

$$aK - G + \pi^* A \equiv_{\mathrm{lin}} D$$

where $A$ is a divisor on $T$, and $D$ and $G$ are fibre-free. Then replace $T$ by a neighborhood of $0$ to ensure $\mathcal{O}_T(A) \cong \mathcal{O}_T$.

$$\mathcal{O}_X\Big(\big(K_{X_0} + (n+1)B_0\big) + (k+1)K_{X_0}\Big) \otimes \mathcal{J}\big(\|\,(k+1)K_{X_0}\,\|\big)$$

is free thanks to Corollary 11.2.13. By virtue of this global generation, it suffices for (**) to establish the corresponding inclusion on global sections. In other words, it is enough to prove that any $(k+a)$-canonical form on $X_0$ vanishing along the ideal sheaf $\mathcal{J}\big(\|\,(k+1)K_{X_0}\,\|\big) \otimes \mathcal{O}_{X_0}(-D_0)$ necessarily vanishes along $\mathcal{J}\big(\|\,(k+a)K\,\|_0\big)$.

To verify this, we use the induction hypothesis (*) to deduce

$$\mathcal{J}\big(\|\,(k+1)K_{X_0}\,\|\big) \otimes \mathcal{O}_{X_0}(-D_0) \ \subseteq\ \mathcal{J}\big(\|\,kK_{X_0}\,\|\big) \otimes \mathcal{O}_{X_0}(-D_0)$$
$$\subseteq\ \mathcal{J}\big(\|\,(k+a-1)K\,\|_0\big).$$

So if $s \in \Gamma\big(X_0, \mathcal{O}_{X_0}\big((k+a)K_{X_0}\big)\big)$ is a section vanishing along

$$\mathcal{J}\big(\|\,(k+1)K_{X_0}\,\|\big) \otimes \mathcal{O}_{X_0}(-D_0),$$

then $s$ vanishes along $\mathcal{J}\big(\|\,(k+a-1)K\,\|_0\big)$. But then Lemma 11.5.5 shows that $s$ extends to a $(k+a)$-canonical form $\tilde{s} \in \Gamma\big(X, \mathcal{O}_X\big((k+a)K\big)\big)$. Therefore $s$ lies in the subspace $\mathrm{im}\phi_{k+a} \subseteq H^0\big(X_0, \mathcal{O}_{X_0}\big((k+a)K_{X_0}\big)\big)$ and consequently vanishes along the corresponding multiplier ideal $\mathcal{J}\big(\|\,(k+a)K\,\|_0\big)$. In summary, we have established the inclusion

$$H^0\Big(X_0,\, \mathcal{O}_{X_0}\big((k+a)K_{X_0}\big) \otimes \mathcal{J}\big(\|\,(k+1)K_{X_0}\,\|\big)(-D_0)\Big)$$
$$\subseteq\ H^0\Big(X_0,\, \mathcal{O}_{X_0}\big((k+a)K_{X_0}\big) \otimes \mathcal{J}\big(\|\,(k+a)K\,\|_0\big)\Big),$$

and this implies (**), as required.  $\square$

**Remark 11.5.7. (Extensions of Theorem 11.5.1).** Kawamata and Nakayama have used some of the ideas from [539] to establish a number of striking results. First, Kawamata shows in [323] that canonical singularities are preserved under deformation. He also proves the deformation invariance of plurigenera of varieties of general type having such singularities. (Along the way [323] cast Siu's arguments from [539] in algebro-geometric language.) Using related ideas, Nakayama [471] proves that if $X_0$ is a variety of general type having only canonical singularities, then any deformation of $X_0$ is again of general type. We recommend Kawamata's paper [324] for a very nice overview of these and related problems. As noted above, Siu [540] has established the deformation-invariance of plurigenera of varieties which may not be of general type.  $\square$

# Notes

At several points, we have drawn on [144] and [323]. In particular, the proof of 11.1.18 follows [144]. Statements (ii) and (iii) of Theorem 11.1.8 were worked out in discussions with K. Smith. Graded linear systems play an important role in [158].

Corollary 11.2.13 was established by Siu in [539] via an analytic argument. Remark 11.2.5 and Theorem 11.2.16 were worked out with Ein. Theorems 11.2.1 and 11.2.8 are taken from [161].

Theorem 11.4.11 appears in [130], but it is implicit in Tsuji's paper [574], and was undoubtably known to Fujita as well. Theorem 11.4.9 is new. Section 11.4.C is based on an early draft of [68]: we thank the authors of that paper for giving us access to this draft, and for their permission to include the result.

# References

1. Shreeram Abhyankar, *Tame coverings and fundamental groups of algebraic varieties. I. Branch loci with normal crossings; Applications: Theorems of Zariski and Picard*, Amer. J. Math. **81** (1959), 46–94.

2. Dan Abramovich and A. Johan de Jong, *Smoothness, semistability, and toroidal geometry*, J. Algebraic Geom. **6** (1997), no. 4, 789–801.

3. Allen Altman and Steven Kleiman, *Introduction to Grothendieck Duality Theory*, Lect. Notes in Math., vol. 146, Springer-Verlag, Berlin, 1970.

4. A. Alzati and G. Ottaviani, *A linear bound on the t-normality of codimension two subvarieties of* $\mathbf{P}^n$, J. Reine Angew. Math. **409** (1990), 35–40.

5. Marco Andreatta and Andrew J. Sommese, *On the projective normality of the adjunction bundles*, Comment. Math. Helv. **66** (1991), no. 3, 362–367.

6. Marco Andreatta and Jarosław A. Wiśniewski, *On manifolds whose tangent bundle contains an ample subbundle*, Invent. Math. **146** (2001), no. 1, 209–217.

7. Aldo Andreotti and Theodore Frankel, *The Lefschetz theorem on hyperplane sections*, Ann. of Math. (2) **69** (1959), 713–717.

8. Aldo Andreotti and Hans Grauert, *Théorème de finitude pour la cohomologie des espaces complexes*, Bull. Soc. Math. France **90** (1962), 193–259.

9. Aldo Andreotti and Alan Mayer, *On period relations for abelian integrals on algebraic curves*, Ann. Scuola Norm. Sup. Pisa (3) **21** (1967), 189–238.

10. Urban Angehrn and Yum-Tong Siu, *Effective freeness and point separation for adjoint bundles*, Invent. Math. **122** (1995), no. 2, 291–308.

11. Flavio Angelini, *An algebraic version of Demailly's asymptotic Morse inequalities*, Proc. Amer. Math. Soc. **124** (1996), no. 11, 3265–3269.

12. Marian Aprodu and Claire Voisin, *Green–Lazarsfeld's conjecture for generic curves of large gonality*, C. R. Math. Acad. Sci. Paris **336** (2003), no. 4, 335–339.

13. Donu Arapura, *Higgs line bundles, Green–Lazarsfeld sets, and maps of Kähler manifolds to curves*, Bull. Amer. Math. Soc. (N.S.) **26** (1992), no. 2, 310–314.

14. _____, *Frobenius amplitude and strong vanishing theorems for vector bundles*, Duke Math. J. **121** (2004), no. 2, 231–267, With an appendix by Dennis S. Keeler.

15. Enrico Arbarello, Maurizio Cornalba, Phillip Griffiths, and Joe Harris, *Geometry of Algebraic Curves. Vol. I*, Grundlehren Math. Wiss., vol. 267, Springer-Verlag, New York, 1985.

16. Enrico Arbarello and Corrado De Concini, *Another proof of a conjecture of S. P. Novikov on periods of abelian integrals on Riemann surfaces*, Duke Math. J. **54** (1987), no. 1, 163–178.

17. V. I. Arnol'd, *Mathematical Methods of Classical Mechanics*, Graduate Texts in Mathematics, vol. 60, Springer-Verlag, New York, 1989.

18. V. I. Arnol'd, S. M. Guseĭn-Zade, and A. N. Varchenko, *Singularities of Differentiable Maps. Vol. II: Monodromy and Asymptotics of Integrals*, Birkhäuser Boston Inc., Boston, MA, 1988.

19. Michael Atiyah and Jurgen Berndt, *Projective planes, Severi varieties and spheres*, preprint.

20. Lucian Bădescu, *Projective Geometry and Formal Geometry*, forthcoming book.

21. _____, *Special chapters of projective geometry*, Rend. Sem. Mat. Fis. Milano **69** (1999/00), 239–326 (2001).

22. _____, *Algebraic Surfaces*, Universitext, Springer-Verlag, New York, 2001.

23. Edoardo Ballico, *Spanned and ample vector bundles with low Chern numbers*, Pacific J. Math. **140** (1989), no. 2, 209–216.

24. _____, *On ample and spanned rank-3 bundles with low Chern numbers*, Manuscripta Math. **68** (1990), no. 1, 9–16.

25. Daniel Barlet, *À propos d'une conjecture de R. Hartshorne*, J. Reine Angew. Math. **374** (1987), 214–220.

26. Daniel Barlet, L. Doustaing, and J. Magnússon, *La conjecture de R. Hartshorne pour les hypersurfaces lisses de* $\mathbf{P}^n$, J. Reine Angew. Math. **457** (1994), 189–202.

27. Daniel Barlet, Thomas Peternell, and Michael Schneider, *On two conjectures of Hartshorne's*, Math. Ann. **286** (1990), no. 1-3, 13–25.

28. Wolf Barth, *Verallgemeinerung des Bertinischen Theorems in abelschen Mannigfaltigkeiten*, Ann. Scuola Norm. Sup. Pisa (3) **23** (1969), 317–330.

29. _____, *Der Abstand von einer algebraischen Mannigfaltigkeit im komplex-projektiven Raum*, Math. Ann. **187** (1970), 150–162.

30. _____, *Transplanting cohomology classes in complex-projective space*, Amer. J. Math. **92** (1970), 951–967.

31. _____, *Larsen's theorem on the homotopy groups of projective manifolds of small embedding codimension*, Algebraic Geometry – Arcata 1974, Proc. Symp. Pure Math., Amer. Math. Soc., Providence, R.I., 1975, pp. 307–313.

32. Wolf Barth and Mogens Larsen, *On the homotopy groups of complex projective algebraic manifolds*, Math. Scand. **30** (1972), 88–94.

33. Wolf Barth and A. Van de Ven, *A decomposability criterion for algebraic 2-bundles on projective spaces*, Invent. Math. **25** (1974), 91–106.

34. Charles M. Barton, *Tensor products of ample vector bundles in characteristic p*, Amer. J. Math. **93** (1971), 429–438.

35. Alexander Barvinok, *A Course in Convexity*, Graduate Studies in Mathematics, vol. 54, American Mathematical Society, Providence, RI, 2002.

36. Thomas Bauer, *Seshadri constants of quartic surfaces*, Math. Ann. **309** (1997), no. 3, 475–481.

37. _____, *On the cone of curves of an abelian variety*, Amer. J. Math. **120** (1998), no. 5, 997–1006.

38. _____, *Seshadri constants and periods of polarized abelian varieties*, Math. Ann. **312** (1998), no. 4, 607–623, With an appendix by the author and Tomasz Szemberg.

39. _____, *Seshadri constants on algebraic surfaces*, Math. Ann. **313** (1999), no. 3, 547–583.

40. Thomas Bauer, Alex Küronya, and Tomasz Szemberg, *Zariski chambers, volume and stable base loci*, preprint, 2003.

41. Thomas Bauer and Tomasz Szemberg, *Local positivity of principally polarized abelian threefolds*, J. Reine Angew. Math. **531** (2001), 191–200.

42. David Bayer and David Mumford, *What can be computed in algebraic geometry?*, Computational Algebraic Geometry and Commutative Algebra (Cortona, 1991), Cambridge Univ. Press, Cambridge, 1993, pp. 1–48.

43. David Bayer and Michael Stillman, *A criterion for detecting m-regularity*, Invent. Math. **87** (1987), no. 1, 1–11.

44. _____, *On the complexity of computing syzygies*, J. Symbolic Comput. **6** (1988), no. 2–3, 135–147.

45. Arnaud Beauville, *Annulation du $H^1$ et systèmes paracanoniques sur les surfaces*, J. Reine Angew. Math. **388** (1988), 149–157.

46. _____, *Annulation du $H^1$ pour les fibrés en droites plats*, Complex Algebraic Varieties (Bayreuth, 1990), Lect. Notes in Math., vol. 1507, Springer, Berlin, 1992, pp. 1–15.

47. _____, *Vector bundles on curves and generalized theta functions: recent results and open problems*, Current Topics in Complex Algebraic Geometry, Cambridge Univ. Press, Cambridge, 1995, pp. 17–33.

48. Alexander Beilinson, Joseph Bernstein, and Pierre Deligne, *Faisceaux Pervers*, Astérisque, vol. 100, Soc. Math. France, Paris, 1982.

49. Mauro Beltrametti, Michael Schneider, and Andrew Sommese, *Chern inequalities and spannedness of adjoint bundles*, Proceedings of the Hirzebruch 65 Conference on Algebraic Geometry (Ramat Gan, 1993), 1996, pp. 97–107.

50. Mauro Beltrametti and Andrew Sommese, *The Adjunction Theory of Complex Projective Varieties*, Walter de Gruyter & Co., Berlin, 1995.

51. Carlos Berenstein, Roger Gay, Alekos Vidras, and Alain Yger, *Residue Currents and Bezout Identities*, Progr. Math., vol. 114, Birkhäuser Verlag, Basel, 1993.

52. P. Berthelot, Alexander Grothendieck, and Luc Illusie, *Théorie des Intersections et Théorème de Riemann–Roch* (SGA6), Lect. Notes in Math., vol. 225, Springer-Verlag, Berlin, 1971.

53. Aaron Bertram, *An existence theorem for Prym special divisors*, Invent. Math. **90** (1987), no. 3, 669–671.

54. Aaron Bertram, Lawrence Ein, and Robert Lazarsfeld, *Surjectivity of Gaussian maps for line bundles of large degree on curves*, Algebraic Geometry (Chicago, 1989), Lect. Notes in Math., vol. 1479, Springer, Berlin, 1991, pp. 15–25.

55. _____, *Vanishing theorems, a theorem of Severi, and the equations defining projective varieties*, J. Amer. Math. Soc. **4** (1991), no. 3, 587–602.

56. Edward Bierstone and Pierre D. Milman, *Canonical desingularization in characteristic zero by blowing up the maximum strata of a local invariant*, Invent. Math. **128** (1997), no. 2, 207–302.

57. Paul Biran, *Constructing new ample divisors out of old ones*, Duke Math. J. **98** (1999), no. 1, 113–135.

58. _____, *A stability property of symplectic packing*, Invent. Math. **136** (1999), no. 1, 123–155.

59. _____, *From symplectic packing to algebraic geometry and back*, European Congress of Mathematics, Vol. II (Barcelona, 2000), Progr. in Math., vol. 202, Birkhäuser, Basel, 2001, pp. 507–524.

60. Spencer Bloch and David Gieseker, *The positivity of the Chern classes of an ample vector bundle*, Invent. Math. **12** (1971), 112–117.

61. Fedor Bogomolov, *Holomorphic tensors and vector bundles on projective manifolds*, Izv. Akad. Nauk SSSR Ser. Mat. **42** (1978), no. 6, 1227–1287, 1439.

62. _____, *Unstable vector bundles and curves on surfaces*, Proceedings of the International Congress of Mathematicians (Helsinki), Acad. Sci. Fennica, 1980, pp. 517–524.

63. Fedor Bogomolov and Bruno De Oliveira, *Convexity of coverings of algebraic varieties and vanishing theorems*, Preprint, 2003.

64. Fedor Bogomolov and Tony Pantev, *Weak Hironaka theorem*, Math. Res. Lett. **3** (1996), no. 3, 299–307.

65. Ciprian Borcea, *Homogeneous vector bundles and families of Calabi–Yau threefolds. II*, Several Complex Variables and Complex Geometry, Part 2, Proc. Symp. Pure Math., vol. 52, Amer. Math. Soc., Providence, RI, 1991, pp. 83–91.

66. Sébastien Boucksom, *On the volume of a line bundle*, Internat. J. Math. **13** (2002), no. 10, 1043–1063.

67. _____, *Higher dimensional Zariski decompositions*, preprint, 2003.

68. Sébastien Boucksom, Jean-Pierre Demailly, Mihai Paun, and Thomas Peternall, *The pseudo-effective cone of a compact Kähler manifold and varieties of negative Kodaira dimension*, preprint, 2004.

69. Nicolas Bourbaki, *Commutative Algebra. Chapters 1–7*, Elements of Mathematics, Springer-Verlag, Berlin, 1998.

70. Ana Bravo and Orlando Villamayor, *A strong desingularization theorem*, preprint.

71. Glen E. Bredon, *Topology and Geometry*, Graduate Texts in Mathematics, vol. 139, Springer-Verlag, New York, 1993.

72. Markus Brodmann and Rodney Sharp, *Local Cohomology: An Algebraic Introduction with Geometric Applications*, Cambridge studies in advanced mathematics, vol. 60, Cambridge University Press, Cambridge, 1998.

73. W. Dale Brownawell, *Bounds for the degrees in the Nullstellensatz*, Ann. of Math. (2) **126** (1987), no. 3, 577–591.

74. Winfried Bruns and Jürgen Herzog, *Cohen–Macaulay Rings*, Cambridge Studies in Advanced Mathematics, vol. 39, Cambridge University Press, Cambridge, 1993.

75. Nero Budur and Morihiko Saito, *Multiplier ideals, V-filtration, and spectrum*, preprint, 2003.

76. P. Buser and Peter Sarnak, *On the period matrix of a Riemann surface of large genus*, Invent. Math. **117** (1994), no. 1, 27–56.

77. Frédéric Campana, *Ensembles de Green–Lazarsfeld et quotients résolubles des groupes de Kähler*, J. Algebraic Geom. **10** (2001), no. 4, 599–622.

78. Frédéric Campana and Thomas Peternell, *Algebraicity of the ample cone of projective varieties*, J. Reine Angew. Math. **407** (1990), 160–166.

79. ———, *Projective manifolds whose tangent bundles are numerically effective*, Math. Ann. **289** (1991), no. 1, 169–187.

80. ———, *On the second exterior power of tangent bundles of threefolds*, Compositio Math. **83** (1992), no. 3, 329–346.

81. ———, *Rational curves and ampleness properties of the tangent bundle of algebraic varieties*, Manuscripta Math. **97** (1998), no. 1, 59–74.

82. James Carlson, Mark Green, Phillip Griffiths, and Joe Harris, *Infinitesimal variations of Hodge structure. I*, Compositio Math. **50** (1983), no. 2-3, 109–205.

83. James Carlson, Stefan Müller-Stach, and Chris Peters, *Period Mappings and Period Domains*, Cambridge Studies in Advanced Mathematics, vol. 85, Cambridge University Press, Cambridge, 2003.

84. Karen Chandler, *Regularity of the powers of an ideal*, Comm. Algebra **25** (1997), no. 12, 3773–3776.

85. Pierre-Emmanuel Chaput, *Severi varieties*, Math. Z. **240** (2002), no. 2, 451–459.

86. Marc Chardin and Clare D'Cruz, *Castelnuovo-Mumford regularity: examples of curves and surfaces*, J. Algebra **270** (2003), no. 1, 347–360.

87. Marc Chardin and Bernd Ulrich, *Liaison and Castelnuovo–Mumford regularity*, Amer. J. Math. **124** (2002), no. 6, 1103–1124.

88. Jungkai Chen and Christopher Hacon, *Characterization of abelian varieties*, Invent. Math. **143** (2001), no. 2, 435–447.

89. ———, *Pluricanonical maps of varieties of maximal Albanese dimension*, Math. Ann. **320** (2001), no. 2, 367–380.

90. Jaydeep V. Chipalkatti, *A generalization of Castelnuovo regularity to Grassmann varieties*, Manuscripta Math. **102** (2000), no. 4, 447–464.

91. Koji Cho and Ei-ichi Sato, *Smooth projective varieties with the ample vector bundle $\lambda^2 T_X$ in any characteristic*, J. Math. Kyoto Univ. **35** (1995), no. 1, 1–33.

92. Ciro Ciliberto and Alexis Kouvidakis, *On the symmetric product of a curve with general moduli*, Geom. Dedicata **78** (1999), no. 3, 327–343.

93. Herbert Clemens, *Curves on generic hypersurfaces*, Ann. Sci. École Norm. Sup. (4) **19** (1986), no. 4, 629–636.

94. _____, *A local proof of Petri's conjecture at the general curve*, J. Differential Geom. **54** (2000), no. 1, 139–176.

95. Herbert Clemens and Christopher Hacon, *Deformations of the trivial line bundle and vanishing theorems*, Amer. J. Math. **124** (2002), no. 4, 769–815.

96. John Conway and Neil Sloane, *Sphere Packings, Lattices and Groups*, third ed., Springer-Verlag, New York, 1999.

97. S. Dale Cutkosky, *Zariski decomposition of divisors on algebraic varieties*, Duke Math. J. **53** (1986), no. 1, 149–156.

98. _____, *Irrational asymptotic behaviour of Castelnuovo–Mumford regularity*, J. Reine Angew. Math. **522** (2000), 93–103.

99. S. Dale Cutkosky, Lawrence Ein, and Robert Lazarsfeld, *Positivity and complexity of ideal sheaves*, Math. Ann. **321** (2001), no. 2, 213–234.

100. S. Dale Cutkosky, Jürgen Herzog, and Ngô Viêt Trung, *Asymptotic behaviour of the Castelnuovo–Mumford regularity*, Compositio Math. **118** (1999), no. 3, 243–261.

101. S. Dale Cutkosky and V. Srinivas, *On a problem of Zariski on dimensions of linear systems*, Ann. of Math. (2) **137** (1993), no. 3, 531–559.

102. _____, *Periodicity of the fixed locus of multiples of a divisor on a surface*, Duke Math. J. **72** (1993), no. 3, 641–647.

103. Mark de Cataldo, *Singular Hermitian metrics on vector bundles*, J. Reine Angew. Math. **502** (1998), 93–122.

104. Mark de Cataldo and Luca Migliorini, *The hard Lefschetz theorem and the topology of semismall maps*, Ann. Sci. École Norm. Sup. (4) **35** (2002), no. 5, 759–772.

105. _____, *The Hodge theory of algebraic maps*, preprint, 2003.

106. Tommaso de Fernex, Lawrence Ein, and Mircea Mustaţă, *Multiplicities and log canonical threshold*, J. Alg. Geometry, to appear.

107. _____, *Bounds for log canonical thresholds with applications to birational rigidity*, Math. Res. Lett. **10** (2003), no. 2–3, 219–236.

108. A. Johan de Jong, *Smoothness, semi-stability and alterations*, Inst. Hautes Études Sci. Publ. Math. (1996), no. 83, 51–93.

109. Olivier Debarre, *Fulton–Hansen and Barth–Lefschetz theorems for subvarieties of abelian varieties*, J. Reine Angew. Math. **467** (1995), 187–197.

110. _____ , *Théorèmes de connexité et variétés abéliennes*, Amer. J. Math. **117** (1995), no. 3, 787–805.

111. _____ , *Théorèmes de connexité pour les produits d'espaces projectifs et les grassmanniennes*, Amer. J. Math. **118** (1996), no. 6, 1347–1367.

112. _____ , *Tores et Variétés Abéliennes Complexes*, Société Mathématique de France, Paris, 1999.

113. _____ , *Théorèmes de Lefschetz pour les lieux de dégénérescence*, Bull. Soc. Math. France **128** (2000), no. 2, 283–308.

114. _____ , *Higher-Dimensional Algebraic Geometry*, Springer-Verlag, New York, 2001.

115. _____ , *Varieties with ample cotangent bundle*, preprint, 2003.

116. Olivier Debarre, Klaus Hulek, and J. Spandaw, *Very ample linear systems on abelian varieties*, Math. Ann. **300** (1994), no. 2, 181–202.

117. Pierre Deligne, *Cohomologie Étale* (SGA4.5), Lect. Notes in Math., vol. 569, Springer-Verlag, Berlin, 1977.

118. _____ , *Le groupe fondamental du complément d'une courbe plane n'ayant que des points doubles ordinaires est abélien (d'après W. Fulton)*, Séminaire Bourbaki, Année 1979/80, Lect. Notes in Math., vol. 842, Springer, Berlin, 1981, pp. 1–10.

119. Jean-Pierre Demailly, *Complex Analytic and Algebraic Geometry*, book in preparation.

120. _____ , *Champs magnétiques et inégalités de Morse pour la $d''$-cohomologie*, C. R. Acad. Sci. Paris Sér. I Math. **301** (1985), no. 4, 119–122.

121. _____ , *Vanishing theorems for tensor powers of an ample vector bundle*, Invent. Math. **91** (1988), no. 1, 203–220.

122. _____ , *Regularization of closed positive currents and intersection theory*, J. Algebraic Geom. **1** (1992), no. 3, 361–409.

123. _____ , *Singular Hermitian metrics on positive line bundles*, Complex Algebraic Varieties (Bayreuth, 1990), Lect. Notes in Math., vol. 1507, 1992, pp. 87–104.

124. _____ , *A numerical criterion for very ample line bundles*, J. Differential Geom. **37** (1993), no. 2, 323–374.

125. _____ , *Effective bounds for very ample line bundles*, Invent. Math. **124** (1996), no. 1-3, 243–261.

126. _____ , *$L^2$ vanishing theorems for positive line bundles and adjunction theory*, Transcendental Methods in Algebraic Geometry (Cetraro, 1994), Lect. Notes in Math., vol. 1464, Springer, Berlin, 1996, pp. 1–97.

127. _____ , *Algebraic criteria for Kobayashi hyperbolic projective varieties and jet differentials*, Algebraic Geometry, Santa Cruz 1995, Proc. Symp. Pure Math., vol. 62, Amer. Math. Soc., Providence, RI, 1997, pp. 285–360.

128. _____ , *Méthodes $L^2$ et résultats effectifs en géométrie algébrique*, Séminaire Bourbaki, Année 1998/99, Exp. No. 852, no. 266, 2000, pp. 59–90.

129. _____ , *Multiplier ideal sheaves and analytic methods in algebraic geometry*, School on Vanishing Theorems and Effective Results in Algebraic Geometry

(Trieste, 2000), ICTP Lect. Notes, vol. 6, Abdus Salam Int. Cent. Theoret. Phys., Trieste, 2001, pp. 1–148.

130. Jean-Pierre Demailly, Lawrence Ein, and Robert Lazarsfeld, *A subadditivity property of multiplier ideals*, Michigan Math. J. **48** (2000), 137–156.

131. Jean-Pierre Demailly and János Kollár, *Semi-continuity of complex singularity exponents and Kähler–Einstein metrics on Fano orbifolds*, Ann. Sci. École Norm. Sup. (4) **34** (2001), no. 4, 525–556.

132. Jean-Pierre Demailly and Mihai Paun, *Numerical characterization of the Kähler cone of a compact Kähler manifold*, to appear.

133. Jean-Pierre Demailly, Thomas Peternell, and Michael Schneider, *Compact complex manifolds with numerically effective tangent bundles*, J. Algebraic Geom. **3** (1994), no. 2, 295–345.

134. Jean-Pierre Demailly and Henri Skoda, *Relations entre les notions de positivités de P. A. Griffiths et de S. Nakano pour les fibrés vectoriels*, Séminaire Pierre Lelong–Henri Skoda. Année 1978/79, Lect. Notes in Math, vol. 822, 1980, pp. 304–309.

135. Michel Demazure, *Caractérisations de l'espace projectif (conjectures de Hartshorne et de Frankel) (d'après Shigefumi Mori)*, Séminaire Bourbaki, Année 1979/80, Lect. Notes in Math., vol. 842, Springer, Berlin, 1981, pp. 11–19.

136. Jan Denef and François Loeser, *Germs of arcs on singular algebraic varieties and motivic integration*, Invent. Math. **135** (1999), no. 1, 201–232.

137. Harm Derksen and Jessica Sidman, *A sharp bound for the Castelnuovo–Mumford regularity of subspace arrangements*, Adv. Math. **172** (2002), no. 2, 151–157.

138. Mireille Deschamps, *Courbes de genre géométrique borné sur une surface de type général (d'après F. A. Bogomolov)*, Séminaire Bourbaki, Année 1977/78, Lect. Notes in Math., vol. 710, Springer, 1979, pp. 233–247.

139. S. K. Donaldson, *Anti self-dual Yang–Mills connections over complex algebraic surfaces and stable vector bundles*, Proc. London Math. Soc. (3) **50** (1985), no. 1, 1–26.

140. Lawrence Ein, *An analogue of Max Noether's theorem*, Duke Math. J. **52** (1985), no. 3, 689–706.

141. _____, *Varieties with small dual varieties. II*, Duke Math. J. **52** (1985), no. 4, 895–907.

142. _____, *Varieties with small dual varieties. I*, Invent. Math. **86** (1986), no. 1, 63–74.

143. _____, *Subvarieties of generic complete intersections*, Invent. Math. **94** (1988), no. 1, 163–169.

144. _____, *Unpublished notes on multiplier ideals*, 1995.

145. _____, *Multiplier ideals, vanishing theorems and applications*, Algebraic Geometry—Santa Cruz 1995, Proc. Symp. Pure Math., vol. 62, Amer. Math. Soc., Providence, RI, 1997, pp. 203–219.

146. _____, *Notes from lectures at Catania*, unpublished, 1999.

147. _____, *Linear systems with removable base loci*, Comm. Algebra **28** (2000), no. 12, 5931–5934.

148. Lawrence Ein, Bo Ilic, and Robert Lazarsfeld, *A remark on projective embeddings of varieties with non-negative cotangent bundles*, Complex Analysis and Algebraic Geometry, de Gruyter, Berlin, 2000, pp. 165–171.

149. Lawrence Ein, Oliver Küchle, and Robert Lazarsfeld, *Local positivity of ample line bundles*, J. Differential Geom. **42** (1995), no. 2, 193–219.

150. Lawrence Ein and Robert Lazarsfeld, *Stability and restrictions of Picard bundles, with an application to the normal bundles of elliptic curves*, Complex Projective Geometry (Trieste, 1989/Bergen, 1989), London Math. Soc. Lecture Note Series, vol. 179, Cambridge Univ. Press, Cambridge, 1992, pp. 149–156.

151. _____, *Global generation of pluricanonical and adjoint linear series on smooth projective threefolds*, J. Amer. Math. Soc. **6** (1993), no. 4, 875–903.

152. _____, *Seshadri constants on smooth surfaces*, Journées de Géométrie Algébrique d'Orsay (Orsay, 1992), Astérisque, no. 218, 1993, pp. 177–186.

153. _____, *Syzygies and Koszul cohomology of smooth projective varieties of arbitrary dimension*, Invent. Math. **111** (1993), no. 1, 51–67.

154. _____, *Singularities of theta divisors and the birational geometry of irregular varieties*, J. Amer. Math. Soc. **10** (1997), no. 1, 243–258.

155. _____, *A geometric effective Nullstellensatz*, Invent. Math. **137** (1999), no. 2, 427–448.

156. Lawrence Ein, Robert Lazarsfeld, and Mircea Mustaţă, *Contact loci in arc spaces*, to appear.

157. Lawrence Ein, Robert Lazarsfeld, Mircea Mustaţă, Michael Nakamaye, and Mihnea Popa, *Asymptotic invariants of base loci*, preprint, 2003.

158. Lawrence Ein, Robert Lazarsfeld, and Michael Nakamaye, *Zero-estimates, intersection theory, and a theorem of Demailly*, Higher-Dimensional Complex Varieties (Trento, 1994), de Gruyter, Berlin, 1996, pp. 183–207.

159. Lawrence Ein, Robert Lazarsfeld, and Karen E. Smith, *Uniform bounds and symbolic powers on smooth varieties*, Invent. Math. **144** (2001), no. 2, 241–252.

160. _____, *Uniform approximation of Abhyankar valuation ideals in smooth function fields*, Amer. J. Math. **125** (2003), no. 2, 409–440.

161. Lawrence Ein, Robert Lazarsfeld, Karen E. Smith, and Dror Varolin, *Jumping coefficients of multiplier ideals*, to appear.

162. Lawrence Ein, Mircea Mustaţă, and Takehiko Yasuda, *Jet schemes, log discrepancies and inversion of adjunction*, Invent. Math. **153** (2003), no. 3, 519–535.

163. David Eisenbud, *The Geometry of Syzygies*, forthcoming.

164. _____, *Commutative Algebra, with a View Toward Algebraic Geometry*, Graduate Texts in Math, no. 150, Springer-Verlag, New York, 1995.

165. David Eisenbud and Shiro Goto, *Linear free resolutions and minimal multiplicity*, J. Algebra **88** (1984), no. 1, 89–133.

166. Santiago Encinas and Orlando Villamayor, *Good points and constructive resolution of singularities*, Acta Math. **181** (1998), no. 1, 109–158.

167. _____, *A course on constructive desingularization and equivariance*, Resolution of Singularities (Obergurgl, 1997), Progr. Math., vol. 181, Birkhäuser, Basel, 2000, pp. 147–227.

168. Hélène Esnault and Eckart Viehweg, *Sur une minoration du degré d'hypersurfaces s'annulant en certains points*, Math. Ann. **263** (1983), no. 1, 75–86.

169. _____, *Dyson's lemma for polynomials in several variables (and the theorem of Roth)*, Invent. Math. **78** (1984), no. 3, 445–490.

170. _____, *Logarithmic de Rham complexes and vanishing theorems*, Invent. Math. **86** (1986), no. 1, 161–194.

171. _____, *Rêvetements cycliques. II (autour du théorème d'annulation de J. Kollár)*, Géométrie algébrique et applications, II (La Rábida, 1984), Travaux en cours, no. 23, Hermann, Paris, 1987, pp. 81–96.

172. _____, *Effective bounds for semipositive sheaves and for the height of points on curves over complex function fields*, Compositio Math. **76** (1990), no. 1-2, 69–85, Algebraic geometry (Berlin, 1988).

173. _____, *Ample sheaves on moduli schemes*, Algebraic Geometry and Analytic Geometry (Tokyo, 1990), ICM-90 Satell. Conf. Proc., Springer, Tokyo, 1991, pp. 53–80.

174. _____, *Lectures on Vanishing Theorems*, DMV Seminar, vol. 20, Birkhäuser Verlag, Basel, 1992.

175. E. Graham Evans and Phillip Griffith, *The syzygy problem*, Ann. of Math. (2) **114** (1981), no. 2, 323–333.

176. Gerd Faltings, *Formale Geometrie und homogene Räume*, Invent. Math. **64** (1981), no. 1, 123–165.

177. _____, *Verschwindungssätze und Untermannigfaltigkeiten kleiner Kodimension des komplex-projektiven Raumes*, J. Reine Angew. Math. **326** (1981), 136–151.

178. _____, *Diophantine approximation on abelian varieties*, Ann. of Math. (2) **133** (1991), no. 3, 549–576.

179. _____, *The general case of S. Lang's conjecture*, Barsotti Symposium in Algebraic Geometry (Abano Terme, 1991), Academic Press, San Diego, CA, 1994, pp. 175–182.

180. _____, *Mumford-Stabilität in der algebraischen Geometrie*, Proceedings of the International Congress of Mathematicians, (Zürich, 1994), Birkhäuser, 1995, pp. 648–655.

181. Gerd Faltings and Gisbert Wüstholz, *Diophantine approximations on projective spaces*, Invent. Math. **116** (1994), no. 1-3, 109–138.

182. Fuquan Fang, Sérgio Mendonça, and Xiaochung Rong, *A connectedness principle in the geometry of positive curvature*, preprint.

183. Gavril Farkas and Angela Gibney, *The Mori cones of moduli spaces of pointed curves of small genus*, Trans. Amer. Math. Soc. **355** (2003), no. 3, 1183–1199.

184. Charles Favre and Mattias Jonsson, *Valuative criterion for integrability*, preprint.

185. Guillermo Fernández del Busto, *A Matsusaka-type theorem on surfaces*, J. Algebraic Geom. **5** (1996), no. 3, 513–520.

186. Hubert Flenner, Liam O'Carroll, and Wolfgang Vogel, *Joins and Intersections*, Monographs in Mathematics, Springer-Verlag, Berlin, 1999.

187. Hubert Flenner and Wolfgang Vogel, *Joins, tangencies and intersections*, Math. Ann. **302** (1995), no. 3, 489–505.

188. Jens Franke, Yuri Manin, and Yuri Tschinkel, *Rational points of bounded height on Fano varieties*, Invent. Math. **95** (1989), no. 2, 421–435.

189. Klaus Fritzsche, *Pseudoconvexity properties of complements of analytic subvarieties*, Math. Ann. **230** (1977), no. 2, 107–122.

190. Osamu Fujino, *Notes on toric varieties from Mori theoretic viewpoint*, Tohoku Math. J. (2) **55** (2003), no. 4, 551–564.

191. Takao Fujita, *Defining equations for certain types of polarized varieties*, Complex Analysis and Algebraic Geometry, Cambridge University Press, 1977, pp. 165–173.

192. _____, *On Kähler fiber spaces over curves*, J. Math. Soc. Japan **30** (1978), no. 4, 779–794.

193. _____, *On Zariski problem*, Proc. Japan Acad. Ser. A Math. Sci. **55** (1979), no. 3, 106–110.

194. _____, *Semipositive line bundles*, J. Fac. Sci. Univ. Tokyo Sect. IA Math. **30** (1983), no. 2, 353–378.

195. _____, *Vanishing theorems for semipositive line bundles*, Algebraic Geometry (Tokyo/Kyoto, 1982), Lect. Notes in Math., no. 1016, Springer, Berlin, 1983, pp. 519–528.

196. _____, *Triple covers by smooth manifolds*, J. Fac. Sci. Univ. Tokyo Sect. IA Math. **35** (1988), no. 1, 169–175.

197. _____, *Classification Theories of Polarized Varieties*, London Math. Soc. Lect. Note Series, vol. 155, Cambridge University Press, Cambridge, 1990.

198. _____, *The Kodaira energy and the classification of polarized manifolds*, Sūgaku **45** (1993), no. 3, 244–255.

199. _____, *Approximating Zariski decomposition of big line bundles*, Kodai Math. J. **17** (1994), no. 1, 1–3.

200. _____, *On Kodaira energy of polarized log varieties*, J. Math. Soc. Japan **48** (1996), no. 1, 1–12.

201. William Fulton, *Ample vector bundles, Chern classes, and numerical criteria*, Invent. Math. **32** (1976), no. 2, 171–178.

202. _____, *On the fundamental group of the complement of a node curve*, Ann. of Math. (2) **111** (1980), no. 2, 407–409.

203. _____, *On the topology of algebraic varieties*, Algebraic Geometry, Bowdoin, 1985, Proc. Symp. Pure Math., vol. 46, Amer. Math. Soc., Providence, RI, 1987, pp. 15–46.

204. _____, *Flags, Schubert polynomials, degeneracy loci, and determinantal formulas*, Duke Math. J. **65** (1992), no. 3, 381–420.

205. _____, *Introduction to Toric Varieties*, Annals of Math. Studies, vol. 131, Princeton University Press, Princeton, NJ, 1993.

206. _____, *Positive polynomials for filtered ample vector bundles*, Amer. J. Math. **117** (1995), no. 3, 627–633.

207. _____, *Young Tableaux*, London Mathematical Society Student Texts, vol. 35, Cambridge University Press, Cambridge, 1997.

208. _____, *Intersection Theory*, second ed., Ergebnisse der Math. und ihrer Grenzgebiete (3), vol. 2, Springer-Verlag, Berlin, 1998.

209. William Fulton and Johan Hansen, *A connectedness theorem for projective varieties, with applications to intersections and singularities of mappings*, Ann. of Math. (2) **110** (1979), no. 1, 159–166.

210. William Fulton and Joe Harris, *Representation Theory, a First Course*, Graduate Texts in Math., vol. 129, Springer-Verlag, New York, 1991.

211. William Fulton and Robert Lazarsfeld, *Connectivity and its applications in algebraic geometry*, Algebraic Geometry (Chicago, 1980), Lect. Notes in Math., vol. 862, Springer, Berlin, 1981, pp. 26–92.

212. _____, *On the connectedness of degeneracy loci and special divisors*, Acta Math. **146** (1981), no. 3-4, 271–283.

213. _____, *Positivity and excess intersection*, Enumerative Geometry and Classical Algebraic Geometry (Nice, 1981), Prog. Math., vol. 24, Birkhäuser Boston, Mass., 1982, pp. 97–105.

214. _____, *Positive polynomials for ample vector bundles*, Ann. of Math. (2) **118** (1983), no. 1, 35–60.

215. William Fulton and Piotr Pragacz, *Schubert Varieties and Degeneracy Loci*, Lect. Notes in Math., vol. 1689, Springer-Verlag, Berlin, 1998.

216. Terence Gaffney and Robert Lazarsfeld, *On the ramification of branched coverings of* $\mathbf{P}^n$, Invent. Math. **59** (1980), no. 1, 53–58.

217. Francisco Gallego and B. P. Purnaprajna, *Projective normality and syzygies of algebraic surfaces*, J. Reine Angew. Math. **506** (1999), 145–180.

218. _____, *Syzygies of projective surfaces: an overview*, J. Ramanujan Math. Soc. **14** (1999), no. 1, 65–93.

219. Jørgen Anders Geertsen, *Push-forward of degeneracy classes and ampleness*, Proc. Amer. Math. Soc. **129** (2001), no. 7, 1885–1890.

220. Anthony V. Geramita, Alessandro Gimigliano, and Yves Pitteloud, *Graded Betti numbers of some embedded rational n-folds*, Math. Ann. **301** (1995), no. 2, 363–380.

221. Franco Ghione, *Un problème du type Brill–Noether pour les fibrés vectoriels*, Algebraic Geometry—Open Problems (Ravello, 1982), Lecture Notes in Math., vol. 997, Springer, Berlin, 1983, pp. 197–209.

222. Angela Gibney, Seán Keel, and Ian Morrison, *Towards the ample cone of* $\overline{M}_{g,n}$, J. Amer. Math. Soc. **15** (2002), no. 2, 273–294.

223. David Gieseker, *p-ample bundles and their Chern classes*, Nagoya Math. J. **43** (1971), 91–116.

224. _____, *On a theorem of Bogomolov on Chern classes of stable bundles*, Amer. J. Math. **101** (1979), no. 1, 77–85.

225. _____, *Stable curves and special divisors: Petri's conjecture*, Invent. Math. **66** (1982), no. 2, 251–275.

226. Jacob Eli Goodman, *Affine open subsets of algebraic varieties and ample divisors*, Ann. of Math. (2) **89** (1969), 160–183.

227. Mark Goresky and Robert MacPherson, *Stratified Morse Theory*, Ergebnisse der Math. und ihrer Grenzgebiete (3), vol. 14, Springer-Verlag, Berlin, 1988.

228. Gerd Gotzmann, *Eine Bedingung für die Flachheit und das Hilbertpolynom eines graduierten Ringes*, Math. Z. **158** (1978), no. 1, 61–70.

229. Russell Goward, *A simple algorithm for principalization of monomial ideals*, preprint, 2003.

230. William Graham, *Non-emptiness of skew symmetric degeneracy loci*, preprint, 2003.

231. _____, *Non-emptiness of symmetric degeneracy loci*, preprint, 2003.

232. Hans Grauert, *Über Modifikationen und exzeptionelle analytische Mengen*, Math. Ann. **146** (1962), 331–368.

233. _____, *Mordells Vermutung über rationale Punkte auf algebraischen Kurven und Funktionenkörper*, Inst. Hautes Études Sci. Publ. Math. (1965), no. 25, 131–149.

234. Hans Grauert and Oswald Riemenschneider, *Verschwindungssätze für analytische Kohomologiegruppen auf komplexen Räumen*, Invent. Math. **11** (1970), 263–292.

235. Mark Green, *Koszul cohomology and the geometry of projective varieties*, J. Differential Geom. **19** (1984), no. 1, 125–171.

236. _____, *A new proof of the explicit Noether–Lefschetz theorem*, J. Differential Geom. **27** (1988), no. 1, 155–159.

237. _____, *Koszul cohomology and geometry*, Lectures on Riemann surfaces (Trieste, 1987), World Sci. Publishing, Teaneck, NJ, 1989, pp. 177–200.

238. _____, *Restrictions of linear series to hyperplanes, and some results of Macaulay and Gotzmann*, Algebraic Curves and Projective Geometry (Trento, 1988), Lect. Notes in Math., vol. 1389, Springer, Berlin, 1989, pp. 76–86.

239. _____, *Infinitesimal methods in Hodge theory*, Algebraic Cycles and Hodge Theory (Torino, 1993), Lect. Notes in Math., vol. 1594, Springer, Berlin, 1994, pp. 1–92.

240. _____, *Generic initial ideals*, Six Lectures on Commutative Algebra (Bellaterra, 1996), Progr. Math., vol. 166, Birkhäuser, Basel, 1998, pp. 119–186.

241. Mark Green and Robert Lazarsfeld, *On the projective normality of complete linear series on an algebraic curve*, Invent. Math. **83** (1985), no. 1, 73–90.

242. _____, *Deformation theory, generic vanishing theorems, and some conjectures of Enriques, Catanese and Beauville*, Invent. Math. **90** (1987), no. 2, 389–407.

243. _____, *Some results on the syzygies of finite sets and algebraic curves*, Compositio Math. **67** (1988), no. 3, 301–314.

244. _____, *Higher obstructions to deforming cohomology groups of line bundles*, J. Amer. Math. Soc. **4** (1991), no. 1, 87–103.

245. Phillip Griffiths, *Hermitian differential geometry and the theory of positive and ample holomorphic vector bundles*, J. Math. Mech. **14** (1965), 117–140.

246. _____, *The extension problem in complex analysis. II. Embeddings with positive normal bundle*, Amer. J. Math. **88** (1966), 366–446.

247. _____, *Hermitian differential geometry, Chern classes, and positive vector bundles*, Global Analysis (Papers in Honor of K. Kodaira), Univ. Tokyo Press, Tokyo, 1969, pp. 185–251.

248. Phillip Griffiths and Joseph Harris, *Principles of Algebraic Geometry*, Wiley-Interscience, New York, 1978.

249. _____, *Algebraic geometry and local differential geometry*, Ann. Sci. École Norm. Sup. (4) **12** (1979), no. 3, 355–452.

250. _____, *On the variety of special linear systems on a general algebraic curve*, Duke Math. J. **47** (1980), no. 1, 233–272.

251. Mikhael Gromov, *Pseudoholomorphic curves in symplectic manifolds*, Invent. Math. **82** (1985), no. 2, 307–347.

252. _____, *Partial Differential Relations*, Ergebnisse der Math. und ihrer Grenzgebiete (3), vol. 9, Springer-Verlag, Berlin, 1986.

253. _____, *Convex sets and Kähler manifolds*, Advances in Differential Geometry and Topology, World Sci. Publishing, Teaneck, NJ, 1990, pp. 1–38.

254. _____, *Systoles and intersystolic inequalities*, Actes de la Table Ronde de Géométrie Différentielle (Luminy, 1992), Soc. Math. France, Paris, 1996, pp. 291–362.

255. Alexander Grothendieck, *Éléments de Géométrie Algébrique. II. Étude Globale Élémentaire de Quelques Classes de Morphismes*, Inst. Hautes Études Sci. Publ. Math. (1961), no. 8, 222.

256. _____, *Éléments de Géométrie Algébrique. III. Étude Cohomologique des Faisceaux Cohérents. I*, Inst. Hautes Études Sci. Publ. Math. (1961), no. 11, 167.

257. _____, *Éléments de Géométrie Algébrique. IV. Étude Locale des Schémas et des Morphismes de Schémas IV*, Inst. Hautes Études Sci. Publ. Math. (1967), no. 32, 361.

258. _____, *Cohomologie Locale des Faisceaux Cohérents et Théorèmes de Lefschetz Locaux et Globaux* (SGA 2), Advanced Stud. in Pure Math, vol. 2, North-Holland, Amsterdam, 1968.

259. Laurent Gruson, Robert Lazarsfeld, and Christian Peskine, *On a theorem of Castelnuovo, and the equations defining space curves*, Invent. Math. **72** (1983), no. 3, 491–506.

260. Christopher Hacon, *Examples of spanned and ample vector bundles with small numerical invariants*, C. R. Acad. Sci. Paris Sér. I Math. **323** (1996), no. 9, 1025–1029.

261. _____, *Divisors on principally polarized abelian varieties*, Compositio Math. **119** (1999), no. 3, 321–329.

262. ———, *Fourier transforms, generic vanishing theorems and polarizations of abelian varieties*, Math. Z. **235** (2000), no. 4, 717–726.

263. ———, *A derived category approach to generic vanishing*, preprint, 2003.

264. Helmut Hamm, *Lokale topologische Eigenschaften komplexer Räume*, Math. Ann. **191** (1971), 235–252.

265. ———, *On the vanishing of local homotopy groups for isolated singularities of complex spaces*, J. Reine Angew. Math. **323** (1981), 172–176.

266. Helmut Hamm and Lê Dung Tráng, *Rectified homotopical depth and Grothendieck conjectures*, The Grothendieck Festschrift, Vol. II, Birkhäuser Boston, Boston, MA, 1990, pp. 311–351.

267. ———, *Relative homotopical depth and a conjecture of Deligne*, Math. Ann. **296** (1993), no. 1, 87–101.

268. ———, *Vanishing theorems for constructible sheaves. I*, J. Reine Angew. Math. **471** (1996), 115–138.

269. Johan Hansen, *A connectedness theorem for flagmanifolds and Grassmannians*, Amer. J. Math. **105** (1983), no. 3, 633–639.

270. Nobuo Hara, *Geometric interpretation of tight closure and test ideals*, Trans. Amer. Math. Soc. **353** (2001), no. 5, 1885–1906.

271. Nobuo Hara and Ken-Ichi Yoshida, *A generalization of tight closure and multiplier ideals*, Trans. Amer. Math. Soc. **355** (2003), no. 8, 3143–3174.

272. Joe Harris, *On the Severi problem*, Invent. Math. **84** (1986), no. 3, 445–461.

273. Joe Harris and Loring W. Tu, *The connectedness of symmetric degeneracy loci: odd ranks*, Topics in Algebra, Part 2 (Warsaw, 1988), PWN, Warsaw, 1990, pp. 249–256.

274. Robin Hartshorne, *Ample vector bundles*, Inst. Hautes Études Sci. Publ. Math. No. **29** (1966), 63–94.

275. ———, *Cohomological dimension of algebraic varieties*, Ann. of Math. (2) **88** (1968), 403–450.

276. ———, *Ample Subvarieties of Algebraic Varieties*, Lect. Notes in Math., vol. 156, Springer-Verlag, Berlin, 1970.

277. ———, *Ample vector bundles on curves*, Nagoya Math. J. **43** (1971), 73–89.

278. ———, *Equivalence relations on algebraic cycles and subvarieties of small codimension*, Proc. Symp. Pure Math., vol. 29, pp. 129–164, Amer. Math. Soc., Providence, R.I., 1973.

279. ———, *Varieties of small codimension in projective space*, Bull. Amer. Math. Soc. **80** (1974), 1017–1032.

280. ———, *Algebraic Geometry*, Graduate Texts in Math., vol. 52, Springer-Verlag, New York, 1977.

281. Robin Hartshorne (ed.), *Algebraic Geometry, Arcata 1974*, Proc. Symp. Pure Math., vol. 29, Providence, R.I., American Mathematical Society, 1979.

282. Brendan Hassett, Hui-Wen Lin, and Chin-Lung Wang, *The weak Lefschetz principle is false for ample cones*, Asian J. Math. **6** (2002), no. 1, 95–99.

283. Brendan Hassett and Yuri Tschinkel, *Rational curves on holomorphic symplectic fourfolds*, Geom. Funct. Anal. **11** (2001), no. 6, 1201–1228.

284. ———, *On the effective cone of the moduli space of pointed rational curves*, Topology and geometry: commemorating SISTAG, Contemp. Math., vol. 314, Amer. Math. Soc., Providence, RI, 2002, pp. 83–96.

285. Stefan Helmke, *On Fujita's conjecture*, Duke Math. J. **88** (1997), no. 2, 201–216.

286. ———, *On global generation of adjoint linear systems*, Math. Ann. **313** (1999), no. 4, 635–652.

287. Michel Hickel, *Solution d'une conjecture de C. Berenstein–A. Yger et invariants de contact à l'infini*, Ann. Inst. Fourier (Grenoble) **51** (2001), no. 3, 707–744.

288. Heisuke Hironaka, *Resolution of singularities of an algebraic variety over a field of characteristic zero. I, II*, Ann. of Math. (2) 79 (1964), 109–203; ibid. (2) **79** (1964), 205–326.

289. André Hirschowitz, *Une conjecture pour la cohomologie des diviseurs sur les surfaces rationnelles génériques*, J. Reine Angew. Math. **397** (1989), 208–213.

290. Melvin Hochster and Craig Huneke, *Comparison of symbolic and ordinary powers of ideals*, Invent. Math. **147** (2002), no. 2, 349–369.

291. Audun Holme and Michael Schneider, *A computer aided approach to codimension 2 subvarieties of $\mathbf{P}_n$, $n \geq 6$*, J. Reine Angew. Math. **357** (1985), 205–220.

292. Jason Howald, *Computations with monomial ideals*, Ph.D. thesis, University of Michigan, 2001.

293. ———, *Multiplier ideals of monomial ideals*, Trans. Amer. Math. Soc. **353** (2001), no. 7, 2665–2671.

294. ———, *Multiplier ideals of sufficiently general polynomials*, preprint, 2003.

295. Reinhold Hübl and Irena Swanson, *Discrete valuations centered on local domains*, J. Pure Appl. Algebra **161** (2001), no. 1-2, 145–166.

296. Craig Huneke, *Uniform bounds in Noetherian rings*, Invent. Math. **107** (1992), no. 1, 203–223.

297. ———, *Tight Closure and Its Applications*, CBMS Regional Conference Series in Math., vol. 88, Conference Board of the Mathematical Sciences, Washington, DC, 1996.

298. Daniel Huybrechts, *Compact hyper-Kähler manifolds: basic results*, Invent. Math. **135** (1999), no. 1, 63–113.

299. ———, *Erratum: "Compact hyper-Kähler manifolds: basic results"*, Invent. Math. **152** (2003), no. 1, 209–212.

300. Daniel Huybrechts and Manfred Lehn, *The Geometry of Moduli Spaces of Sheaves*, Aspects of Mathematics, E31, Friedr. Vieweg & Sohn, Braunschweig, 1997.

301. Jun-Muk Hwang and Jong Hae Keum, *Seshadri-exceptional foliations*, Math. Ann. **325** (2003), no. 2, 287–297.

302. Jun-Muk Hwang and Wing-Keung To, *On Seshadri constants of canonical bundles of compact complex hyperbolic spaces*, Compositio Math. **118** (1999), no. 2, 203–215.

303. _____, *On Seshadri constants of canonical bundles of compact quotients of bounded symmetric domains*, J. Reine Angew. Math. **523** (2000), 173–197.

304. _____, *Volumes of complex analytic subvarieties of Hermitian symmetric spaces*, Amer. J. Math. **124** (2002), no. 6, 1221–1246.

305. Eero Hyry and Karen E. Smith, *On a Non-Vanishing Conjecture of Kawamata and the Core of an Ideal*, preprint, 2002.

306. Bo Ilic and J. M. Landsberg, *On symmetric degeneracy loci, spaces of symmetric matrices of constant rank and dual varieties*, Math. Ann. **314** (1999), no. 1, 159–174.

307. Shuzo Izumi, *A measure of integrity for local analytic algebras*, Publ. Res. Inst. Math. Sci. **21** (1985), no. 4, 719–735.

308. Jean-Pierre Jouanolou, *Théorèmes de Bertini et Applications*, Progr. Math., vol. 42, Birkhäuser Boston, Boston, MA, 1983.

309. Morris Kalka, Bernard Shiffman, and Bun Wong, *Finiteness and rigidity theorems for holomorphic mappings*, Michigan Math. J. **28** (1981), no. 3, 289–295.

310. M. Kapranov, *Veronese curves and Grothendieck–Knudsen moduli space $\overline{M}_{0,n}$*, J. Algebraic Geom. **2** (1993), no. 2, 239–262.

311. K. K. Karčjauskas, *A generalized Lefschetz theorem*, Funkcional. Anal. i Priložen. **11** (1977), no. 4, 80–81.

312. _____, *Homotopy properties of algebraic sets*, Zap. Nauchn. Sem. Leningrad. Otdel. Mat. Inst. Steklov. (LOMI) **83** (1979), 67–72, 103, Studies in topology, III.

313. Masaki Kashiwara, *B-functions and holonomic systems. Rationality of roots of B-functions*, Invent. Math. **38** (1976/77), no. 1, 33–53.

314. Ludger Kaup, *Eine topologische Eigenschaft Steinscher Räume*, Nachr. Akad. Wiss. Göttingen (1966), 213–224.

315. Yujiro Kawamata, *Characterization of abelian varieties*, Compositio Math. **43** (1981), no. 2, 253–276.

316. _____, *A generalization of Kodaira–Ramanujam's vanishing theorem*, Math. Ann. **261** (1982), no. 1, 43–46.

317. _____, *The cone of curves of algebraic varieties*, Ann. of Math. (2) **119** (1984), no. 3, 603–633.

318. _____, *Pluricanonical systems on minimal algebraic varieties*, Invent. Math. **79** (1985), no. 3, 567–588.

319. _____, *The Zariski decomposition of log-canonical divisors*, Algebraic Geometry, Bowdoin, 1985, Amer. Math. Soc., Providence, RI, 1987, pp. 425–433.

320. _____, *On Fujita's freeness conjecture for 3-folds and 4-folds*, Math. Ann. **308** (1997), no. 3, 491–505.

321. _____, *Subadjunction of log canonical divisors for a subvariety of codimension 2*, Birational Algebraic Geometry (Baltimore, 1996), Contemp. Math., vol. 207, Amer. Math. Soc., Providence, RI, 1997, pp. 79–88.

322. _____, *Subadjunction of log canonical divisors. II*, Amer. J. Math. **120** (1998), no. 5, 893–899.

323. _____, *Deformations of canonical singularities*, J. Amer. Math. Soc. **12** (1999), no. 1, 85–92.

324. _____, *On the extension problem of pluricanonical forms*, Algebraic Geometry: Hirzebruch 70 (Warsaw, 1998), Amer. Math. Soc., Providence, RI, 1999, pp. 193–207.

325. _____, *On a relative version of Fujita's freeness conjecture*, Complex Geometry (Göttingen, 2000), Springer, Berlin, 2002, pp. 135–146.

326. Yujiro Kawamata, Katsumi Matsuda, and Kenji Matsuki, *Introduction to the minimal model problem*, Algebraic Geometry, Sendai, 1985, North-Holland, Amsterdam, 1987, pp. 283–360.

327. Yujiro Kawamata and Eckart Viehweg, *On a characterization of an abelian variety in the classification theory of algebraic varieties*, Compositio Math. **41** (1980), no. 3, 355–359.

328. Seán Keel, *Intersection theory of moduli space of stable n-pointed curves of genus zero*, Trans. Amer. Math. Soc. **330** (1992), no. 2, 545–574.

329. _____, *Basepoint freeness for nef and big line bundles in positive characteristic*, Ann. of Math. (2) **149** (1999), no. 1, 253–286.

330. _____, *Polarized pushouts over finite fields*, Comm. Algebra **31** (2003), no. 8, 3955–3982, Special issue in honor of Steven L. Kleiman.

331. Dennis Keeler, *Ample filters of invertible sheaves*, J. Algebra **259** (2003), no. 1, 243–283.

332. George Kempf, *Varieties with rational singularities*, The Lefschetz Centennial Conference, Part I, Comtemp. Math., vol. 58, Amer. Math. Soc., Providence, R.I., 1986, pp. 179–182.

333. _____, *Projective coordinate rings of abelian varieties*, Algebraic Analysis, Geometry, and Number Theory, Johns Hopkins Univ. Press, Baltimore, MD, 1989, pp. 225–235.

334. A. G. Khovanskiĭ, *Analogues of Aleksandrov–Fenchel inequalities for hyperbolic forms*, Dokl. Akad. Nauk SSSR **276** (1984), no. 6, 1332–1334.

335. Meeyoung Kim, *A Barth–Lefschetz type theorem for branched coverings of Grassmannians*, J. Reine Angew. Math. **470** (1996), 109–122.

336. _____, *On branched coverings of quadrics*, Arch. Math. (Basel) **67** (1996), no. 1, 76–79.

337. Meeyoung Kim and Laurent Manivel, *On branched coverings of some homogeneous spaces*, Topology **38** (1999), no. 5, 1141–1160.

338. Meeyoung Kim and Andrew J. Sommese, *Two results on branched coverings of Grassmannians*, J. Math. Kyoto Univ. **38** (1998), no. 1, 21–27.

339. Meeyoung Kim and Jon Wolfson, *Theorems of Barth–Lefschetz type on Kähler manifolds of non-negative bisectional curvature*, Forum Math. **15** (2003), no. 2, 261–273.

340. Steven Kleiman, *A note on the Nakai–Moisezon test for ampleness of a divisor*, Amer. J. Math. **87** (1965), 221–226.

341. _____, *Toward a numerical theory of ampleness*, Ann. of Math. (2) **84** (1966), 293–344.

342. _____, *Ample vector bundles on algebraic surfaces*, Proc. Amer. Math. Soc. **21** (1969), 673–676.

343. _____, *The transversality of a general translate*, Compositio Math. **28** (1974), 287–297.

344. _____, *Misconceptions about $K_X$*, Enseign. Math. (2) **25** (1979), no. 3-4, 203–206.

345. _____, *Bertini and his two fundamental theorems*, Rend. Circ. Mat. Palermo (2) Suppl. (1998), no. 55, 9–37.

346. _____, *Cartier divisors versus invertible sheaves*, Comm. Algebra **28** (2000), no. 12, 5677–5678.

347. Steven Kleiman and Dan Laksov, *On the existence of special divisors*, Amer. J. Math. **94** (1972), 431–436.

348. _____, *Another proof of the existence of special divisors*, Acta Math. **132** (1974), 163–176.

349. Steven Kleiman and Anders Thorup, *A geometric theory of the Buchsbaum–Rim multiplicity*, J. Algebra **167** (1994), no. 1, 168–231.

350. _____, *Mixed Buchsbaum–Rim multiplicities*, Amer. J. Math. **118** (1996), no. 3, 529–569.

351. Shoshichi Kobayashi, *Negative vector bundles and complex Finsler structures*, Nagoya Math. J. **57** (1975), 153–166.

352. _____, *Differential Geometry of Complex Vector Bundles*, Princeton University Press, Princeton, NJ, 1987.

353. K. Kodaira, *On a differential-geometric method in the theory of analytic stacks*, Proc. Nat. Acad. Sci. U.S.A. **39** (1953), 1268–1273.

354. Vijay Kodiyalam, *Asymptotic behaviour of Castelnuovo–Mumford regularity*, Proc. Amer. Math. Soc. **128** (2000), no. 2, 407–411.

355. Shoji Koizumi, *Theta relations and projective normality of Abelian varieties*, Amer. J. Math. **98** (1976), no. 4, 865–889.

356. János Kollár, *Higher direct images of dualizing sheaves. I*, Ann. of Math. (2) **123** (1986), no. 1, 11–42.

357. _____, *Higher direct images of dualizing sheaves. II*, Ann. of Math. (2) **124** (1986), no. 1, 171–202.

358. _____, *Subadditivity of the Kodaira dimension: fibers of general type*, Algebraic Geometry, Sendai, 1985, North-Holland, Amsterdam, 1987, pp. 361–398.

359. _____, *Sharp effective Nullstellensatz*, J. Amer. Math. Soc. **1** (1988), no. 4, 963–975.

360. _____, *Flips and Abundance for Algebraic Threefolds*, Astérisque, vol. 211, 1992.

361. _____, *Shafarevich maps and plurigenera of algebraic varieties*, Invent. Math. **113** (1993), no. 1, 177–215.

362. _____, *Shafarevich Maps and Automorphic Forms*, Princeton University Press, Princeton, NJ, 1995.

363. _____, *Rational Curves on Algebraic Varieties*, Ergebnisse der Math. und ihrer Grenzgebiete (3), vol. 32, Springer-Verlag, Berlin, 1996.

364. _____, *Singularities of pairs*, Algebraic Geometry, Santa Cruz 1995, Proc. Symp. Pure Math., vol. 62, Amer. Math. Soc., Providence, RI, 1997, pp. 221–287.

365. _____, *Effective Nullstellensatz for arbitrary ideals*, J. Eur. Math. Soc. (JEMS) **1** (1999), no. 3, 313–337.

366. János Kollár and T. Matsusaka, *Riemann–Roch type inequalities*, Amer. J. Math. **105** (1983), no. 1, 229–252.

367. János Kollár, Yoichi Miyaoka, and Shigefumi Mori, *Rational curves on Fano varieties*, Classification of Irregular Varieties (Trento, 1990), Lect. Notes in Math., vol. 1515, Springer, Berlin, 1992, pp. 100–105.

368. János Kollár and Shigefumi Mori, *Birational Geometry of Algebraic Varieties*, Cambridge Tracts in Math., vol. 134, Cambridge University Press, Cambridge, 1998.

369. Jian Kong, *Seshadri constants on Jacobian of curves*, Trans. Amer. Math. Soc. **355** (2003), no. 8, 3175–3180.

370. Alexis Kouvidakis, *Divisors on symmetric products of curves*, Trans. Amer. Math. Soc. **337** (1993), no. 1, 117–128.

371. Sándor J. Kovács, *The cone of curves of a K3 surface*, Math. Ann. **300** (1994), no. 4, 681–691.

372. Henrik Kratz, *Compact complex manifolds with numerically effective cotangent bundles*, Doc. Math. **2** (1997), 183–193.

373. Oliver Küchle and Andreas Steffens, *Bounds for Seshadri constants*, New Trends in Algebraic Geometry (Warwick, 1996), London Math. Soc. Lect. Note Series, vol. 264, Cambridge Univ. Press, Cambridge, 1999, pp. 235–254.

374. Alex Küronya, *A divisorial valuation with irrational volume*, J. Algebra **262** (2003), no. 2, 413–423.

375. Sijong Kwak, *Castelnuovo regularity for smooth subvarieties of dimensions 3 and 4*, J. Algebraic Geom. **7** (1998), no. 1, 195–206.

376. Klaus Lamotke, *The topology of complex projective varieties after S. Lefschetz*, Topology **20** (1981), no. 1, 15–51.

377. J. M. Landsberg, *Differential-geometric characterizations of complete intersections*, J. Differential Geom. **44** (1996), no. 1, 32–73.

378. _____, *On degenerate secant and tangential varieties and local differential geometry*, Duke Math. J. **85** (1996), no. 3, 605–634.

379. _____, *Algebraic Geometry and Projective Differential Geometry*, Seoul National University Research Institute of Mathematics Global Analysis Research Center, Seoul, 1999.

380. Serge Lang, *Hyperbolic and Diophantine analysis*, Bull. Amer. Math. Soc. (N.S.) **14** (1986), no. 2, 159–205.

381. _____, *Introduction to Complex Hyperbolic Spaces*, Springer-Verlag, New York, 1987.

382. _____ , *Number Theory, III. Diophantine Geometry*, Encyclopedia of Math. Sciences, vol. 60, Springer-Verlag, Berlin, 1991.

383. Herbert Lange and Christina Birkenhake, *Complex Abelian Varieties*, Grundlehren Math. Wiss., vol. 302, Springer-Verlag, Berlin, 1992.

384. Mogens Larsen, *On the topology of complex projective manifolds*, Invent. Math. (1973), 251–260.

385. F. Laytimi, *On degeneracy loci*, Internat. J. Math. **7** (1996), no. 6, 745–754.

386. F. Laytimi and W. Nahm, *A generalization of Le Potier's vanishing theorem*, preprint, 2002.

387. Robert Lazarsfeld, *A Barth-type theorem for branched coverings of projective space*, Math. Ann. **249** (1980), no. 2, 153–162.

388. _____ , *Branched coverings of projective space*, Ph.D. thesis, Brown University, 1980.

389. _____ , *Some applications of the theory of positive vector bundles*, Complete Intersections (Acireale, 1983), Lect. Notes in Math, vol. 1092, Springer, Berlin, 1984, pp. 29–61.

390. _____ , *Brill–Noether–Petri without degenerations*, J. Differential Geom. **23** (1986), no. 3, 299–307.

391. _____ , *A sharp Castelnuovo bound for smooth surfaces*, Duke Math. J. **55** (1987), no. 2, 423–429.

392. _____ , *A sampling of vector bundle techniques in the study of linear series*, Lectures on Riemann Surfaces (Trieste, 1987), World Sci. Publishing, Teaneck, NJ, 1989, pp. 500–559.

393. _____ , *Lengths of periods and Seshadri constants of abelian varieties*, Math. Res. Lett. **3** (1996), no. 4, 439–447.

394. _____ , *Lectures on linear series*, Complex Algebraic Geometry (Park City, UT, 1993), IAS/Park City Math. Series, vol. 3, Amer. Math. Soc., Providence, RI, 1997, pp. 161–219.

395. Robert Lazarsfeld and A. Van de Ven, *Topics in the Geometry of Projective Space: Recent Work of F. L. Zak*, DMV Seminar, vol. 4, Birkhäuser Verlag, Basel, 1984.

396. Joseph Le Potier, *"Vanishing theorem" pour un fibré vectoriel holomorphe positif de rang quelconque*, Journées de Géométrie Analytique (Poitiers, 1972), Soc. Math. France, Paris, 1974, pp. 107–119. Bull. Soc. Math. France Suppl. Mém., No. 38.

397. _____ , *Annulation de la cohomologie à valeurs dans un fibré vectoriel holomorphe positif de rang quelconque*, Math. Ann. **218** (1975), no. 1, 35–53.

398. Marc Levine, *Pluri-canonical divisors on Kähler manifolds*, Invent. Math. **74** (1983), no. 2, 293–303.

399. Anatoly Libgober, *Alexander invariants of plane algebraic curves*, Singularities, Part 2 (Arcata, Calif., 1981), Proc. Symp. Pure Math., vol. 40, Amer. Math. Soc., Providence, RI, 1983, pp. 135–143.

400. Ben Lichtin, *Poles of $|f(z, w)|^{2s}$ and roots of the b-function*, Ark. Mat. **27** (1989), no. 2, 283–304.

401. David Lieberman and David Mumford, *Matsusaka's big theorem*, Algebraic Geometry – Arcata 1974, Proc. Symp. Pure Math., vol. 29, Amer. Math. Soc., 1975, pp. 513–530.

402. Joseph Lipman, *Unique factorization in complete local rings*, Algebraic Geometry – Arcata, 1974, Proc. Symp. Pure Math., vol. 29, Amer. Math. Soc., Providence, R.I., 1975, pp. 531–546.

403. _____, *Adjoints of ideals in regular local rings*, Math. Res. Lett. **1** (1994), no. 6, 739–755.

404. Joseph Lipman and Bernard Teissier, *Pseudorational local rings and a theorem of Briançon–Skoda about integral closures of ideals*, Michigan Math. J. **28** (1981), no. 1, 97–116.

405. Joseph Lipman and Kei-ichi Watanabe, *Integrally closed ideals in two-dimensional regular local rings are multiplier ideals*, Math. Res. Lett. **10** (2003), no. 4, 423–434.

406. François Loeser and Michel Vaquié, *Le polynôme d'Alexander d'une courbe plane projective*, Topology **29** (1990), no. 2, 163–173.

407. Eduard Looijenga, *Cohomology and intersection homology of algebraic varieties*, Complex Algebraic Geometry (Park City, UT, 1993), IAS/Park City Math. Series, vol. 3, Amer. Math. Soc., Providence, RI, 1997, pp. 221–263.

408. _____, *Motivic measures*, Séminaire Bourbaki, Année 1999/2000, Exp. 276, 2002, pp. 267–297.

409. Martin Lübke, *Beweis einer Vermutung von Hartshorne für den Fall homogener Mannigfaltigkeiten*, J. Reine Angew. Math. **316** (1980), 215–220.

410. Tie Luo, *Riemann–Roch type inequalities for nef and big divisors*, Amer. J. Math. **111** (1989), no. 3, 457–487.

411. _____, *A note on the Hodge index theorem*, Manuscripta Math. **67** (1990), no. 1, 17–20.

412. Gennady Lyubeznik, *Étale cohomological dimension and the topology of algebraic varieties*, Ann. of Math. (2) **137** (1993), no. 1, 71–128.

413. B. Malgrange, *Sur les polynômes de I. N. Bernstein*, Séminaire Goulaouic–Schwartz 1973–1974: Équations aux dérivées partielles et analyse fonctionnelle, Exp. No. 20, Centre de Math., École Polytech., Paris, 1974, p. 10.

414. _____, *Le polynôme de Bernstein d'une singularité isolée*, Fourier Integral Operators and Partial Differential Equations, Lect. Notes in Math., vol. 459, Springer, Berlin, 1975, pp. 98–119.

415. Yuri Manin, *Rational points on algebraic curves over function fields*, Izv. Akad. Nauk SSSR Ser. Mat. **27** (1963), 1395–1440.

416. Laurent Manivel, *Théorèmes d'annulation sur certaines variétés projectives*, Comment. Math. Helv. **71** (1996), no. 3, 402–425.

417. _____, *Vanishing theorems for ample vector bundles*, Invent. Math. **127** (1997), no. 2, 401–416.

418. Mireille Martin-Deschamps, *Propriétés de descente des variétés à fibré cotangent ample*, Ann. Inst. Fourier (Grenoble) **34** (1984), no. 3, 39–64.

419. Kenji Matsuki, *Introduction to the Mori Program*, Universitext, Springer-Verlag, New York, 2002.

420. T. Matsusaka, *Polarized varieties with a given Hilbert polynomial*, Amer. J. Math. **94** (1972), 1027–1077.

421. Arthur Mattuck, *Symmetric products and Jacobians*, Amer. J. Math. **83** (1961), 189–206.

422. Dusa McDuff and Leonid Polterovich, *Symplectic packings and algebraic geometry*, Invent. Math. **115** (1994), no. 3, 405–434.

423. Dusa McDuff and Dietmar Salamon, *Introduction to Symplectic Topology*, Oxford University Press, New York, 1995.

424. John Milnor, *Morse Theory*, Annals of Math. Studies, vol. 51, Princeton University Press, Princeton, N.J., 1963.

425. Rick Miranda, *Triple covers in algebraic geometry*, Amer. J. Math. **107** (1985), no. 5, 1123–1158.

426. _____, *Linear systems of plane curves*, Notices Amer. Math. Soc. **46** (1999), no. 2, 192–201.

427. Yoichi Miyaoka, *On the Mumford–Ramanujam vanishing theorem on a surface*, Journées de Géometrie Algébrique d'Angers, Sijthoff & Noordhoff, Alphen aan den Rijn, 1980, pp. 239–247.

428. _____, *Algebraic surfaces with positive indices*, Classification of Algebraic and Analytic Manifolds (Katata, 1982), Birkhäuser Boston, Boston, Mass., 1983, pp. 281–301.

429. _____, *The Chern classes and Kodaira dimension of a minimal variety*, Algebraic Geometry, Sendai, 1985, North-Holland, Amsterdam, 1987, pp. 449–476.

430. _____, *Deformations of a morphism along a foliation and applications*, Algebraic Geometry, Bowdoin, 1985, Proc. Symp. Pure Math., vol. 46, Amer. Math. Soc., Providence, RI, 1987, pp. 245–268.

431. Yoichi Miyaoka and Shigefumi Mori, *A numerical criterion for uniruledness*, Ann. of Math. (2) **124** (1986), no. 1, 65–69.

432. Yoichi Miyaoka and Thomas Peternell, *Geometry of Higher-Dimensional Algebraic Varieties*, DMV Seminar, vol. 26, Birkhäuser Verlag, Basel, 1997.

433. Boris Moishezon, *Projective imbeddings of algebraic manifolds*, Dokl. Akad. Nauk SSSR **141** (1961), 555–557.

434. _____, *Remarks on projective imbeddings of algebraic manifolds*, Dokl. Akad. Nauk SSSR **145** (1962), 996–999.

435. Ngaiming Mok, *The uniformization theorem for compact Kähler manifolds of nonnegative holomorphic bisectional curvature*, J. Differential Geom. **27** (1988), no. 2, 179–214.

436. Frank Morgan, *Geometric Measure Theory*, second ed., Academic Press Inc., San Diego, CA, 1995.

437. Shigefumi Mori, *Projective manifolds with ample tangent bundles*, Ann. of Math. (2) **110** (1979), no. 3, 593–606.

438. _____, *Threefolds whose canonical bundles are not numerically effective*, Ann. of Math. (2) **116** (1982), no. 1, 133–176.

439. _____, *Classification of higher-dimensional varieties*, Algebraic Geometry, Bowdoin, 1985, Proc. Symp. Pure Math., vol. 46, Amer. Math. Soc., Providence, RI, 1987, pp. 245–268.

440. _____, *Flip theorem and the existence of minimal models for 3-folds*, J. Amer. Math. Soc. **1** (1988), no. 1, 117–253.

441. Atsushi Moriwaki, *Geometric height inequality on varieties with ample cotangent bundles*, J. Algebraic Geom. **4** (1995), no. 2, 385–396.

442. _____, *Remarks on rational points of varieties whose cotangent bundles are generated by global sections*, Math. Res. Lett. **2** (1995), no. 1, 113–118.

443. Christophe Mourougane and Francesco Russo, *Some remarks on nef and good divisors on an algebraic variety*, C. R. Acad. Sci. Paris Sér. I Math. **325** (1997), no. 5, 499–504.

444. Shigeru Mukai, *Duality between $D(X)$ and $D(\hat{X})$ with its application to Picard sheaves*, Nagoya Math. J. **81** (1981), 153–175.

445. David Mumford, *Lectures on Curves on an Algebraic Surface*, Annals of Math. Studies, vol. 59, Princeton University Press, Princeton, N.J., 1966.

446. _____, *Pathologies. III*, Amer. J. Math. **89** (1967), 94–104.

447. _____, *Abelian Varieties*, Tata Institute Studies in Mathematics, vol. 5, Oxford University Press, 1970.

448. _____, *Varieties defined by quadratic equations*, Questions on Algebraic Varieties (C.I.M.E., III Ciclo, Varenna, 1969), Edizioni Cremonese, Rome, 1970, pp. 29–100.

449. _____, *Theta characteristics of an algebraic curve*, Ann. Sci. École Norm. Sup. (4) **4** (1971), 181–192.

450. _____, *Curves and Their Jacobians*, The University of Michigan Press, Ann Arbor, Mich., 1975, (reprinted in The Red Book of Varieties and Schemes, 2nd Edition, Lect. Notes in Math., Volume 1358).

451. _____, *Hilbert's fourteenth problem–the finite generation of subrings such as rings of invariants*, Mathematical Developments Arising from Hilbert Problems, Proc. Symp. Pure Math., vol. 28, Amer. Math. Soc.., 1976, pp. 431–444.

452. _____, *Some footnotes to the work of C. P. Ramanujam*, C. P. Ramanujam—a tribute, Springer, Berlin, 1978, pp. 247–262.

453. _____, *Algebraic Geometry, I. Complex Projective Varieties*, Grundlehren Math. Wiss., vol. 221, Springer-Verlag, Berlin, 1981.

454. Vicente Muñoz and Francisco Presas, *Semipositive bundles and Brill–Noether theory*, Bull. London Math. Soc. **35** (2003), no. 2, 179–190.

455. Mircea Mustaţă, *Jet schemes of locally complete intersection canonical singularities*, Invent. Math. **145** (2001), no. 3, 397–424.

456. _____, *The multiplier ideals of a sum of ideals*, Trans. Amer. Math. Soc. **354** (2002), no. 1, 205–217.

457. _____, *On multiplicities of graded sequences of ideals*, J. Algebra **256** (2002), no. 1, 229–249.

458. _____, *Singularities of pairs via jet schemes*, J. Amer. Math. Soc. **15** (2002), no. 3, 599–615.

459. _____ , *Vanishing theorems on toric varieties*, Tohoku Math. J. (2) **54** (2002), no. 3, 451–470.

460. Alan Nadel, *Multiplier ideal sheaves and existence of Kähler–Einstein metrics of positive scalar curvature*, Proc. Nat. Acad. Sci. U.S.A. **86** (1989), no. 19, 7299–7300.

461. _____ , *Multiplier ideal sheaves and Kähler–Einstein metrics of positive scalar curvature*, Ann. of Math. (2) **132** (1990), no. 3, 549–596.

462. _____ , *The boundedness of degree of Fano varieties with Picard number one*, J. Amer. Math. Soc. **4** (1991), no. 4, 681–692.

463. Masayoshi Nagata, *On the 14-th problem of Hilbert*, Amer. J. Math. **81** (1959), 766–772.

464. Yoshikazu Nakai, *Non-degenerate divisors on an algebraic surface*, J. Sci. Hiroshima Univ. Ser. A **24** (1960), 1–6.

465. _____ , *A criterion of an ample sheaf on a projective scheme*, Amer. J. Math. **85** (1963), 14–26.

466. _____ , *Some fundamental lemmas on projective schemes*, Trans. Amer. Math. Soc. **109** (1963), 296–302.

467. Michael Nakamaye, *Multiplicity estimates and the product theorem*, Bull. Soc. Math. France **123** (1995), no. 2, 155–188.

468. _____ , *Stable base loci of linear series*, Math. Ann. **318** (2000), no. 4, 837–847.

469. _____ , *Base loci of linear series are numerically determined*, Trans. Amer. Math. Soc. **355** (2003), no. 2, 551–566.

470. _____ , *Seshadri constants at very general points*, preprint, 2003.

471. Noboru Nakayama, *Invariance of plurigenera of algebraic varieties*, preprint.

472. _____ , *Zariski-decomposition and abundance*, unpublished preprint, 1997.

473. Terrence Napier and Mohan Ramachandran, *The $L^2$ $\bar{\partial}$-method, weak Lefschetz theorems, and the topology of Kähler manifolds*, J. Amer. Math. Soc. **11** (1998), no. 2, 375–396.

474. M. S. Narasimhan and C. S. Seshadri, *Stable and unitary vector bundles on a compact Riemann surface*, Ann. of Math. (2) **82** (1965), 540–567.

475. Raghavan Narasimhan, *On the homology groups of Stein spaces*, Invent. Math. **2** (1967), 377–385.

476. Lei Ne and Jon Wolfson, *The Lefschetz theorem for CR submanifolds and the nonexistence of real analytic Levi flat submanifolds*, preprint, 2003.

477. P. E. Newstead, *Introduction to Moduli Problems and Orbit Spaces*, Tata Institute of Fundamental Research, Bombay, 1978.

478. Peter A. Nielsen, *A sharper Riemann–Roch inequality for big, semi-ample divisors*, Japan. J. Math. (N.S.) **17** (1991), no. 2, 267–284.

479. Junjiro Noguchi, *A higher-dimensional analogue of Mordell's conjecture over function fields*, Math. Ann. **258** (1981/82), no. 2, 207–212.

480. Junjiro Noguchi and Toshikazu Sunada, *Finiteness of the family of rational and meromorphic mappings into algebraic varieties*, Amer. J. Math. **104** (1982), no. 4, 887–900.

481. Madhav V. Nori, *Zariski's conjecture and related problems*, Ann. Sci. École Norm. Sup. (4) **16** (1983), no. 2, 305–344.

482. _____, *Algebraic cycles and Hodge-theoretic connectivity*, Invent. Math. **111** (1993), no. 2, 349–373.

483. _____, *Constructible sheaves*, Algebra, Arithmetic and Geometry, Part I, II (Mumbai, 2000), Tata Inst. Fund. Res. Stud. Math., vol. 16, Tata Inst. Fund. Res., Bombay, 2002, pp. 471–491.

484. Joseph Oesterlé, *Empilements de sphères*, Astérisque (1990), no. 189-190, Exp. No. 727, 375–397, Séminaire Bourbaki, Vol. 1989/90.

485. Keiji Oguiso, *Seshadri constants in a family of surfaces*, Math. Ann. **323** (2002), no. 4, 625–631.

486. Arthur Ogus, *Local cohomological dimension of algebraic varieties*, Ann. of Math. (2) **98** (1973), 327–365.

487. Christian Okonek, *Barth–Lefschetz theorems for singular spaces*, J. Reine Angew. Math. **374** (1987), 24–38.

488. Christian Okonek, Michael Schneider, and Heinz Spindler, *Vector Bundles on Complex Projective Spaces*, Progr. Math., vol. 3, Birkhäuser Boston, Mass., 1980.

489. Andrei Okounkov, *Why would multiplicities be log-concave?*, preprint.

490. Gianluca Pacienza, *On the nef cone of symmetric products of a generic curve*, Amer. J. Math. **125** (2003), no. 5, 1117–1135.

491. Roberto Paoletti, *Seshadri constants, gonality of space curves, and restriction of stable bundles*, J. Differential Geom. **40** (1994), no. 3, 475–504.

492. _____, *Free pencils on divisors*, Math. Ann. **303** (1995), no. 1, 109–123.

493. _____, *The asymptotic growth of equivariant sections of positive and big line bundles*, preprint, 2002.

494. Kapil H. Paranjape, *The Bogomolov–Pantev resolution, an expository account*, New Trends in Algebraic Geometry (Warwick, 1996), London Math. Soc. Lecture Note Ser., vol. 264, Cambridge Univ. Press, Cambridge, 1999, pp. 347–358.

495. Giuseppe Pareschi, *Syzygies of abelian varieties*, J. Amer. Math. Soc. **13** (2000), no. 3, 651–664.

496. Giuseppe Pareschi and Mihnea Popa, *Regularity on abelian varieties. I*, J. Amer. Math. Soc. **16** (2003), no. 2, 285–302.

497. _____, *Regularity on abelian varieties. II. Basic results on linear series and defining equations*, J. Algebraic Geom. **13** (2004), no. 1, 167–193.

498. Sam Payne, *Fujita's very ampleness conjecture for singular toric varieties*, preprint, 2004.

499. Irena Peeva and Bernd Sturmfels, *Syzygies of codimension 2 lattice ideals*, Math. Z. **229** (1998), no. 1, 163–194.

500. Mathias Peternell, *Ein Lefschetz-Satz für Schnitte in projektiv-algebraischen Mannigfaltigkeiten*, Math. Ann. **264** (1983), no. 3, 361–388.

501. _____, *Algebraische Varietäten und q-vollständige komplexe Räume*, Math. Z. **200** (1989), no. 4, 547–581.

502. Thomas Peternell, Joseph Le Potier, and Michael Schneider, *Vanishing theorems, linear and quadratic normality*, Invent. Math. **87** (1987), no. 3, 573–586.

503. Thomas Peternell and Andrew Sommese, *Ample vector bundles and branched coverings, II*, preprint.

504. _____, *Ample vector bundles and branched coverings*, Comm. Algebra **28** (2000), no. 12, 5573–5599.

505. Henry Pinkham, *A Castelnuovo bound for smooth surfaces*, Invent. Math. **83** (1986), no. 2, 321–332.

506. Piotr Pragacz, *Symmetric polynomials and divided differences in formulas of intersection theory*, Parameter Spaces (Warsaw, 1994), Polish Acad. Sci., Warsaw, 1996, pp. 125–177.

507. Yuri Prokhorov, *On Zariski decomposition problem*, preprint, 2002.

508. S. Ramanan, *A note on C. P. Ramanujam*, C. P. Ramanujam—a Tribute, Tata Inst. Fund. Res. Studies in Math., vol. 8, Springer, Berlin, 1978, pp. 11–13.

509. C. P. Ramanujam, *Remarks on the Kodaira vanishing theorem*, J. Indian Math. Soc. (N.S.) **36** (1972), 41–51.

510. Ziv Ran, *On projective varieties of codimension 2*, Invent. Math. **73** (1983), no. 2, 333–336.

511. _____, *Local differential geometry and generic projections of threefolds*, J. Differential Geom. **32** (1990), no. 1, 131–137.

512. Ziv Ran and Herbert Clemens, *A new method in Fano geometry*, Internat. Math. Res. Notices (2000), no. 10, 527–549.

513. M. S. Ravi, *Regularity of ideals and their radicals*, Manuscripta Math. **68** (1990), no. 1, 77–87.

514. David Rees and Rodney Sharp, *On a theorem of B. Teissier on multiplicities of ideals in local rings*, J. London Math. Soc. (2) **18** (1978), no. 3, 449–463.

515. Miles Reid, *Canonical 3-folds*, Journées de Géometrie Algébrique d'Angers, Sijthoff & Noordhoff, Alphen aan den Rijn, 1980, pp. 273–310.

516. _____, *Young person's guide to canonical singularities*, Algebraic Geometry, Bowdoin, 1985, Proc. Symp. Pure Math., vol. 46, Amer. Math. Soc., Providence, RI, 1987, pp. 345–414.

517. _____, *Twenty-five years of 3-folds—an old person's view*, Explicit Birational Geometry of 3-Folds, London Math. Soc. Lecture Note Ser., vol. 281, Cambridge Univ. Press, Cambridge, 2000, pp. 313–343.

518. Igor Reider, *Vector bundles of rank 2 and linear systems on algebraic surfaces*, Ann. of Math. (2) **127** (1988), no. 2, 309–316.

519. Paul C. Roberts, *Multiplicities and Chern Classes in Local Algebra*, Cambridge Tracts in Mathematics, vol. 133, Cambridge University Press, Cambridge, 1998.

520. Francesco Russo, *A characterization of nef and good divisors by asymptotic multiplier ideals*, unpublished manuscript.

521. Bernard Saint-Donat, *Projective models of K-3 surfaces*, Amer. J. Math. **96** (1974), 602–639.

522. Michael Schneider, *Ein einfacher Beweis des Verschwindungssatzes für positive holomorphe Vektorraumbündel*, Manuscripta Math. **11** (1974), 95–101.

523. _____, *Symmetric differential forms as embedding obstructions and vanishing theorems*, J. Algebraic Geom. **1** (1992), no. 2, 175–181.

524. Michael Schneider and Alessandro Tancredi, *Almost-positive vector bundles on projective surfaces*, Math. Ann. **280** (1988), no. 4, 537–547.

525. Michael Schneider and Jörg Zintl, *The theorem of Barth–Lefschetz as a consequence of Le Potier's vanishing theorem*, Manuscripta Math. **80** (1993), no. 3, 259–263.

526. Richard Schoen and Jon Wolfson, *Theorems of Barth–Lefschetz type and Morse theory on the space of paths*, Math. Z. **229** (1998), no. 1, 77–89.

527. Frank-Olaf Schreyer, *Green's conjecture for general p-gonal curves of large genus*, Algebraic Curves and Projective Geometry (Trento, 1988), Lect. Notes in Math., vol. 1389, Springer, Berlin, 1989, pp. 254–260.

528. Stefan Schröer, *Remarks on the existence of Cartier divisors*, Arch. Math. (Basel) **75** (2000), no. 1, 35–38.

529. Tsutomu Sekiguchi, *On the normal generation by a line bundle on an Abelian variety*, Proc. Japan Acad. Ser. A Math. Sci. **54** (1978), no. 7, 185–188.

530. Jean-Pierre Serre, *Faisceaux algébriques cohérents*, Ann. of Math. (2) **61** (1955), 197–278.

531. _____, *Revêtements ramifiés du plan projectif (d'après S. Abhyankar)*, Séminaire Bourbaki, Année 1959/60, Exp. 204, Soc. Math. France, Paris, 1995, pp. 483–489.

532. Igor R. Shafarevich, *Basic Algebraic Geometry. 1. Varieties in Projective Space*, second ed., Springer-Verlag, Berlin, 1994.

533. Bernard Shiffman and Andrew Sommese, *Vanishing Theorems on Complex Manifolds*, Progr. Math., vol. 56, Birkhäuser Boston Inc., Boston, MA, 1985.

534. V. V. Shokurov, *A nonvanishing theorem*, Izv. Akad. Nauk SSSR Ser. Mat. **49** (1985), no. 3, 635–651.

535. Jessica Sidman, *On the Castelnuovo–Mumford regularity of products of ideal sheaves*, Adv. Geom. **2** (2002), no. 3, 219–229.

536. Carlos Simpson, *Subspaces of moduli spaces of rank one local systems*, Ann. Sci. École Norm. Sup. (4) **26** (1993), no. 3, 361–401.

537. Yum-Tong Siu, *An effective Matsusaka big theorem*, Ann. Inst. Fourier (Grenoble) **43** (1993), no. 5, 1387–1405.

538. _____, *Effective very ampleness*, Invent. Math. **124** (1996), no. 1–3, 563–571.

539. _____, *Invariance of plurigenera*, Invent. Math. **134** (1998), no. 3, 661–673.

540. _____, *Extension of twisted pluricanonical sections with plurisubharmonic weight and invariance of semipositively twisted plurigenera for manifolds not necessarily of general type*, Complex Geometry (Göttingen, 2000), Springer, Berlin, 2002, pp. 223–277.

541. _____, *A new bound for the effective Matsusaka big theorem*, Houston J. Math. **28** (2002), no. 2, 389–409, Special issue for S. S. Chern.

542. Yum-Tong Siu and Shing-Tung Yau, *Compact Kähler manifolds of positive bisectional curvature*, Invent. Math. **59** (1980), no. 2, 189–204.

543. Karen E. Smith, *Fujita's freeness conjecture in terms of local cohomology*, J. Algebraic Geom. **6** (1997), no. 3, 417–429.

544. _____, *The multiplier ideal is a universal test ideal*, Comm. Algebra **28** (2000), no. 12, 5915–5929.

545. Roy Smith and Robert Varley, *Multiplicity g points on theta divisors*, Duke Math. J. **82** (1996), no. 2, 319–326.

546. Ernst Snapper, *Polynomials associated with divisors*, J. Math. Mech. **9** (1960), 123–139.

547. Martín Sombra, *A sparse effective Nullstellensatz*, Adv. in Appl. Math. **22** (1999), no. 2, 271–295.

548. Andrew Sommese, *On manifolds that cannot be ample divisors*, Math. Ann. **221** (1976), no. 1, 55–72.

549. _____, *Theorems of Barth–Lefschetz type for complex subspaces of homogeneous complex manifolds*, Proc. Nat. Acad. Sci. U.S.A. **74** (1977), no. 4, 1332–1333.

550. _____, *Concavity theorems*, Math. Ann. **235** (1978), no. 1, 37–53.

551. _____, *Submanifolds of Abelian varieties*, Math. Ann. **233** (1978), no. 3, 229–256.

552. _____, *Complex subspaces of homogeneous complex manifolds. I. Transplanting theorems*, Duke Math. J. **46** (1979), no. 3, 527–548.

553. _____, *Complex subspaces of homogeneous complex manifolds. II. Homotopy results*, Nagoya Math. J. **86** (1982), 101–129.

554. _____, *A convexity theorem*, Singularities (Arcata, 1981), Proc. Symp. Pure Math., vol. 40, Amer. Math. Soc., Providence, RI, 1983, pp. 497–505.

555. _____, *On the density of ratios of Chern numbers of algebraic surfaces*, Math. Ann. **268** (1984), no. 2, 207–221.

556. Andrew Sommese and A. Van de Ven, *Homotopy groups of pullbacks of varieties*, Nagoya Math. J. **102** (1986), 79–90.

557. Frauke Steffen, *A generalized principal ideal theorem with an application to Brill–Noether theory*, Invent. Math. **132** (1998), no. 1, 73–89.

558. _____, *Connectedness theorems for determinantal schemes*, J. Algebraic Geom. **8** (1999), no. 1, 169–179.

559. Andreas Steffens, *Remarks on Seshadri constants*, Math. Z. **227** (1998), no. 3, 505–510.

560. Irena Swanson, *Powers of ideals. Primary decompositions, Artin–Rees lemma and regularity*, Math. Ann. **307** (1997), no. 2, 299–313.

561. _____, *Linear equivalence of ideal topologies*, Math. Z. **234** (2000), no. 4, 755–775.

562. L. Szpiro, *Lectures on Equations Defining Space Curves*, Tata Institute of Fundamental Research, Bombay, 1979.

352     References

563. Shunsuke Takagi, *An interpretation of multiplier ideals via tight closure*, preprint.

564. Abdelhak Taraffa, *Conditions explicites d'amplitude liées au théorème de Faltings*, C. R. Acad. Sci. Paris Sér. I Math. **322** (1996), no. 12, 1209–1212.

565. Bernard Teissier, *Cycles évanescents, sections planes et conditions de Whitney*, Singularités à Cargèse, Soc. Math. France, Paris, 1973, pp. 285–362. Astérisque, Nos. 7 et 8.

566. _____, *Sur une inégalité à la Minkowski pour les multiplicités*, Ann. Math. (2) **106** (1977), no. 1, 38–44.

567. _____, *Bonnesen-type inequalities in algebraic geometry. I. Introduction to the problem*, Seminar on Differential Geometry, Annals of Math. Studies, vol. 102, Princeton Univ. Press, Princeton, N.J., 1982, pp. 85–105.

568. _____, *Variétés polaires. II. Multiplicités polaires, sections planes, et conditions de Whitney*, Algebraic Geometry (La Rábida, 1981), Lect. Notes in Math., vol. 961, Springer, Berlin, 1982, pp. 314–491.

569. _____, *Résultats récents d'algèbre commutative effective*, Séminaire Bourbaki, Année 1989/90, no. 189-190, 1990, pp. Exp. No. 718, 107–131.

570. Evgueni Tevelev, *Projectively dual varieties*, preprint, 2001.

571. Siu-Kei Tin, *Numerically positive polynomials for k-ample vector bundles*, Math. Ann. **294** (1992), no. 4, 579–590.

572. I-Hsun Tsai, *Dominating the varieties of general type*, J. Reine Angew. Math. **483** (1997), 197–219.

573. Yuri Tschinkel, *Fujita's program and rational points*, Higher dimensional varieties and rational points (Budapest, 2001), Bolyai Soc. Math. Stud., vol. 12, Springer, Berlin, 2003, pp. 283–310.

574. Hajime Tsuji, *On the structure of pluricanonical systems of projective varieties of general type*, preprint, 1999.

575. _____, *Numerically trivial fibration*, preprint, 2000.

576. Loring W. Tu, *The connectedness of symmetric and skew-symmetric degeneracy loci: even ranks*, Trans. Amer. Math. Soc. **313** (1989), no. 1, 381–392.

577. _____, *The connectedness of degeneracy loci*, Topics in Algebra, Part 2 (Warsaw, 1988), PWN, Warsaw, 1990, pp. 235–248.

578. Kenji Ueno, *Introduction to classification theory of algebraic varieties and compact complex spaces*, Classification of Algebraic Varieties and Compact Complex Manifolds, Lect. Notes in Math., vol. 412, Springer, Berlin, 1974, pp. 288–332.

579. Karen Uhlenbeck and Shing-Tung Yau, *On the existence of Hermitian–Yang–Mills connections in stable vector bundles*, Comm. Pure Appl. Math. **39** (1986), no. S, suppl., S257–S293.

580. Sampei Usui and Hiroshi Tango, *On numerical positivity of ample vector bundles with additional condition*, J. Math. Kyoto Univ. **17** (1977), no. 1, 151–164.

581. Bert van Geemen and Gerard van der Geer, *Kummer varieties and the moduli spaces of abelian varieties*, Amer. J. Math. **108** (1986), no. 3, 615–641.

582. Michel Vaquié, *Irrégularité des revêtements cycliques*, Singularities (Lille, 1991), London Math. Soc. Lecture Note Ser., vol. 201, Cambridge Univ. Press, Cambridge, 1994, pp. 383–419.

583. A. N. Varchenko, *Asymptotic Hodge structure on vanishing cohomology*, Izv. Akad. Nauk SSSR Ser. Mat. **45** (1981), no. 3, 540–591, 688.

584. ———, *Semicontinuity of the complex singularity exponent*, Funktsional. Anal. i Prilozhen. **17** (1983), no. 4, 77–78.

585. Wolmer Vasconcelos, *Arithmetic of Blowup Algebras*, London Mathematical Society Lecture Note Series, vol. 195, Cambridge University Press, Cambridge, 1994.

586. Jean-Louis Verdier, *Stratifications de Whitney et théorème de Bertini–Sard*, Invent. Math. **36** (1976), 295–312.

587. Peter Vermeire, *A counterexample to Fulton's conjecture on $\overline{M}_{0,n}$*, J. Algebra **248** (2002), no. 2, 780–784.

588. Eckart Viehweg, *Die Additivität der Kodaira Dimension für projektive Faserräume über Varietäten des allgemeinen Typs*, J. Reine Angew. Math. **330** (1982), 132–142.

589. ———, *Vanishing theorems*, J. Reine Angew. Math. **335** (1982), 1–8.

590. ———, *Weak positivity and the additivity of the Kodaira dimension for certain fibre spaces*, Algebraic Varieties and Analytic Varieties (Tokyo, 1981), North-Holland, Amsterdam, 1983, pp. 329–353.

591. ———, *Weak positivity and the stability of certain Hilbert points*, Invent. Math. **96** (1989), no. 3, 639–667.

592. ———, *Weak positivity and the stability of certain Hilbert points. II*, Invent. Math. **101** (1990), no. 1, 191–223.

593. ———, *Weak positivity and the stability of certain Hilbert points. III*, Invent. Math. **101** (1990), no. 3, 521–543.

594. ———, *Quasi-Projective Moduli for Polarized Manifolds*, Ergebnisse der Math. und ihrer Grenzgebiete (3), vol. 30, Springer-Verlag, Berlin, 1995.

595. Kari Vilonen, *Cohomology of constructible sheaves*, Master's thesis, Brown University, 1979.

596. Claire Voisin, *Courbes tétragonales et cohomologie de Koszul*, J. Reine Angew. Math. **387** (1988), 111–121.

597. ———, *On a conjecture of Clemens on rational curves on hypersurfaces*, J. Differential Geom. **44** (1996), no. 1, 200–213.

598. ———, *Green's generic syzygy conjecture for curves of even genus lying on a K3 surface*, J. Eur. Math. Soc. **4** (2002), no. 4, 363–404.

599. ———, *Hodge Theory and Complex Algebraic Geometry. I*, Cambridge Studies in Advanced Mathematics, vol. 76, Cambridge University Press, Cambridge, 2002, Translated from the French original by Leila Schneps.

600. ———, *Théorie de Hodge et Géométrie Algébrique Complexe*, Cours Spécialisés, vol. 10, Société Mathématique de France, Paris, 2002.

601. ———, *Green's canonical syzygy conjecture for generic curves of odd genus*, preprint, 2003.

602. Paul Vojta, *Mordell's conjecture over function fields*, Invent. Math. **98** (1989), no. 1, 115–138.

603. Jonathan Wahl, *A cohomological characterization of* $\mathbf{P}^n$, Invent. Math. **72** (1983), no. 2, 315–322.

604. R. O. Wells, Jr., *Differential Analysis on Complex Manifolds*, second ed., Graduate Texts in Mathematics, vol. 65, Springer-Verlag, New York, 1980.

605. Gerald E. Welters, *The surface* $C - C$ *on Jacobi varieties and 2nd order theta functions*, Acta Math. **157** (1986), no. 1–2, 1–22.

606. Burkhard Wilking, *Torus actions on manifolds of positive sectional curvature*, preprint.

607. P. M. H. Wilson, *On the canonical ring of algebraic varieties*, Compositio Math. **43** (1981), no. 3, 365–385.

608. _____, *Calabi–Yau manifolds with large Picard number*, Invent. Math. **98** (1989), no. 1, 139–155.

609. _____, *The Kähler cone on Calabi–Yau threefolds*, Invent. Math. **107** (1992), no. 3, 561–583.

610. Jarosław A. Wiśniewski, *On contractions of extremal rays of Fano manifolds*, J. Reine Angew. Math. **417** (1991), 141–157.

611. _____, *On deformation of nef values*, Duke Math. J. **64** (1991), no. 2, 325–332.

612. Bun Wong, *A class of compact complex manifolds with negative tangent bundles*, Complex Analysis of Several Variables (Madison, Wis., 1982), Proc. Sympos. Pure Math., vol. 41, Amer. Math. Soc., Providence, RI, 1984, pp. 217–223.

613. Geng Xu, *Curves in* $\mathbf{P}^2$ *and symplectic packings*, Math. Ann. **299** (1994), no. 4, 609–613.

614. _____, *Subvarieties of general hypersurfaces in projective space*, J. Differential Geom. **39** (1994), no. 1, 139–172.

615. Tamaki Yano, *b-functions and exponents of hypersurface isolated singularities*, Singularities (Arcata, 1981), Proc. Symp. Pure Math., vol. 40, Amer. Math. Soc., Providence, RI, 1983, pp. 641–652.

616. Sai-Kee Yeung, *Very ampleness of line bundles and canonical embedding of coverings of manifolds*, Compositio Math. **123** (2000), no. 2, 209–223.

617. _____, *Effective estimates on the very ampleness of the canonical line bundle of locally Hermitian symmetric spaces*, Trans. Amer. Math. Soc. **353** (2001), no. 4, 1387–1401.

618. M. G. Zaidenberg and V. Ya. Lin, *Finiteness theorems for holomorphic mappings*, Several Complex Variables III, Encyclopedia of Math. Sciences, vol. 9, Springer Verlag, Berlin, 1989, pp. 113–172.

619. F. L. Zak, *Projections of algebraic varieties*, Mat. Sb. (N.S.) **116(158)** (1981), no. 4, 593–602, 608.

620. _____, *Linear systems of hyperplane sections on varieties of small codimension*, Funktsional. Anal. i Prilozhen. **19** (1985), no. 3, 1–10, 96.

621. _____, *Severi varieties*, Mat. Sb. (N.S.) **126(168)** (1985), no. 1, 115–132, 144.

622. _____, *Tangents and Secants of Algebraic Varieties*, American Mathematical Society, Providence, RI, 1993.

623. Oscar Zariski, *The theorem of Riemann–Roch for high multiples of an effective divisor on an algebraic surface*, Ann. of Math. (2) **76** (1962), 560–615.

# Glossary of Notation

## Notation Introduced in Volume I

| | |
|---|---|
| $f^{-1}\mathfrak{a}$ | Pullback of ideal sheaf, I: 1 |
| $S^k E$ | Symmetric power of vector space or bundle, I: 1 |
| $\mathrm{Sym}E$ | Symmetric algebra of vector space or bundle, I: 1 |
| $E^*$ | Dual of vector space or bundle, I: 1 |
| $\mathbf{P}(E)$ | Projective bundle of one-dimensional quotients, I: 1 |
| $\mathbf{P}_{\mathrm{sub}}(E)$ | Projective bundle of one-dimensional subspaces, I: 1 |
| $\mathrm{pr}_1, \mathrm{pr}_2$ | The two projections of $X_1 \times X_2$ onto its factors, I: 2 |
| $O(m^k)$ | Rate of growth of function of $m$, I: 2 |
| $\mathcal{M}_X$ | Sheaf of rational functions on $X$, I: 8 |
| $\mathrm{Div}(X)$ | Additive group of Cartier divisors, I: 8 |
| $\succcurlyeq$ | Effectivity of difference of divisors, I: 8 |
| $Z_k(X)$ | Group of $k$-cycles on $X$, I: 9 |
| $\mathrm{WDiv}(X)$ | Group of Weil divisors on $X$, I: 9 |
| $\mathrm{div}(f)$ | Divisor of rational function, I: 9 |
| $\mathrm{Princ}(X)$ | Group of principal divisors on $X$, I: 9 |
| $\equiv_{\mathrm{lin}}$ | Linear equivalence of divisors, I: 9 |
| $\mathrm{Pic}(X)$ | Picard group of $X$, I: 10 |
| $\omega_X$ | Canonical bundle of $X$, I: 11 |
| $K_X$ | Canonical divisor of $X$, I: 11 |
| $\mathfrak{b}(|V|)$ | Base ideal of linear series, I: 13 |
| $\mathrm{Bs}(|V|)$ | Base locus or scheme of linear series, I: 13 |
| $\phi_{|V|}$ | Morphism determined by linear series, I: 14 |

| | |
|---|---|
| $(D_1 \cdot \ldots \cdot D_k \cdot V)$ | Intersection number of divisors with subvariety, I: 15 |
| $\int_V D_1 \cdot \ldots \cdot D_k$ | Alternative notation for intersection number, I: 15 |
| $(D_1 \cdot \ldots \cdot D_n)$ | Intersection number of $n$ divisors on $n$-fold, I: 16 |
| $[V]$ | Cycle of a scheme $V$, I: 16 |
| $\equiv_{\text{num}}$ | Numerical equivalence of divisors, I: 18 |
| $\text{Num}(X)$ | Subroup of numerically trivial divisors, I: 18 |
| $N^1(X)$ | Néron–Severi group of $X$, I: 18 |
| $\rho(X)$ | Picard number of $X$, I: 18 |
| $(\delta_1 \cdot \ldots \cdot \delta_k \cdot [V])$ | Intersection of numerical equivalence classes, I: 19 |
| $\text{rank}(\mathcal{F})$ | Rank of coherent sheaf, I: 20 |
| $Z_n(\mathcal{F})$ | Cycle of coherent sheaf, I: 20 |
| $\omega_{\text{FS}}$ | Fubini–Study form on $\mathbf{P}^n$, I: 40 |
| $\omega_{\text{std}}$ | Standard symplectic form on $\mathbf{C}^n$, I: 42 |
| $\Theta(L, h)$ | Curvature form of Hermitian line bundle, I: 42 |
| $\text{Div}_{\mathbf{Q}}(X)$ | Group of $\mathbf{Q}$-divisors on $X$, I: 44 |
| $N^1(X)_{\mathbf{Q}}$ | Numerical equivalence classes of $\mathbf{Q}$-divisors, I: 45 |
| $\text{WDiv}_{\mathbf{Q}}(X)$ | Group of Weil $\mathbf{Q}$-divisors, I: 47 |
| $\text{Div}_{\mathbf{R}}(X)$ | Group of $\mathbf{R}$-divisors, I: 48 |
| $N^1(X)_{\mathbf{R}}$ | Numerical equivalence classes of $\mathbf{R}$-divisors, I: 48 |
| $\text{mult}_x C$ | Multiplicity of curve at a point, I: 55 |
| $\text{Amp}(X)$ | Ample cone of $X$, I: 59 |
| $\text{Nef}(X)$ | Nef cone of $X$, I: 59 |
| $N_1(X)_{\mathbf{R}}$ | Numerical equivalence classes of real one-cycles, I: 61 |
| $\text{NE}(X)$ | Cone of curves on $X$, I: 61 |
| $\overline{\text{NE}}(X)$ | Closed cone of curves, I: 62 |
| $D^{\perp}, D_{>0}, D_{\leq 0}$ | Hyperplane, half-spaces determined by divisor, I: 62 |
| $\mathcal{N}_V$ | Null cone determined by subvariety, I: 82 |
| $\mathcal{B}_X$ | Nef boundary of $X$, I: 83 |
| $\text{Kahler}(X)$ | Cone of Kähler classes on $X$, I: 84 |
| $\overline{\text{NE}}(X)_{D\geq 0}$ | Subset of $\overline{\text{NE}}(X)$ in $D$-non-negative halfspace, I: 86 |
| $\text{cont}_{\mathbf{r}}$ | Contraction determined by extremal ray, I: 87 |
| $e(\mathfrak{a}_1; \ldots; \mathfrak{a}_n)$ | Mixed multiplicity of ideals, I: 91 |
| $\text{reg}(\mathcal{F})$ | Regularity of coherent sheaf, I: 103 |
| $d(\mathcal{I})$ | Generating degree of ideal sheaf on $\mathbf{P}^r$, I: 111 |
| $d(I)$ | Generating degree of homogeneous ideal, I: 111 |
| $\text{reg}(I)$ | Regularity of homogeneous ideal, I: 111 |

| | |
|---|---|
| $\mathrm{Cliff}(A)$ | Clifford index of line bundle on a curve,  I: 117 |
| $\mathrm{Cliff}(X)$ | Clifford index of curve,  I: 117 |
| $\mathbf{N}(L)$, $\mathbf{N}(X,L)$ | Semigroup of a line bundle,  I: 122 |
| $e(L)$ | Exponent of a line bundle,  I: 122 |
| $\kappa(L)$, $\kappa(X,L)$ | Iitaka dimension of a line bundle or divisor,  I: 123 |
| $\kappa(X)$ | Kodaira dimension of a variety,  I: 123 |
| $R(L)$, $R(X,L)$ | Section ring of line bundle or divisor,  I: 126 |
| $\mathbf{B}(D)$ | Stable base locus of a divisor,  I: 127 |
| $\mathrm{Big}(X)$ | Big cone of $X$,  I: 147 |
| $\overline{\mathrm{Eff}}(X)$ | Pseudoeffective cone of $X$,  I: 147 |
| $\mathrm{vol}(L)$, $\mathrm{vol}_X(L)$ | Volume of line bundle or divisor,  I: 148 |
| $\mathrm{vol}$, $\mathrm{vol}_X$ | Volume function on $N^1(X)_{\mathbf{R}}$,  I: 153 |
| $\mathrm{mult}_x |V|$ | Multiplicity of linear series at a point,  I: 165 |
| $V_\bullet$ | Graded linear series,  I: 172 |
| $R(V_\bullet)$ | Section ring of graded linear series,  I: 173 |
| $V_\bullet \cdot W_\bullet$ | Product of graded linear series,  I: 174 |
| $V_\bullet \cap W_\bullet$ | Intersection of graded linear series,  I: 174 |
| $\mathrm{Span}(V_\bullet, W_\bullet)$ | Span of graded linear series,  I: 174 |
| $V_\bullet^{(p)}$ | Veronese of graded linear series,  I: 174 |
| $\mathbf{N}(V_\bullet)$ | Semigroup of graded linear series,  I: 174 |
| $e(V_\bullet)$ | Exponent of graded linear series,  I: 174 |
| $\kappa(V_\bullet)$ | Iitaka dimension of graded linear series,  I: 175 |
| $\mathbf{B}(V_\bullet)$ | Stable base locus of graded linear series,  I: 175 |
| $\mathrm{vol}(V_\bullet)$ | Volume of graded linear series,  I: 176 |
| $\mathfrak{a}_\bullet$ | Graded family of ideal sheaves,  I: 176 |
| $\mathrm{Rees}(\mathfrak{a}_\bullet)$ | Rees algebra of graded system,  I: 176 |
| $\mathrm{Pow}(\mathfrak{a})$ | Graded system of powers of ideal,  I: 177 |
| $\mathfrak{q}^{<k>}$ | Symbolic power of radical ideal,  I: 177 |
| $\mathfrak{a}_\bullet \cap \mathfrak{b}_\bullet$ | Intersection of graded families of ideals,  I: 180 |
| $\mathfrak{a}_\bullet \cdot \mathfrak{b}_\bullet$ | Product of graded families of ideals,  I: 180 |
| $\mathfrak{a}_\bullet + \mathfrak{b}_\bullet$ | Sum of graded families of ideals,  I: 180 |
| $\mathfrak{a}_\bullet^{(p)}$ | Veronese of graded system of ideals,  I: 181 |
| $\mathrm{mult}(\mathfrak{a}_\bullet)$ | Multiplicity of graded system of ideals,  I: 182 |
| $L_\omega$ | Cup product with Kähler form $\omega$,  I: 199 |
| $\mathrm{Sec}(X)$ | Secant variety of projective variety,  I: 215 |
| $\mathrm{Tan}(X)$ | Tangent variety of projective variety,  I: 215 |

| | |
|---|---|
| $e_f(x)$ | Local degree of branched covering $f$, I: 216 |
| $\mathbf{T}_x$ | Embedded tangent space to projective variety, I: 219 |
| $\mathrm{Trisec}(X)$ | Variety of tri-secant two-planes to $X$, I: 223 |
| $\mathrm{except}(\mu)$ | Exceptional locus of birational map $\mu$, I: 241 |
| $\Omega_X^1(\log D)$ | Bundle of one-forms with log poles along $D$, I: 250 |
| $\Omega_X^p(\log D)$ | Bundle of $p$-forms with log poles along $D$, I: 250 |
| $\mathcal{K}_X$ | Grauert–Riemenschneider canonical sheaf, I: 258 |
| $\mathrm{Pic}^0(X)$ | Identity component of $\mathrm{Pic}(X)$, I: 261 |
| $\mathrm{Alb}(X)$ | Albanese variety of $X$, I: 262 |
| $\mathrm{alb}_X$ | Albanese mapping of $X$, I: 262 |
| $\varepsilon(L;x)$ | Seshadri constant of $L$ at $x$, I: 270 |
| $s(B;x)$ | Number of jets separated by $B$ at $x$, I: 273 |
| $B(\lambda)$ | Open ball of radius $\lambda$ in $\mathbf{C}^g$, I: 276 |
| $w_G(M,\omega)$ | Gromov width of symplectic manifold, I: 276 |
| $\mathrm{mult}_x(F)$ | Multiplicity of divisor $F$ at a point, I: 282 |
| $\mathrm{mult}_Z(F)$ | Multiplicity of divisor $F$ along subvariety $Z$, I: 282 |
| $(d_1,\dots,d_g)$ | Type of polarization on abelian variety, I: 291 |
| $m(A,L)$ | Buser–Sarnak invariant of abelian variety, I: 291 |
| $(A,\Theta)$ | Principally polarized abelian variety (PPAV), I: 292 |
| $\mathcal{A}_g$ | Moduli space of PPAVs, I: 292 |
| $(JC,\Theta_C)$ | Polarized Jacobian of Riemann surface $C$, I: 293 |
| $\varepsilon(A,L)$ | Seshadri constant of polarized abelian variety, I: 293 |
| $s_L(\mathcal{I})$ | $s$-invariant of ideal sheaf, I: 303 |
| $d_L(\mathcal{I})$ | Generating degree of ideal sheaf w.r.t. line bundle, I: 304 |
| $\deg_L(V)$ | Degree of variety w.r.t. line bundle, I: 308 |
| $\mathrm{reg}_L(\mathcal{I})$ | Regularity of ideal sheaf w.r.t. line bundle, I: 310 |
| $\mathcal{H}_i$ | Homology sheaf of complex, I: 318 |
| $K_\bullet(E,s)$ | Koszul complex of section of vector bundle, I: 320 |

## Notation Introduced in Volume II

| | |
|---|---|
| $\mathbf{P}(E)$, $\mathbf{P}_X(E)$ | Projective bundle over $X$, II: 7 |
| $\xi$, $\xi_E$ | Class of Serre line bundle on $\mathbf{P}(E)$, II: 8 |
| $V_X$ | Trivial vector bundle modeled on vector space $V$, II: 8 |
| $\Gamma^\lambda E$ | Bundle associated to $E$ via representation $\lambda$, II: 15 |
| $E{<}\delta{>}$, $E{<}D{>}$ | Bundle twisted by $\mathbf{Q}$-divisor class, II: 21 |

| | |
|---|---|
| $\delta(X, E, h)$ | Barton invariant of a bundle,  II: 25 |
| $N_{X/M}$ | Normal bundle to subvariety $X \subseteq M$,  II: 27 |
| $\widehat{M}_{/X}$ | Formal completion of $M$ along subvariety $X$,  II: 31 |
| $E_f$ | Bundle associated to branched covering,  II: 48 |
| $\mu(E)$ | Slope of vector bundle on curve,  II: 57 |
| $\mathrm{HN}_\bullet(E)$ | Harder–Narasimhan filtration of bundle $E$,  II: 59 |
| $E_{\mathrm{norm}}$ | Normalized $\mathbf{Q}$-twist of vector bundle on curve,  II: 60 |
| $\mathrm{Zeroes}(s)$ | Zeroes of section of vector bundle,  II: 65 |
| $D_k(u)$ | Degeneracy locus of vector bundle map,  II: 74 |
| $W_d^r(C)$ | Variety of special divisors on a curve,  II: 82 |
| $W^r(C; V)$ | Special divisors associated to vector bundle,  II: 84 |
| $S^i(f)$ | Singularity locus of a mapping,  II: 86 |
| $c_i(E{<}\delta{>})$ | Chern class of $\mathbf{Q}$-twisted vector bundle,  II: 102 |
| $z(C, E)$ | Cone class of cone in vector bundle,  II: 106 |
| $\mathbf{P}_{\mathrm{sub}}(C)$ | Projectivization of cone,  II: 107 |
| $C_1 \oplus C_2$ | Direct sum of cones,  II: 107 |
| $\overline{C} \subset \overline{E}$ | Projective closure of cone in vector bundle,  II: 107 |
| $z(C{<}\delta{>}, E{<}\delta{>})$ | $\mathbf{Q}$-twisted cone class,  II: 110 |
| $s_\lambda(c_1, \ldots, c_e)$ | Schur polynomial,  II: 118 |
| $\ulcorner D \urcorner$ | Round-up of $\mathbf{Q}$-divisor,  II: 140 |
| $\llcorner D \lrcorner, [D]$ | round-down or integral part of $\mathbf{Q}$-divisor,  II: 140 |
| $\{D\}$ | Fractional part of $\mathbf{Q}$-divisor,  II: 141 |
| $K_{X'/X}$ | Relative canonical divisor,  II: 146 |
| $\mathcal{J}(D), \mathcal{J}(X, D)$ | Multiplier ideal of a $\mathbf{Q}$-divisor,  II: 152 |
| $\mathcal{J}(\mathfrak{a}^c), \mathcal{J}(c \cdot \mathfrak{a})$ | Multiplier ideal associated to ideal sheaf,  II: 152 |
| $\mathcal{J}(\mathfrak{a}_1^{c_1} \cdot \ldots \cdot \mathfrak{a}_t^{c_t})$ | Mixed multiplier ideal,  II: 153 |
| $\mathcal{J}(c \cdot |V|)$ | Multiplier ideal of linear series,  II: 154 |
| $\mathcal{J}(f, c \cdot |L|)$ | Multiplier ideal of relative linear series,  II: 158 |
| $\mathrm{mult}_x(D)$ | Multiplicity of $\mathbf{Q}$-divisor,  II: 162 |
| $\mathrm{ord}_Z(\mathfrak{a})$ | Order of vanishing of ideal along subvariety,  II: 164 |
| $\Sigma_k(D)$ | Multiplicity locus of $\mathbf{Q}$-divisor,  II: 165 |
| $\mathrm{lct}(D; x), \mathrm{lct}(D)$ | Log-canonical threshold of $\mathbf{Q}$-divisor,  II: 166 |
| $\mathrm{lct}(\mathfrak{a}; x), \mathrm{lct}(\mathfrak{a})$ | Log-canonical threshold of an ideal sheaf,  II: 166 |
| $b_f(s)$ | Bernstein–Sato polynomial,  II: 169 |
| $P(\mathfrak{a})$ | Newton polyhedron of monomial ideal,  II: 171 |
| $\mathbf{1} = (1, \ldots, 1)$ | Vector of 1's,  II: 171 |

| | |
|---|---|
| $\mathcal{J}(\varphi)$ | Multiplier ideal associated to PSH function, II: 176 |
| $\mathrm{adj}(D)$ | Adjoint ideal of reduced divisor, II: 179 |
| $\mathrm{Jacobian}(D)$ | Jacobian ideal of divisor, II: 181 |
| $\mathrm{Jacobian}_m(\mathfrak{a})$ | Jacobian ideals of ideal sheaf, II: 181 |
| $(X, \Delta)$ | A pair, II: 182 |
| $\mathcal{J}((X, \Delta); D)$ | Multiplier ideal of $\mathbf{Q}$-divisor on pair, II: 183 |
| $\mathcal{J}(h)$ | Multiplier ideal of singular Hermitian metric, II: 192 |
| $D_H$ | Restriction of $\mathbf{Q}$-divisor to hypersurface, II: 195 |
| $\mathfrak{a}_H$ | Restriction of ideal sheaf to hypersurface, II: 197 |
| $\mathfrak{a}_1 \overset{\circ}{+} \mathfrak{a}_2$ | Exterior sum of two ideals, II: 206 |
| $\overline{\mathfrak{a}}$ | Integral closure of ideal, II: 216 |
| $\mathbf{B}_+(L)$ | Augmented stable base locus, II: 247 |
| $\mathrm{Null}(L)$ | Null locus of a nef divisor, II: 248 |
| $\mathrm{LC}(D; x)$ | Locus of log-canonical singularities, II: 255 |
| $\mathcal{J}(c \cdot \|L\|)$ | Asymptotic multiplier ideal of linear series, II: 271 |
| $\mathcal{J}(c \cdot \mathfrak{a}_\bullet), \mathcal{J}(\mathfrak{a}_\bullet^c)$ | Asymptotic multiplier ideal of graded system, II: 276 |
| $\mathrm{lct}(\mathfrak{a}_\bullet; x)$ | Log-canonical threshold of graded family, II: 280 |
| $\mathcal{J}(\mathfrak{a}_\bullet^c \cdot \mathfrak{b}_\bullet^d)$ | Mixed asymptotic multiplier ideal, II: 281 |
| $\mathcal{J}(c \cdot \|V_m\|)$ | Asymptotic multiplier ideal of graded linear series, II: 281 |
| $\mathcal{J}(c \cdot V_\bullet)$ | Alternate notation for $\mathcal{J}(c \cdot \|V_1\|)$, II: 281 |
| $P_m(X)$ | $m^{\mathrm{th}}$ plurigenus of $X$, II: 292 |
| $\overline{\mathrm{Mov}}(X)$ | Cone of movable curves, II: 307 |

# Index

The small Roman numbers I and II refer to Volume I and II, respectively.

CPSIA information can be obtained
at www.ICGtesting.com
Printed in the USA
LVOW01s1535150316
479258LV00006B/110/P